On-Farm Drying and Storage Systems

Otto J. Loewer
Thomas C. Bridges
Ray A. Bucklin

Library of Congress Card Number (LCCN) 94-72087
International Standard Book Number (ISBN) 0-929355-53-9
ASAE Publication 9

Pam DeVore-Hansen, Acquisitions
Books & Journals

Preface

The book "On-farm Drying and Storage Systems" had its beginnings in 1973 when I came to the Agricultural Engineering Department at the University of Kentucky as an assistant extension professor working in the area of grain drying and storage. An upturn in grain and energy prices had triggered an increased interest in on-farm grain systems throughout the grain growing areas of the United States. In western Kentucky, which was an established corn and soybean area similar to the midwest, producers were primarily interested in learning how to expand their existing facilities and operate them more efficiently while producing a quality product. Central Kentucky began to grow significant quantities of shelled corn for the first time, and their primary interest was primarily related to the design and operation of new facilities. At that time, the use of computers was a new and somewhat untested concept in providing extension assistance. Microcomputers, as we now know them, were not even on the horizon. Thus, we were presented with a broad range of grain drying and storage situations and a new tool for addressing the needs of our clientele.

A group of faculty in the Agricultural Engineering Department at the University of Kentucky, along with cooperators at other universities, coordinated their grain drying and storage research and extension programs in a way that was somewhat unique at that time. We continued the traditional method of providing extension publications based on field and laboratory research results, but we coupled this activity with the development of computer software that was designed to meet the individual needs of our farm clientele. Our first workshop, held in 1975 at the University of Kentucky, was directed nationwide for research and extension faculty who had an interest in extending grain drying and storage information to farmers through computers. Our first workshop for farmers was held early the next year. Although there were skeptics, we offered a three-day workshop, charged the farmers a fee to cover our out-of-pocket expenses, and offered them hands-on experience with our mainframe computer system using the grain drying and storage computer software that we had developed. Typically over the years, 20 to 40 farmers would attend, often accompanied by their county agents. Usually eight to ten faculty would assist with the workshops along with an equal number of graduate students. The feedback we received from the farmers helped us to develop additional software to meet their needs. We continued the workshop for 10 years.

Computer capability has changed rapidly since we offered our first workshop. Nevertheless, the major principles for layout, design, and operation of grain drying and storage systems remain much the same. The goal of this book is to provide its user with much of the information needed in the design and management of on-farm grain drying and storage systems. One unique aspect of the book is that it is combined with computer software designed to run on a microcomputer and help its user make important management decisions. Information related to prices for various items quickly becomes obsolete. However, the principles used in making decisions remain much the same. It is our hope that this book may be used to meet the needs of a broad range of individuals. It contains numerous tables and charts so that it may be used as a reference book for farmers, contractors, and engineers. The equations, example problems, and problem sets make it an ideal textbook for students in Agricultural Operations Management and similar programs.

The material in this textbook is the product of many individuals besides the authors. We have borrowed liberally from their publications, software, knowledge and counsel with their full cooperation and assistance. Accordingly, the authors would like to especially acknowledge the contributions of the following: Drs. I. J. Ross, Gerald M. White, Douglas G. Overhults, Samuel G. McNeill, Robert L. Fehr, Larry W. Turner and John N. Walker - University of Kentucky Agricultural Engineering Department; Dr. Terry J. Siebenmorgen - University of Arkansas Biological and Agricultural Engineering Department; Drs. Michael Kocher and the late Thomas L. Thompson - University of Nebraska Biological Systems Engineering Department; Mr. Bruce A. McKenzie - Purdue University Agricultural Engineering Department; Mr. Larry Miller and Mr. Art Taylor - University of Florida Agricultural Engineering Department; Dr. David H. Loewer and Mr. Carl A. Loewer - Wynne, Ark.; Mr. Gerald T. Benock - Punta Gorda, Fla.

Otto J. Loewer

Dedication

This book is dedicated to some of my greatest teachers: my wife Betty, my daughters Cynthia and Sarah, my brothers Carl and David, my mother Barbara and my father Otto, the one who taught me most about how to teach.

Otto J. Loewer

Table of Contents

Chapter 2

Chapter 3

1

Systems Analysis, Computer Models, and Grain Systems

The material in this text was generated over a 20-year period. The selection of topics was based on questions that were asked repeatedly by farmers, grain systems designers, and extension personnel. This textbook is intended to serve the needs of a broad range of clientele that includes farmers, salespersons, extension personnel, engineers, and students. It may be used as:

1. A handbook that contains tables of information and associated equations.
2. A manual that carries the user through the design process including the utilization of computer models to address individual situations.
3. A reference or text for workshop or formal classroom settings where fundamental principles of certain processes are presented.

The depth of coverage varies considerably with the topic. While the text may be used effectively by engineers, it is not intended to be an engineering textbook. Although some complex mathematical expressions are used, most of the mathematics is straightforward and relatively easy to understand. To serve the broad range of clientele, material in the text is presented in several ways including text discussion, example situations, equations, tables, figures, and computer program software. Problem sets designed for formal classroom use are presented at the end of each chapter except the first.

Each chapter is somewhat independent of the others. The reader is encouraged to first seek the information that is of greatest interest regardless of the order in which the material is presented. The text contains considerable economic information. Prices of products and services tend to become dated very quickly. Thus, specific costs are best used in making comparisons among alternatives rather than taken at face value.

Computer skills will enhance the utility of the text for those wishing to use the computer programs directly on a microcomputer. However, these skills are certainly not a requirement for using the text effectively.

In overview, design and utilization of grain drying, handling, and storage systems can be enhanced by having a greater understanding of how these systems function and interact. We begin a text geared to provide insight into grain drying and storage with a discussion about systems analysis and computer models. Why? What do systems analysis and computer models have to do with grain systems? Grain drying and storage facilities have been around for a long time, certainly longer than computer models, and there have been many excellent texts written on the subject such as Henderson and Perry (1976), Johnson and Lamp (1966), and Brooker et al. (1992). Yet, today we have the technological skill and insight necessary to build the most efficient systems ever if we are willing to take advantage of the concepts of systems analysis and the technological capabilities offered to us in the form of computer models. However, we should develop an understanding of the jargon associated with systems analysis and computer modeling if we are to use these tools successfully and to better understand the nature of some the material discussed later in the book.

Principles of Systems Analysis

The term "systems analysis", as used in this text, refers to analyzing systematically the whole system as a collection of individual parts (subsystems) rather than to concentrate on the individual components. This infers that trade-offs will often have to be made so as to obtain "global" efficiency rather than "local" efficiency.

The Primary Components

The primary components of the systems analysis procedure are shown in figure 1.1. These include:

Environment. The environment is the aggregate of all external conditions and influences affecting the system. No particular individual or group establishes the environment totally, but its influence is felt throughout the system.

Decision Maker. The decision maker is the authority that decides which alternative(s) should be used to accomplish the objective(s) of the system. This person is the goal setter of any organization and the "final word" in terms of selecting among alternative courses of action.

Goals. Goals tend to be broadly defined aspirations, aims or desires for the system that are usually not quantified. Goals give direction and guidance to the system and should be established at the beginning of any process. The decision maker is the goal setter for the system.

SYSTEMS APPROACH TO PROBLEM SOLVING

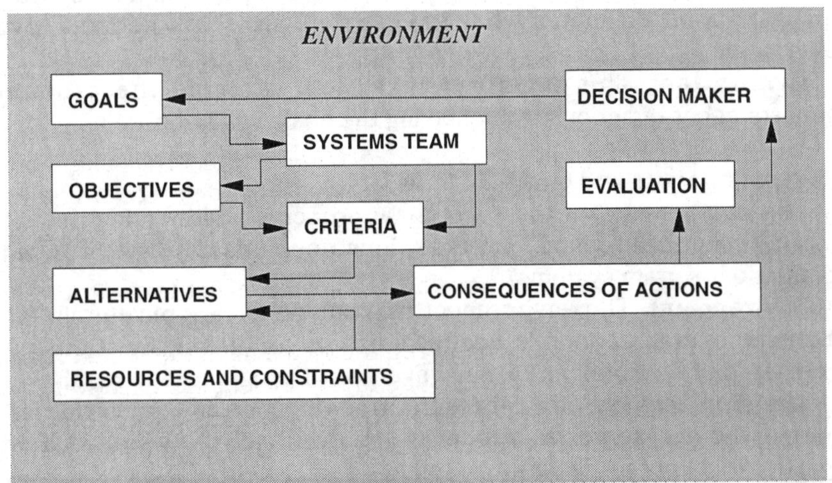

Figure 1.1–Flow chart describing the systems approach to problem solving.

Objectives. Objectives are aspirations, aims or desires that are quantified so as to give reference points for possible alternatives. The systems analysis team usually establishes the objectives, perhaps in concert with the decision maker.

Systems Analysis Team. The systems analysis team is given goals by the decision maker with the directive being to provide alternative courses of action. The team is often interdisciplinary so as to give a broad perspective when addressing the goals of the system.

Resources and Constraints. Resources become constraints when they become limiting for an alternative course of action. Resources are most often provided by the decision maker and include available material in the form of labor and capital. Constraints, however, may be imposed from outside the system. Resources and constraints must be considered when selecting alternatives.

Alternatives. An alternative is a feasible means of accomplishing an objective. The systems analysis team establishes the alternatives. Alternatives are considered by the decision maker when determining which change(s), if any, should be made to the system. The only viable alternative may be to retain the status quo.

Criteria for Evaluation. Criteria are standards of judgement applied to a possible alternative to determine if it can be classified as a feasible alternative. The systems analysis team usually establishes the criteria to ensure feasible alternatives immediately after objectives are established.

Consequences of Actions. Consequences of actions are the expected results of the implementation of a given alternative. The systems analysis team estimates what the consequences will be for a given alternative.

Evaluation. The decision maker will examine the probable consequences of actions when making the final evaluation.

Systems Analysis and Grain Systems

The components of the systems approach to addressing decision making have been given. However, how does this concept fit in with evaluation of grain systems?

Environment. The environment that surrounds a typical grain farm includes overall economic health, business and family relationships, standards of the community, and government programs and regulations.

Decision Maker. The representative decision maker is the farm owner who also serves as farm manager. Eventually, this individual will decide what courses of action to follow.

Goals. Most often, profit maximization is the stated goal of a business including a grain farm. However, factors such as risk preference and convenience influence the goal of profit maximization. For example, as the decision maker ages, this individual may wish to be assured of an adequate income that does not require as much outside labor even though the farm might produce greater profits given another strategy.

The goals for any business may include a "hidden agenda" that is not well-defined. For example, the farmer may want to consider the welfare of neighbors to the extent that the grain facility does not adversely impact them, or there may be a desire to pass the family farm on to other family members even though this decision cannot be justified strictly on short-term economics. Recognize, however, that economic value for one individual may include factors that are not as highly valued by another individual. For example, the fact that a particular land area has "been in the family for generations" may mean more to its owner than to a prospective land buyer.

Objectives. Objectives of the farmer who is considering purchasing a grain system may include a specific quantity of grain to be stored and dried in a given time period. For example, the farmer may wish to be able to dry 2,000 bushels of corn and 1,000 bushels of soybeans per day and have sufficient capacity to store 30,000 bushels of corn and 20,000 additional bushels of soybeans.

Interdisciplinary Team. The interdisciplinary team that advises the farm decision maker includes: the banker that will provide the short-term financial resources for purchasing the system; the county agent and extension agricultural engineer who will provide technical input into what types of drying and storage systems will function correctly; the extension farm management specialist who analyzes the potential profitability of adding drying and storage; the salespeople who represent various companies that provide the grain drying and storage equipment; and the builder(s) who may become involved in construction of the facility. These individuals may work as a team at some level. Recognize, also, that these people have varying degrees of what might be conflicting and vested interests in the project. For example, a salesperson who sells only one type of drying equipment will definitely want that line to be included in the design. Similarly, the builder may specialize in one particular type of layout and may not want to look at alternative designs. The banker wants to make sure that the loan will be repaid in a timely manner. The county agent and the agricultural engineer want the facility to be designed and constructed correctly, and the extension farm management specialist wants the farmer's investment to be profitable. These different views can provide the decision maker a full spectrum of attitudes, all of which are important. However, unlike an industrial setting the decision maker may have to serve as "chairman" of this interdisciplinary team.

Resources and Constraints. Resources include savings and potential loans. As with all resources, these become constraints if they limit the type of facility that may be built. On-farm land, labor, energy supplies, management, and experience should also be considered. Publications and expertise from the Extension Service and the private sector can prove very beneficial in the planning, design, and construction of a grain facility.

Alternatives. There are many different drying methods, material handling techniques, and storage layouts that may be technologically and economically feasible. The most promising of these need to be identified.

Criteria for Evaluation. The criteria must be referenced to the goals and objectives for the grain system. That is, does the facility design allow for all the features stated in the goals and objectives? The ranking of alternatives may vary depending on the criteria used. For example, an annual cost ranking may be somewhat different from a purchase cost ranking.

Consequences of Actions. Each alternative design has an associated consequence of actions. For example, a given alternative may be of relatively low cost but require manual unloading of delivery vehicles. Another alternative may meet the stated criteria but fail to allow for any future expansion. Yet another alternative may seem to be ideal but is much more costly than was expected. And, perhaps another alternative will use fans that are noisy and may bother the neighbors.

Evaluation. Evaluation of alternatives may lead the decision maker to redefine personal objectives which, in turn, may lead to a refining of resources and constraints. In fact, many of the items that were in the hidden agenda may become part of the goals and objectives for the facility. For example, the decision maker may state that designs requiring manual unloading of delivery vehicles are unacceptable, that all designs must allow for doubling the initial storage capacity, that purchase cost may not exceed $100,000, and that only fans that are relatively quiet may be used. Of course, restating the objectives and criteria will reduce further the number of viable alternatives, and, in fact, may result in no feasible design. That is, it may be impossible to build a facility that satisfies all the other criteria for a cost of $100,000 or less. However, remember that doing nothing is always an alternative.

Systems Analysis and Economics

Eventually, economic considerations in the broadest sense will govern any system. This text limits its discussion strictly to capital costs and returns. During periods of inflation, costs and returns are subject to rapid change that, in turn, tends to date quickly specific economic values in a publication. Thus, the reader is encouraged to understand the principles associated with economic comparisons (Chapter 9) rather than to take the absolute calculations stated in this text at face value.

Systems Analysis and Computer Models

Computer models provide insight into how portions of a given system function. This text often uses computer models to generate a greater understanding of portions of the grain harvesting/delivery/drying/storage system. The remainder of this chapter provides some insight into the limitations and capabilities of computer models in addressing grain systems.

Conceptual Utilization of Computer Models

Computer models are used primarily for design, analysis, and management. Each of these applications may utilize one or more models of differing structure and complexity. The key to efficient utilization of computer models is in defining one's objectives and selecting the model that is applicable within acceptable limits of structure, complexity, and availability. For example, two models may provide the same output. However, the assumptions that are incorporated into one of the models may not be sufficiently valid for addressing the particular problem at hand. Similarly, one of the models may run only on "main frame" computers and not be readily accessible for use.

Design, analysis, and management models are interrelated. However, the following discussion will address the differences in these types of computer programs.

Design Models

Design models are generally static in structure; that is, simulated time remains constant. Design models tend to give "exact" or "point" solutions to problems. They are associated primarily with determining proper "sizes" (lengths, widths, heights, angles, direction, weight, volume, cost, etc.) of certain structures (buildings and/or machines) given a performance criteria. For example, the BNDZN (pronounced bin design) program developed by Loewer et al. (1976a) specifies the total equipment set including cost and location for a grain drying and storage facility given model inputs such as capacity, drying method, and the number of bins to be built. Design models do much of the mathematical computation associated with planning but in much less time. They allow their user to quickly evaluate the consequences of changes in design specifications. For example, if the storage capacity is changed from 30,000 to 40,000 bushels, how much larger will the bins be? What is the cost increment? Will an existing bucket elevator still be tall enough? Design models may be used directly by designers, contractors and builders so as to enhance communication and iensure proper layout and design within clearly identified cost constraints.

Design models are not used as frequently as the other types of models. However, they are extremely valuable in helping to put together a physical system that is sized properly within the input specification guidelines as to performance and cost. Microcomputer spreadsheet programs are usually design models.

Models Used for Analysis and Management

Models used for purposes of analysis are very closely related to management type models. The primary differences lie in their application. Models used for analyses are usually directed toward a broader set of problems and are used to develop a greater understanding of the functioning of the system. An example is in Chapter 7 where two models are used together to define the relative importance of materials handling equipment on combine and delivery vehicle performance. Used in that manner, the models helped establish broad principles to be used in facility design.

Management usage of models is usually site specific and directed towards the solving of relatively specialized problems. For example, a farmer has a particular fan and bin combination and wishes to use it as a batch-in-bin system for corn drying in October. One model may be used to determine fan performance and another to evaluate the effects of drying air temperature and grain depth on drying rate. The "solution" to the drying problem is very much directed toward an individual situation and has relatively limited application to other farmers.

Realistic Expectations

The computer model user, whether for purposes of design, analysis or management, must have realistic expectations of model performance. A model is an abstraction of reality and not reality itself. Hence, a model cannot be completely "accurate" in all possible situations. Likewise, not all models are created equal! Therefore, the user needs to become familiar with the inherent assumptions of the model so as to better judge its output. Generally, simpler models have relatively more "simplifying" assumptions. All models become less accurate as the number of "violations" of inherent assumptions increases.

The "accuracy" of a model is a term used to compare model output with observed values. There are two primary comparisons that may be made. The first is a comparison between measured and computed outputs. The second is a comparison between predicted and observed trends.

For example, suppose one wishes to evaluate a dryer performance model with regard to energy efficiency expressed as Btu per pound of water evaporated. In the "real" dryer, temperature would be varied incrementally and measurements taken of the Btu usage. These same temperatures would be entered into the dryer performance model with one of the outputs being Btu per pound of water evaporated. How do we judge the "accuracy" of this model, and can we incorporate the two comparisons mentioned previously into a single measure? The answers to these questions lie in developing a procedure capable of comparing statistically both the value and trend between computed and observed values. We know the following. If the model were "perfect", a plot of observed versus computed values would yield a straight line sloped at 45° and passing through the origin (fig. 1.2). The statistical measure of "closeness of fit" is called the "correlation coefficient," usually referred to as the "R^2" value. The correlation coefficient may be determined by the following expression:

$$R^2 = \frac{\left(\text{sum of xi} * \text{yi}\right) **2}{\left[\left(\text{sum of } x_i{}^2\right) / \left(\text{sum of } y_i{}^2\right)\right]} \tag{1.1}$$

where

x_i = an individual "observed" value for a given set of input conditions

y_i = an individual "computed" value for a given set of input conditions

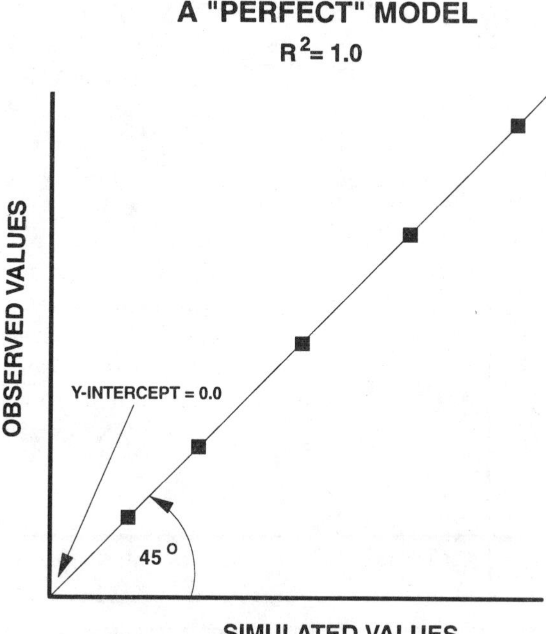

Figure 1.2–A comparison between simulated and observed values that produces perfect correlation (R^2 = 1.0) with no bias (a 45° slope regression line with its Y-intercept passing through the origin).

A "perfect" correlation has an "R^2" value of 1.0 while an "R^2" value of 0.0 implies that there is no correlation between the observed and computed values.

We will now examine several situations that influence the acceptability of a model as measured by goodness of fit; that is, how closely do the observed and computed values compare. Suppose we have the situation shown in figure 1.3 where R^2 is equal to 0.95. Visually, we would conclude that this model very closely matches observed values. The "intercept" is equal to zero indicating that the model is not "biased" in its predictions. Suppose the comparison is as shown in figure 1.4. Again, the model shows no bias. Visually, however, the correlation coefficient appears to be substantially lower indicating less agreement between observed and computed values. In figure 1.5, we see the situation where the model is biased "high"; that is, the model consistently predicts a higher value than observed as evidenced by a line that is perfectly parallel to the "perfect" condition and has a negative Y-intercept value. In figure 1.6, the model underpredicts progressively for higher observed values even though the Y-intercept does pass through the origin. In figure 1.7, the model predicts relatively

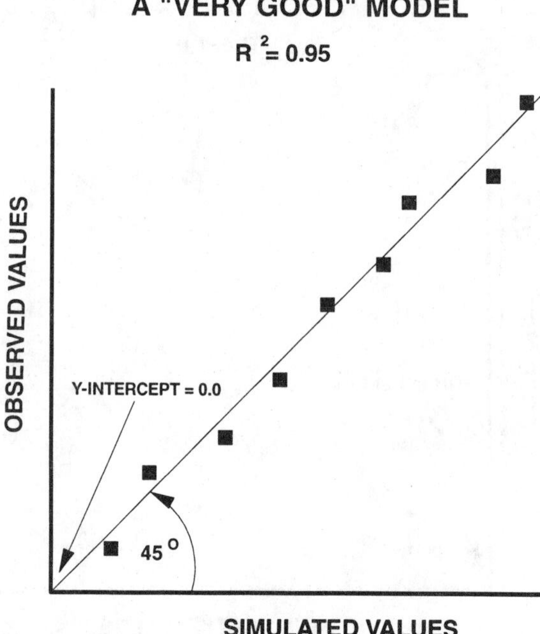

Figure 1.3–A comparison between simulated and observed values that produces relatively close correlation (R^2 = 0.95) with no bias (a 45° slope regression line with its Y-intercept passing through the origin).

well in the center range of observed values while over and under-predicting at the extremes.

From the above discussion, one can obtain a statistical measure of how closely observed and modeled values compare. The conclusions that may be drawn, however, are subject to qualitative rather than quantitative considerations. For example, observed values are simply that, and as such are subject to all the errors associated with measurement and experimentation. Hence, it is possible that the model may more closely reflect reality than the observed values. Similarly, experiments are conducted by fixing all input variables except the "independent" variable which is to be entered over a range of values. The "dependent" variable is measured at each level of the independent variable. The model performs the same way, with the inference being that the "fixed" variables are the same in both the model and the observed values. However, this may not be so. For the grain drying example, grain variety may be different as may be ambient temperature. These factors influence the predictability of the model, but do not mean that the model is wrong. Rather, the model is constructed with a different set of assumptions (fixed variables) than the experiment.

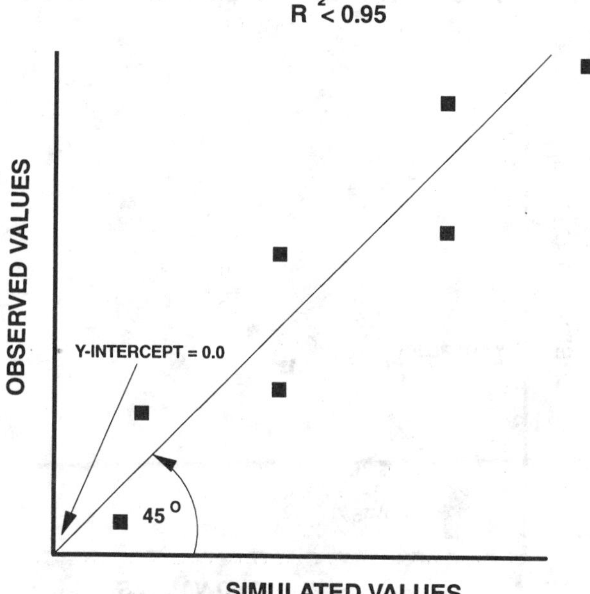

Figure 1.4–A comparison between simulated and observed values that produces less correlation than in figure 1.3 ($R^2 < 0.95$) with no bias (a 45° slope regression line with its Y-intercept passing through the origin).

Comparisons between observed and computed values usually are made by model developers rather than model users. Hence, the user of a model can only assume that the model is "acceptably" accurate to the model developer. However, this says nothing about how "acceptable" the model is to its user. Of course, design models must be more accurate than management or analysis models. Model users generally are more interested in "correct" trends than with "exact" answers. Models generally are regarded to be more accurate in predicting differences in strategies rather than absolute values.

Summary

The model user must recognize that models are reflections of reality and not reality themselves. However, mathematical models, as a group, generally are the best predictors of cause-effect relationships. Eventually, the user will develop a type of "faith" in the model based on how well the model reflects personal concepts of reality. Individual "faith" will increase if the model is "acceptably" accurate in simulating situations for which the user believes to know the correct answer. Again, it is very important that the inherent assumptions used in

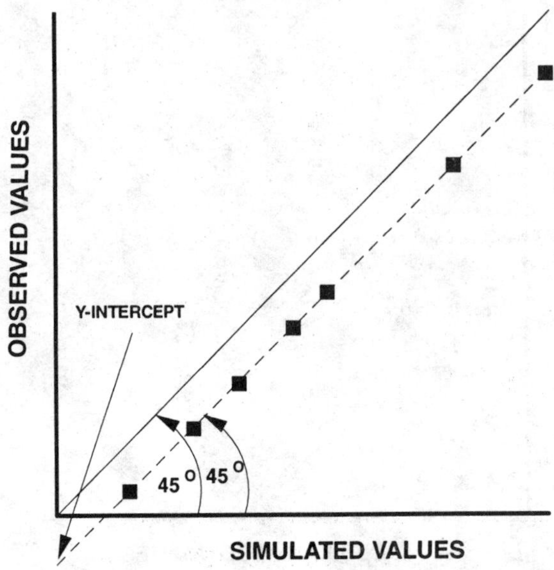

Figure 1.5–A comparison between simulated and observed values that produces perfect correlation (R^2 = 1.0) with a constant overprediction bias (a 45° slope regression line with its Y-intercept passing below the origin).

constructing the model are well known and understood if model utilization is to be maximized.

Computer Model Terminology

Computers are being used extensively in evaluating, designing, and managing grain drying and storage systems as evidenced by many of the relationships presented in this text. The increase in microcomputer capabilities has resulted in even greater utilization of computer software. The following discussion is directed toward some of the considerations associated with utilizing computer models for typical agricultural situations.

The overriding consideration in selecting equipment is to obtain an optimum set of components with respect to cost. This involves the design and evaluation of the subsystems that comprise the total equipment system in order to obtain the greatest return for a given set of resources. Design and evaluation involve the use of "models" or "simulations", the terms being used synonymously in this chapter. The primary objective of this discussion is to address many of the factors that should be considered when using a computer simulation model. In

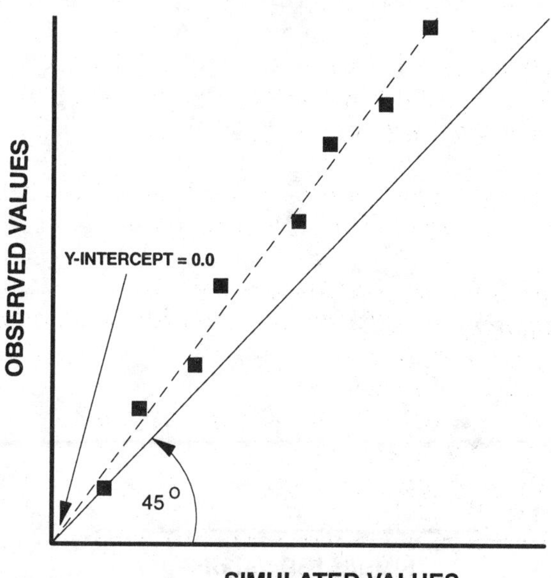

Figure 1.6–A comparison between simulated and observed values that has a progressively under-predicting bias (a greater than 45° slope regression line with its Y-intercept passing through the origin).

addition, many of the terms commonly used in computer simulation modeling will be defined.

Models and Model Building

A model may be defined as an abstraction of reality whereas model building is the process by which concepts of reality are refined and developed further. Models may be mental, physical, or mathematical. Mathematical models are precise quantitative expressions of the functioning of our mental and physical models; that is, a mathematical model is a logical extension of how one supposes a system to operate. When individuals say that the mathematical models they have developed are "bad" or "unacceptable", they are really saying that they have failed to express adequately the functioning of the system under study in an adequate mathematical form. Within defined boundaries a mathematical model should represent the "state of the art" with respect to the functioning of the system under study. Inadequate models are the result of inadequate information and/or inadequate modeling techniques, the latter being the subject of further discussion.

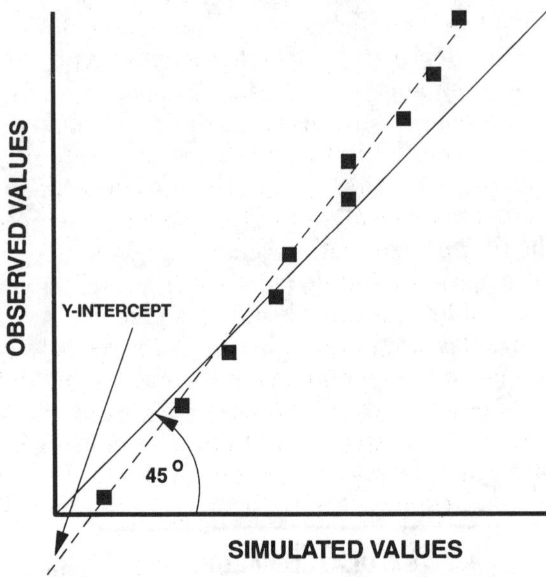

Figure 1.7–A comparison between simulated and observed values that has an overpredictive and then underpredictive bias.

What Are Your Objectives?

The first step in effective utilization of a mathematical model is to define the objectives of the study; that is, what are the questions that should be answered using the simulation model? All the remaining decisions concerning the simulation will depend in large part on this initial decision. One way of defining objectives closely is to list the desired output from a model. Each output relationship not known prior to utilization of the model represents an objective of the study.

The Model and Its Environment

Each simulation model operates in an environment composed of limitations and assumptions as specified by the model developer. Limitations define the boundaries of the system under investigation and include factors relating to breadth and depth.

Breadth refers to limitations imposed by physical, economic, geographical, political, climatological or biological boundaries. For example, in the extreme one might wish to model the performance of a single combine having a 12-ft header, costing between $50,000 and $60,000, operating in Kentucky on flat terrain with a clay soil during October, and harvesting 'Dare' soybeans at 13% moisture content. Each of the factors serves to limit both the breadth of the model and its applicability.

Depth is the degree of detail or aggregation desired in the simulation in order to accomplish its objectives. Using our previous example, does one desire sufficient depth to model each moving part of the combine, or is combine speed an adequate measure of machine performance? As with questions associated with breadth, the depth of the model must only be sufficient to accomplish the objectives of the study.

Assumptions refer to the suppositions made concerning the functional relationships considered in the model. Assumptions may be expressed explicitly through mathematical expressions or implicitly by the lack of a mathematical relationship. Explicit assumptions are defined by two types of variables: exogenous and endogenous.

Exogenous variables are determined outside the boundaries of the system. They influence the functioning of the system but are not in turn influenced by it. In essence, there is no feedback from the system to the exogenous variable. For example, in harvesting management systems, weather may be considered an exogenous variable in that it may influence field operations while field operations do not determine weather.

Endogenous variables are computed internally and may be a function of both exogenous and other endogenous variables. Endogenous variables affect the system and in turn are affected by it. For example, drying air temperature both influences and is influenced by thermostat operation.

Entities and Attributes

Having defined the objectives of the simulation study and its environment, a particular model may require that certain entities and their associated attributes be identified. An entity is an individual or object that maintains its identity throughout the simulation. The entity is described by its attributes. Attributes are characteristics that may or may not change over the simulation period and are either exogenous or endogenous variables. In a harvesting simulation study, a combine might be considered an entity. Certain attributes of the combine would be exogenous such as horsepower, grain tank capacity, and purchase price. Other attributes would be endogenous such as grain present in the grain tank, number of unloadings, and amount of grain harvested. An attribute is used in a model only if it is required to identify the entity, compute its performance, or provide pertinent output information. Using the combine example, horsepower will be required as an attribute only if it serves one of the three aforementioned functions; that is, to determine if horsepower is needed to identify the combine, compute combine performance, or provide output information essential to the model user.

Simulated Time and Model Selection

Simulated time is the chronological time reference used in simulation models and provides the model user with insight into the capabilities of a particular model. Simulated time may start or stop under conditions specified by the model user. As with chronological time, simulated time always advances. Unlike chronological time, simulated time may "stand still" while mathematical computations are being made. Simulation models are usually classified as being either static or dynamic, depending on the way in which simulated time is considered.

Static Models. Simulated time remains constant in static models. Endogenous variables are computed for a single point in time rather than recomputed periodically during the course of the simulation. This is not to say that time is not used in static models, but rather that it does not advance. Examples of static simulation models include the design of stationary structures (such as grain storage facilities) and the computation of depreciation for tax purposes. Most optimization models are static in that they give an optimum solution for one point in time. Linear programming (LP), for example, involves the solution to a single set of static equations. Although many iterations of the LP model may be required to obtain an optimum solution, simulated time does not advance. Likewise, many statistical regression models are static in that time does not appear as an independent variable.

Dynamic Models. Simulated time advances in a dynamic model, and endogenous variables are updated as required. Dynamic models are categorized as being discrete, continuous, or combinations of both.

Discrete Models. Development of a discrete model requires identifying the events that may be associated with the activities of an entity. Simulated time advances from one event to another. By definition, nothing occurs between the events of a discrete model, and the event occurs instantaneously; that is, simulated time "stands still" during the course of an event. In a discrete model, the entity is either involved in the event, or it is waiting for an event to occur.

To illustrate the above concepts, suppose that one wishes to simulate a harvesting system composed of one combine and one delivery truck. The world, as viewed by the combine operator, would in simple form consist of two events, start harvesting and stop harvesting to unload. This simple world, as seen through the eyes of the delivery truck operator, would consist of three events; receive grain from the unloading combine, exit the field for the grain facility, and return from the grain facility to the field. The two entities (combine and delivery truck) interface (or share) in the event where grain is unloaded from the combine onto the truck. If the truck arrives from the field and the combine has not completed filling its tank, the truck must wait. Likewise, if the combine holding capacity is reached and the truck has not yet arrived in the field, the combine must wait. Each of the events, i.e., arriving in the field, exiting the field, etc., occurs instantaneously in a discrete model, and nothing considered important to the results of the model occurs between these explicit events. For example, grain enters the tank between the time the combine starts and stops harvesting. However, for purposes of this example situation, the manner in which grain enters the grain tank is not required to accomplish the objectives of the study. If flow rates are needed for this purpose, a continuous model should be used.

Continuous Models. Simulated time advances in relatively small uniform increments in most continuous models, and the endogenous variables are updated at each time update. Continuous models are, in effect, a means of numeric integration over time. Waiting lines are not used in continuous models, and entities tend to be treated in the aggregate rather than as individuals.

Referring to the previous example, suppose one wishes to study the rate at which grain filled the combine receiving tank. One option would be to treat each grain kernel as an entity and to identify the series of discrete events associated with movement of the kernel from the plant to the tank. However, the more efficient method would be to use a continuous model and simulate grain flow in the aggregate, that is, to compute the flow rate of grain into the tank as a function of yield, combine speed and header width, and integrate the flow with respect to time.

Combined Models. Combined simulation models (not to be confused with the combine used in previous examples!) include both discrete and continuous features in combination. A combined simulation model has both time events and state events. Time events occur at a point in time and are associated with a discrete simulation model. State events are triggered by the status (or state) of the system. Both types of events are assumed to occur instantaneously.

One of the events used in the previous example was the stopping of the combine to initiate unloading after the grain tank filled. This event could be either a time or state event depending on how the model was constructed. For example, suppose the grain tank held 100 bu and the harvesting rate was 5 bu/min. If the "stop to unload" event was scheduled 20 min after harvesting began, it would be a "time" event. On the other hand, if the level of grain in the grain tank is computed periodically, say each minute, and the "stop to unload" event is stated to occur when the grain level reaches 100 bu, this would be a "state" event.

Flow Charts

Flow charts provide the model user with a visual display of model logic, and material and information flow, usually in simplified form. Well-constructed flow charts are especially effective in providing the model user a greater understanding of the overall complexity of the model and its inherent assumptions.

Input and Output

A well-written program will give the model user the option of displaying all input, intermediate and output variables. In this way, the total program logic may be examined readily.

Input data for the program may be supplied externally by the model user or internally by the program developer. For example, the program user might specify a certain type of tractor by model name. The program developer may have defined previously the data required to describe the tractor, i.e., horsepower, cost, etc. Usually it is desirable to list the user input for further reference. However, the model user should also have the option of seeing the complete data input.

In many models, the user may specify the types and quantity of output that is desired. As a minimum, output from the program should provide a relatively brief set of summary data that contains the primary factors to be computed from the simulation study. However, an abbreviated output often leads to additional questions concerning how the primary computations were determined. For example, harvesting efficiency of a combine may have been the calculation of primary interest. But, after seeing this efficiency in the primary factor output section, most individuals then wish to know particular details such as (1) how many times did the combine have to wait to unload, (2) what caused the delays in unloading, and (3) how much grain was actually harvested? Thus, it is often better to have a complete and detailed output section for future reference, even if it is not displayed automatically with every output.

Simulation Languages

Simulation languages are special purpose languages designed to help the model builder use programming time more efficiently. Generally, simulation languages are used in more complex models and are not of concern to the model user, the exception being a "spreadsheet" type simulation model used on a microcomputer. Typically, a simulation language is nothing more than a collection of subroutines that are able to perform special functions such as manipulation of data files, computing statistics, preparing histograms, and constructing graphical and tabular displays. Some simulation languages may only be used with either a continuous or a discrete model, and may be limited to one type of computer operating system.

Sensitivity Analysis

A sensitivity analysis is the process by which exogenous variables are evaluated with regard to their importance relative to simulation results.

A sensitivity analysis also provides a mechanism for testing the simulation in the extremes; that is, using extreme values of the exogenous variables will test rigorously the model in terms of mathematical logic and stability.

The sensitivity analysis begins with selection of model output results that are considered most crucial to the study. A set of "base" conditions is then established. Base conditions are comprised of the set of the best estimates of each exogenous variable. The base results are the simulation output obtained when using base condition values. Typically, a range of values is selected representing the extreme conditions associated with each exogenous variable to be evaluated. Simulation runs are made changing each value within the extreme range while holding all other base conditions constant. A comparison between changes in the base condition value as compared to changes in the base results provide an indication of the relative importance of the variable. Base condition variables may also be tested in larger groups to determine their interactive significance.

For example, in a model of a grain harvesting system, an exogenous input could be combine hopper size, and the primary simulation output could be the number of bushels of grain harvested in one day. First, the base conditions composed of all the exogenous variables, including hopper size, are defined. Then, the extreme and intermediate values to be tested are specified. For example, suppose the base value for hopper size was 100 bu. The alternative values for testing might be 25, 50, 75, 125, 150, 175, and 200 bu. A simulation run would be made for each of these values while holding the remaining base conditions constant. Results from each of the runs would be compared to the base results, and comparisons could take the form of ratios, percentages or absolute differences.

The sensitivity analysis may also be used to compare base conditions that have similar units of measure. For example, which is most important, an increase in delivery vehicle carrying capacity or combine hopper capacity? Although much of the interpretation of a sensitivity analysis is subjective, it does assist the model user in gaining a better understanding of the overall system performance and gaining an indication of the capabilities of a particular model.

Is the Model Accurate?

The question of whether a model is accurate is of great concern to the model user but must be followed by the question "as compared to what?". In the final analysis, one must ascertain whether the cost of obtaining an additional unit of accuracy is greater than the benefits to be gained in terms of the study objectives. Using our previous examples, a model that simulates the movement of each kernel of grain through the combine into the grain tank might provide greater accuracy than using an average filling rate to describe the same phenomenon. However, the cost of obtaining this additional accuracy, both in terms of user and computer time, may exceed its value if the objectives of the study are simply to determine the number of times that the combine fills and unloads during the course of a harvesting day. Also, using a more detailed model may reduce rather than increase accuracy. Again, detail should be a function of the study objectives.

Accuracy may be defined in terms of three progressive stages: verification, calibration, and validation. These stages are often defined differently in the literature and become the source of much discussion and confusion.

Verification is the process by which the programming logic is compared with the intentions of the model builder. In other words, does the programming logic of the model do "accurately" what the model builders intended for it to do? Stated differently, is the programming logic and the flow chart description exactly the same? For example, suppose that in a multi-delivery vehicle harvesting model, one wishes the combine to unload into the waiting delivery vehicle that has the most grain already in it. If the model carries out this desired logic correctly, we would say this portion of the simulation has been verified.

Calibration refers to the adjustments made to a model by its builder to give the most "accurate" comparison between simulated results and results obtained from field measurements. In other words, calibration involves the adjustment of certain model parameters by systematically comparing simulated results to field observations that did not necessarily measure directly the parameter in question. For example, suppose that one has reason to believe that corn harvesting rate is a function of variety, everything else being equal. It is possible through calibration to determine an appropriate parameter modifier so that the simulated results and field observations more closely correspond. Calibration should be conducted only within the confines of a given data set, and the model is said to be more "accurate" the closer the simulated and observed data. After the model has been calibrated, it may then be validated.

Validation is the process by which a simulation model is compared to field data not used previously in the calibration process. Validation involves subjective judgement both by the model builder and the model user. First, the areas for comparison must be selected. Then, a measure of "accuracy" or "closeness of fit" must be established such as the mean value of harvesting efficiency using the earlier example. It is also possible that the model will yield conflicting signals concerning validation. For example, in a dynamic model, two common measures of model validity are the mean and directional trend over time. It may be possible that the mean predicted by the model will be exactly that observed in the field while the trend is exactly opposite. Likewise, results from the model may be offset by a constant time factor. In other words, the model predictions are similar to field observations but occur some time units apart. Is this model valid? Again, this is a relative question and depends on subjective judgements and the objectives of the study. Although there is no single measurement that distinguishes between a valid and invalid model, a common criterion may be developed to determine if one model is superior to another. One would expect that only "verified, calibrated, and valid" models are released to general users. However, it is impossible to test every possible input to a large and complex simulation model, and the degree to which a model is "valid" may be viewed differently by the builder as compared to the user.

Summary

Simulation is a viable tool for the design and management of machinery systems in that it represents a logical extension of the thought processes used to make management decisions. Inadequate models are the result of inadequate information and/or inadequate model building techniques.

Effective model utilization begins by defining the objectives of the study and the environment under which a model is assumed to function. The model user must provide accurate input information if the model is to provide accurate output information; that is, "garbage in, garbage out" is very true when using computer models.

Models may be classified with respect to simulated time. If simulated time does not change, the model is said to be static. If simulated time advances, the model is said to be dynamic. Dynamic models may be further classified as being discrete, continuous, or combinations of both.

Modeling technique may be enhanced by using flow charts, proper programming procedures, input and output options, and simulation languages. Models of systems may be tested rigorously by the model user using a sensitivity analysis. Accuracy may refer to verification, calibration, or validation. Validity of the model is a subjective judgement based in part on the objectives of the modeling study.

The remaining chapters in this text often use computer models to generate design and system performance information. The collection of software that accompanies this text may be used to evaluate individual systems. It is important to understand how to use these models as well as to know their limitations.

Computer Models of Grain Systems

Beginning in the late 1960s, the following computer models were developed at the University of Kentucky, the University of Arkansas, the

University of Nebraska, and Purdue University. The developers of the models worked directly with farmers, extension personnel, and industry representatives, often in workshop settings (Loewer et al., 1977; Loewer and Bridges, 1986a), in determining their design needs in terms of grain drying and storage systems. Many of these models have been converted from mainframe to microcomputer programs. More recent models have utilized microcomputer spreadsheet programs.

The following discussion is designed to provide (1) a brief overview of these programs using some of the terminology presented earlier, and (2) to give the reader some understanding of how to use best a number of these programs. Many of these programs were used to generate the information contained in the remainder of the text and, in some instances, a more detailed description of a particular computer model is provided in later chapters. These programs are varied in size, type and function. Insofar as possible, they are included as part of an executable software collection on a diskette that may be acquired with this text. The reader who wishes to use these programs directly should examine the "README" file on the software diskettes. Note that the following programs may have different names on the diskette and as noted in the "README" file.

Grain Drying

FANMATCH. The FANMATCH model (Thompson, 1975a) determines the fan horsepower requirements for delivering a given volume of air per bushel of grain. It also computes the performance of a selected fan given a user defined bin and type of grain.

CROSSFLOW. The CROSSFLOW model (Thompson et al., 1968) projects the time to dry and drying rate, cost, and energy requirements for drying either corn or grain sorghum. The user must supply grain type, drying air temperature, fuel cost, incoming grain moisture content, the desired average final moisture content of the grain, and the air delivery capability of the fan. Crossflow drying is assumed; that is, the drying air moves across the flow of grain rather than in the same (concurrent) or opposite (countercurrent) direction. CROSSFLOW, programmed in FORTRAN, was originally a mainframe interactive program. It has been modified slightly from its original form and may now be run interactively using a microcomputer. Portions of this model have been included as a component of several of the models discussed below that require estimates of drying performance.

NATAIR. The NATAIR model (Thompson, 1975) projects the time to dry the top layer of grain to a specified final moisture content given either natural air or low temperature drying conditions. This model may be used also to evaluate aeration. Either corn or grain sorghum may be evaluated. The user must supply incoming drying air temperature and relative humidity, initial grain temperature and moisture content, desired final moisture content of the grain and the air delivery capability of the fan. NATAIR provides an estimate of dry matter decomposition losses over the drying period. As with CROSSFLOW, NATAIR is programmed in FORTRAN and was originally a mainframe interactive program. It too has been modified slightly from its original form and may now be run interactively using a microcomputer. Portions of this model also have been included in several of the models discussed below that require estimates of drying performance.

CONTNBN. The CONTNBN model (Bridges et al., 1983) evaluates the drying performance of continuous in-bin drying systems. Inputs include a schedule of incoming grain deliveries, bin size, fan performance data, and drying air conditions. CONTNBN models the dynamics of the grain drying process to determine if drying capacity is a bottleneck for a given grain delivery strategy. CONTNBN is a relatively large dynamic simulation batch model currently limited to the mainframe computer. It utilizes the GASP IV simulation language (Pritsker and Allen 1974).

LAYERD. The LAYERD model (Bridges et al., 1982) provides a filling schedule for a layer drying bin. Input information includes type of grain (corn or grain sorghum), ambient air temperature and relative humidity, drying air temperature, initial and desired grain moisture content, fan performance data, drying bin diameter and height, and whether the system is to be controlled by a thermostat or humidistat. A filling schedule is generated taking into account dry matter decomposition and the possibility of aflatoxin contamination. LAYERD is interactive and may be run on a microcomputer.

STIRDRY. The STIRDRY model (Bridges et al., 1984) was developed to provide management information concerning the use of stirring devices in individual layer and batch-in-bin corn drying systems. The model user specifies drying fan and bin information, drying temperatures, and ambient air conditions to compare stirred and non-stirred drying strategies. The model can be used as a decision-making tool to aid the producer in determining the benefits (if any) that a stirring device will provide to the individual drying system. The model may be used to compare average daily rates with and without stirring for several fan and bin combinations in a layer drying situation. STIRDRY is interactive and may be run on a microcomputer.

DUCT. The DUCT model (Bridges et al., 1988) determines duct sizes and duct spacing for aeration of rectangular storages. Design is based on the configuration of the grain mass, design airflow rate, grain volume, and a duct spacing criteria that will provide a relatively uniform distribution of air throughout the grain mass. DUCT is interactive and may be run on a microcomputer.

Facility Layout and Design

BNDZN. The BNDZN (pronounced "bin design") model (Loewer et al., 1976a) provides the layout dimensions, bill of materials, and estimated fixed and annual costs for centralized grain storage facilities. Most inputs are defaulted so that required information pertains only to drying method and rate, amount of storage capacity, and number of bins. An extensive set of input data contains the cost information for various components. All default input information may be modified by editing the input file directly. BNDZN is a static design model programmed in FORTRAN that may be run on a microcomputer.

CIRCLE. The CIRCLE model (Loewer et al., 1986b) provides the layout for a circular arrangement of storage bins. Inputs include storage capacity and the number of bins. No economic output information is provided. CIRCLE is a spreadsheet model that runs on a microcomputer.

FLATSTOR. The FLATSTOR model (Loewer et al., 1989) provides the geometric and purchase cost information for a flat storage system. Inputs include external dimensions, the grain angle of repose, and the cost of flooring and cover material. FLATSTOR is a spreadsheet model that runs on a microcomputer.

Total System Evaluation

CHASE. The CHASE model (Bridges et al., 1979) is a static design model that optimizes the delivery, drying, and storage components of 60 different types of systems and then ranks them as to purchase and annual cost. Physical layout information is provided also. Inputs include acreage, yield per acre, row width, harvest season (days), harvest time (hours/day), energy cost, portable drying time (hours/day), travel distance from the field to the facility, labor cost ($/hour), and initial and desired grain moisture content. CHASE is a batch FORTRAN model that may be run on a microcomputer.

SQUASH. The SQUASH model (Benock et al., 1981) is a dynamic simulation that describes the harvesting, delivery, handling, drying, and storage of grain at any point in time over the course of a day for almost any type of system. Inputs include descriptions of harvesters, delivery vehicles, drying, materials handling capacities, and management strategies. SQUASH provides a listing of events and a graphical output of grain flow and vehicle activities that occur over the day. The primary purpose of SQUASH is to identify bottlenecks in the total system given a set of equipment and management strategies. SQUASH is a batch FORTRAN model that is limited currently to the mainframe and utilizes the GASP IV simulation language.

EXSQUASH. The EXSQUASH model (Loewer et al., 1990) is an expert system that guides the user through a series of questions that helps identify the most probable bottleneck in the harvesting/delivery handling/drying/storage system. EXSQUASH is a complement to

SQUASH in that SQUASH information may be used in conjunction with EXSQUASH to better evaluate actual and potential bottlenecks. EXSQUASH utilizes the EXSYS expert system (EXSYS, 1985) and may be run on a microcomputer.

Economics of Systems

CACHE. The CACHE model (Loewer et al., 1976b) provides an economic analysis of having a grain drying and storage facility with or without feed processing. Inputs include harvesting strategies with and without a facility, volume of grain, expected prices of corn at harvest and when sold after storage, drying fuel cost, and feed processing information. The model gives insight into the economics of the total system and provides a "bottom line" estimate of whether on-farm grain drying and storage will be profitable. CACHE is an interactive FORTRAN program that may be run on a microcomputer.

HARVEST. The HARVEST model (Loewer et al., 1984a) is a dynamic simulation model that projects corn field and machine losses given a set of weather conditions, a harvesting rate, and quantity of grain to be harvested. It is a batch mainframe FORTRAN program that utilizes the GASP IV simulation language. Its primary utility is in establishing the quantities of energy associated with field and machine losses that serve as inputs to the OPTMC model.

OPTMC. The OPTMC model (Loewer et al., 1984a) is a static regression model that calculates the optimum moisture content to begin harvesting of corn given the value of corn, the cost of drying fuel, the quantity of grain to be harvested, and the harvesting rate. Weather conditions in western Kentucky were used to generate the regression equations. OPTMC is a FORTRAN program that may be run interactively on a microcomputer.

STIRECON. The STIRECON program (Loewer et al., 1984b) computes the economics of adding a stirring device to an in-bin drying system. Inputs include the diameter and eave height of the bin, the drying air temperature and relative humidity, fan performance information, and the capital investment associated with the stirring device. STIRECON is an interactive FORTRAN program that may be run on a microcomputer.

Seed Processing

PACASACS. The PACASACS model (Bucklin et al., 1989) is a dynamic simulation of seed flow through a processing plant including cleaning, bagging and delivery. Inputs include the processing rates of cleaners and spiral separators, delivery capacity of materials handling equipment, filling capacity of bagging machines, and work rates for individuals involved in bagging, stacking, and delivery. PACASACS is a batch mainframe FORTRAN program that utilizes the GASP IV simulation language.

JAWS. The JAWS model (Bucklin et al., 1982) is a static design simulation that provides an optimum equipment set and associated purchase and annual costs for a seed processing facility. Inputs include overall flow rates and the costs for the range of equipment available. JAWS is a batch FORTRAN program that may be run on a microcomputer.

Summary

The previous computer models were developed over many years and on many different computer systems. The diskette that was developed to accompany this text contains many of these programs. Each program was used in some way to generate portions of the information contained within this text. The skills needed to use these programs include a fundamental understanding of how to manipulate and edit files on a microcomputer. Some of the programs require an understanding of commercial spreadsheet programs.

Specific information about program availability and execution may be obtained from the "README" file on the diskette and the reference publications. These programs were not developed by commercial software producers for the general public. Hence, they may require some study if they are to be used effectively. However, almost all of them have been used very effectively in workshop and classroom settings by individuals with a broad range of backgrounds including farmers, county agents, students, and engineers. Economic inputs used as default values quickly become dated. Therefore, the user of programs that contain economic defaults (BNDZN for example) are encouraged to either update the economic defaults as necessary, use only the physical design portions of the programs, or to use the default economic inputs to compare systems rather than project current economic costs.

References for all chapters begin on page 541.

2

Principles of Drying

Drying of grain is usually essential for successful storability. The manager of on-farm drying systems needs to understand the principles of drying and how an individual situation may dictate the desired final moisture content for storage of grain and the selection of a particular drying method.

Purpose of Drying Systems

Storage of grain on the farm has become increasingly important to grain producers and feeders. Marketing and commercial storage situations have made on-the-farm drying and storage of grain advantageous from the standpoint of economics and the overall efficiency of grain harvesting and processing operations.

One of the advantages for a farmer having a drying system is the choice of whether to sell grain immediately after harvest or to store it for later sale or for feeding. Artificial drying provides a means of reducing grain moisture content so that grain can be stored safely, and so that there will be no price dockage because of high moisture when the grain is sold. A low storage moisture content also reduces the potential for insect and mold damage.

By having a grain drying system, a farmer can harvest grain much earlier, thus giving more individual control over the grain harvesting operation and providing many potential advantages. Among these are:

1. There is reduced potential for weather and pest related field losses. A drying system can improve both the quality and the quantity of the harvested grain.
2. Harvesting operations can be scheduled to obtain more efficient use of labor and available equipment. Harvest rates and the length of the harvest can be varied to accommodate the resources that the farmer possesses.
3. Preparation of land for double cropping or fall planting can begin sooner. This can be very important, especially in some double cropping situations where a few days difference in planting dates can have a significant impact on crop yields.

The primary disadvantages of on-farm drying systems relate to the cost of purchase, operation, and maintenance. In addition, the farmer assumes the risks associated with management and maintenance of the quality of stored grain.

Functional Requirements

The use of the word drying implies moisture removal. After growth ceases and a grain crop has reached maturity, a considerable amount of moisture must still be removed from the grain before it can be safely stored. Prior to use of mechanized harvesting techniques, essentially all the excess moisture was removed in the field by natural drying processes, one result being excessive field losses. With mechanized harvesting, grain can be harvested at moisture contents that are much too high for safe storage. The primary function of a farmstead drying system is to dry high moisture content grain to a safe storage moisture content, thus reducing the probability of excessive field losses associated with allowing the grain to dry in the field. Table 2.1 lists moisture contents at which various grains can be harvested and stored. Actual safe storage moisture contents will vary depending on the storage temperature, the condition of the grain and the length of the storage period. Figure 2.1, for example, contains curves that indicate allowable storage times for shelled corn for various combinations of temperature and moisture content. The lower the moisture content and temperature, the longer the grain can be stored safely. High moisture grain will deteriorate rapidly at high storage temperatures. In addition, wet grain will undergo chemical changes and is more susceptible to insect damage. However, the most severe problems encountered in storing high moisture grain are usually associated with the growth of various storage fungi and resulting mycotoxin production.

In most drying situations, warm air at a high relative humidity exits the drying zone. Grain above the drying front is exposed to this air before being dried. The exposure time should not exceed the maximum

Table 2.1. Commonly suggested moisture contents for harvesting and "safe storage" of various grains

Type of Grain	Moisture Content at Harvest (%)	Moisture Content for Safe Storage (%)
Shelled Corn	14 to 30	Up to 13.0
Soft Red Winter Wheat	9 to 17	Up to 13.5
Oats	10 to 18	Up to 13.0
Grain Sorghum	10 to 25	Up to 13.0
Soybeans	9 to 20	Up to 11.0
Rough Rice	18 to 22	Up to 13.5

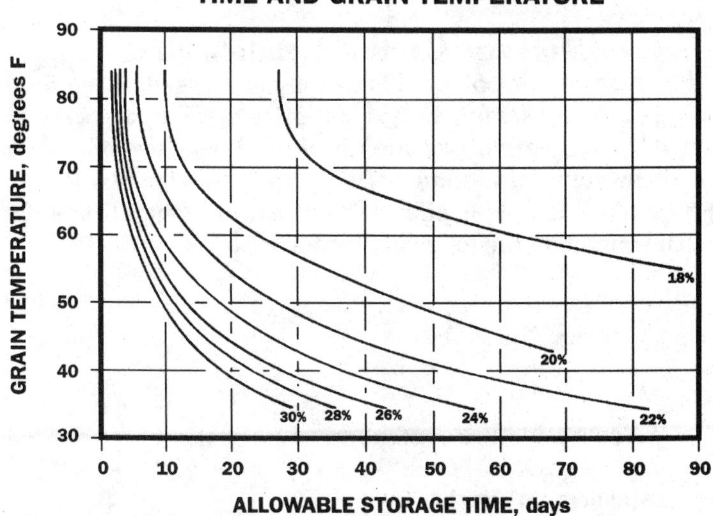

Figure 2.1–Allowable storage times for shelled corn for various grain moisture contents (Data from the USDA Grain Research Laboratory, Ames, Iowa).

allowable storage time for existing grain moisture and temperature. This performance restriction must be taken into account in designing any grain drying system.

A drying system should be capable of drying high moisture grain to safe storage moisture contents without a significant reduction in grain quality. Quality for commercial grains may be defined by government grading standards. Depending on the type of grain, quality may include such factors as germination levels, stress crack occurrence, breakage susceptibility, heat damage, and milling or processing indices. These factors can be affected by the dryer design, operational procedures, initial and final grain moisture contents, drying air temperature, airflow rates, and grain cooling procedures after drying. In addition, quality restrictions are often dependent on the intended use of the grain. The relationship of these parameters to grain quality and the economic value of the dried grain must be considered in designing any dryer.

In addition to the above considerations, a grain dryer should be functionally compatible with the farmer's grain harvesting, handling, storage, and processing facilities. It needs to be sized to accommodate the farmer's harvesting rate and total production of grain, and must consider labor and financial resources as well as management abilities. The sizing of grain dryers begins with determining the quantity of moisture that must be removed which, in turn, is dependent on grain moisture content.

Moisture Content Relationships

The moisture content of grain is usually expressed as the percentage of total weight of the sample that is water (referred to as "wet basis"). It is assumed that the grain is made up of only moisture and dry matter. Thus, if the weight of moisture (W_w) and the weight of dry matter (W_{dm}) in a given sample is known, its moisture content on a wet basis can be determined by the following formula:

$$M_{wb} = W_w * 100/W_t \tag{2.1}$$

where

M_{wb} = percent moisture content on a wet basis

W_t = $W_w + W_{dm}$

= total sample weight

(Weight and mass will be used interchangeably throughout the text.)

The oven method is one of the most accurate methods for determining the moisture content of grain. The usual procedure is to weigh a known quantity of grain and place it in a convection oven heated to the recommended temperature for a prescribed time. This procedure is designed to remove essentially all the moisture from the grain leaving only the dry matter. The dry matter is then weighed to determine W_{dm}. The weight of moisture in the original sample is equal to the initial sample weight (W_t) minus the final sample weight (W_{dm}). The moisture content can then be determined from equation 2.1. For example, if 100 g of grain yield 75 g of dry matter after the recommended oven treatment, the sample moisture content would be:

$$\begin{aligned} M_{wb} &= W_w * (100)/W_t = (W_t - W_{dm}) * (100)/W_t \\ &= (100g - 75g) * 100/100g = 25\% \end{aligned}$$

The moisture content of grain is sometimes expressed as a percentage by weight on a dry basis as follows:

$$M_{db} = W_w * (100) / W_{dm} \tag{2.2}$$

where

M_{db} = percent moisture content on a dry basis

Dry basis moisture content is always higher than wet basis moisture content. This can be illustrated using the values from the previous example to give:

$$M_{db} = W_w * 100 / W_{dm} = (W_t - W_{dm}) * 100 / W_{dm}$$
$$= (100g - 75g) * 100 / 75g = 33.3\%$$

Dry basis readings provide an easier method for making moisture calculations because the basis (W_{dm}) does not change as a material dries. Dry basis readings are used extensively in scientific research and professional journals; however, they are rarely used in commercial grain operations. Farmers will seldom encounter this method of specifying moisture content since all moisture meters are calibrated on a wet basis. All references to moisture content in this text assume wet basis unless stated specifically otherwise. Conversion from one basis to another is made using the following equations:

$$M_{db} = [M_{wb} / (100 - M_{wb})] * 100 \tag{2.3}$$

$$M_{wb} = [M_{db} / (100 + M_{db})] * 100 \tag{2.4}$$

Moisture Tester Considerations

The management of grain drying and storage systems requires an accurate measure of moisture content and a knowledge of how it relates to volume and quality. There are many different types of moisture testers available (a detailed discussion of their performance will follow later in the chapter). However, the oven method is considered the most accurate method. Unfortunately, it usually takes several days to complete. For example, the recommended test for corn specifies that the grain be placed in a 217° F oven for 72 h to remove completely all moisture. Obviously, this procedure is not satisfactory whenever a rapid determination of moisture content is needed.

Many moisture meters or testers are commercially available for use in making rapid determinations of grain moisture. These are secondary instruments that do not sense moisture directly. Instead, they usually measure an electrical property that is related to moisture content. Most modern testers measure the dielectric (capacitance) properties or the electrical resistance of the grain. Oven moisture readings are used to calibrate the output of the various instruments over a range of moisture contents. Separate calibrations are required for each type of grain.

Many factors can affect the moisture reading obtained from a specific moisture tester. Nelson (1980) states that temperature, grain bulk density, foreign material, kernel shape and size, and chemical composition can all affect the dielectric constant of grain. Hurburgh et al. (1981) compared moisture meter-oven readings for six commonly used meters to quantify their accuracy in measuring the moisture content of shelled corn. They found that in the 15 to 20% moisture content range it was not unusual for meter readings to differ as much as 1% from that found using a standard oven test. Differences at higher moistures (25 to 30% moisture) were found to be even higher (plus or minus 2 to 2.5% moisture). Much of this difference was attributed to sample-to-sample variability. Other investigators (Paulsen et al., 1983) have found similar variations in meter readings as compared to oven test results.

In operating any grain moisture tester, the user should carefully follow the manufacturer's instructions in order to achieve representative grain moisture readings. In addition, several basic rules should be adhered to including:

1. When testing the moisture content of grain, make sure representative samples of grain are obtained by drawing them from several locations within the grain bulk. Samples should not be taken from just one location such as the top of the grain mass.
2. Test as many samples as is reasonably possible within the time allowed, and average the results. The more samples taken, the more nearly the average value will represent the true average moisture content of the grain. Increasing the number of samples does not, however, correct for inherent inaccuracies in the tester or for improper measurement procedures by the operator.
3. If necessary, correct the moisture readings for temperature effects. Electrical properties of grain vary with temperature. Therefore, manufacturers often supply temperature correction charts or include automatic temperature compensation in the electrical circuit of the instrument.
4. When testing grain soon after it has been dried, take into account that the outside surface of the grain will have a lower moisture content than its center; therefore, the tester will indicate a moisture content that is lower than the true average moisture content of the grain. The approximate magnitude of this deviation can be determined by testing the same grain immediately after drying and cooling, and repeating the procedure again several days later with the same sample of grain. This information can be used to develop an approximate correction factor that can be applied to future moisture test readings taken under similar circumstances. One should use caution in applying this approach, however, since

the observed deviations in moisture readings will depend on a number of grain and drying system conditions, such as initial and final grain moisture content, type of grain, type of dryer, drying air temperatures, rate of drying, cooling time, etc.

Farmers should regularly compare their moisture meter readings with those from a grain elevator moisture tester. This will give them an indication of the accuracy of their moisture meters and allow them to compensate for any consistent meter bias. Although elevator moisture meters may not always be calibrated properly, they will generally be more accurate and consistent than the typical farm-type meter.

Definition of Grain Quantity

In producing, marketing and utilizing grain, the amount of grain involved in any operation is usually stated in terms of "bushels"; however, the term "bushel" can have different meanings depending on the situation in which it is being applied. It is very important that one be aware of which "bushel" designation is being used when discussing quantitative measures of grain.

By definition, the bushel is a volume measure containing 1.25 ft^3. Thus, when referring to bushels as measures of volume, one is speaking of the number of 1.25 ft^3 units of volume contained in a given quantity of grain. This does not indicate the weight of the grain unless the test weight of the grain in pounds per bushel is known.

Through general usage, the term "bushel" has come to have several other meanings. One of these is related to the buying and selling of grain by commercial elevators. For trading purposes, a bushel is designated as a certain weight of grain. This weight is usually based on the minimum test weight (lb/bu) listed in the USDA Grain Standards for that grain at its designated trading moisture content, and is defined as a bushel regardless of the grain's actual moisture content or test weight. For example, a bushel of corn is given as 56 lb, a bushel of wheat as 60 lb, and a bushel of oats as 32 lb. A load of shelled corn that contains 56,000 lb will be equal to 1,000 bu for trading purposes regardless of its actual volume. In fact, grain often becomes "packed" meaning that its weight per unit volume increases over time. These bushels are classified as "weight bushels".

Another interpretation of "bushel" is referred to as a "dry-matter bushel". It is defined as the quantity of grain that contains the same weight of dry matter as that contained in a "volume bushel" of standard test weight (as listed by USDA Grain Standards) at the designated trading moisture content for that grain. It is the weight of grain at any moisture content that will yield a standard bushel weight after its moisture content has been adjusted to its standard moisture content for trading purposes. This type of bushel is used in adjusting crop yields to a standard moisture content or for determining the dry matter weight of grain for the purpose of establishing its relative feeding value. For example, suppose a farmer harvested 10,000 lb of 25% moisture corn per acre and wishes to determine yield in terms of 15.5% moisture bushels. A 15.5% moisture standard bushel weighs 56 lb and contains:

$$\frac{56 \text{ lb} * (100\% - 15.5\%)}{100\%} = 56 \text{ lb} * (0.845)$$
$$= 47.3 \text{ lb of dry matter}$$

The 10,000 lb of corn contains:

$$\frac{10{,}000 \text{ lb / acre} * (100\% - 25\%)}{100\%} = 7{,}500 \text{ lb of dry matter / acre}$$

The number of 15.5% moisture content bushels per acre will then equal:

$$\frac{7{,}500 \text{ lb dry matter / acre}}{47.3 \text{ lb dry matter / bu}} = 158.5 \text{ bu / acre}$$

Another approach would be to establish the weight of 25% moisture corn required to yield 56 lb of 15.5% corn and divide this value into 10,000 lb. The weight of 25% moisture corn required to yield a standard bushel can be calculated by determining the weight of corn associated with the weight of dry matter (47.3 lb) in the standard bushel. This equals:

$$(47.3 \text{ lb / bu}) / [(100\%\text{-}25\%) / 100\%] =$$
$$(47.3 \text{ lb / bu}) / 0.75 = 63.1 \text{ lb / bu}$$

The yield then equals:

$$\frac{(10{,}000 \text{ lb / acre})}{(63.1 \text{ lb / bu})} = 158.5 \text{ bu / acre}$$

which is the same as the previous answer.

Grain Shrinkage During Drying

When grain is dried, the number of "dry matter bushels" does not change (except for certain "invisible" losses generally associated with handling, such as dust and trash, and ignored for purposes of this discussion). However, the overall weight and volume of the grain will be reduced during the drying process. This is commonly referred to as "shrinkage". In buying and selling grain, it is important to understand shrinkage and how drying can affect the value of grain.

Weight Shrinkage

The weight lost when grain is dried depends on the initial and final moisture content of the grain. It represents a reduction in the number of bushels (by weight) available for sale. The value per bushel must be sufficiently increased to off-set this shrinkage and to pay for the cost of drying the grain. Otherwise, a drying system cannot be justified economically. To be able to evaluate this situation, a farmer must determine the weight shrinkage expected from the drying process. This can be calculated by determining the weight of water lost during the drying process. In determining shrinkage, the weight of dry matter in the grain remains essentially constant during drying (ignoring invisible losses), but the total weight of the grain decreases with reductions in moisture content. Since the grain is assumed to be made up of only moisture and dry matter, the percentage of dry matter in the grain will always be equal to ($100\%-M_{wb}$). For example, if grain has a 20% moisture content, it will contain (100% - 20%) or 80% dry matter. Using this concept, the determination of weight shrinkage during drying is illustrated in the following example.

Suppose a farmer has 56,000 lb of 25% shelled corn. If the grain is sold without drying, the payment would be for 1,000 bu (each 56 lb by weight) at a reduced price per bushel based on the prevailing moisture dockage rate. If the farmer dries the grain to 15.5% moisture (the moisture content used in specifying the 56 lb/bu weight), payment would be received based on the market price per bushel but for a reduced number of bushels because of the weight lost from drying. In evaluating alternatives, the farmer must determine how many 56 lb bushels will remain to sell after drying. Since the weight of dry matter will be the same before and after drying, it is known that:

$$W_{dm} = \text{initial grain weight} * (100 - \text{initial moisture content})/100$$

and also

$$W_{dm} = \text{final grain weight} * (100 - \text{final moisture content})/100$$

where

W_{dm} = weight of the dry matter

Setting these two expressions equal to each other and inserting known values,

$$56{,}000 \text{ lb} * (100\% - 25\%)/100\% = \text{final grain weight} * (100\% - 15.5\%)/100\%$$

which yields

$$\text{final grain weight} = 56{,}000 \text{ lb} * 0.75/0.845 = 49{,}704 \text{ lb}$$

Dividing by 56 lb/bu, 888 bu remain at 15.5% moisture.

The shrinkage will be:

$$(56{,}000 \text{ lb} - 49{,}704 \text{ lb}) = 6{,}296 \text{ lb}$$

or

$$1{,}000 \text{ bu} - 888 \text{ bu} = 112 \text{ bu}$$

It is important to know how much grain will be left after drying. The above procedures can be expressed in equation form as:

$$\text{Final Quantity} = \frac{(100 - \text{initial \% moisture}) * (\text{initial quantity})}{(100 - \text{final \% moisture})} \qquad (2.5)$$

where grain quantity can be expressed in terms of either grain weight or bushels by weight.

Charts such as that in table 2.2 can be used to determine the weight fraction of grain left after drying from various initial moisture contents to certain specific final moisture contents. For the previous example, the table indicates that 88.3% or 883 bushels remain after drying from 25 to 15.5% moisture. This value is slightly less than the 888 bushels arrived at in the example because the table also includes a 0.5% invisible (or not accounted for) loss in addition to the calculated grain moisture losses. This additional reduction in weight is based on weight losses that typically occur in actual drying installations.

Volume Shrinkage

The volume of grain also decreases during the drying process. The amount of shrinkage volume can be computed by taking into account the reduction in the grain's weight during drying and its changing test weight (lb/volume bu) at different moisture contents. Typical test weights for shelled corn over a range of moisture contents are presented in table 2.3. These values will vary depending on variety, degree of mechanical damage, and other factors.

Table 2.2. Selected base and field moisture content relationships

Field Moisture Content	Fraction of Wet Weight after Drying*		Units of Wet Grain at Various Field Moisture Contents†	
	14.0%‡	15.5%‡	14.0%‡	15.5%‡
14.0	0.995	–	1.005	0.988
14.5	0.989	–	1.011	0.993
15.0	0.983	–	1.017	0.999
15.5	0.978	0.995	1.023	1.005
16.0	0.972	0.989	1.029	1.011
16.5	0.966	0.983	1.035	1.017
17.0	0.960	0.977	1.041	1.023
17.5	0.954	0.971	1.047	1.029
18.0	0.948	0.965	1.054	1.035
18.5	0.943	0.959	1.060	1.042
19.0	0.937	0.954	1.067	1.048
19.5	0.931	0.948	1.073	1.055
20.0	0.925	0.942	1.080	1.061
20.5	0.919	0.936	1.087	1.068
21.0	0.914	0.930	1.094	1.075
21.5	0.908	0.924	1.101	1.081
22.0	0.902	0.918	1.108	1.088
22.5	0.896	0.912	1.115	1.095
23.0	0.890	0.906	1.122	1.102
23.5	0.885	0.900	1.129	1.110
24.0	0.879	0.894	1.137	1.117
24.5	0.873	0.888	1.144	1.124
25.0	0.867	0.883	1.152	1.132
25.5	0.861	0.877	1.159	1.139
26.0	0.855	0.871	1.167	1.147
26.5	0.850	0.865	1.175	1.155
27.0	0.844	0.859	1.183	1.163
27.5	0.838	0.853	1.191	1.171
28.0	0.832	0.847	1.199	1.179
28.5	0.826	0.841	1.208	1.187
29.0	0.821	0.835	1.216	1.195
29.5	0.815	0.829	1.225	1.204
30.0	0.809	0.823	1.234	1.212
30.5	0.803	0.817	1.242	1.221
31.0	0.797	0.812	1.251	1.230
31.5	0.792	0.806	1.260	1.239
32.0	0.786	0.800	1.270	1.248
32.5	0.780	0.794	1.279	1.257
33.0	0.774	0.788	1.289	1.266
33.5	0.768	0.782	1.298	1.276
34.0	0.762	0.776	1.308	1.285
34.5	0.757	0.770	1.318	1.295
35.0	0.751	0.764	1.328	1.305

* From various field moisture contents to the indicated base moisture content.
† Required to obtain one unit of grain at the indicated base moisture content.
‡ Includes ½% invisible losses.

To find the percentage reduction in volume (volume shrinkage) associated with drying, the volume of grain before and after drying must be known. For example, a "volume" bushel (1.25 ft^3) of 26% moisture shelled corn will weigh 50.6 lb before it is dried (table 2.3). After drying to 15.5% moisture it will weigh (using equation 2.5):

$$\frac{50.6\text{ lb} * (100\%\text{-}26\%)}{(100\%\text{-}15.5\%)} = \frac{50.6\text{ lb} * 74}{84.5} = 44.3\text{ lb}$$

Assuming its test weight at 15.5% moisture is 56 lb/bu, its new volume will be:

$$\frac{44.3\text{ lb}}{56\text{ lb / bu}} = 79.1\%\text{ of the original volume or}$$

Thus, the original bushel of corn now occupies

$$0.791 * 1.25\text{ ft}^3\text{/bu} = 0.99\text{ ft}^3$$

The percent reduction in volume or volume shrinkage from the drying process is therefore equal to:

$$(1.25\text{ ft}^3\text{/bu} - 0.99\text{ ft}^3\text{/bu}) * 100/1.25\text{ ft}^3\text{/bu} = 20.8\%$$

Table 2.3. Variation in shelled corn test weight (pounds per bushel) with changes in moisture content*

Moisture Content (% wet basis)	Test Weight (lb / bu)
10.0	57.5
12.0	57.3
14.0	56.7
15.5	56.0
16.0	55.8
18.0	54.7
20.0	53.1
22.0	52.0
24.0	51.3
26.0	50.6

* Based on data from Lorenzen, 1958.

Table 2.4. Approximate volume shrinkage (percent decrease in volume) of several grains when dried to 12% moisture content

	Initial Moisture Content (% wet basis)			
Grain Type	17	20	25	30
Milo	7.0	12.3	–	–
Rough Rice	3.8	6.5	11.5	–
Shelled Corn	9.0	15.8	24.2	30.9
Wheat	11.5	18.7	–	–

Thus, 26% moisture shelled corn will occupy 20.8% less volume after drying to a 15.5% moisture than it did before. Another interpretation is that the grain depth in a bin-type drying system will be reduced by approximately 20.8% after drying 26% moisture shelled corn down to 15.5%.

Table 2.4 presents typical volume shrinkage values that one can expect from drying several common grains to 12% moisture content from selected initial moisture contents.

Air-Water Vapor Mixtures

Grain drying depends on air-water vapor mixtures as well as grain moisture content.

Properties and Relationships

Ambient air is composed of a mixture of various gases (primarily nitrogen, oxygen, argon, carbon dioxide, and hydrogen) and water vapor. An understanding of the physical and thermal properties of air-vapor mixtures is essential to the analysis of grain drying processes. The psychrometric chart is a graphical representation of these properties and their interrelationships, and can be used in the analysis of grain drying systems.

Definition of Terms

It is necessary to become familiar with the meaning of the various terms used to describe the properties of such mixtures in order to understand air-vapor mixtures and the psychrometric chart. The most commonly used terms are:

Dry Bulb Temperature. This temperature is usually referred to as ambient air temperature or drying air temperature in a drying system. It is the air temperature measured by using any of a number of general purpose temperature measuring devices or thermometers.

Wet Bulb Temperature. This temperature is measured by using a wet bulb thermometer. A wet bulb thermometer is similar to that used to measure dry bulb temperature except that its bulb is covered with a water-moistened wick. When placed in a moving air stream, the thermometer bulb is cooled by the evaporation of water from the wick. After several minutes, a steady-state condition is attained; the heat transferred from the surrounding air to the thermometer bulb is equal to the heat being utilized to evaporate moisture from the wick. At this point, the thermometer will indicate a minimum temperature which is referred to as the wet bulb temperature. This temperature is also the temperature to which an air-vapor mixture will cool when saturated with moisture and without the addition of any external energy. The heat required to evaporate the mixture that is added to the air as it is being saturated must come from the air itself. The wet bulb depression (difference between the wet and dry bulb temperatures) depends on the dry bulb temperature and the relative humidity of the air.

Dew Point Temperature. If an air-vapor mixture is cooled without adding or taking away any moisture, its relative humidity will increase until it reaches 100%. Any further cooling will cause moisture to condense out of the air. The temperature at which this condensation begins is called the dew point temperature. Whenever air next to a cold surface is cooled below its dew point temperature, moisture will condense on the surface. This occurs on the inside surface of window panes when the outside temperature drops below the dew point temperature of the inside room air.

Dew point temperature can be measured using any of several available dew point sensors; however, these are usually more complicated and expensive than wet bulb thermometers. For this reason, dew point temperatures are not used as often as wet bulb temperatures in establishing the psychrometric properties of air-vapor mixtures.

Humidity Ratio. The ratio of the weight of water vapor in an air-vapor mixture to the weight of dry air associated with that moisture is defined as the humidity ratio. Usually it is expressed in terms of weight of water per unit weight of dry air.

Vapor Pressure. Vapor pressure is the pressure exerted by the water vapor in a given sample of air. If the air is saturated with water vapor (that is, it contains all the water vapor it can hold under the existing conditions), the pressure is referred to as the saturated vapor pressure.

Both the dry air and the water vapor components of any air-vapor mixture produce partial pressures related to the mixture temperature and the relative concentration of the components. Total vapor pressure is equal to the sum of the component partial pressures and, in an unpressurized environment, is equal to the prevailing atmospheric pressure.

Relative Humidity. Relative humidity is defined for a given dry bulb temperature as the ratio of the vapor pressure of the water vapor contained in an air-vapor mixture to the vapor pressure of an air-vapor mixture that is completely saturated (air having a relative humidity of 100%).

Enthalpy (Heat Content). Enthalpy is the amount of heat energy contained in an air-vapor mixture per unit weight of dry air. It is usually expressed in terms of Btu/lb dry air and includes both sensible and latent heat components. Sensible heat is that heat associated with a dry bulb temperature increase of an air mixture. Latent heat, as used in air-vapor mixtures, is the heat required to change the state of water (liquid to vapor or vapor to liquid) without changing its temperature.

Heat content is not usually expressed on an absolute basis but rather as the difference in energy between the condition being considered and reference condition. The usual reference temperatures for dry air and water are 0° F and 32° F, respectively. The enthalpy, h, of an air vapor mixture can be expressed as:

$$h = 0.24\, T_{db} + W * \left[1{,}075.2 + 0.45\left(t_{db} - 32\right)\right] \tag{2.6}$$

where

h	=	enthalpy in Btu/lb of dry air
0.24	=	specific heat of dry air (Btu/lb-°F)
W	=	humidity ratio (lb water vapor/lb of dry air)
T_{db}	=	dry bulb temperature (°F)
1,075.2	=	heat of vaporization of water at 0° F (Btu/lb water)
0.45	=	specific heat of water vapor (Btu/lb-°F)

Volume. In psychrometric relationships, two terms are used to define volume. "Humid" volume is the volume occupied by an air-vapor mixture per pound of dry air at a given temperature. Specific volume is the volume of dry air per pound of dry air. The differences between humid volume and specific volume are sufficiently small so that they are used interchangeably in this text. The base of one pound of dry air is used since the pounds of dry air involved in any psychrometric process remains constant under steady state conditions. The pounds of dry air in any psychrometric process equals the total volume of air being considered, divided by the humid or specific volume.

Psychrometric Chart

The psychrometric chart is a graphic representation of the various psychrometric properties of moist air (see Appendix A, page 527). A skeletonized psychrometric chart is shown in figure 2.2 while a more

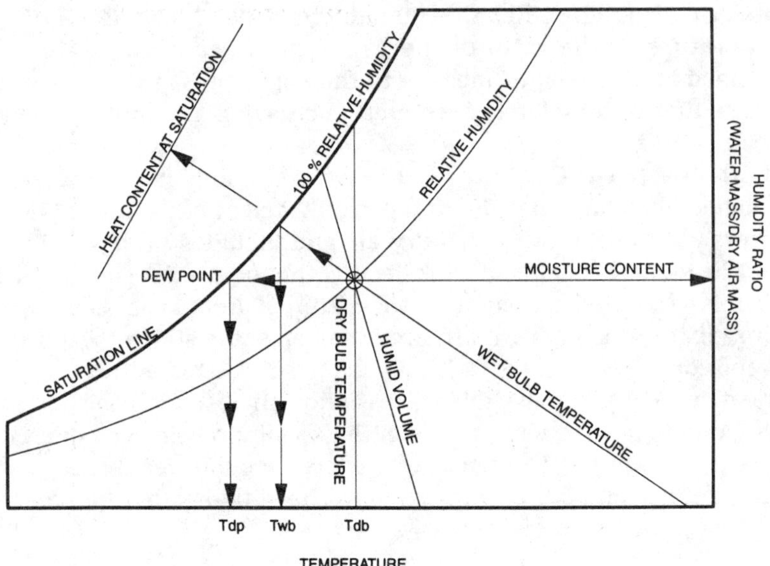

Figure 2.2–Skeletonized psychrometric chart.

detailed chart for standard atmospheric pressure is presented in figure 2.3. An understanding of this chart, the meanings of its lines, and its use in working with various psychrometric processes is important when studying crop drying processes.

The psychrometric chart provides a graphical means of specifying the theoretical relationships that define the properties of air-vapor mixtures. If any two psychrometric properties are known, all other psychrometric properties can be determined.

Psychrometric processes, such as the heating, cooling, drying and humidifying of air-vapor mixtures, can also be traced on the chart. The tracing procedure may be used to determine the changes in psychrometric properties as a result of different conditioning operations.

In figures 2.2 and 2.3, dry bulb temperature (T_{db}) is plotted along the bottom of the chart with vertical lines representing constant dry bulb temperatures. Horizontal lines represent conditions of constant humidity ratios (W) and constant dew point temperatures. Humidity ratio can be read from the vertical scale along the right edge of the chart. Dew point temperature (T_{dp}) can be measured by moving horizontally to the left until the saturation (100% relative humidity) line is reached. The dew point temperature can be read either from the temperature scale on the saturation line or by moving vertically down to the temperature scale along the bottom of the chart.

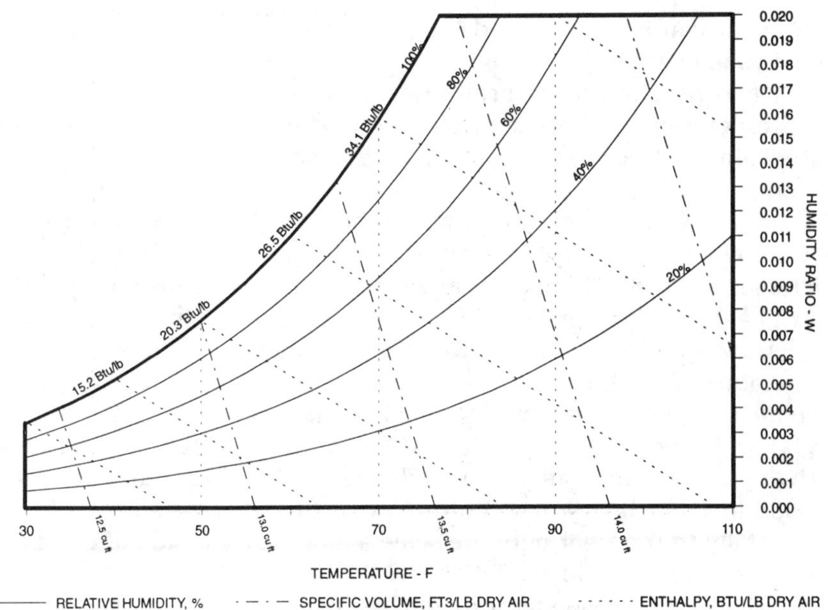

Figure 2.3–Psychrometric chart.

The curved lines that extend from the lower left side of the chart to the upper right side of the chart are relative humidity (Rh) lines. The curved line on the left side of the chart represents the saturated or 100% relative humidity line. Movement to the right represents progressively lower relative humidities.

Wet bulb temperature (T_{wb}) lines slant up from the lower right to the upper left of the chart and, for practical purposes, may be considered to be constant enthalpy lines. They are distinguished from the constant humid volume (HV) lines in that their slopes are not nearly as steep. Wet bulb temperature can be determined for any point on the chart by proceeding along or parallel to the nearest wet bulb line until the saturation line is reached. The wet bulb temperature then can be read from the temperature scale on the saturation line or from the dry bulb temperature scale at the bottom of the chart in a manner similar to that used to determine the dew point temperature.

Enthalpy (h) of any air-vapor mixture can be determined by considering wet bulb lines to be lines of constant heat. By extending these lines past the saturation line until they intersect the heat content (or enthalpy) scale, one can determine the heat content per pound of dry air.

Humid volume (HV) lines slope from the lower right to the upper left at a steeper angle than the wet bulb lines. The humid volume for any state point can be determined by interpolation between the different humid volume lines.

Any two psychrometric properties of an air-vapor mixture may be used to determine all other properties of the mixture by locating the state point of the mixture on a psychrometric chart (fig. 2.4). For example, suppose an air-vapor mixture has a dry bulb temperature (T_{db}) of 80° F and a relative humidity of 40%. The state point can be located by moving up the constant 80° F dry bulb temperature line until it intersects the 40% Rh line. The wet bulb temperature (63.5° F), Point A, can be determined by moving from the state point parallel to the wet bulb lines until the saturation line is reached. The dew point temperature (54° F), Point B, can be established by moving horizontally from the state point to the saturation line. The heat content (29.0 Btu/lb dry air) can be established by extending the wet bulb line to the enthalpy scale and reading the indicated value. The humidity ratio (0.0088 lb H_2O/lb dry air), Point C, can be determined by moving horizontally to the right until the scale along the right side of the chart is reached. The humid volume (13.8 ft^3/lb dry air), Point D, can be established at the given state point by interpolating between the two adjacent humid volume lines.

As an additional exercise, assume that an air-vapor mixture has a dry bulb temperature of 90° F and a wet bulb temperature of 65° F. Following the previous procedure results in the following values:

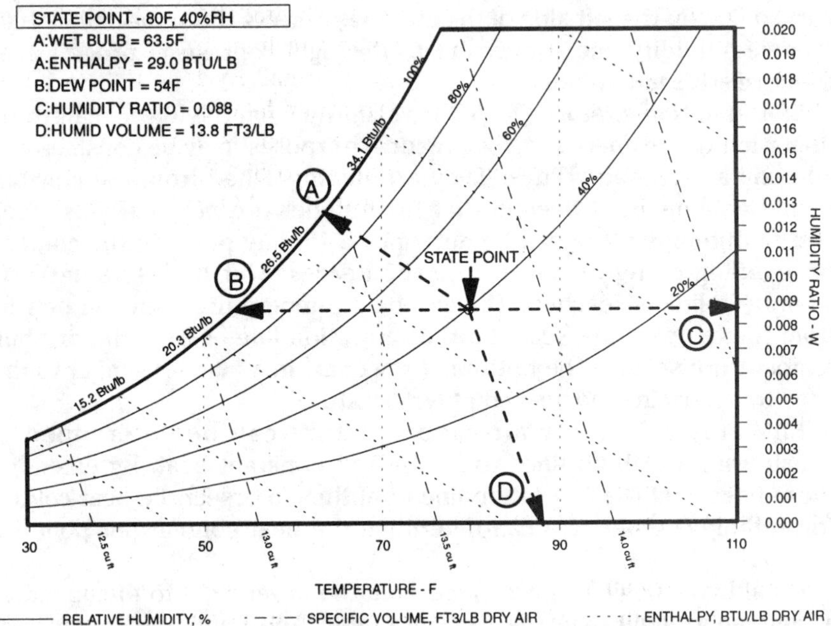

Figure 2.4–Properties of air given state point conditions.

T_{dp} = 49.3° F
h = 30.0 Btu/lb dry air
Rh = 25%
W = 0.0074 lb H_2O/lb dry air
HV = 14.0 ft^3/lb dry air

Psychrometric Processes

Psychrometric processes include heating, cooling, humidifying or dehumidifying air-vapor mixtures, and mixing various air volumes. Many of these processes are important in the operation of a crop dryer and must be understood in analyzing the performance of different drying systems.

Heating or Cooling

These processes are similar, as long as the air temperature during cooling always remains above the dew point temperature. Sufficient information must be known about the heating or cooling system so that initial and final air conditions can be determined. In heating (or cooling without condensation), any change in temperature is assumed to occur without a change in moisture content; hence, the air will maintain a constant humidity ratio. Therefore, the final and initial air conditions will be on the same horizontal line (constant dew point temperature and constant humidity ratio). For example, suppose ambient air at 60° F and 70% Rh is to be heated to 80° F for crop drying (fig. 2.5). What is the relative humidity of the heated air and the heat input required per pound of dry air passing through the system? By locating the initial state point on the psychrometric chart, initial heat content of the air is determined to be 22.7 Btu/lb dry air, and its humid volume is 13.25 ft^3/lb dry air. By moving horizontally to 80° F, the state point of the heated air is located. Its heat content is 27.8 Btu/lb dry air, its relative humidity 35% and its humid volume 13.75 ft^3/lb dry air. The heat added per pound of drying air, dh, equals the increase in enthalpy between the two state points.

dh = (27.8 Btu/lb dry air – 22.7 Btu/lb dry air)
= 5.1 Btu/lb dry air

If the airflow rate through the heating section is known, the total heat input required can be computed. Suppose the fan is supplying 5,000 cfm (ft^3/min) of ambient air to the dryer.

pounds of dry air moved per hour = (5,000 ft^3/min*60 min/h)/(13.25 ft^3/lb dry air)
= 22,642 lb dry air/h

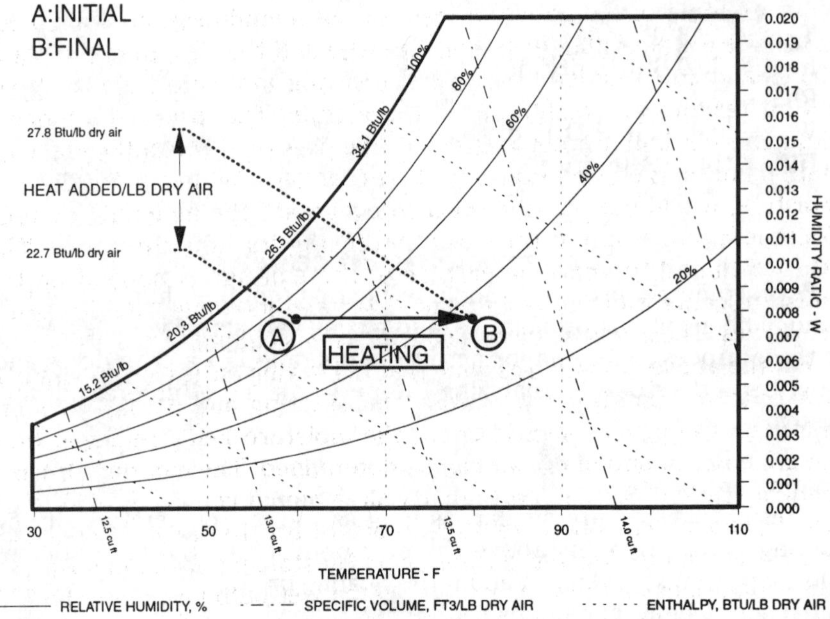

Figure 2.5–Heating example.

total heat input
per hour = (lb dry air/h) * (Btu input/lb dry air)
= (22,642 lb dry air/h) * (5.1 Btu/lb dry air)
= 115,474 Btu/h

Therefore, the burner must supply heat at the rate of 115,474 Btu/h to accomplish the indicated temperature rise in the drying air. Heat transferred in a cooling application can be computed in the same manner; however, it is important to note that if the air temperature drops below its dew point temperature, some dehumidification must occur. In that case, the problem involves a combination of cooling and condensation, and the state point of the cooled air will be below the horizontal line through the original state point. It will then have a lower dew point temperature and a lower humidity ratio.

Drying or Wetting

In grain drying, most drying and wetting processes are assumed to be adiabatic; that is, they occur under essentially constant heat conditions. The water vapor is either added or removed from the air without any change in the heat content of the air. The ratio of latent to sensible heat changes, but the total heat does not. Such a process is then assumed to take place along a constant enthalpy (constant heat) line on the psychrometric chart.

An example of the above process is the humidification of air being used to dry a bin full of high moisture wheat. If the air enters the bin at 100° F and 30% relative humidity, it will gain moisture from the grain until it reaches an equilibrium with the grain. This process will occur along the wet bulb (or constant heat) line passing through the initial air state point on the psychrometric chart (fig. 2.6). As the air increases in moisture, it will use sensible heat to evaporate the moisture, thereby decreasing air temperature and increasing its humidity ratio. The relative humidity and temperature of the exiting air depend on the moisture content of the grain being dried and its equilibrium moisture properties. These will be discussed later in this chapter.

For the above situation, assume that the drying air leaves the grain at 85% relative humidity. By locating the entering and exiting air state points on the psychrometric chart, the moisture being removed from the grain per pound of dry air can be determined. The entering air has a humidity ratio of 0.0124 lb H_2O/lb dry air, a humid volume of 14.37 ft^3/lb dry air, and a heat content (enthalpy) of 38.1 Btu/lb dry air. The exit air conditions can be located on the chart by starting at the entering air state conditions and moving parallel to the wet bulb lines until the 85% relative humidity line is reached. The exit air temperature is 78° F, its humidity ratio is 0.0176 lb H_2O/lb dry air and its heat content is still

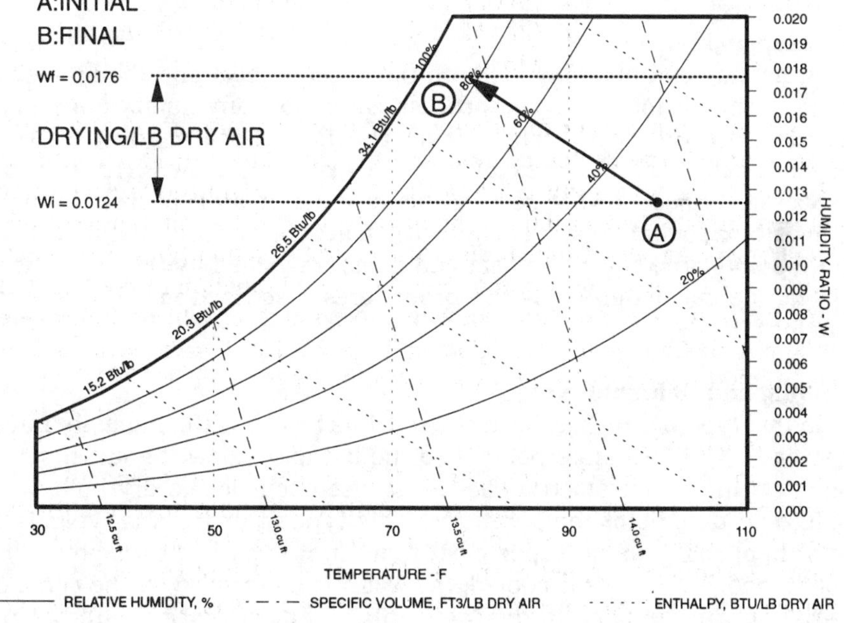

Figure 2.6–Drying example.

38.1 Btu/lb dry air. The moisture gained as the drying air passes through the grain mass, dW, can be calculated as follows:

$$
\begin{aligned}
dW &= W_{out} - W_{in} \\
&= (0.0176 \text{ lb } H_2O/\text{lb dry air} - 0.0124 \text{ lb } H_2O/\text{lb dry air}) \\
&= 0.0052 \text{ lb } H_2O/\text{lb dry air}
\end{aligned}
$$

The moisture removed from the grain per hour, dM/dt, can also be calculated if the airflow rate is known. For 5,000 cfm of incoming air the moisture removal rate will be:

$$
\begin{aligned}
\frac{dM}{dt} &= \left(5{,}000 \text{ ft}^3 / \text{min}\right) / \left(14.37 \text{ ft}^3 / \text{lb dry air}\right) * \\
&\quad \left(60 \text{ min} / \text{h}\right) * \left(0.0052 \text{ lb } H_2O / \text{lb dry air}\right) \\
&= 108.48 \text{ lb water} / \text{h}
\end{aligned}
$$

If the total moisture to be removed in the drying process is known, the drying time can be estimated by dividing the total weight of moisture to be removed by the rate of moisture removal. This calculation assumes that the drying rate remains relatively constant over the entire drying period. If the rate of moisture removal decreases later in the drying process, the drying time will increase accordingly.

Heating and Humidifying

In heating and humidifying, both the sensible heat and the humidity ratio of the air increase. The process can follow any number of paths moving up and to the right from the initial state point on the psychrometric chart. The exact path is usually not known. This type of process is typical of what occurs when ventilation air is passed through a livestock shelter or a plant growth structure. If the initial and final state points are known, the heat and moisture gained by the ventilation air can be calculated using the procedures discussed in the previous sections.

Cooling and Dehumidifying

In this type of process, both the sensible heat and the humidity ratio decrease. The final state point is to the left and below the initial state point on the psychrometric chart. Changes occur in the dry bulb, wet bulb, and dew point temperatures. This type of process is typical of what happens to an air-vapor mixture when it is passed through an air conditioner. As the air is cooled, the water is condensed on the cooling coils. Again, if the final state point can be established, the psychrometric chart can be used to determine the heat and moisture removed from the conditioned air.

Summary and Discussion

Dry bulb, wet bulb and dew point temperature, along with the other psychrometric properties of air-vapor mixtures, are related so that if any two of these properties are known, the rest of the properties may be determined from the psychrometric chart.

Heating and cooling of air without any change in its moisture content involves a change in sensible heat only and takes place along a horizontal line on the psychrometric chart. This process occurs when air is heated prior to its use for grain drying.

In a constant heat-humidification process, air picks up moisture without any change in its total heat content. This type of process occurs when air is passed through an evaporative cooling pad or through moist grain in a drying system. The process can be assumed to be a constant wet bulb temperature process. The air temperature is lowered as sensible heat from the air is used to evaporate moisture from the cooling pad or the moist grain. The sensible heat used to vaporize moisture remains in the air as latent heat; therefore, no significant amount of heat is added or removed from the process air. Dew point temperature, relative humidity, and humidity ratio all increase during this type of process. In grain drying, the relative humidity of the air leaving the grain will depend on the type of grain, its moisture content, the depth of the grain, and the airflow rate.

Equilibrium Moisture Properties

Drying and storage relationships for various grains are directly related to their equilibrium moisture properties. Because grain is hygroscopic, it will exchange moisture with the surrounding air until the vapor pressure of the moisture in the grain and that of the air reach a state of equilibrium. If grain comes to equilibrium with air maintained at a relatively constant environmental condition, the grain moisture content is referred to as the equilibrium moisture content (EMC) corresponding to the existing air conditions. On the other hand, if the grain is surrounded by a limited amount of air (such as occurs in the interstitial spaces of a grain mass in a storage bin), the air will reach moisture equilibrium with the grain without any significant change in the grain moisture content. The relative humidity of the air in this situation is referred to as the equilibrium relative humidity (ERH) corresponding to the existing grain moisture content at the prevailing temperature. All equilibrium moisture properties are a function of temperature; that is, the properties change with changes in temperature.

Table 2.5. Equilibrium moisture content values (wet basis) for shelled corn at various moisture contents and relative humidities*

Air °F	Air Relative Humidity† % 45	50	55	60	65	70	75	80	85
30	12.5	13.4	14.3	15.2	16.2	17.2	18.3	19.5	21.0
40	11.9	12.7	13.6	14.5	15.4	16.4	17.5	18.7	20.0
50	11.3	12.2	13.0	13.9	14.8	15.7	16.7	17.9	19.2
60	10.9	11.7	12.5	13.3	14.2	15.1	16.1	17.2	18.5
70	10.5	11.2	12.0	12.8	13.6	14.5	15.5	16.6	17.8
80	10.1	10.8	11.6	12.4	13.2	14.0	15.0	16.0	17.2
90	9.7	10.5	11.2	12.0	12.7	13.6	14.5	15.5	16.7
100	9.4	10.1	10.9	11.6	12.4	13.2	14.1	15.1	16.2
110	9.2	9.8	10.5	11.3	12.0	12.8	13.7	14.6	15.8
120	8.9	9.6	10.2	10.9	11.7	12.5	13.3	14.3	15.4
130	8.7	9.3	10.0	10.7	11.4	12.1	13.0	13.9	15.0
140	8.5	9.1	9.7	10.4	11.1	11.8	12.7	13.6	14.6

* Modified Henderson-Perry equation, ASAE, 1985.

† Boxed areas indicate moisture contents usually considered safe for long-term storage. Conditions to the left of the boxes represent overdried grain while those to the right will require additional heat supplied to the air for drying to safe long term storage.

Table 2.6. Equilibrium moisture content values (wet basis) for grains at various relative humidities, all at 77° F*

Type of Grain	Relative Humidity (%) 10	20	30	40	50	60	70	80
Barley	4.7	6.9	8.4	9.6	10.6	11.9	13.4	15.7
Buckwheat	5.6	7.7	9.2	10.2	11.2	12.4	13.9	15.9
Corn								
shelled, YD	5.3	7.4	8.8	9.8	10.8	12.2	13.9	16.3
shelled, WD	5.0	7.1	8.8	10.0	11.0	12.4	14.0	16.1
Popcorn, Y	4.2	6.1	7.6	9.1	10.5	12.2	14.1	16.5
Oats	4.5	6.6	8.2	9.4	10.3	11.4	12.8	15.0
Rice								
whole	5.9	8.0	9.5	10.9	12.2	13.3	14.1	15.2
milled	4.9	7.7	9.5	10.3	11.0	12.0	13.4	15.3
Rye	5.3	7.4	8.8	9.8	10.8	12.2	13.9	16.3
Sorghum	5.0	7.4	9.2	10.4	11.6	12.7	13.8	15.0
Soybeans	3.8	5.3	6.1	6.9	7.8	9.7	12.1	15.8
Wheat								
Duram	5.3	7.3	8.5	9.4	10.4	11.7	13.4	15.8
soft red, w.	4.8	7.0	8.6	9.8	10.8	12.1	13.6	15.8
hard red, w.	5.0	7.2	8.2	9.2	10.9	12.1	13.8	16.0
hard red, s.	5.3	7.2	8.4	9.5	10.7	12.2	14.0	16.3

* ASAE, 1985.

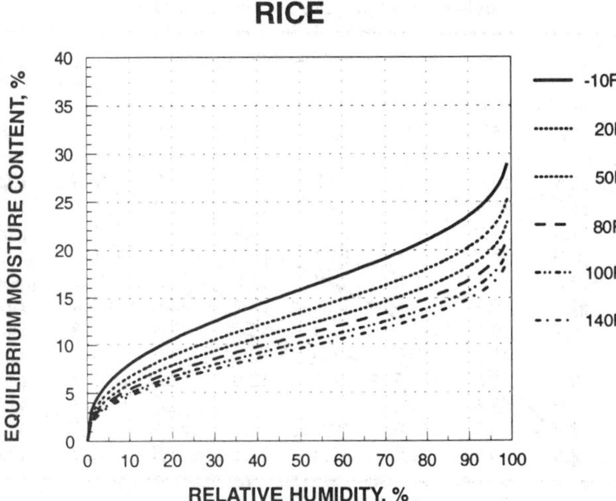

Figure 2.7–Equilibrium moisture content for rice (ASAE, 1985).

Equilibrium moisture properties are specific for each type of grain and are important in developing storage and drying recommendations. Typical equilibrium moisture content (EMC) values for shelled corn are presented in table 2.5 for a range of temperatures and relative humidities. Equilibrium moisture contents for other grains at various relative humidities for one temperature (77° F) are presented in table 2.6. These values can be used to compare EMC data for various grains under the same environmental conditions. The EMC graphs for

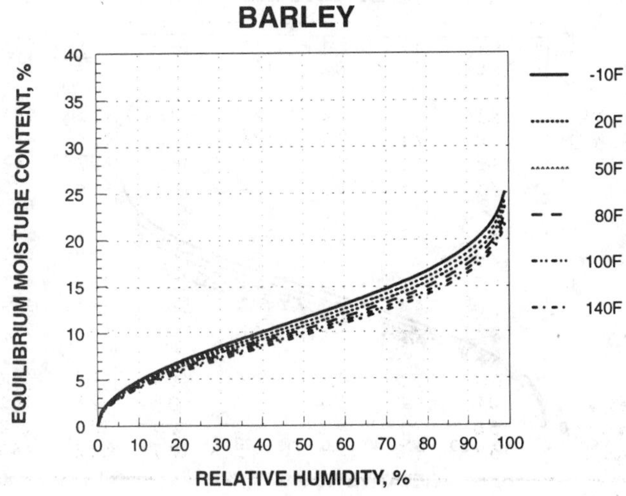

Figure 2.8–Equilibrium moisture content for barley (ASAE, 1985).

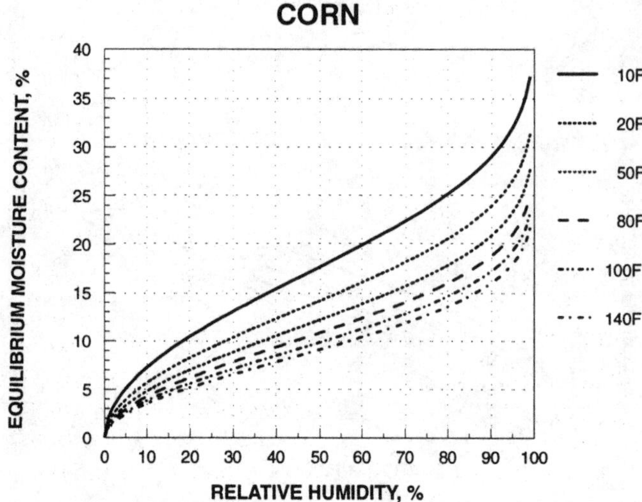

Figure 2.9–Equilibrium moisture content for corn (ASAE, 1985).

several different grains are shown in figures 2.7 to2.14 using the modified Henderson-Perry equation (ASAE Standards, 1985).

Equilibrium moisture properties are useful in analyzing drying and storage systems. Grain will dry when the air that surrounds it has a relative humidity less than the equilibrium relative humidity (ERH) corresponding to the moisture content of the grain. The reverse will occur if the air has a relative humidity that is greater than the ERH corresponding to the moisture content of the grain. In any drying

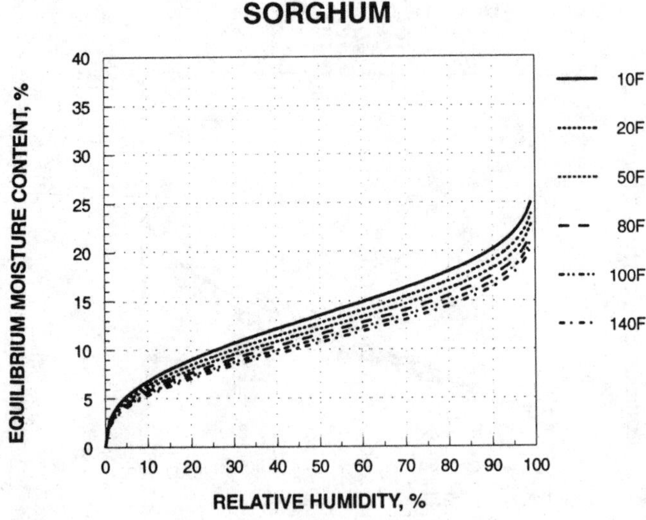

Figure 2.10–Equilibrium moisture content for sorghum (ASAE, 1985).

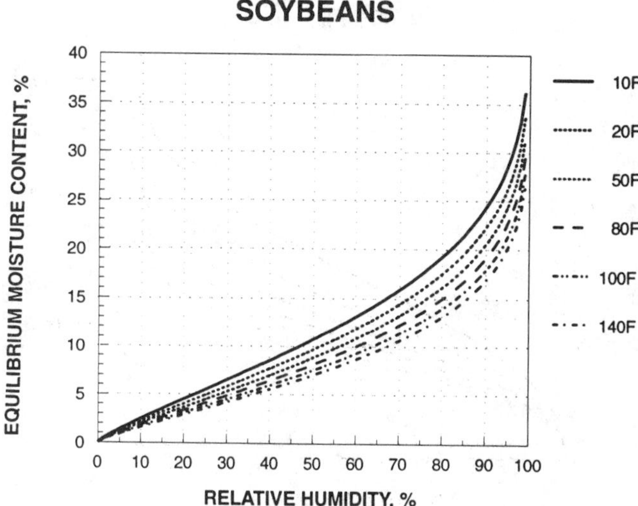

Figure 2.11–Equilibrium moisture content for soybeans (ASAE, 1985).

system, the minimum moisture content to which the grain can be dried is the EMC corresponding to the drying air conditions being used. For example, from table 2.5, grain can be dried to no lower than 12.7% moisture when using air at 90° F and 65% Rh. On the other hand, grain will dry to 8.7% moisture when using 130° F and 45% Rh air. It can also be established that when 13.9% moisture corn is stored at 50° F, the air contained within the grain will have a relative humidity of 60%. This

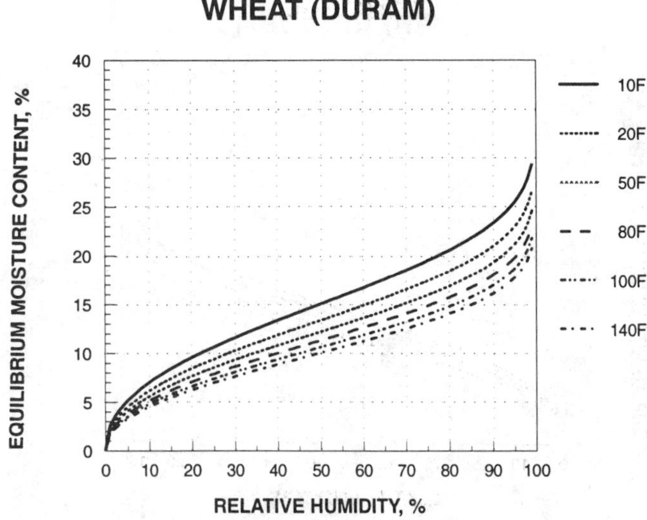

Figure 2.12–Equilibrium moisture content for Duram wheat (ASAE, 1985).

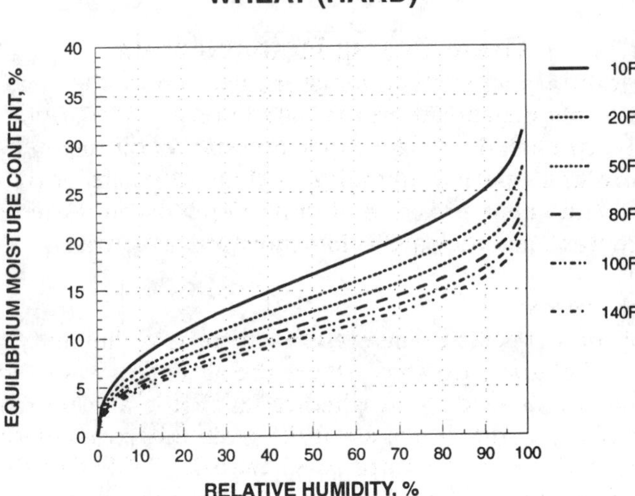

Figure 2.13–Equilibrium moisture content for hard wheat (ASAE, 1985).

type of information can be used to predict the growth of microorganisms in the stored grain and the potential for deterioration in storage. Most storage fungi cannot grow and reproduce in grain that is in equilibrium with air at a relative humidity less than 65%. In addition, the activity of storage insects will greatly decrease at relative humidities below 50%.

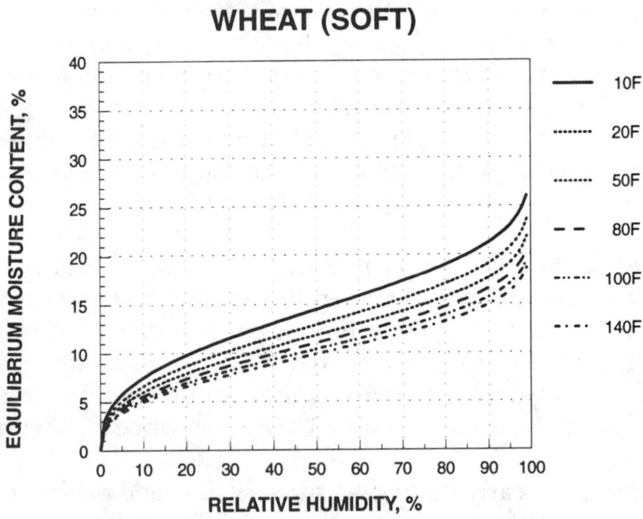

Figure 2.14–Equilibrium moisture content for soft wheat (ASAE, 1985).

Grain Drying Fundamentals

In any grain drying system, moisture will move from one point to another at a rate dependent on the difference in water vapor pressure between the two locations. The vapor pressure of air depends upon its temperature and relative humidity as defined by its psychrometric properties. The vapor pressure of grain depends on its temperature, moisture content, and the equilibrium moisture properties.

The Drying Process

When air of a specified temperature and relative humidity is passed through a bed of grain, the grain where the air enters will either lose or absorb moisture depending on whether its ERH is above or below the relative humidity of the drying air. If the grain ERH is above that of the air, the grain will lose moisture (dry) until its ERH (and thereby its vapor pressure) equals that of the air. At this point the grain will be in moisture equilibrium with the air and no further drying can occur. The grain will also be in thermal equilibrium (at the same temperature) with the drying air.

"Dry" air passing through "wet" grain absorbs moisture. Its humidity ratio (and thereby its relative humidity) increases while its drying potential decreases. If the grain depth is great enough, the air will approach moisture equilibrium with the grain. Its relative humidity will be the same as the grain's ERH, and no further drying will occur as the air travels the rest of the way through the grain mass. Initially, the drying air is removing the maximum amount of moisture possible per unit weight of air being moved by the drying fan (unless the airflow rate is sufficiently high so that there is not enough time for the air to completely absorb the moisture). For the same airflow, the drying (moisture removal) rate can be increased by removing moisture from the air prior to the drying process. However, in grain drying, the usual approach is to increase drying potential by raising air temperature to decrease its relative humidity and increase its moisture holding capacity.

Application of the Psychrometric Chart

An abbreviated psychrometric chart (fig. 2.3) has been modified to include equilibrium temperature-relative humidity relationships for several different corn moisture contents. The modified chart (fig. 2.15) describes changes in air conditions that occur during a grain drying process, and can be used to analyze the performance of different drying systems.

Suppose it is an early fall or cool summer day, and natural air is to be used for drying at the conditions shown at Point A. If there is potential

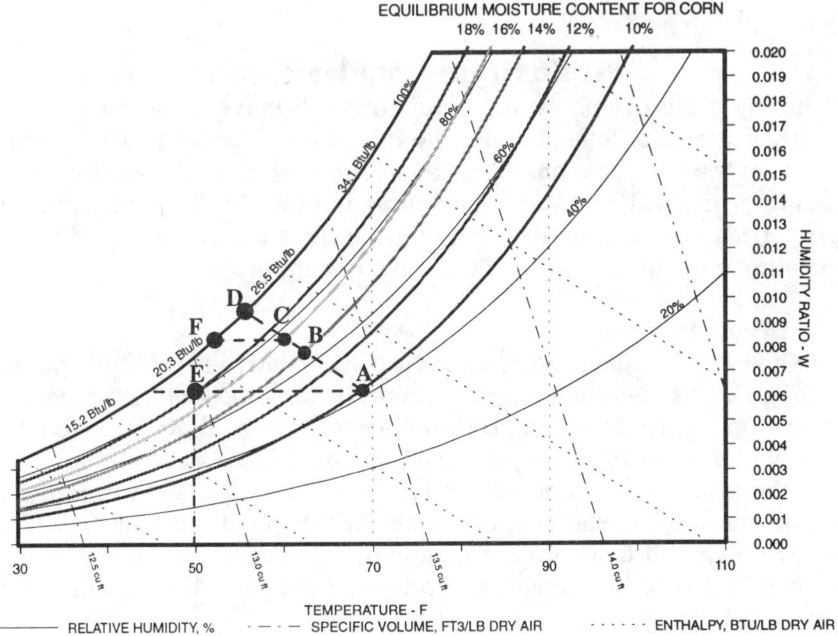

Figure 2.15–Drying example using EMC imposed on the psychrometric chart.

for drying, moisture will be evaporated from the grain into the air as air moves through the grain mass. The latent heat of vaporization required to evaporate this moisture will cool both the air and the grain. The cooling process can be represented by a constant heat (enthalpy) or constant wet bulb line on the psychrometric chart if it is assumed that the heat used to vaporize the moisture comes primarily from the air. Thus, as the air increases in moisture, its state conditions move along a constant wet bulb line on the psychrometric chart as represented by the dashed line through Points A-D. If the air remains in the grain long enough, it will exit in moisture and temperature equilibrium with the grain. The higher the grain moisture content, the closer the exiting drying air will be to the wet bulb temperature and a relative humidity of 100% (Point D). This represents the minimum temperature and the maximum humidity ratio that the stated drying air can attain through the drying process alone. The observed increase in the humidity ratio from Point A to Point D represents the maximum amount of moisture that can be removed from the grain per pound of drying air.

The drying air will never become saturated if the grain in the above drying system is at a moisture content sufficiently low so that its ERH is below 100%. The air will increase in moisture only to the point where it reaches equilibrium with the grain. For example, given sufficient time, 14% moisture content shelled corn would result in the drying air

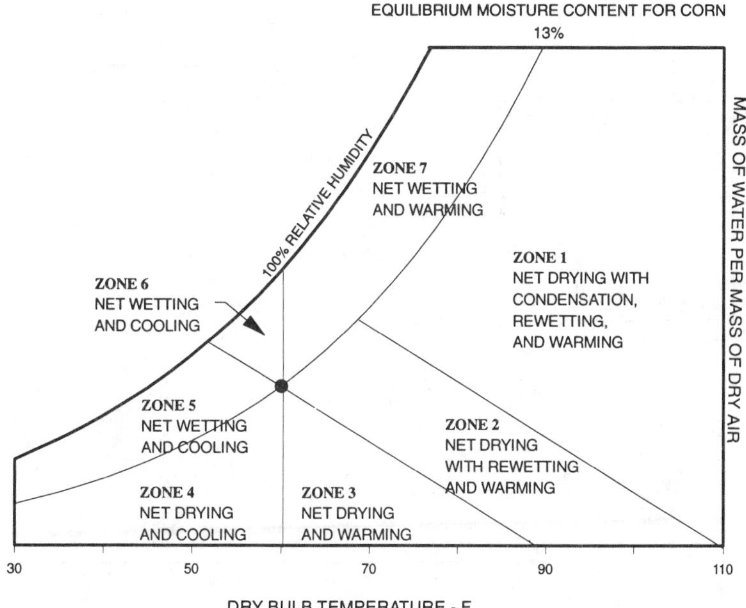

Figure 2.16–Zones within the psychrometric chart.

exhausting at Point B (64% Rh), while air passing through 16% corn would exhaust at Point C (76% Rh). Once moisture equilibrium has been reached, no additional moisture can be removed from the grain regardless of its depth. Similarly, if initial drying air conditions are represented by Point B, no drying could occur in 14% moisture shelled corn.

Suppose drying began in the late fall with air at the conditions represented by Point E (50° F and 80% Rh). The relative humidity of this air is on the 18% ERH line for corn; thus, 14% corn could not be dried (in fact, it would gain moisture from the drying air). Adding supplemental heat would increase the dry bulb temperature of the air without increasing its humidity ratio. If sufficient heat is added, the air state condition would move along the horizontal line to Point A on the psychrometric chart. This situation represents a temperature and humidity condition that correspond to a 10% EMC for shelled corn. This air can now dry shelled corn to the point where overdrying (drying below the desired moisture content) can be a problem in certain types of drying systems.

There are some conditions under which drying air that is in moisture equilibrium with grain may result in moisture condensation. This can occur at the beginning of a drying process when supplemental heat is being used to dry very cold grain. For example, consider the air in equilibrium with 16% moisture shelled corn as represented by Point C in

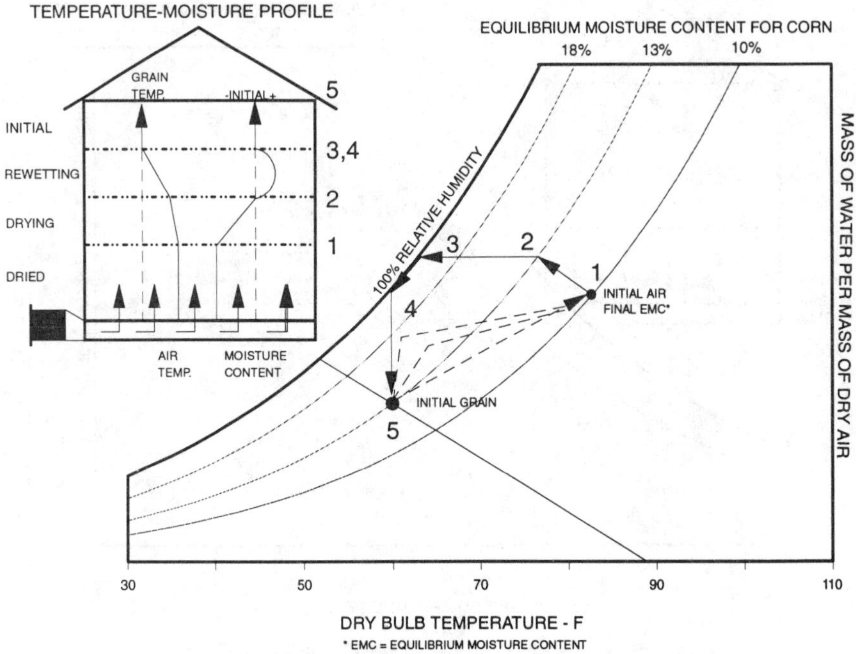

Figure 2.17–Zone 1 conditions.

figure 2.15. The dew point of this air, as determined by following a horizontal line to Point F, is approximately 53° F. If the air is cooled to below 53° F before it exits the grain, moisture will condense. No further condensation will occur once the grain reaches thermal equilibrium with the air at Point C. Grain spoilage problems do not usually occur if condensation conditions exist for a relatively short period of time. If these conditions persist, however, spoilage can occur in either the dryer or during storage after drying. Condensation does not usually occur in high temperature, high airflow, drying systems but should be considered when supplemental heat is used with relatively low temperature, low airflow systems to dry relatively cold grain. Condensation can also occur if warm humid air is used to aerate cold grain.

Grain can also adsorb moisture from drying air that has a relative humidity higher than the grain ERH. This is quite different from the condensation process in that relatively larger volumes of air and longer periods of time are required to transfer significant amounts of moisture to the grain. The moisture gained is adsorbed as a vapor rather than as a condensed liquid, and is distributed over a larger volume of grain. With condensation, the moisture is usually concentrated in a localized area and is, therefore, more likely to result in spoilage.

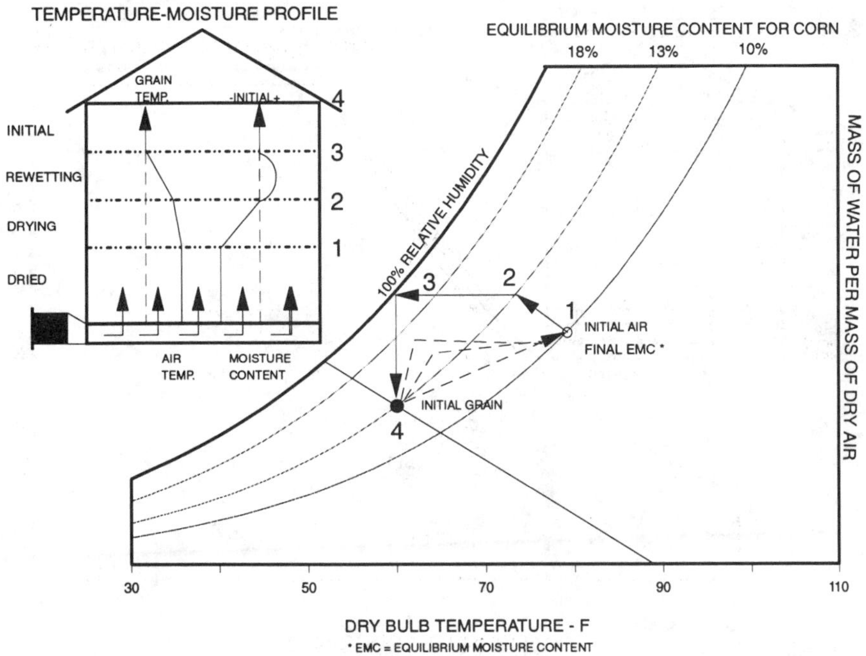

Figure 2.18–Zone 2 conditions.

Development of Drying Zones

When drying air is forced through grain, all the grain does not dry uniformly and at the same time. This is particularly true in bin-type dryers where airflow rates per unit volume (cfm/bu) are relatively low. Usually, in these dryers the grain is considered to be in one of three zones: (1) the dried zone, (2) the drying zone, and (3) the undried or wet zone. However, the process is considerably more complex in that there may be zones of rewetting, cooling, and warming, all along the same air path. The zones that develop are dependent upon the temperature and relative humidity of the incoming air as related to the equilibrium temperature and moisture conditions of the grain along the air path. There are seven zones on the modified psychrometric chart (fig. 2.16) that depict possible relationships between grain and incoming air for a situation where grain and air are both at 60° F and the grain is in moisture equilibrium at 13%. Each zone will be discussed in some detail. Recognize that there are many processes occurring simultaneously along the air path. Figures 2.17 to 2.23 use 10%, 13%, and 18% equilibrium moisture contents for corn to illustrate low, intermediate, and high moisture conditions. Initial and final grain moisture content and temperature conditions, with associated pathways, are shown. The "bin" to the left of the modified psychrometric chart depicts a profile of grain moisture content and

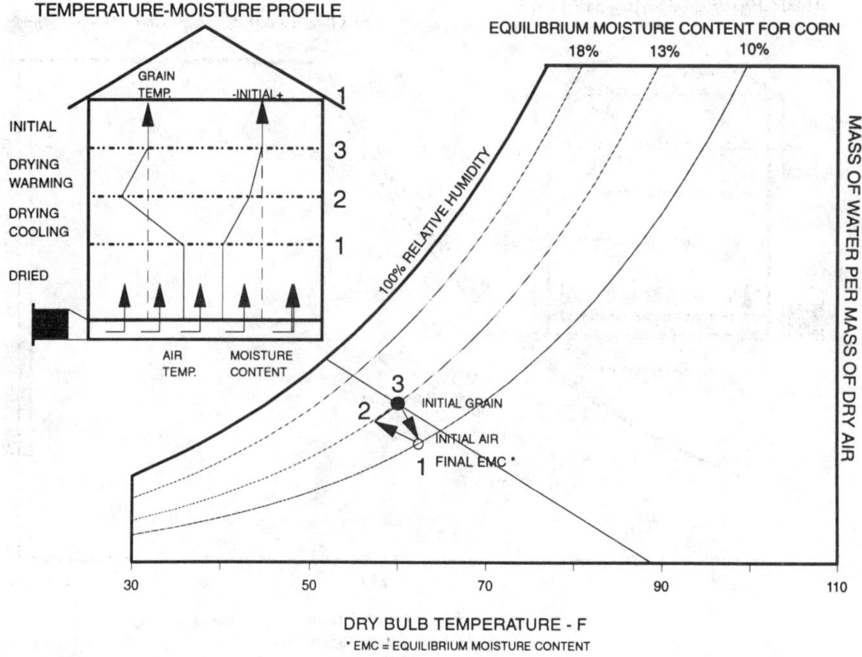

Figure 2.19–Zone 3 conditions.

entering air temperature as compared to initial grain moisture content and grain temperature, respectively. Depths shown in the bin are for purposes of illustration and are not intended to depict relative sizes of the zones.

An example of Zone 1 conditions (fig. 2.16) is shown in figure 2.17. The entering air is at a temperature and relative humidity such that the grain will dry from 13 to 10% given sufficient time. This is the most typical grain drying situation, especially when heat is added. Note that the air reaches its highest drying potential humidity ratio at Point 2. However, the drying air continues to be cooled providing sensible heat to the grain mass. At Point 3, the dew point temperature is reached that results in a condensation (rewetting) zone in the grain mass. Condensed water and both latent and sensible heat additions to the grain mass continue until the air reaches its exit temperature at Point 4. Thus, the "rewetting" zone precedes the drying zone. In all instances, however, the grain temperature is approaching constantly that of the entering air.

Zone 2 conditions are illustrated in figure 2.18 and differ from Zone 1 only in that the grain temperature is greater than the dew point temperature of the drying air. Thus, there is no condensation. However, rewetting still occurs because of the high relative humidity of the air.

Zone 3 conditions are shown in figure 2.19. The loss of heat from the drying air results in the air becoming cooler than the grain and then

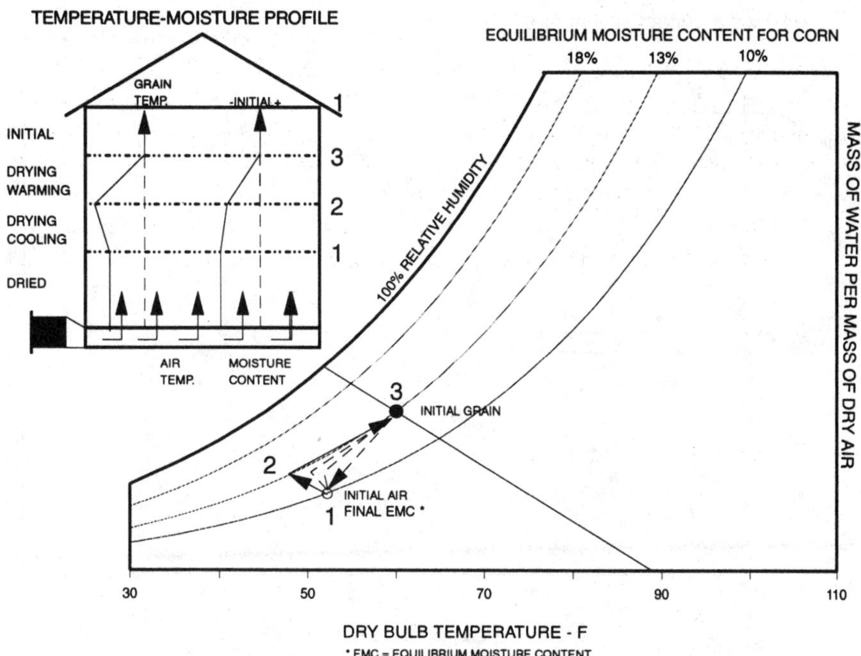

Figure 2.20–Zone 4 conditions.

warming back up to the grain temperature before exiting the grain mass. Thus, a cooling front precedes the drying front. However, no condensation occurs because the dew point temperature is not reached. Eventually the grain dries to equilibrium with the incoming air becoming warmer. In this zone, the energy for drying is first supplied by the cooling of the drying air (Point 2). After the drying air reaches its lowest temperature, its drying capacity is increased by using sensible heat from the grain mass to evaporate water from the grain (Point 3).

Zone 4 conditions, shown in figure 2.20, differ from Zone 3 only in that the grain is cooled eventually. The energy exchange resulting from drying still results in a cooling front (Point 2) that is lower in temperature than the drying air (Point 1).

Whereas conditions in Zones 1 through 4 always result in a net drying of the grain mass, Zones 5 through 7 conditions always result in a net rewetting. In Zone 5, as illustrated in figure 2.21, the incoming air is being warmed constantly using the sensible heat exchanged with the grain. However, its humidity ratio decreases initially (Point 2). Eventually, the grain mass cools to the same temperature as the incoming air (Point 1) and, as a result, a rewetting zone is established as the grain adsorbs moisture as it comes into equilibrium.

Zone 6 conditions are illustrated in figure 2.22 and differ from Zone 5 in that the temperature of the entering air temporarily will exceed that

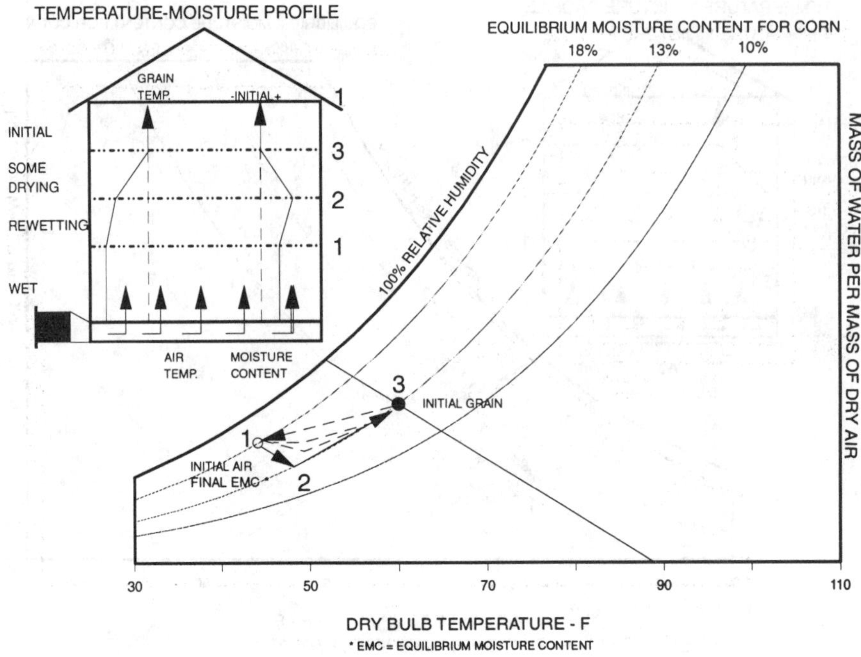

Figure 2.21–Zone 5 conditions.

of the grain (Point 2). Thus, a warm front will be established with temperatures in excess of grain or incoming air temperatures. The net effect remains that given sufficient time the grain will absorb moisture until it is in equilibrium with the initial air conditions.

In the Zone 7 situation, shown in figure 2.23, the air is warmed initially to the equilibrium moisture content conditions (Point 2) and then cooled to the temperature of the grain mass. Thus, a warm front with temperatures in excess of the entering air temperature is created. Eventually, the grain is warmed and adsorbs moisture as it equilibrates with the entering air. This is reflected by a rewetting zone that begins at the point where the air enters the grain.

When drying air enters the grain in a bin-type dryer, the region nearest the inlet dries first (fig. 2.24). In both natural and heated air drying, this region generally will dry below the desired final moisture content. The degree of overdrying depends on the EMC corresponding to the input air conditions. This region of dry grain (the so-called dried zone) gradually will move upward as drying proceeds. The temperature of the grain in the dried zone will approach the drying air temperature, and its moisture content will approach equilibrium with the drying air.

The situations represented by Zones 1 through 7 are for uniform grain and incoming air conditions. In reality, grain temperature and moisture content will vary within the grain mass. In addition, the

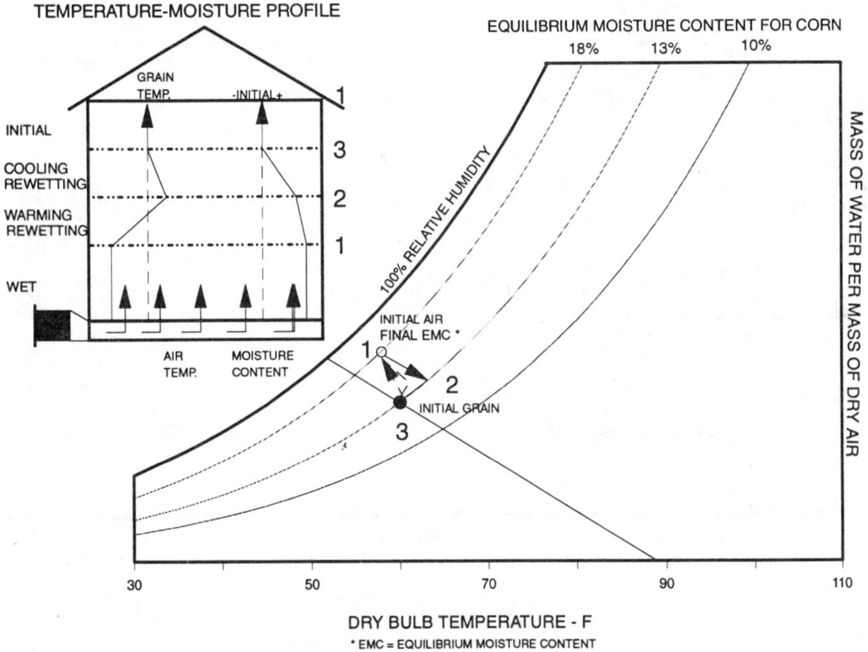

Figure 2.22–Zone 6 conditions.

temperature and relative humidity of the incoming air will vary throughout the day, especially if natural air is used. Thus, the grain mass may simultaneously contain many of the Zone 1 through Zone 7 situations. Although the grain mass will reach equilibrium with the incoming air given sufficient time, changes in incoming air conditions present a "moving target" to the grain. When drying high moisture content grain, however, the exiting air will be nearly saturated. The depth of the various zones (drying, rewetting, warming, cooling), depends on initial grain and air conditions, the airflow rate, the rate at which moisture can be exchanged with the grain, and the net amount of moisture to be exchanged between the air and the grain. In situations where drying is the goal and sufficient heat and airflow are utilized, the dominant consideration is the drying zone. The upper edge of the drying zone is at the interface with the undried grain and is called the "drying front". The region above the drying front is called the undried zone since essentially no drying will take place in any location until the drying front reaches that point.

The term "moisture spread" refers to the moisture content difference of the grain between where the drying air enters and exits. The amount of moisture spread depends on the initial grain moisture content, the EMC corresponding to drying air conditions, and the width of the drying zone relative to the overall grain depth. At high airflow rates, the drying

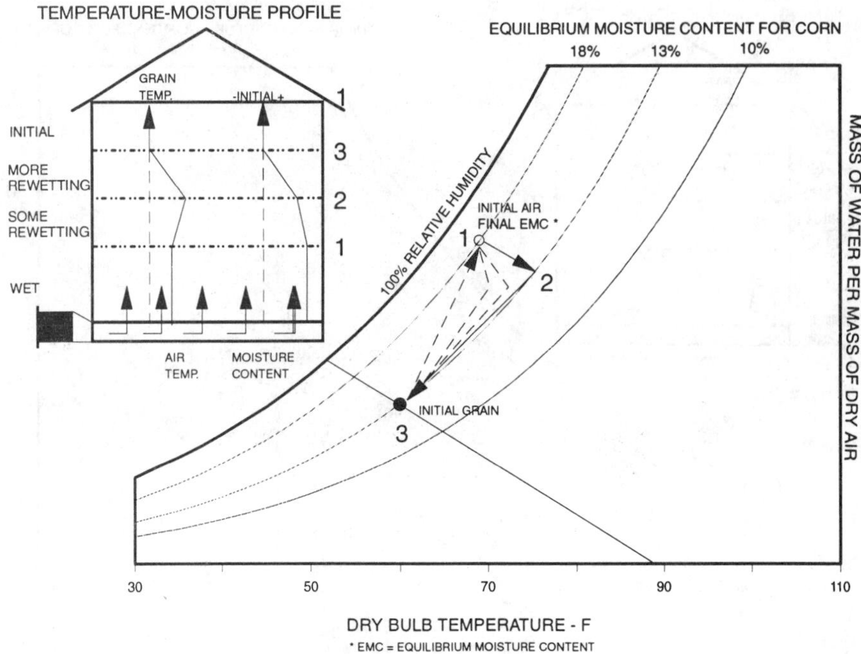

Figure 2.23–Zone 7 conditions.

zone may extend through the entire grain depth. This reduces the moisture spread in the dried grain but also lowers the efficiency of energy utilization in the drying process. Controlling the drying air relative humidity can be used to minimize moisture spread and reduce overdrying in low temperature drying methods; however, this approach is ineffective with high temperature drying because the heated drying air always has a very low relative humidity. Control of moisture spread and overdrying in this case can only be achieved by drying in relatively shallow layers using high airflow rates and/or mixing the grain during the drying process.

A drying front is rarely as uniform as that shown in figure 2.24. A uniform drying front occurs only when there is parallel airflow with a uniform velocity across the entire cross-section of the dryer. A non-uniform grain depth in the dryer, variability in the grain resistance to airflow at different locations in the drying bin, and the use of ducts to distribute the drying air can all result in non-uniform drying patterns. Localized concentrations of trash and broken kernels in the grain can modify airflow patterns, thereby affecting drying rates at different locations in the bin. A non-uniform drying pattern may result in excessive moisture content variation across a horizontal section of the grain mass. It is therefore advisable that the grain be as clean as

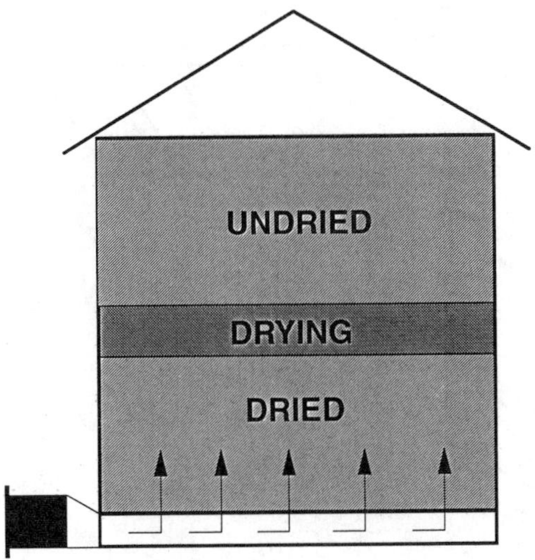

Figure 2.24–Three zones of drying.

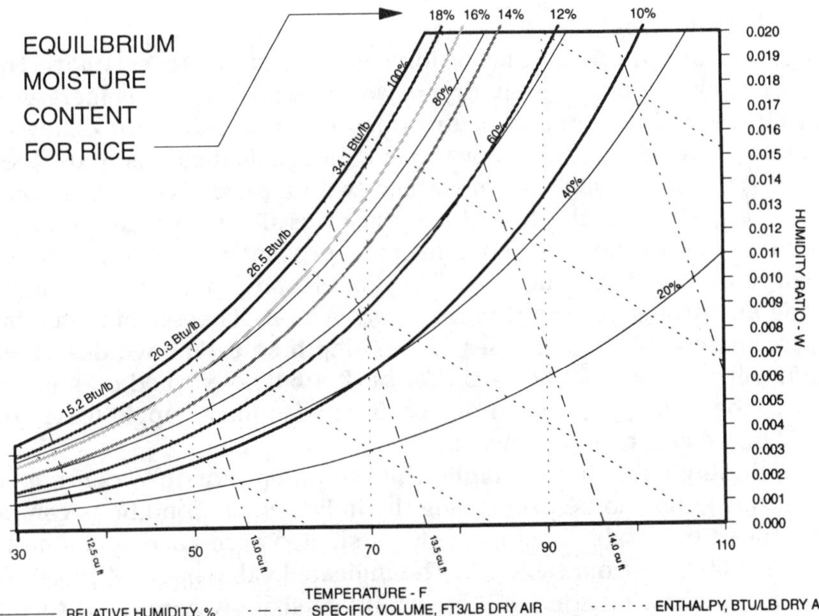

Figure 2.25–EMC-psychrometric chart for rice.

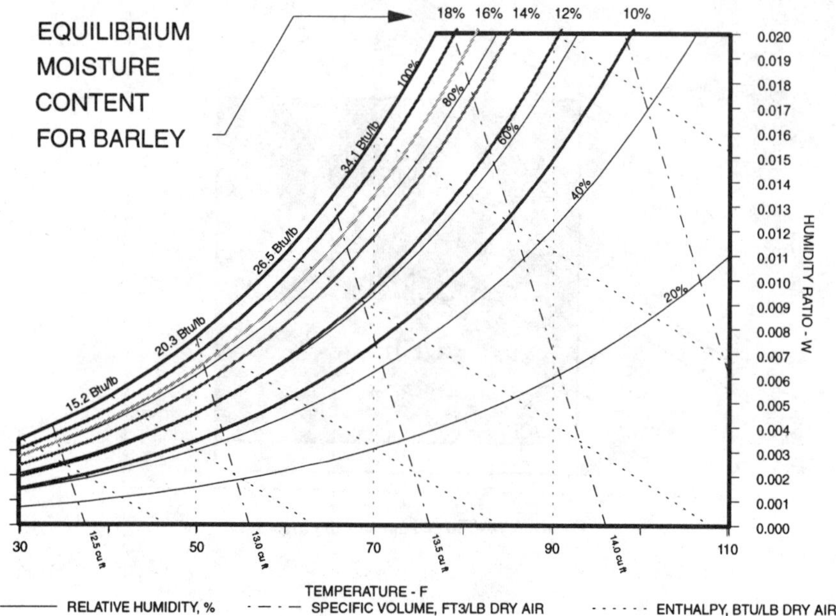

Figure 2.26–EMC-psychrometric chart for barley.

possible prior to drying, and that any broken or fine material remaining in the grain be uniformly distributed throughout the grain mass.

Grain Drying Calculations

Certain simplifying calculations can be made to estimate the performance of bin-type grain dryers where relatively low temperatures and airflow rates are used to dry deep beds of grain. For example, suppose a 6 ft depth of 15% moisture content shelled corn is to be dried in a 21-ft-diameter cylindrical metal bin (with a perforated drying floor) until the grain near the top of the bin dries down to a safe storage condition of 13% moisture content. For purposes of example, assume the drying fan is delivering 4 cfm/bu of 60° F and 50% relative humidity drying air throughout the drying period. A 21-ft-diameter bin contains approximately 277 bushels per foot of depth so the above described batch will contain 1,662 bushels (277 bu/ft * 6 ft) of shelled corn at the start of the drying period. On this basis, the fan must supply 6,648 cfm (1,662 bu * 4 cfm/bu) of drying air.

Continuing with this example, the minimum moisture content to which this grain can be dried using the indicated air conditions can be determined by establishing the EMC of shelled corn corresponding to 60° F and 50% Rh from table 2.5. The indicated value is 11.7% moisture content. The drying zone will be relatively shallow compared to the entire grain depth for the airflow in this example. Thus, the air leaving

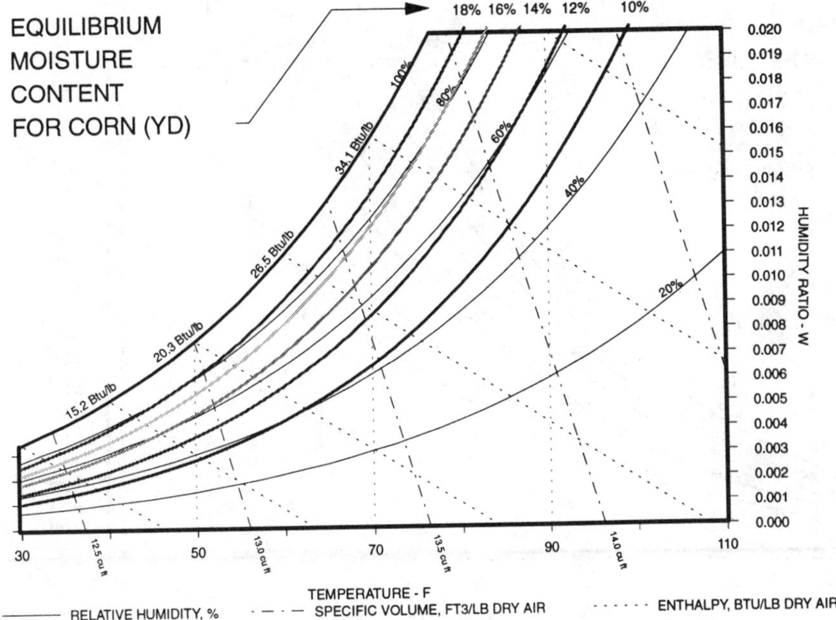

Figure 2.27–EMC-psychrometric chart for corn.

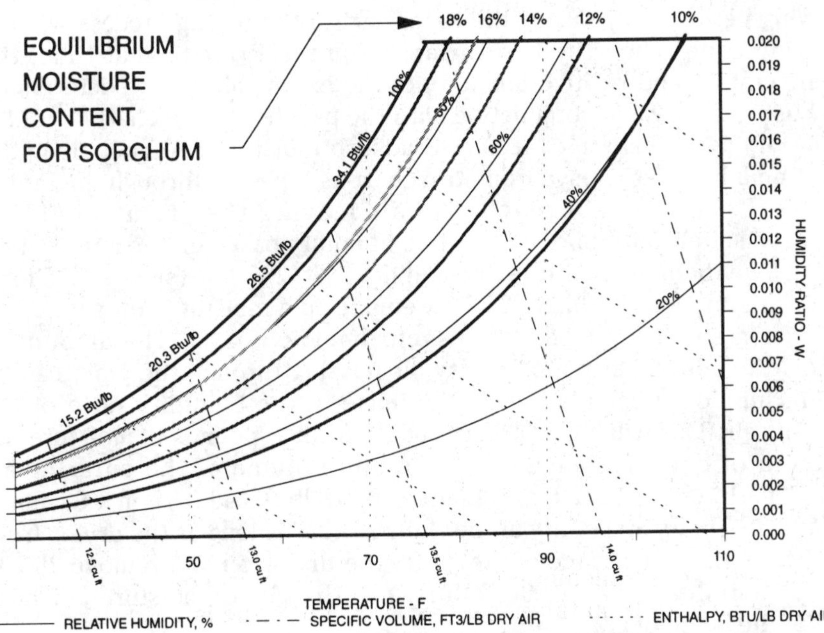

Figure 2.28–EMC-psychrometric chart for sorghum.

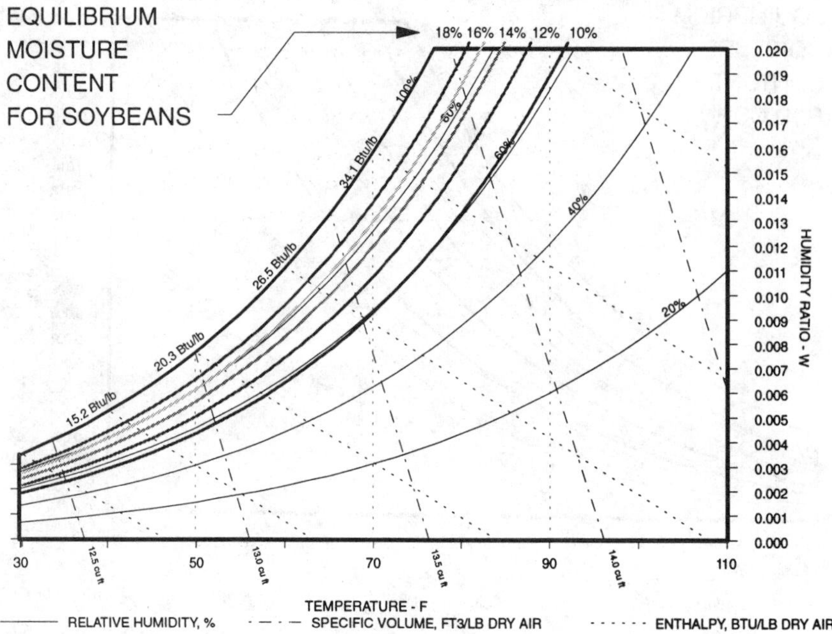

Figure 2.29–EMC-psychrometric chart for soybeans.

the grain will be in equilibrium with the undried grain for most of the drying period. Only during the latter part of the drying process will the drying zone reach the top surface of grain. Prior to that time, the temperature and relative humidity of the exiting air can be determined by using table 2.5 in conjunction with the psychrometric chart (fig. 2.3). The drying air can be expected to adsorb moisture and cool along the wet-bulb line on the psychrometric chart as it passes through the drying zone. Using this approach, the air leaving the grain should be approximately 56° F and 68% relative humidity based on the intersection of the wet bulb temperature line and the 15% EMC line (see fig. 2.15 for a similar situation and figure 2.27 for equilibrium moisture content curves for corn superimposed on a psychrometric chart). The amount of moisture added per pound of dry air passing through the grain can be determined by calculating the increase in the humidity ratio of the drying air. This will be 0.00095 (0.00645 - 0.0055) pounds of moisture per pound of dry air. Based on the humid volume of the entering air (13.2 ft^3/lb dry air) the fan will move 30,218 lb (6,648 ft^3/min * 60 min/h ÷ 13.2 ft^3/lb dry air) of dry air per hour. Thus, as long as the drying front is below the top surface of the grain, the drying air will remove 28.7 lb (30,218 lb dry air/h * 0.00095 lb H_2O/lb dry air) of moisture per hour from the batch of grain being dried.

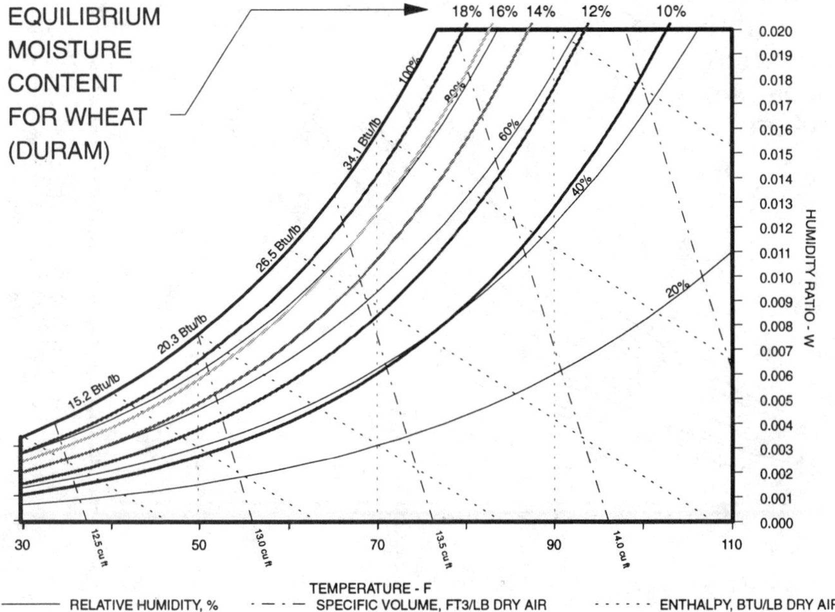

Figure 2.30–EMC-psychrometric chart for Duram wheat.

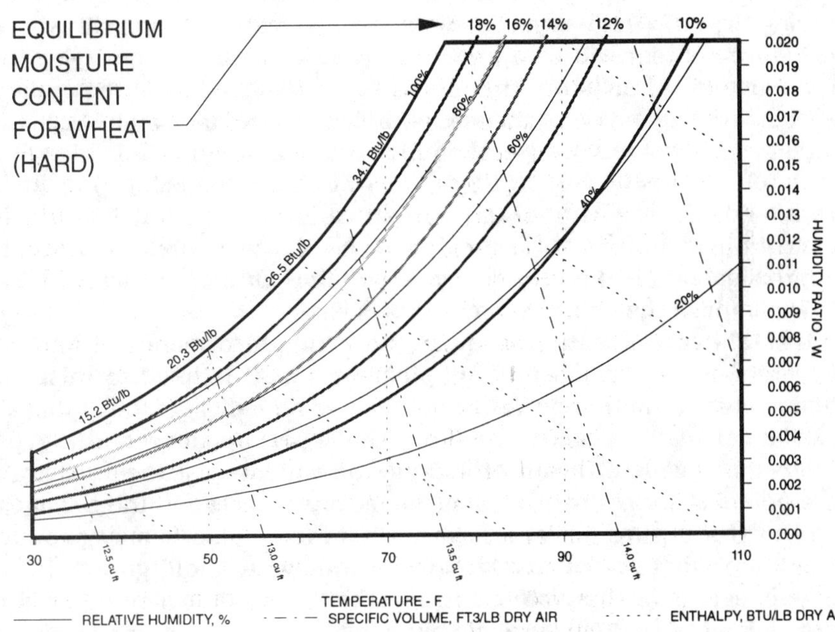

Figure 2.31–EMC-psychrometric chart for hard wheat.

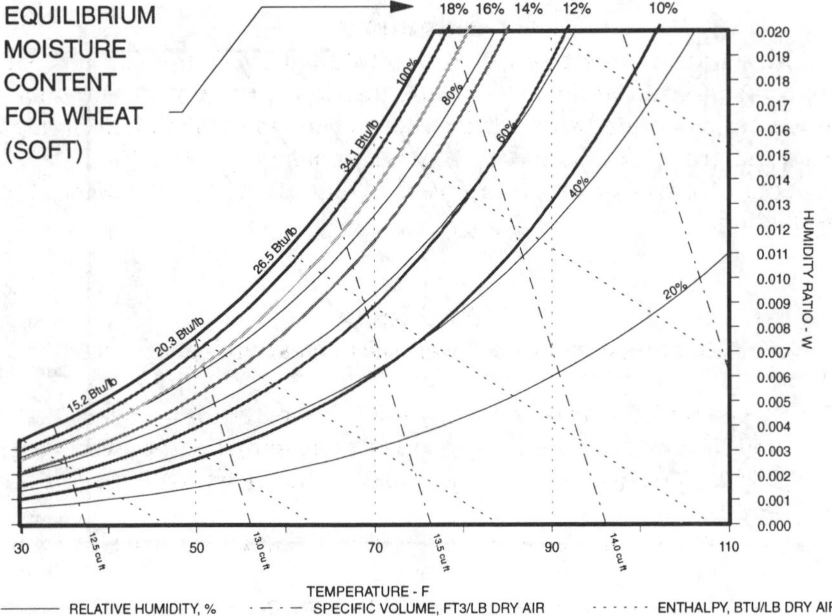

Figure 2.32–EMC-psychrometric chart for soft wheat.

When the grain on the top surface of the batch reaches 13% moisture content, the rest of the bin will have dried to near equilibrium with the drying air, 11.7% moisture content. If it is assumed that the final average moisture content will be approximately 12% at the end of drying, the total amount of moisture that needs to be removed from the grain to accomplish the indicated amount of drying can be calculated. Assuming an initial bushel test weight of 56 lb and using equation 2.5, 3,173 lb of moisture [1 – (100 – 15)/(100 – 12) ∗ 56 lb/bu ∗ 1,662 bu) must be removed. If, on the average, moisture is removed at 95% of the maximum drying rate (as determined above), then it will take approximately 116.4 h or 4.85 days [3,173 lb H_2O/(28.7 lb H_2O/h ∗ 0.95)] to accomplish the indicated amount of drying.

The above calculations indicate how the performance of bin-type dryers can be evaluated. If supplemental heat is used to raise the temperature (and thereby lower the relative humidity) of the drying air, similar calculations can be used to evaluate performance of other types of dryers. In general, heating the drying air will increase the drying rate, the potential for overdrying, and the moisture spread in the bin at the end of the drying period. In some instances, also it may produce conditions that are relatively more conducive to mold growth in the upper regions of the grain mass. Figures 2.25 to 2.32 may be used to determine the final moisture condition after drying for several different grain types.

Summary

This chapter has been devoted to examining the fundamental concepts of grain drying. Obviously, the drying of grain involves many processes associated with the transfer of heat and moisture both to and from the grain. The following chapter presents more of the practical aspects of grain drying as related to equipment performance and evaluation.

Problems

NOTE: Moisture contents are wet basis unless otherwise stated.

2.1: A sample of wheat weighing 105 g is heated in an oven kept at 130° C for 19 h. The sample when removed from the oven weighs 88.5 g. Assuming that all the moisture has been removed from the grain:
- (a) What is the moisture content of the grain expressed on a wet basis?
- (b) What is the moisture content of the grain expressed on a dry basis?

2.2: A sample of corn has a moisture content of 11.0% on a dry basis. What is its moisture content on a wet basis?

2.3: A bin is 12 ft in diameter and has an eave height of 10 ft. If the bin is filled with grain leveled at the eave height, what is the volumetric capacity of the bin in bushels?

2.4: A truckload of grain contains corn at a moisture content of 18%. The truck-bed is 12 ft long, 8 ft wide, and has 6-ft tall sidewalls. If the truck-bed is filled full flush with the top of the sidewalls, what amount of grain is contained in the bed expressed in 15.5% dry bushels.

2.5: A bin contains 4,000 bu of corn with a moisture content of 20.0%:
- (a) What will the shrinkage in weight be if the grain is dried to 14% moisture content?
- (b) What will the shrinkage in volume be if the grain is dried to 14% moisture content?
- (c) What will the shrinkage in weight be if the grain is dried to 12% moisture content?
- (d) What will the shrinkage in volume be if the grain is dried to 12% moisture content?

2.6: Using a psychrometric chart, find the enthalpy (h) for the following conditions:
(a) 95° F T_{db} and 85° T_{wb}
(b) 80° F DP and 40% Rh
(c) 70° F DP and 60% Rh
(d) 0.011 lb moisture/lb dry air W and 80% Rh
(e) 13.5 ft^3/lb dry air HV and 60°F T_{dp}

2.7: Using a psychrometric chart, find the humidity ratio (W) for the following conditions:
(a) 85° F T_{db} and 60° F T_{wb}
(b) 75° F T_{db} and 40% Rh
(c) 60° F T_{db} and 40° T_{dp}

2.8: Using a psychrometric chart, find the relative humidity (Rh) for the following conditions:
(a) 80° F T_{db} and 65° F T_{wb}
(b) 70° F T_{db} and 65° F T_{dp}
(c) 70° F T_{db} and 0.014 lb moisture/lb dry air W

2.9: Using a psychrometric chart, find the Specific Volume (SV) for the following conditions:
(a) 60° F T_{db} and 45° F T_{wb}
(b) 75° F T_{db} and 55° F T_{dp}
(c) 0.008 lb moisture per lb dry air W and 40% Rh

2.10: The dry bulb temperature (T_{db}) of an air moisture mixture is 75° F. If the humidity ratio (W) is 0.006 lb moisture/lb dry air:
(a) What is the enthalpy (h) calculated from equation 2.6?
(b) What is the enthalpy (h) read from a psychrometric chart?

2.11: A fan supplies 4,000 cfm of air to a dryer. If ambient conditions are 80° F dry bulb temperature (T_{db}) and 40% relative humidity (Rh), what size heater is required to heat the air to 90° F dry bulb temperature (T_{db})?

2.12: Heated air enters a dryer at 85° F dry bulb temperature (T_{db}) and 20% relative humidity (Rh) at an airflow rate of 6,000 cfm. If the air exiting the dryer has a dry bulb temperature (T_{db}) of 70° F and a relative humidity (Rh) of 50%, how much water is removed from the grain per hour?

References for all chapters begin on page 541.

3

On-Farm Drying Methods

Drying and storage are interrelated in terms of overall system economics. This chapter provides the basic information needed to evaluate a major component of the total on-farm grain drying and storage system. Drying systems usually are compared with regard to daily and seasonal drying capacity and annual and purchase costs. The following sections compare the commonly used on-farm drying systems often using the terms "low, medium, and high" to reflect ranking of comparative performance and costs. There is no absolute standard for these terms. However, for purposes of this text, low capacity systems refer generally to those that process less than a total of 10,000 bushels per year at an approximate drying rate of 500 bu/day. Medium capacity systems process from 10,000 to 30,000 bu/year with drying rates ranging from 500 to 1,500 bu/day. High capacity systems exceed 30,000 bu in total processing capacity with drying rates that exceed 1,500 bu/day. Economic comparisons are made relative to other systems. A more complete discussion of economic considerations is given in Chapter 9.

Natural Air Drying

Natural air drying is a process used to dry grain by forcing unheated air through the grain mass until the grain comes into an equilibrium moisture condition with the air. The only fossil-based energy supplied for drying is the electricity used to operate the drying fan. Energy for evaporating the moisture from the grain comes, for the most part, from the energy contained within the ambient air as supplied by the sun. In one sense, natural air drying is the most basic form of solar drying.

Drying with natural or ambient air can be accomplished only if the air temperature and relative humidity conditions allow a net moisture transfer from the grain to the air. The drying fan removes the moisture laden air and supplies a fresh supply of ambient air. The process continues until sufficient moisture has been removed from the grain so that no additional net moisture transfer may be made to the air. When this occurs, the grain is said to be in equilibrium with the drying air. The moisture content of the grain when no moisture transfer is possible is referred to as the equilibrium moisture content (EMC) for a given air temperature and relative humidity. The ENC also depends on the grain type in addition to temperature and relative humidity as is shown in tables 2.5 to 2.6 from Chapter 2 and figures 2.7 to 2.14 from Chapter 2.

Moisture may be transferred from the air to the grain (referred to as "rewetting") during natural air drying. However, the potential for transfer usually becomes less as the grain dries.

Equipment Requirements

The equipment required for natural air drying consists of a bin equipped with a perforated floor and drying fan. Typically, a grain spreader, sweep auger, and unloading auger are also used (fig. 3.1). Stirring devices may also be added. However, they are not economical for natural air drying systems in most grain producing areas because overdrying is usually not a significant problem (Loewer et al., 1984).

Loading the drying bin may be accomplished by either a portable auger or a bucket elevator. The loading rate should be geared to accommodate the desired harvesting rate, and should be sufficient to empty a large trailer truck in no more than 2 h, thus a minimum of approximately 500 bu/h capacity.

Advantages, Disadvantages, and Limitations

Successful grain drying with natural air is usually the most energy efficient method of drying. It is also the slowest method and has the greatest potential for grain spoilage. Consequently, natural air drying

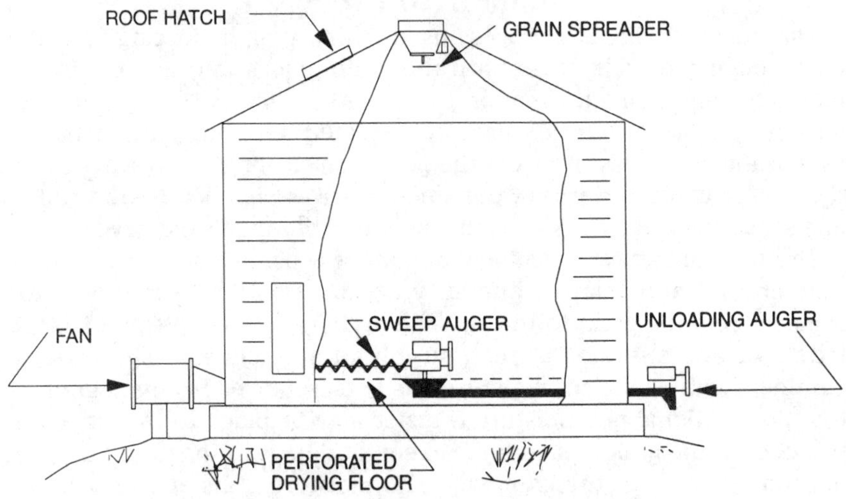

Figure 3.1–Natural air drying equipment.

requires the highest level of management if spoilage and/or aflatoxin problems are to be prevented. Much of the risk is because there is little "reserve" capacity for speeding up the drying process in that the inlet air conditions vary with the weather. Typically, the bin is filled only once each harvest season. If spoilage begins, the mid-course corrections are limited to either (1) drying immediately using another drying method or (2) selling before the grain degrades because of unacceptable damage levels.

Natural air drying requires relatively little direct labor except to enter, level, and remove the grain from the bin. However, a major effort should be made to inspect the grain regularly for spoilage and insect infestations.

Performance Expectations

The rate of drying when using natural air is related directly to the airflow delivered by the fan and the ambient air temperature and relative humidity. Airflow is usually expressed as cubic feet per minute (cfm) of air per bushel of grain, a bushel being defined volumetrically as 1.25 ft^3. Airflow (see Chapter 4) is influenced by type, quantity, and total height of grain in the bin (especially the latter). Grain cleanliness and the extent to which the grain has settled are also very important factors. Each type and size of drying fan delivers a different airflow rate for similar bin conditions. Drying fan performance is specified by the airflow rate that the fan will deliver when subjected to a given amount of resistance. Grain moisture content is usually not considered to influence airflow resistance significantly. All fans of the same horsepower do not deliver the same airflow rate. Airflow rate does generally increase with increases in fan horsepower. However, a doubling of the desired airflow rate per bushel of grain requires approximately a five-fold increase in fan horsepower, other factors being equal. This makes relatively large airflows in filled bins impractical, and limits severely the practical use of natural air drying in warmer climates.

Airflow delivery in corn, as influenced by bin diameter and fan horsepower for a constant grain depth, is shown in figure 3.2 using a representative line of grain drying fans. Large increases in fan horsepower show diminishing increases in cfm delivered per bushel of grain per horsepower. Thus, the design of a natural air drying system will have limited auxiliary drying capacity.

Ambient air temperature and relative humidity vary over the harvesting season. This becomes especially important over the long drying periods associated with natural air drying. Later in the season, the air temperature-relative humidity combination may reach a point at which the EMC corresponding to this combination is above that of the grain. In this situation, little drying (or perhaps even rewetting) of the grain will occur if further drying is attempted. This is why grain farmers

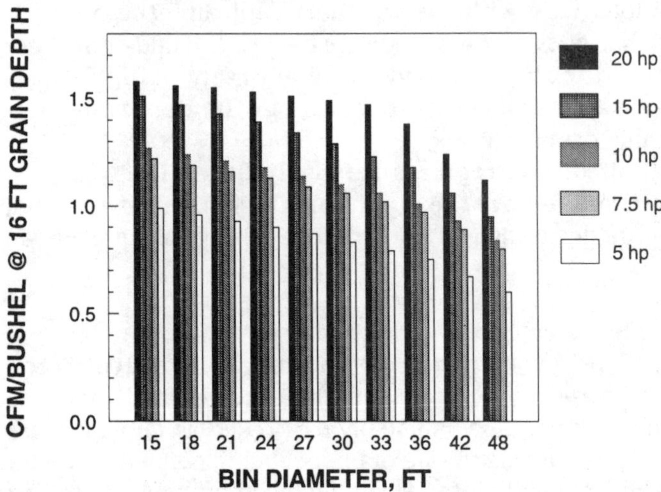

Figure 3.2–Representative airflows for axial fans of various horsepowers operating with corn at a depth of 16 ft over a range of bin diameters (Thompson, 1975b).

in the northern Midwest may dry partially in the fall and then complete the process in the spring. The grain, in this case, is cooled sufficiently so that relatively little spoilage occurs so long as there is sufficient aeration.

The time required for drying may be estimated by assuming a constant air temperature, relative humidity, airflow rate and quantity of moisture removed. The approximate drying times are given in figure 3.3 for a range of conditions resulting in the equilibrium grain moisture contents shown in figure 3.4. Each drying situation and weather condition is unique. Although increasing drying air temperature and airflow will decrease drying time (fig. 3.3), the drying time required when using natural air must be considered as it relates to the potential for spoilage. This is demonstrated by comparing the relationships among final equilibrium grain moisture content after drying (fig. 3.4), the expected allowable storage time as influenced by grain temperature and moisture content (fig. 2.1 from Chapter 2) and approximate dry matter decomposition over the drying period (fig. 3.5). Figure 2.1 (Chapter 2) illustrates the importance of harvesting relatively low moisture content corn if ambient air temperatures are high and a slow drying process is being used. This concept also applies to other grains.

The probability of drying grain successfully with natural air improves if layers of grain are added periodically over the harvest in order to maximize airflow per undried bushel. If the desired harvest rate is

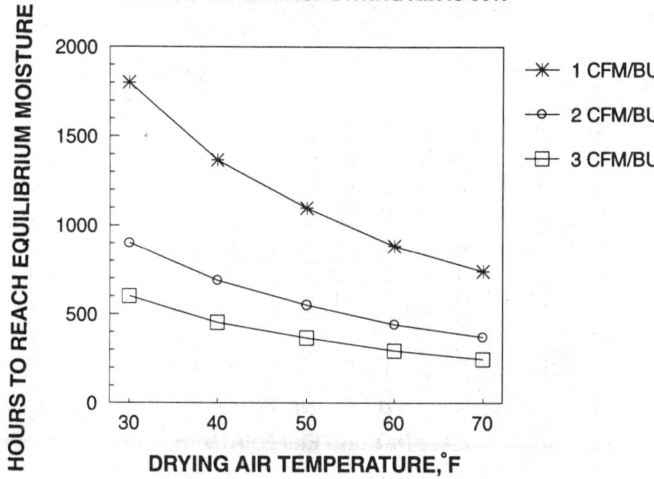

Figure 3.3–Approximate time for 22% moisture content corn to reach equilibrium moisture given various airflow rates and drying air temperatures (Thompson et al., 1975a).

sufficiently high, bin filling may have to be rotated if spoilage and/or the occurrence of aflatoxin are to be avoided. If the desired moisture content of corn after drying is to be close to but no higher than a certain

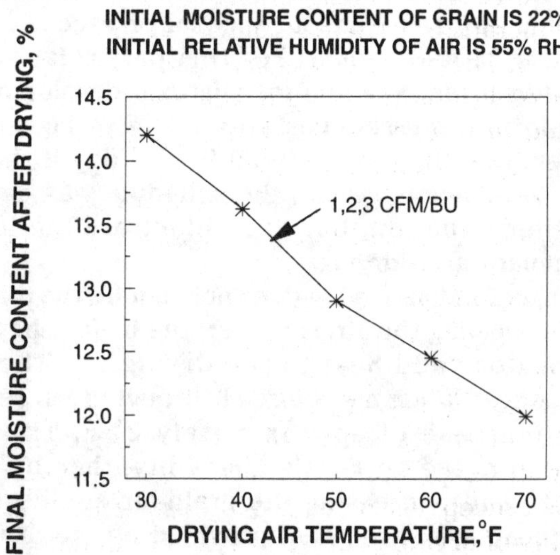

Figure 3.4–Final moisture content for corn when dried with heated air that is initially at 55% relative humidity (Thompson, 1975a).

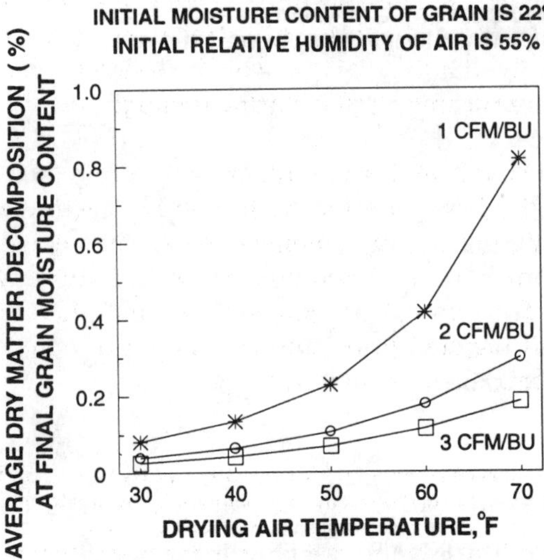

Figure 3.5–Approximate dry matter decomposition expected when drying with a range of airflow rates using heated air that is initially at 55% relative humidity (Thompson, 1968).

value, say 13%, a range of effective temperature and relative humidity values may be used (table 2.5 from Chapter 2). If these temperature-relative humidity values are assumed constant, a filling schedule may be developed for natural air drying using a computer program developed by Bridges et al. (1982). An example schedule is shown in table 3.1 for a representative bin and fan combination over a range of initial moisture contents. The model assumes that aflatoxin may occur if temperature and relative humidity simultaneously equal or exceed 55° F and 85%, respectively, for 72 h or longer. For this particular set of ambient weather conditions, the potential for aflatoxin development limits the bin filling rate for initial grain moistures of 20% or higher (Schedules 1 to 3). However, the bin may be filled in one day if the initial grain moisture is 18% or below. Again, the schedule was developed for a specific weather situation, but does illustrate the importance of management in natural air drying.

By definition, natural air drying does not require the direct burning of fossil energy for heating the air. However, the inefficiency of the drying fan and motor does add heat to the drying air. The friction and turbulence produced when air is forced through a fan results in an air temperature increase of approximately 2° F. This increase in temperature lowers the air relative humidity, thereby increasing its drying capacity and decreasing the grain-air equilibrium moisture content. The recapture of the heat provided by the drying fan makes natural air drying an even more efficient utilizer of energy assuming the grain can be dried safely.

One of the primary management mistakes associated with natural air drying is turning off the fan before the grain is cooled and the potential for drying still exists. Drying potential certainly decreases with increases in relative humidity. However, the following factors should be considered before turning off fans during damp periods:

1. The added heat from the fan will increase drying potential.
2. If grain in the lower portion of the bin has been overdried, it may rewet during periods of dampness. When this moisture is removed from the air, the air becomes drier and the drying potential increases. Thus, the higher moisture grain in the upper portions of the bin will become drier, and the moisture content of the grain mass will become more uniform.

Table 3.1. Schedules for natural air corn drying given 60° F and 55% Rh ambient air conditions with 1.5° supplied by the fan to obtain a desired final moisture content of 14%

In-Bin Shelled Corn Drying Model
(without stirring)
Developed by Tom Bridges, University of Kentucky

Grain = Corn	Fan ID XL430-7 (10HP)
Bin Diameter = 24.0 ft.	Bin Eave = 16.0 ft
Drying Temp. Added (°) = 1.5	Drying Air Rel. Hum. = 52.1%
Specified Final Moisture = 14%	Thermostatic Control

Ambient Conditions

Avg. Temp (° F)	Avg. Rh (%)	M.E. of Drying Air (% w.b.)
60.0	55.0	12.0

Fan Performance Information
Static Pressure (in. water)-CFM delivered: 0.028400.

0.5	28,400.	1.0	25,900.	1.5	23,130.	2.0	19,900.
2.5	17,000.	3.0	13,730.	3.5	10,700.	4.0	7,800.

Drying Schedule No. 1
Initial Moisture = 24% Layer Size Limited by Aflatoxin

Layer No.	Layer Depth (ft)	Total Depth (ft)	Bu/ Layer	Bu Total	Drying Days/ Layer	Drying Days Total	CFM/ Undried Bu
1	3.5	3.5	1,267	1,267	6.6	6.6	13.1
2	2.5	6.0	905	2,171	6.3	12.9	14.9
3	2.0	8.0	724	2,895	6.1	19.1	16.4
4	2.0	10.0	724	3,619	6.3	25.4	14.8
5	2.0	12.0	724	4,343	6.6	32.0	13.6
6	1.5	13.5	543	4,886	6.0	38.0	17.0
7	1.5	15.0	543	5,429	6.2	44.2	16.0
8	1.0	16.0	362	5,791	5.4	49.6	23.1

Average Drying Rates = 116.7 Bu/day; 0.32 ft/day

Table 3.1. Continued

Drying Schedule No. 2
Initial Moisture – 22% Layer Size Limited by Aflatoxin

Layer No.	Layer Depth (ft)	Total Depth (ft)	Bu/ Layer	Bu Total	Drying Days/ Layer	Drying Days Total	CFM/ Undried Bu
1	4.5	4.5	1,629	1,629	7.2	7.2	9.2
2	3.5	8.0	1,267	2,895	7.2	14.5	9.4
3	3.0	11.0	1,086	3,981	7.2	21.7	9.4
4	2.5	13.5	905	4,886	6.9	28.6	10.2
5	2.5	16.0	905	5,791	7.3	35.9	9.3

Average Drying Rates – 161.3 Bu/day; 0.45 ft/day

Drying Schedule No. 3
Initial Moisture – 20% Layer Size Limited by Aflatoxin

Layer No.	Layer Depth (ft)	Total Depth (ft)	Bu/ Layer	Bu Total	Drying Days/ Layer	Drying Days Total	CFM/ Undried Bu
1	6.0	6.0	2,171	2,171	8.0	8.0	6.1
2	4.5	10.5	1,629	3,800	7.8	15.8	6.4
3	3.5	14.0	1,267	5,067	7.4	23.2	7.1
4	2.0	16.0	724	5,791	5.9	29.0	11.6

Average Drying Rates – 199.7 Bu/day; 0.55 ft/day

Drying Schedule No. 4
Initial Moisture – 18% Layer Size Limited by Eave Height

Layer No.	Layer Depth (ft)	Total Depth (ft)	Bu/ Layer	Bu Total	Drying Days/ Layer	Drying Days Total	CFM/ Undried Bu
1	16.0	16.0	5,791	5,791	18.4	18.4	1.4

Average Drying Rates – 314.7 Bu/day; 0.87 ft/day

3. Relatively high grain temperatures lead to more rapid spoilage (fig. 2.1). If the grain mass can be cooled to near freezing, the rate of spoilage will be reduced significantly, especially if the grain moisture content is low.

As a minimum, it is important that the fans be operated until the entire grain mass is below 40° F if at all possible. Grain is often frozen in colder areas of the United States. Moisture will condense on the grain surface when thawing of frozen grain begins (the Zone 1 condition discussed in Chapter 2). Drying during the summer limits the possibility of reducing the spoilage rate by lowering grain temperature and offers greater drying potential, both in the field and in the bin.

Economic Considerations

In general, natural air drying and storage systems are very economical so long as all the grain can be dried successfully in a single bin without extending the harvest time adversely. As additional bins are needed, drying fans and other associated equipment must also be duplicated, thereby reducing the economic advantage of this method.

The variable cost of drying is primarily associated with operation of the drying fan. Generally, natural air drying is the least costly drying method if the grain can be dried successfully. Some of the energy efficiency advantages are offset because electricity to power the drying fan is generally more expensive per unit of energy than liquid petroleum (LP) gas, the usual source of thermal energy for drying. The relatively slow drying rate associated with natural air drying requires that grain be brought in from the field at a relatively lower moisture content. When harvesting is delayed to allow additional field drying, harvest losses also increase, resulting in economic loss.

Successful natural air grain drying does result in a high quality product. However, the primary economic consideration is whether the grain can be dried successfully. If aflatoxin or substantial spoilage occurs, the economic losses are tremendous when compared to the savings associated with the increased efficiency of drying.

Summary

Natural air drying is a high-risk and management-intensive form of drying. There is little reserve drying capacity, and the success of the system is very dependent on the weather and field drying of the grain. Natural air drying is an energy efficient method in terms of fossil energy. Grain quality is high if dried successfully. This method is best suited to single bin systems. The probable success of using this drying method increases as temperature during harvest decreases. The risk of using this method is sufficiently high so that it is not recommended for relatively warm and humid areas.

Low Temperature Drying

Low temperature drying of grain is the process by which relatively low amounts of energy are added to the drying air raising its temperature up to approximately 10° F above ambient conditions. Usually, electricity is the thermal energy source, hence the term “electric drying” is sometimes used instead of “low temperature” drying. However, LP gas and solar energy may also be used as thermal energy sources.

The low temperature drying method is assumed to always have potential for drying grain within the accepted moisture contents associated with long-term storage. This is contrasted with natural air drying where outside air conditions may not allow further drying. When air is heated, its temperature and volume increase, but its moisture level remains constant. This results in a lowering of the relative humidity of the air and allows for a possible net transfer of moisture from the grain to the air. Moisture transfer continues until the grain and air again come into equilibrium. In grain drying, the drying fan continuously supplies air as moisture is transferred from the grain and removed from the bin. With low temperature drying, sufficient heat is added so that drying can continue until normally acceptable final moisture contents are reached.

Equipment Requirements

The equipment normally associated with low temperature drying is shown in figure 3.6. Filling the bin may be accomplished by either a portable auger or a bucket elevator. A perforated floor is required. A grain spreader, under-floor unloading auger, and sweep auger usually are included. A stirring device may also be added.

In low temperature drying, grain is dried and stored in the same bin, thus minimizing handling and labor costs. Generally, the comparative total cost for drying decreases as less energy is used to heat the drying air even though more energy is required to operate the drying fans. Thus, successful low temperature drying is relatively economical in terms of energy cost when compared to most higher temperature methods. Some of this advantage is lost when electricity is used as the

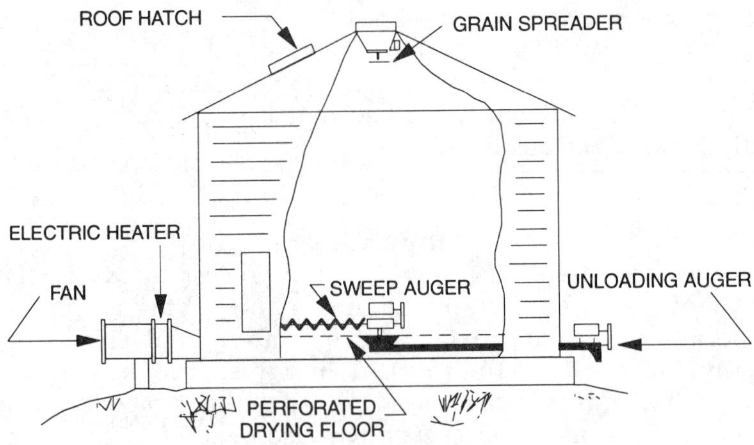

Figure 3.6–Low temperature drying equipment.

thermal energy source because it is usually more expensive than LP gas on a per unit of energy basis.

Successful low temperature drying is defined as drying the grain to a desired moisture content without excessive economic losses through either energy cost or grain spoilage. The storage time limits for corn as influenced by moisture content and temperatures are shown in figure 2.1 (from Chapter 2). The potential for drying increases when adding heat to increase the drying air temperature. Unfortunately, the rate of grain spoilage also increases with higher temperature. Thus, low temperature drying is generally restricted to conditions where the grain moisture is relatively low, near 20% for corn.

Low temperature drying is a high-risk drying method that requires a high level of management if grain is to be dried successfully every year. Each drying bin is usually filled once during the harvest season. Therefore, the manager has few opportunities to make adjustments in drying strategy. Low temperature drying has little reserve capacity; that is, the system dries grain at a slow steady rate that cannot be altered greatly. The dependability of the system is reduced further if solar energy is the heat energy source.

Perhaps the greatest risk associated with low temperature drying is the year-to-year variability of the weather. The manager uses the same drying strategy for several years only to find that changes in normal weather patterns have reduced both field drying and in-bin drying potential. This change possibly can result in aflatoxin contamination and/or grain spoilage. Large quantities of non-usable grain can alter radically the otherwise favorable economics of low temperature drying. The margin for error increases with increasing ambient temperature. Thus, low temperature drying is best suited for colder climates.

Performance

The rate at which low temperature drying systems perform is dependent upon the type of grain and airflow, air temperature, and relative humidity. Resistance to airflow usually is viewed as being independent of grain moisture content. The drying rate is directly proportional to airflow and follows the same principles as discussed previously with natural air drying and shown in figure 3.3.

As with natural air systems, there is sufficient heat generated by the drying fan to raise the drying air temperature approximately 2° F. Additional increases in air temperature are provided by a thermal energy source, usually electricity. The desired temperature of the drying air may be estimated by using the following procedure:

1. Select the desired final moisture content of the grain after drying is completed.

2. Estimate the expected average temperature and relative humidity for the day and mark the intersection of these points on figures 2.25 to 2.32 (from Chapter 2) as appropriate for the particular type of grain.
3. Extend a horizontal line to the right from the point selected in Step 2 above until it reaches the equilibrium moisture content line obtained from Step 1. Drying air relative humidity may be read directly at this intersection.
4. Extend a vertical line from this intersection downward to determine the temperature of the drying air.

Zone 1 through 4 conditions are assumed in the above procedure. Otherwise, a net rewetting of the grain will occur. It is also possible that existing natural air conditions will be sufficient for drying, especially when the 2° F temperature addition from the drying fan is considered.

Suppose that corn is to be dried to a final moisture content of 13% (Step 1 above), and the expected average daily temperature and relative humidity are 50° F and 75%, respectively (Step 2), as shown in figure 3.7. The drying air relative humidity (58%) is determined by extending horizontally the ambient air condition to the right until it intersects the 13% equilibrium moisture content curve. Extending the line vertically downward gives the drying air temperature value of approximately 57° F.

Once the drying air temperature is known, the additional heat requirements may be computed using the following equations:

For thermal systems,

$$\text{Btu / h} = 1.1 * \text{temperature rise} \left(^\circ \text{F}\right) * \text{cfm of fan} \tag{3.1}$$

For electrical heat systems,

$$\text{kW} = \frac{\text{cfm of fan} * \text{temperature rise } (^\circ \text{F})}{3{,}000} \tag{3.2}$$

In both equations, the temperature rise is the difference between the drying air temperature and the ambient air temperature minus the 2° F associated with the inefficiency of the fan. The total airflow of the fan may be determined from the fan performance curves, or estimated by cfm/bu values in figure 3.2 for a full bin, times the volume (cubic feet) of grain in the bin divided by 1.25 (the conversion of cubic feet to bushels). For the example situation, the temperature rise is equal to the drying air temperature of 57° F (from Step 4 above) minus 50° F (ambient temperature) minus 2° F (from the fan) for a final value of 5° F. If a 24-ft-

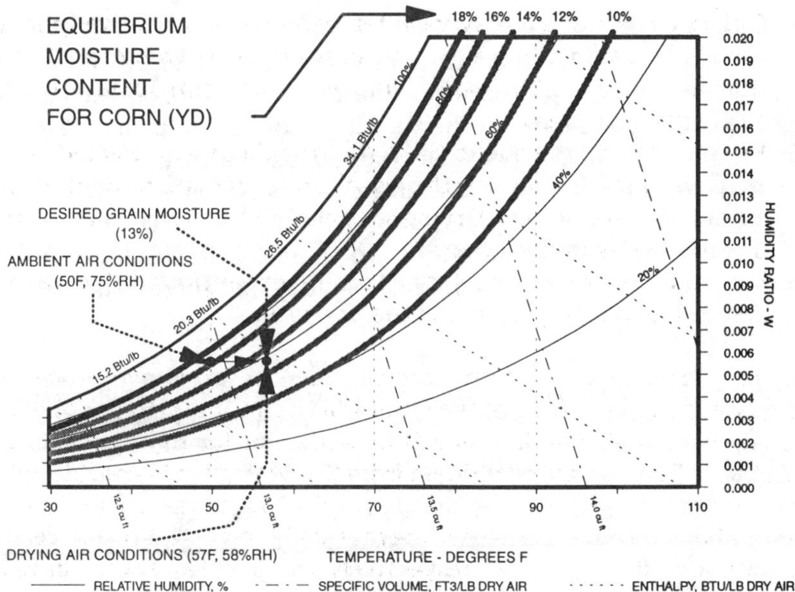

Figure 3.7–Low temperature drying example.

diameter bin is filled to a 16 ft depth, an airflow rate of 1.2 cfm/bu would be expected for corn when a 10-hp fan is used (fig. 3.2). This bin would contain 5,791 bu so that the total airflow would be 6,949 cfm (1.2 cfm/bu * 5,791 bu). The thermal energy requirements would be:

Thermal requirements: Btu/h = 1.1 * 5° F * 6,949 cfm
= 38,220

Electrical requirements: kW = (6,949 cfm * 5° F)/3,000
= 11.6

Filling the Drying Bin

Grain dried using low temperature drying may be subject to spoilage and/or aflatoxin contamination if the drying rate is too slow. The drying rate may be maximized by developing a filling schedule so that spoilage and/or aflatoxin are not likely to occur.

Filling schedules must consider the reduction in airflow as the bin fills as well as air and grain temperature and moisture conditions. Example schedules using a computer simulation model are shown in tables 3.2 to 3.3 for the following average daily weather conditions (Bridges et al., 1982; 1984):

1. 60° F, 85% relative humidity
2. 35° F, 75% relative humidity

In table 3.2, there are four schedules representing 24, 22, 20, and 18% initial grain moisture contents. The relatively warm wet ambient conditions are conducive to the development of aflatoxin. In fact, the temperature rise of 7° F perhaps compounds the situation in that the potential for aflatoxin limits the fill rate, even when initiating harvest with 18% moisture content grain.

The weather situation depicted in table 3.3 is much cooler and slightly drier. The bin can be filled in one day and there is no potential

Table 3.2. Schedules for low temperature corn drying given 60° F and 85% Rh ambient air conditions with 7.0° F added to obtain a desired final moisture content of 15%

In-Bin Shelled Corn Drying Model
(without stirring)
Developed by Tom Bridges, University of Kentucky

Gain = Corn	Fan ID XL420-7 (10 HP)
Bin Diameter = 24.0 ft	Bin Eave = 16.0 ft
Drying Air Rel. Hum. = 66.5%	Temperature Controlled
Drying Temperature Added (°) = 7.0	Specified Final Moisture Content = 15%

Ambient Conditions

Avg. Temp (° F)	Avg. RH (%)	M.E. of Drying Air (% w.b.)
60.0	85.0	13.9

Fan Performance Information
Static Pressure (in. water)-CFM delivered: 0.028400.

0.5	28,400.	1.0	25,900.	1.5	23,130.	2.0	19,900.
2.5	17,000.	3.0	13,730.	3.5	10,700.	4.0	7,800.

Drying Schedule No. 1
Initial Moisture = 24% Layer Size Limited by Aflatoxin

Layer No.	Layer Depth (ft)	Total Depth (ft)	Bu/ Layer	Bu Total	Drying Days/ Layer	Drying Days Total	CFM/ Undried Bu
1	2.5	2.5	905	905	5.7	5.7	20.6
2	2.0	4.5	724	1,629	5.7	11.4	21.1
3	1.5	6.0	543	2,171	5.3	16.7	24.9
4	1.5	7.5	543	2,714	5.5	22.2	22.5
5	1.5	9.0	543	3,257	5.7	27.9	20.7
6	1.0	10.0	362	3,619	5.1	33.0	29.6
7	1.0	11.0	362	3,981	5.2	38.2	28.3
8	1.0	12.0	362	4,343	5.2	43.4	27.1
9	1.0	13.0	362	4,705	5.3	48.7	26.0
10	1.0	14.0	362	5,067	5.4	54.1	25.0
11	1.0	15.0	362	5,429	5.4	59.5	24.0
12	1.0	16.0	362	5,791	5.5	65.0	23.1

Average Drying Rates = 89.1 Bu/day; 0.25 ft/day

Table 3.2. Continued

Drying Schedule No. 2
Initial Moisture = 22% Layer Size Limited by Aflatoxin

Layer No.	Layer Depth (ft)	Total Depth (ft)	Bu/ Layer	Bu Total	Drying Days/ Layer	Drying Days Total	CFM/ Undried Bu
1	3.0	3.0	1,086	1,086	5.8	5.8	16.2
2	2.0	5.0	724	1,810	5.3	11.1	20.2
3	2.0	7.0	724	2,533	5.7	16.8	17.4
4	1.5	8.5	543	3,076	5.3	22.1	21.3
5	1.5	10.0	543	3,619	5.4	27.5	19.7
6	1.5	11.5	543	4,162	5.5	33.0	18.5
7	1.5	13.0	543	4,705	5.7	38.7	17.3
8	1.5	14.5	543	5,248	5.8	44.5	16.3
9	1.0	15.5	362	5,610	5.1	49.7	23.6
10	0.5	16.0	181	5,791	4.3	53.9	46.3

Average Drying Rates = 107.3 Bu/day; 0.30 ft/day

Drying Schedule No. 3
Initial Moisture = 20% Layer Size Limited by Aflatoxin

Layer No.	Layer Depth (ft)	Total Depth (ft)	Bu/ Layer	Bu Total	Drying Days/ Layer	Drying Days Total	CFM/ Undried Bu
1	4.0	4.0	1,448	1,448	6.0	6.0	10.9
2	3.0	7.0	1,086	2,533	5.9	11.9	11.6
3	2.5	9.5	905	3,438	5.8	17.8	12.1
4	2.0	11.5	724	4,162	5.5	23.2	13.9
5	2.0	13.5	724	4,886	5.7	29.0	12.7
6	2.0	15.5	724	5,610	5.9	34.9	11.8
7	0.5	16.0	181	5,791	3.9	38.7	46.3

Average Drying Rates = 149.5 Bu/day; 0.41 ft/day

Drying Schedule No. 4
Initial Moisture = 18% Layer Size Limited by Aflatoxin

Layer No.	Layer Depth (ft)	Total Depth (ft)	Bu/ Layer	Bu Total	Drying Days/ Layer	Drying Days Total	CFM/ Undried Bu
1	7.0	7.0	2,533	2,533	7.7	7.7	4.9
2	5.0	12.0	1,810	4,343	7.3	15.0	5.4
3	4.0	16.0	1,448	5,791	6.9	21.9	5.8

Average Drying Rates = 264.1 Bu/day; 0.73 ft/day

for aflatoxin development, even with the addition of heat to the drying air.

The temperature extremes shown in tables 3.2 to 3.3 are the primary cause of the differences in the schedules. The lower the temperature at harvest, the greater the chance of using low temperature drying successfully without spoilage or aflatoxin contamination.

Economic Considerations

Low temperature drying is most economical when using a single bin drying-storage system. All drying accessories must be duplicated as additional bins are needed to accommodate the harvesting rate, thus eliminating economics of size. Among artificial drying systems, successful low temperature drying is second only to layer drying in terms of energy efficiency and grain quality. Some of the cost advantages of energy efficiency are lost when electricity is converted to thermal energy in that electricity is usually more expensive per unit of energy than LP gas, especially if relatively higher rates are in effect because of the demand schedules for electricity. The major economic

Table 3.3. Schedules for low temperature corn drying given 35° F and 75% Rh ambient air conditions with 7.0° F added to obtain a desired final moisture content of 15%

In-Bin Shelled Corn Drying Model
(without stirring)
Developed by Tom Bridges, University of Kentucky

Grain = Corn	Fan ID XL430-7 (10HP)
Bin Diameter = 24.0 ft.	Bin Eave = 16.0 ft
Drying Air Rel. Hum. = 57.1%	Thermostatic Control = 55%
Drying Temperature Added (°) = 7.0	Specified Final Moisture = 15%

Ambient Conditions

Avg. Temp (° F)	Avg. RH (%)	M.E. of Drying Air (% w.b.)
35.0	75.0	13.7

Fan Performance Information
Static Pressure (in. water)-CFM delivered: 0.0 28400.

0.5	28,400.	1.0	25,900.	1.5	23,130.	2.0	19,900.
2.5	17,000.	3.0	13,730.	3.5	10,700.	4.0	7,800.

Drying Schedule No. 1
Initial Moisture = 24% Layer Size Limited by Eave Height

Layer No.	Layer Depth (ft)	Total Depth (ft)	Bu/ Layer	Bu Total	Drying Days/ Layer	Drying Days Total	CFM/ Undried Bu
1	16.0	16.0	5,791	5,791	49.0	49.0	1.4

Average Drying Rates = 118.2 Bu/day; 0.33 ft/day

concern is, however, whether low temperature drying will work successfully. The chances of spoilage and/or aflatoxin contamination increase greatly with warmer temperatures and higher relative humidities during the harvest season. Non-marketable grain represents the major expense of any unsuccessful drying system.

Summary

Low temperature drying is a relatively high risk drying system requiring substantial management ability. Generally, it is preferable to natural air drying in that drying can occur in all types of weather. When used successfully, low temperature drying results in grain of high quality. However, its susceptibility to failure during high temperature–high relative humidity conditions during harvest limits its application to cooler regions.

Layer Drying

In-storage layer drying is a process whereby the grain is dried in layers in the storage structure with the entire grain depth ultimately being dried in place. The process begins when the initial grain layer is placed in the drying bin. The drying air establishes the drying front that moves through the grain. Additional layers of wet grain are added periodically so that a depth of wet grain always precedes the drying front. The quantity of grain that can be placed in any one layer is limited to that which can be dried before excessive mold growth or aflatoxin develops in the top of the layer. This drying technique is used most successfully in grain systems where relatively slow harvest rates are acceptable and harvest volumes are low to moderate.

Equipment

The desired equipment for a layer drying bin includes a full perforated drying floor, fan and heater unit with a transition, grain spreader, sweep auger, stirring device and unloading auger (fig. 3.8). For typical drying situations, bin diameters and eave heights are no greater than 27 ft and 16 ft, respectively, fan sizes range from 5 to 10 hp, and heater capacities vary from 300,000 to 800,000 Btu/h. It is essential to have sufficient fan capacity to dry the successive grain layers in order to prevent excessive mold growth. Thus, bin eaves are generally no higher than 16 ft because most fans do not deliver sufficient airflow for drying depths above this level.

The capacity for handling wet grain in a layer drying unit is not critical because of the slowness of the drying process. The only requirement is that wet grain should always be ahead of the drying front. Generally, a 6 or 8-in. tube type auger is used for unloading the bin, and its capacity should be designed to match that of the central

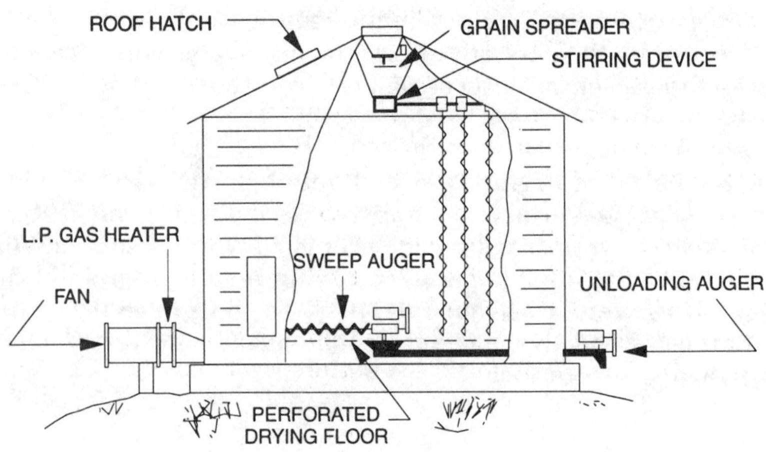

Figure 3.8–Layer drying equipment.

handling unit, whether it be a transport auger or bucket elevator. However, the central handling unit should provide sufficient handling capacity to load out a semi-trailer load in no more than 2 h. This means a minimum handling capacity of 400 to 500 bu/h for layer-drying systems.

Advantages and Disadvantages

Layer drying offers the advantage of low heat input, making it one of the most energy-efficient drying methods in terms of the amount of heat required to remove moisture from the grain. The grain remains in place in the storage structure after being dried, thus requiring a minimum of handling and labor. By combining the dryer and storage unit into one structure, layer drying presents an economical drying alternative for those producers with low harvest rates and harvest volumes, especially those requiring only one bin.

There are several disadvantages that are inherent with layer drying systems. The rate of drying of the system is relatively slow, thus requiring greater system management. The slowness of the system may restrict the rate of harvesting, and normally eliminates the possibility of using the same bin and associated drying equipment more than once during the drying season. The operator is limited to one drying experience per bin per year. This situation requires very careful management in that there is little margin for error for overcoming any adverse conditions. In layer drying, a single mistake may result in the spoilage of an entire bin of grain.

Physical Performance

Characteristically, layer drying is used in grain systems where relatively low harvest rates (200 to 500 bu/day) and low harvest volumes (up to 10,000 bu/year) occur. The drying air temperature should be limited to no more than a 20° F rise above ambient conditions in order to prevent excessive overdrying. A control mechanism for limiting the drying capacity of the air, and hence the final equilibrium moisture content of the grain, is to place a humidistat in the plenum chamber. The control level for the humidistat normally ranges from 50% to 60% relative humidity (Rh) depending on the type of grain, with 55% being a typical setting. Using corn as an example, figure 2.9 (from Chapter 2) shows that equilibrium moisture contents associated with relative humidities below 55% are undesirably low, and there will be a large potential for overdrying in the bottom layers of grain. Care should be taken to limit overdrying as much as possible since no premiums will be paid for grain dried below the base moisture content. Maintaining a humidistat setting of 55% should keep overdrying to a minimum. However, the accuracy of humidistats historically has been low, thus compounding the importance of overall management of layer drying systems. The addition of stirring devices helps reduce overdrying and drying time; however, these gains may be offset by the additional cost of the stirring equipment as is discussed in Chapter 9.

In addition to the normal problems of mold growth associated with grain spoilage, the producer may also be faced with the potential for aflatoxin development, especially during periods of warm temperatures and high humidities. While taking these restraints into account when determining a layer drying schedule, the producer must also consider the following parameters: airflow of the drying fan, outside temperature and humidity, humidistat setting, initial moisture content of the wet grain, and desired final moisture content. A computer model (Bridges et al., 1982; 1984) has been developed to aid grain producers in determining the filling schedule for their individual layer drying systems. The model uses specific drying fan and bin information as well as the projected drying conditions during harvest. Output from the model provides the producer with a tentative filling schedule for a specific set of input conditions while considering the potential for mold growth and aflatoxin development. Table 3.4 presents filling schedules for shelled corn developed by the computer model for a 24-ft diameter drying bin with a 16-ft eave, a 10-hp drying fan and four initial grain moisture contents. Each schedule in table 3.4 was determined for average outside air conditions of 60° F and 65% Rh. The specified final grain moisture content was 16% (w.b.). A humidistat control value of 55% was used to limit drying air temperature to no greater than 20° F (actually 6.2° F). The various schedules provide such management information as the number of fillings or layers required to fill the bin for

Table 3.4. Schedules for corn layer drying given warm high humidity conditions with humidity control to obtain a desired final moisture content of 16%

In-Bin Shelled Corn Drying Model
(without stirring)
Developed by Tom Bridges, University of Kentucky

Grain = Corn — Fan ID XL430-7 (10HP)
Bin Diameter = 24.0 ft. — Bin Eave = 16.0 ft
Controlled Drying Temperature = 66.2 — Drying Air Rel. Hum. = 55.0%
Humidity Control = 55%
Uncontrolled Drying Temperature Added (°) = 20.0
Specified Final Moisture Content = 16%

Ambient Conditions

Avg. Temp (° F)	Avg. RH (%)	M.E. of Drying Air (% w.b.)
60.0	65.0	12.2

Fan Performance Information
Static Pressure (in. water)-CFM delivered: 0.0 28400.

0.5	28,400.	1.0	25,900.	1.5	23,130.	2.0	19,900.
2.5	17,000.	3.0	13,730.	3.5	10,700.	4.0	7,800.

Drying Schedule No. 1
Initial Moisture = 24% Layer Size Limited by Aflatoxin

Layer No.	Layer Depth (ft)	Total Depth (ft)	Bu/ Layer	Bu Total	Drying Days/ Layer	Drying Days Total	CFM/ Undried Bu
1	4.0	4.0	1,448	1,448	3.8	3.8	11.2
2	3.0	7.0	1,086	2,533	3.7	7.5	11.6
3	2.5	9.5	905	3,438	3.6	11.1	12.1
4	2.0	11.5	724	4,162	3.3	14.5	13.9
5	2.0	13.5	724	4,886	3.5	18.0	12.7
6	2.0	15.5	724	5,610	3.7	21.6	11.8
7	0.5	16.0	181	5,791	1.9	23.5	46.3

Average Drying Rates = 246.4 Bu/day; 0.68 ft/day

Drying Schedule No. 2
Initial Moisture = 22% Layer Size Limited by Aflatoxin

Layer No.	Layer Depth (ft)	Total Depth (ft)	Bu/ Layer	Bu Total	Drying Days/ Layer	Drying Days Total	CFM/ Undried Bu
1	4.5	4.5	1,629	1,629	3.6	3.6	9.5
2	3.0	7.5	1,086	2,714	3.3	6.9	11.3
3	3.0	10.5	1,086	3,800	3.6	10.5	9.6
4	2.5	13.0	905	4,705	3.4	14.0	10.4
5	2.0	15.0	724	5,429	3.1	17.1	12.0
6	1.0	16.0	3,652	5,791	2.2	19.3	23.1

Average Drying Rates = 300.5 Bu/day; 0.83 ft/day

Table 3.4. (Continued)

Drying Schedule No. 3
Initial Moisture = 20% Layer Size Limited by Aflatoxin

Layer No.	Layer Depth (ft)	Total Depth (ft)	Bu/ Layer	Bu Total	Drying Days/ Layer	Drying Days Total	CFM/ Undried Bu
1	6.0	6.0	2,171	2,171	3.9	3.9	6.3
2	4.5	10.5	1,629	3,800	3.8	7.7	6.4
3	3.5	14.0	1,267	5,067	3.5	11.3	7.1
4	2.0	16.0	724	5,791	2.5	13.8	11.6

Average Drying Rates = 419.7 Bu/day; 1.16 ft/day

Drying Schedule No. 4
Initial Moisture = 18% Layer Size Limited by Aflatoxin

Layer No.	Layer Depth (ft)	Total Depth (ft)	Bu/ Layer	Bu Total	Drying Days/ Layer	Drying Days Total	CFM/ Undried Bu
1	10.0	10.0	3,619	3,619	5.1	5.1	3.0
2	6.0	16.0	2,171	5,791	3.9	9.1	3.9

Average Drying Rates = 636.4 Bu/day; 1.77 ft/day

a specified harvest moisture content, the volume and approximate drying time for each layer, and the total time required to dry and fill the bin. As each schedule is oriented toward an individual situation, the producer is provided with a guide as to when the drying front will pass through an individual layer and how often one can expect to add a new layer. The various schedules can also be used to determine new layer volumes as the grain moisture decreases during the harvest period.

For the example situation in table 3.4, the initial moisture content is reduced from 24 to 18% w.b. As a result, the total number of layers of fillings per schedule decreases from 7 to 2 while the total drying time decreases from 23.5 to 9.1 days. For all four schedules, there is potential for aflatoxin development so the individual layer sizes are limited to prevent its occurrence.

The example situation in table 3.5 is also for shelled corn using the same drying conditions as those in table 3.4 except for cooler and less humid ambient air conditions. In this example, the average outside air conditions are 55° F and 55% Rh. The humidistat control value is also 55%, thus allowing a minimum temperature rise of only 1.5° F which is provided by the fan. The schedules in table 3.5 indicate that when using layer drying in cooler temperatures and reduced humidities, the size of the layers can be increased significantly over those in table 3.4, because conditions favorable for aflatoxin development are not present and

Table 3.5. Schedules for layer drying of corn given cool and moderate relative humidity conditions with humidity control to obtain a desired final moisture content of 16%

In-Bin Shelled Corn Drying Model
(without stirring)
Developed by Tom Bridges, University of Kentucky

Grain = Corn
Fan ID XL430-7 (10HP)
Bin Diameter = 24.0 ft
Bin Eave = 16.0 ft
Controlled Drying Temperature = 56.5
Drying Air Rel. Hum. = 52.1%
Humidity Control = 55%
Uncontrolled Drying Temperature Added (°) = 20.0
Specified Final Moisture Content = 16%

Ambient Conditions

Avg. Temp (° F)	Avg. RH (%)	M.E. of Drying Air (% w.b.)
55.0	55.0	12.2

Fan Performance Information
Static Pressure (in. water)-CFM delivered: 0.0 28400.

0.5	28,400.	1.0	25,900.	1.5	23,130.	2.0	19,900.
2.5	17,000.	3.0	13,730.	3.5	10,700.	4.0	7,800.

Drying Schedule No. 1
Initial Moisture = 24% Layer Size Limited by Dry Matter Loss

Layer No.	Layer Depth (ft)	Total Depth (ft)	Bu/ Layer	Bu Total	Drying Days/ Layer	Drying Days Total	CFM/ Undried Bu
1	14	14	5,067	5,067	21.3	21.3	1.8
2	2	16	724	5,791	5	26.3	11.6

Average Drying Rates = 220.5 Bu/day; 0.61 ft/day

Drying Schedule No. 2
Initial Moisture = 22% Layer Size Limited by Eave Height

Layer No.	Layer Depth (ft)	Total Depth (ft)	Bu/ Layer	Bu Total	Drying Days/ Layer	Drying Days Total	CFM/ Undried Bu
1	16.0	16.0	5,791	5,791	21.2	21.2	1.5

Average Drying Rates = 273.1 Bu/day; 0.75 ft/day

mold growth is minimal. The trade-off in this situation is that if heat input is limited (in order to minimize overdrying), the drying times are somewhat greater than for the conditions in table 3.4 at comparable initial grain moisture contents. So, while cooler drier climates generally allow more grain to be placed in the individual layers, the drying process may be extended because the drying potential is reduced.

The drying schedules (tables 3.4 and 3.5) generated by the computer model are the maximum filling rates that will prevent grain spoilage for the situations presented. The layer volumes and depths are based on undried grain and will be somewhat less after drying because of shrinkage. Care should be taken in using these schedules for situations other than those shown because each drying situation is unique as to air conditions, grain moisture contents, and drying fans.

Management of layer drying involves tracking of the drying front and adjusting the filling rates accordingly in order to keep potential spoilage development at a minimum. The position of the drying front can be determined at any time by probing the grain mass from above and locating the point where the grain temperature begins to increase significantly. The drying front usually may be detected by feeling the temperature difference along the bin wall. Again, management in layer drying is critical because the operator will have only one learning experience per year.

Economic Considerations

Increases in harvest volume generally are accompanied by an increase in the rate necessary to properly harvest, dry, and store the grain. One method by which the drying capacity of a layer drying system may be increased is to enlarge the fan size or add multiple fan units per storage structure. This additional horsepower may prove to be expensive, and the increased drying capacity may not be significant when related to the cost. Also, duplicating drying fans and associated drying equipment may be more cost prohibitive than selecting another drying method. Layer drying is usually physically and economically undesirable when higher drying rates are needed.

Summary

Layer drying is a good drying technique for low harvest rates and volumes. The method is simple, requires little labor input, but necessitates superior management skills. The system leaves little margin for error because of its relatively low reserve drying capacity. Results can be satisfactory when layer drying is used with appropriate equipment and correct management practices.

Batch-in-bin Drying

Batch-in-bin drying refers to the process where the grain is dried in a drying bin each day in a batch usually 2.5 to 4 ft deep and then cooled and moved to storage bins in time for the next day's harvest. When the storage bins are full, the drying bin may be filled and the grain dried in layers if necessary. No wet grain storage is needed with this technique as the batch size constitutes one day's harvest. The basic principle behind the operation of a batch-in-bin dryer is to force relatively large quantities of air through a shallow grain depth in order to obtain relatively rapid drying, thus allowing the producer to accommodate larger harvest rates than with other in-bin drying methods.

Equipment

Figure 3.9 illustrates the type of equipment necessary for batch-in-bin drying. Generally, the equipment set includes a full perforated floor, fan and heater unit with transition, grain spreader, sweep auger, and an under-bin unloading auger. The drying bin diameter should be selected so that the grain depth resulting from the daily harvest will be no more than the maximum recommended depth of 4 ft. The grain handling equipment, drying fan and floor area associated with the bin diameter should be sized so that the day's harvest can be loaded, dried, cooled, and unloaded in a 24-h period. When selecting the drying bin for this type of system, the eave height may be limited to 4 or 5 rings (10.6 to 13.3 ft for a typical 32-in. ring height) which provides sufficient headroom above the perforated floor for grain distribution and removal. Additional rings may be added if the bin is to also be used for storage. Table 3.6 shows volumetric capacities for various bin diameters and grain depths.

A grain spreader is essential for maintaining a level grain surface and distributing fine material and trash evenly in order to assure uniform drying and airflow. The sweep auger is desirable for rapid and

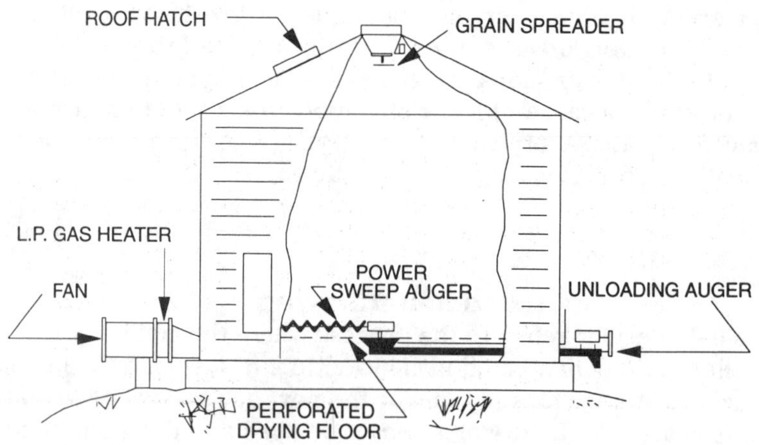

Figure 3.9–Batch-in-bin drying equipment.

Table 3.6. Bushel capacities for round bins

Grain Depth (ft)	Volume in Bushels: Bin Diameter (ft)											
	15	18	21	24	27	30	33	36	39	42	45	48
0.5	71	102	139	181	229	283	342	407	478	554	636	724
1.0	141	204	277	362	458	565	684	814	956	1108	1272	1448
1.5	212	305	416	543	687	848	1026	1221	1433	1662	1908	2171
2.0	283	407	554	724	916	1131	1368	1629	1911	2217	2545	2895
2.5	353	509	693	905	1145	1414	1711	2036	2389	2771	3181	3619
3.0	424	611	831	1086	1374	1696	2053	2443	2867	3325	3817	4343
3.5	495	712	970	1267	1603	1979	2395	2850	3345	3879	4453	5067
4.0	565	814	1108	1448	1832	2262	2737	3257	3823	4433	5089	5790
4.5	636	916	1247	1629	2061	2545	3079	3664	4300	4987	5725	6514
5.0	707	1018	1385	1810	2290	2827	3421	4071	4778	5542	6362	7238
5.5	778	1120	1524	1990	2519	3110	3763	4479	5256	6096	6998	7962
6.0	848	1221	1662	2171	2748	3393	4105	4886	5734	6650	7634	8686
6.5	919	1323	1801	2352	2977	3676	4447	5293	6212	7204	8270	9409
7.0	990	1425	1940	2533	3206	3958	4790	5700	6690	7758	8906	10133
7.5	1060	1527	2078	2714	3435	4241	5132	6107	7167	8312	9542	10857
8.0	1131	1629	2217	2895	3664	4524	5474	6514	7645	8867	10178	11581
8.5	1202	1730	2355	3076	3893	4806	5816	6921	8123	9421	10815	12305
9.0	1272	1832	2494	3257	4122	5089	6158	7328	8601	9975	11451	13028
9.5	1343	1934	2632	3438	4351	5372	6500	7736	9079	10529	12087	13752
10.0	1414	2036	2771	3619	4580	5655	6842	8143	9556	11083	12723	14476
10.5	1484	2137	2909	3800	4809	5937	7184	8550	10034	11637	13359	15200
11.0	1555	2239	3048	3981	5038	6220	7526	8957	10512	12192	13995	15924
11.5	1626	2341	3186	4162	5267	6503	7869	9364	10990	12746	14632	16647
12.0	1696	2443	3325	4343	5496	6786	8211	9771	11468	13300	15268	17371
12.5	1767	2545	3464	4524	5725	7068	8553	10178	11946	13854	15904	18095
13.0	1838	2646	3602	4705	5954	7351	8895	10586	12423	14408	16540	18819
13.5	1908	2748	3741	4886	6183	7634	9237	10993	12901	14962	17176	19543
14.0	1979	2850	3879	5067	6412	7917	9579	11400	13379	15516	17812	20266
14.5	2050	2952	4018	5248	6641	8199	9921	11807	13857	16071	18448	20990
15.0	2121	3054	4156	5429	6870	8482	10263	12214	14335	16625	19085	21714
15.5	2191	3155	4295	5609	7099	8765	10605	12621	14812	17179	19721	22438
16.0	2262	3257	4433	5790	7328	9048	10947	13028	15290	17733	20357	23162
16.5	2333	3359	4572	5971	7558	9330	11290	13436	15768	18287	20993	23885
17.0	2403	3461	4710	6152	7787	9613	11632	13843	16246	18841	21629	24609
17.5	2474	3562	4849	6333	8016	9896	11974	14250	16724	19396	22265	25333
18.0	2545	3664	4987	6514	8245	10178	12316	14657	17202	19950	22902	26057
18.5	2615	3766	5126	6695	8474	10461	12658	15064	17679	20504	23538	26781
19.0	2686	3868	5265	6876	8703	10744	13000	15471	18157	21058	24174	27604
19.5	2757	3970	5403	7057	8932	11027	13342	15878	18635	21612	24810	28228
20.0	2827	4071	5542	7238	9161	11309	13684	16286	19113	22166	25446	28952
20.5	2898	4173	5680	7419	9390	11592	14026	16693	19591	22721	26082	29676
21.0	2969	4275	5819	7600	9619	11875	14369	17100	20069	23275	26718	30400
21.5	3039	4377	5957	7781	9848	12158	14711	17507	20546	23829	27355	31123
22.0	3110	4479	6096	7962	10077	12440	15053	17914	21024	24383	27991	31847
22.5	3181	4580	6234	8143	10306	12723	15395	18321	21502	24937	28627	32571
23.0	3251	4682	6373	8324	10535	13006	15737	18728	21980	25491	29263	33295
23.5	3322	4784	6511	8505	10764	13289	16079	19136	22458	26046	29899	34019
24.0	3393	4886	6650	8686	10993	13571	16421	19543	22935	26600	30535	34742

convenient unloading of the drying bin. The bin unloading auger capacity should be matched to that of the central handling unit. For this type of drying system, the central handling unit (bucket elevator or transport auger) should have sufficient capacity to unload the drying bin in 2 h or less.

After determining the appropriate diameter drying bin for the daily harvest, the drying fan can be selected. Typical fan sizes range from 5 to 10 hp depending on bin diameter and batch height and, as a rule of thumb, should provide an airflow rate of approximately 10 ft^3/min/bu. The heater units generally burn LP gas, although natural gas units are also available, and range in size from 1 million to 2.5 million Btu/h.

Advantages and Disadvantages

One reason for the popularity of batch-in-bin drying is the flexibility available when selecting the drying system equipment. Typical farm size bin diameters range from 18 to 48 ft, and there are several sizes of fans and heaters available to provide for a wide range of drying capacities. Management of a batch-in-bin system is less critical than in other in-bin drying units, since the operator has an opportunity every 24 h to correct any mistakes and to "fine tune" the drying process.

A batch-in-bin system allows drying flexibility in that the drying depth may be varied based on day-to-day operating conditions; thus, the producer is able to adjust the harvesting schedule if necessary. The batch drying bin may also be layer-filled at the end of harvest to provide additional grain storage.

The main disadvantage of a batch-in-bin drying system is that the grain must be handled twice. The drying bin must be leveled and unloaded each day. This requires placing the sweep auger in the drying bin at each unloading unless power sweeps are used. Frequent moving of any portable handling equipment is required unless a second conveyor is purchased to be used exclusively for unloading the dry grain. Another concern is the length of the drying period. This system operates on a 20- to 24-h basis which may require some supervision beyond normal working hours. This schedule may become tiresome over a long harvest season.

Physical Performance

Batch-in-bin drying is used generally used in grain systems where harvest volumes range from 5,000 to 35,000 bushels/year and for harvest rates of 500 to 2,500 bushels/day. For most grains, drying air temperature ranges from 120° to 160° F with 140° F being the recommended value for shelled corn. The heater unit is controlled by a thermostat. When designing batch-in-bin units, it is desirable that the drying fan/drying bin combination dry the daily harvest in about 16 h. Drying may begin when the floor is covered uniformly with approximately 6 in. of grain. After drying is completed, the cooling process will usually require another 2 h. Handling will also require two additional hours for a total of 20 h of activity each drying day. The remaining 4 h provide catch-up time in case of breakdowns, harvesting delays, etc.

Drying is complete when the average moisture of the batch reaches the desired final moisture content for safe storage or marketing. A moisture gradient exists in the batch with grain at the top remaining near its initial moisture content. The cooling period following drying partially will eliminate this gradient as will the mixing that occurs when the grain is moved to storage. While the grain is in storage, moisture diffusion between individual kernels will serve also to reduce this difference. Many producers using batch-in-bin drying will overdry the grain somewhat to offset the possibility of not getting the wet grain to a safe moisture content when mixing and moving it to storage.

Table 3.7 presents the results of a computerized drying simulation to demonstrate how increasing grain depth (decreasing airflow in cfm/bu) affects the drying time and moisture spread of an individual batch of shelled corn. In each example, a 24-ft diameter drying bin and a representative 7.5-hp drying fan are used. An initial moisture of 25% w.b. is assumed, and the grain is dried to an average moisture of 13%. As the grain depth is increased from 3 to 5 ft, the drying time and moisture spread increases significantly as does the efficiency of the drying process (i.e., a decrease in Btu required per pound of water removed). Likewise, when airflow drops below the recommended 10 cfm/bu, the bottom layers of the batch overdry severely, and the drying time is greater than the desired 16 h. These factors are offset somewhat by a gain in drying efficiency as the airflow decreases.

Table 3.7. Batch-in-bin examples showing effects of airflow

Items	Example 1 3-ft Grain Depth 13.84 CFM/Bu	Example 2 4-ft Grain Depth 9.35 CFM/BU	Example 3 5-ft Grain Depth 6.9 CFM/Bu
Drying time (h)	13.30	17.76	22.91
Top moisture (% w.b.)	17.92	19.48	20.92
Bottom moisture (% w.b.)	7.97	6.74	5.76
Fuel Requirements (BTU/lb H_2O)	2,137	1,928	1,835
Fuel Requirements (gal LP gas/bu)	0.253	0.229	0.218

Above data is for the following conditions: 24-ft drying bin; 7.5-hp drying fan; 12 points of moisture removal (25 to 13%); a drying air temperature of 140° F; and outside air conditions of 50° F and 65% Rh.

Table 3.8. Batch-in-bin examples showing effects of drying air temperature

Items	Example 1 4-ft Grain Depth 120° F	Example 2 4-ft Grain Depth 140° F	Example 3 4-ft Grain Depth 160° F
Drying time (h)	24.26	17.76	14.20
Top moisture (% w.b.)	17.69	19.48	21.12
Bottom moisture (% w.b.)	8.49	6.74	4.97
Efficiency (BTU/lb H_2O)	2,048	1,928	1,884
Efficiency (gal LP gas/bu)	0.243	0.229	0.223

Above data is for the following conditions: 24-ft drying bin; 7.5-hp drying fan; 12 points of moisture removal (25 to 13%); an airflow rate of 9.35 cfm/bu; and outside air conditions of 50° F and 65% relative humidity.

Table 3.8 presents results from another computer simulation to illustrate the effect of increasing drying temperature as related to the drying time and moisture spread of an individual batch of shelled corn. As the temperature increases from 120 to 160° F for the fixed grain depth, the drying time decreases and the moisture spread increases as does the degree of overdrying in the bottom layers. The values in table 3.8 illustrate that overdrying of the bottom layers may be severe for temperatures above 140° F. Again there is some trade-off in drying efficiency as drying temperature is increased.

The information presented in tables 3.7 and 3.8 are for the specific examples cited. Using shelled corn, figures 3.10 to 3.15 illustrate the effects of various drying air temperatures and airflow rates on drying performance. The values are the result of a computer simulation (Thompson et al., 1968) for 12 points of moisture removal (25% to 13% w.b.), ambient air conditions of 50° F and 65% Rh, and no stirring devices. The figures show expected trends when the drying temperature or airflow rate is varied in a batch-in-bin system. It can be concluded that:

1. Drying time decreases with increases in air temperature and airflow (fig. 3.10).
2. Drying efficiency decreases as airflow increases but generally increases with drying air temperature (figs. 3.12 to 3.13).
3. Drying rate increases with increases in air temperature and airflow (fig. 3.11).
4. Moisture spread increases with increases in air temperature but decreases with increases in airflow (fig. 3.14).
5. Drying cost for fuel increases with airflow and generally decreases with drying air temperature (fig. 3.15).

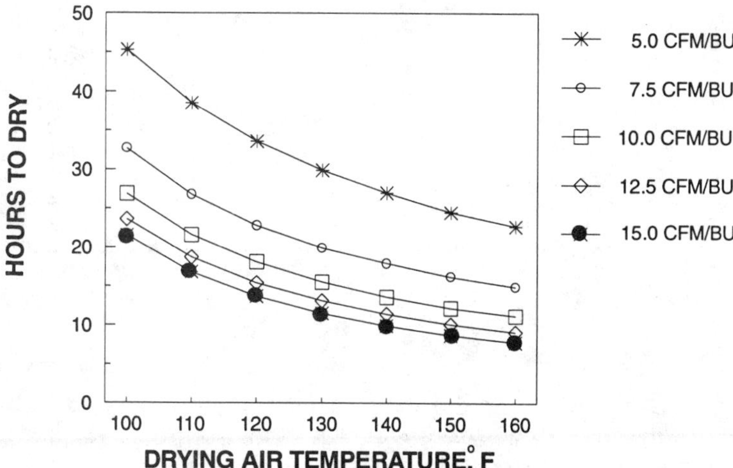

Figure 3.10–Corn drying time, batch-in-bin systems (Thompson et al., 1968).

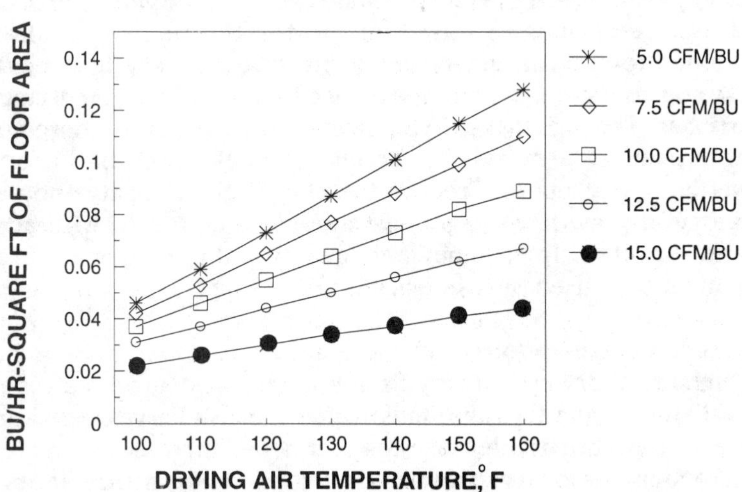

Figure 3.11–Corn drying rate, bu/h/ft² of bin floor area, batch-in-bin systems (Thompson et al., 1968).

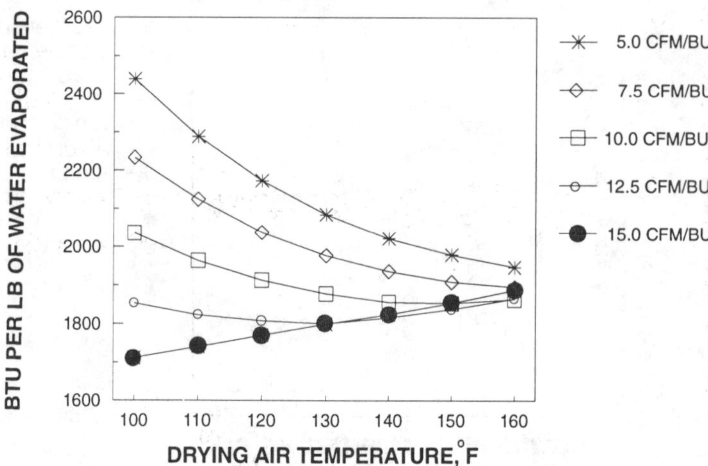

Figure 3.12–Energy required to dry corn, Btu/lb of water evaporated, batch-in-bin systems (Thompson et al., 1968).

Figures 3.10 to 3.15 indicate that drying cost increases as drying rate increases. However, the cost of fuel is only one of many economic considerations when drying as is discussed in Chapter 9.

Several other factors are of concern when managing a batch-in-bin drying system. The temperature and airflow distribution should be as uniform as possible in the plenum chamber. It is desirable that the incoming grain be cleaned to eliminate trash. The drying floor should be kept as free of fines as possible, and the floor supports should be positioned to prevent channeling of the drying air supplied by the fan. These considerations help prevent "hot spots" and uneven drying of the grain mass. The higher the drying temperature, the more thoroughly the grain mass must be mixed after drying. This helps prevent the formation of wet grain pockets that reduce the overall grain quality and possibly cause mold growth. Storage bins should be equipped with adequate aeration fans to help maintain grain quality and to reduce any problems that might be caused by insufficient grain mixing.

Economic Considerations

Batch-in-bin drying is a very flexible drying system with a wide range of available drying bins and fans to meet many harvesting situations. These systems are not flexible, however, when increases in harvest rate require significant expansion of drying capacity in that an enlarged drying bin may be required. One available choice for expansion of drying capacity is the addition of a stirring device to the drying bin. This

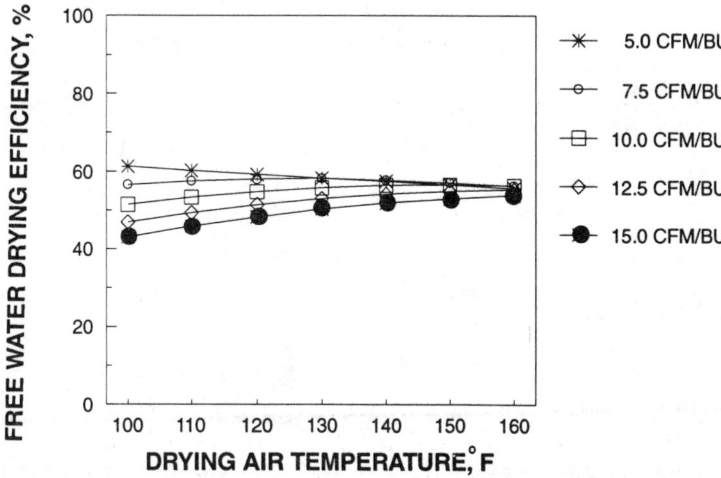

Figure 3.13–Corn drying efficiency compared to free water evaporation when using batch-in-bin systems (Thompson et al., 1968) .

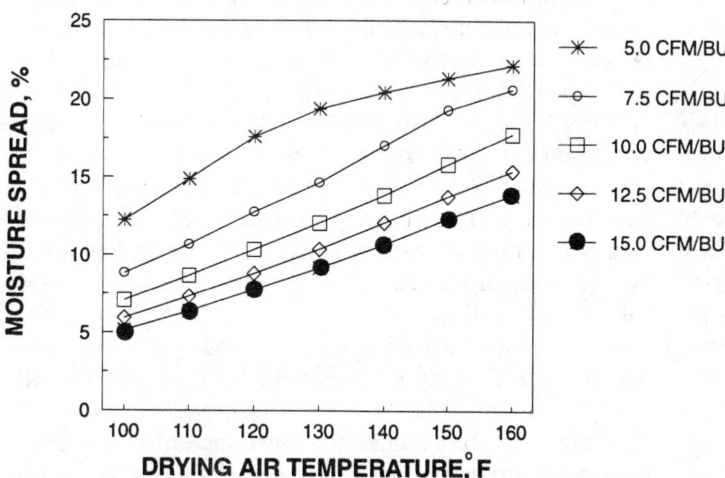

Figure 3.14–Moisture spread across the corn drying column when using batch-in-bin systems (Thompson et al., 1968).

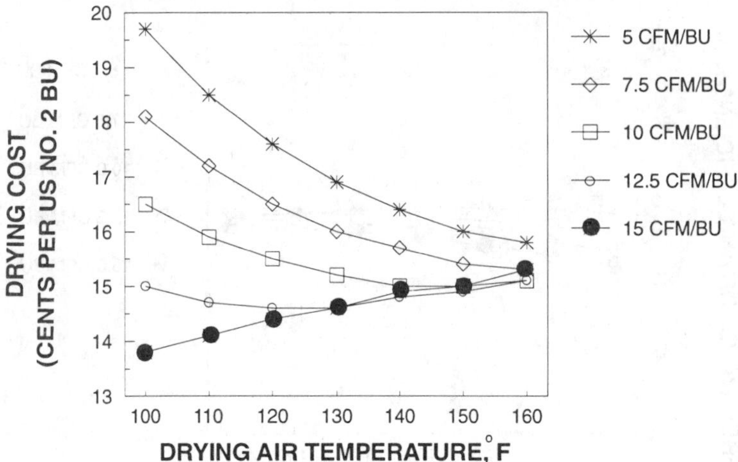

Figure 3.15–Cost of LP gas to dry a bushel of corn when using batch-in-bin systems (Thompson et al., 1968).

allows the producer to increase the drying temperature and drying depth, thereby increasing drying capacity while at the same time helping to reduce the moisture spread in the batch. There would also be some reduction in the drying efficiency. However, a stirring device remains a viable choice if the original system is already at 100% drying capacity.

There are several alternatives for increasing drying capacity in batch-in-bin drying systems. One is to build a larger diameter drying bin to either increase batch size for the same batch depth or increase drying rate for the same grain capacity through the associated decrease in batch depth. Either option probably would require all new drying equipment with the possible exception of the drying fan. Another choice is to increase airflow capacity either by replacing the existing fan or by adding a higher capacity fan. While this may not be as costly as a new drying bin, the system may be designed improperly and result in too much or too little drying capacity. Another possibility is to over-design the original drying bin to allow for future harvest rate increases. This may be the easiest and least expensive choice. However, when drying at lower grain depths, the drying process will be less energy efficient (table 3.7).

Another economic consideration is the purchase of some additional convenience-type equipment to "make life easier" while operating the drying system. One such item is a power sweep auger controlled from outside the drying bin to facilitate daily unloading. If a transport auger is used as the central handling equipment, the purchase of a second auger to unload the drying bin into storage might be considered as a means of eliminating frequent positioning of these units. These equipment items are an additional expense and are not necessary to make the drying system functional; however, they may pay for themselves with long-term labor savings.

Summary

Batch-in-bin is one of the most popular, economical drying techniques available. Management is not as critical as in other in-bin drying methods inasmuch as the operator has many learning experiences per harvest season and can make daily adjustments in system operations. Disadvantages of the system are its labor and unloading time requirements and that grain must be handled twice before going into storage.

Portable Batch-Continuous Flow Drying

Portable batch and continuous flow are two very popular high-speed grain-drying techniques. Both dryer types are similar in appearance and operation and require wet grain storage ahead of the dryer. The basic principle in both dryer types is to force high quantities of air (50 to 125 cfm/bu) through relatively thin grain columns (12 to 24 in.) to obtain relatively high drying rates. The portable batch units usually dry, cool and unload a fixed amount of grain into storage, whereas continuous flow units meter cool-dry grain from the drying chamber at constant flow rate. While the drying units are movable or portable as the name implies, a permanent centralized location is generally desirable.

Equipment Requirements

Portable Batch. A schematic drawing of a typical portable batch dryer is shown in figure 3.16. These dryers are self-contained units with fan sizes generally ranging from 10 to 20 hp. Drying temperatures range from 160° to 200° F with heater capacities of from 2 to 5 million Btu/h. Larger units may have multiple fans and drying chambers that allow grain to be dried in stages. Most portable batch units have a complete set of controls to automatically load, dry, cool, and unload the grain. This eliminates the need for any manual operation of the dryer and allowing extended drying periods after nightfall. Many portable batch units have an optional cooling cycle that can be eliminated when dryeration is practiced. This option increases the capacity of the dryer and will be discussed later in this chapter.

Continuous Flow. Figure 3.17 is a schematic drawing of a typical continuous flow dryer showing the configuration of the heating and cooling chambers. The basic difference between this type of dryer and a

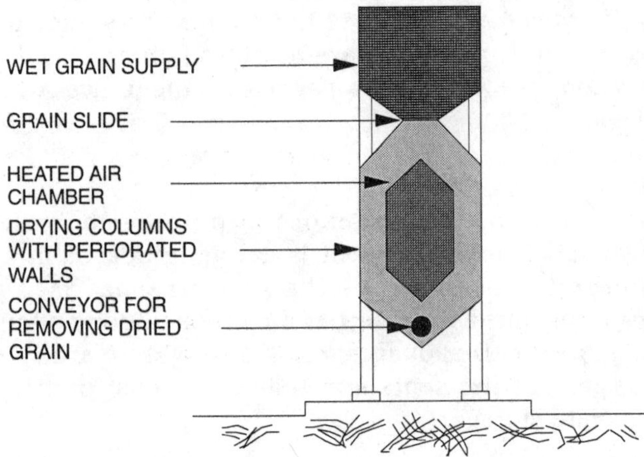

Figure 3.16–Typical column batch dryer.

portable batch unit is that the grain flows continually through the dryer and is regulated by grain meters. Fan sizes for continuous flow dryers are comparable to those of portable batch units, but drying temperatures are generally higher (180 to 220° F) with heater sizes

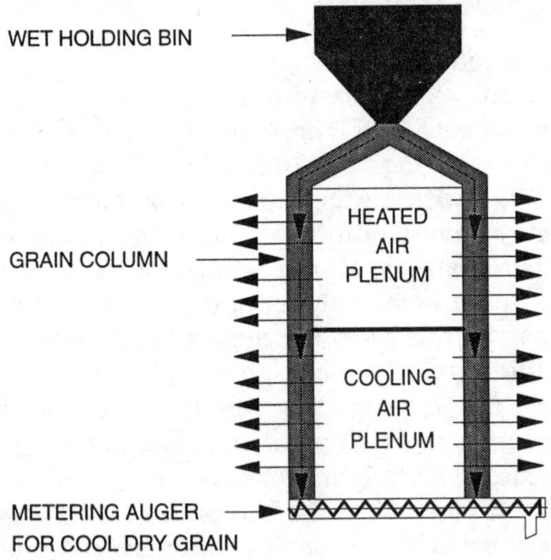

Figure 3.17–Continuous flow dryer without heat recovery.

normally ranging from 3 to 12 million Btu/h. Most continuous flow dryers are completely automated and do not require constant supervision.

Handling. While the name "portable" implies that these drying units are movable, they will function better in a centralized location with permanent grain handling equipment. The central handling unit may be either a bucket elevator or transport auger, with both requiring a minimum handling capacity of 1 to 500 bu/h. In systems where transport augers (or similar portable grain handling equipment) are used, a second auger unit should be available to unload the dry grain into storage. In systems using bucket elevators, a smaller "dry leg" may be installed just to handle dried grain. An important consideration for grain systems requiring high speed dryers is that the drying capacity of these units not be limited by insufficient handling capability for movement of wet and dry grain.

Wet Holding. Both portable batch and continuous flow dryers require wet grain storage preceding the dryer itself. This storage may be a ground-level tank that feeds the dryer by means of a portable auger or an elevated, wet grain bin that provides gravity loading of the dryer. If the elevated tank is chosen, care should be taken to design properly the support structure to accommodate the relatively large stresses that occur when the tank is filled. A rule of thumb is that minimum wet storage requirement for these systems be the maximum of 1½ to 2 times of either (1) the batch size of the dryer or (2) the capacity of the largest delivery vehicle. A good estimate of the maximum size tank may be obtained by using the following relationship:

$$\text{tank capacity} = \text{daily harvest} - (\text{drying time} * \text{effective drying rate}) \quad (3.3)$$

where

tank capacity	=	size of the wet grain tank in bushels
daily harvest	=	amount of grain harvested each day in bushels
drying time	=	hours of dryer operation while harvest is still continuing
drying rate (effective)	=	rate of drying in bu/h, considering the time for loading and unloading plus that of the cooling cycle (if any)

In most cases, the dryer will not be started until after the first or second load reaches the grain facility, thus influencing the effective drying rate. For example, suppose that the producer is harvesting 2,200 bushels in a 10-h harvest period. The batch size of the dryer is 150 bushels that can be dried in approximately 40 min. The cooling cycle requires 15 min, and unloading involves another 5 min for a total cycle time of 1 h. Thus, the effective drying rate is 150 bu/h. The dryer does not begin drying until 1.5 h after harvest has begun because of vehicle travel and loading time. Then:

$$\text{tank capacity} = 2{,}200 - (10.0 - 1.5)*150 = 925 \text{ bu}$$

A detailed discussion of wet holding bins is given in Chapter 5.

Surge Bin. A surge bin is a convenient item of equipment for storing dry grain temporarily that exits a portable drying system. This item generally is used in systems containing bucket elevators and allows the dryer to unload dry grain when ready to facilitate the drying process. In addition, the surge bin unloading conveyor or bucket elevator is required to operate only when the surge bin is unloading rather than for the entire period of dryer operation. The surge bin should have a minimum capacity equal to the batch size of the dryer although it may be much larger if desired. Surge bins generally are equipped with sensing devices that allow automatic unloading of the dry grain to storage. Sizing of surge bins is covered in detail in Chapter 5.

Advantages and Disadvantages

The main advantage of a portable batch or continuous flow drying unit is its greater drying capacity. These units are designed to provide the necessary drying capacity for rapid harvest of large volumes of grain. Most portable drying units of either type are automated completely, thus reducing labor requirements for loading and unloading. They are available in many different sizes to accommodate a wide range of drying needs. The drying units are generally small enough to be moved, if necessary, but generally will operate better in a centralized location. Their portability allows for relatively easy replacement associated with either wear-out or capacity expansion. The main disadvantage of these drying units is the relatively low efficiency in terms of energy utilization. Although the addition of heat recapture devices has improved significantly the energy efficiency of these dryers (fig. 3.18), the trade-off between relatively lower drying energy efficiency and the higher harvesting efficiency (i.e., reduction of harvesting losses) remains a consideration. Rapid harvesting requires rapid drying that comes at the expense of drying energy efficiency. In the short run, portable batch-continuous flow units generally are more expensive to purchase and require more associated handling equipment (wet holding, surge bins, dry legs, etc.) than do other drying methods. Grain must be handled at least twice by systems using these dryers.

Physical Performance

Portable batch and continuous flow dryers are very similar in appearance and performance. Portable batch dryers are classified as

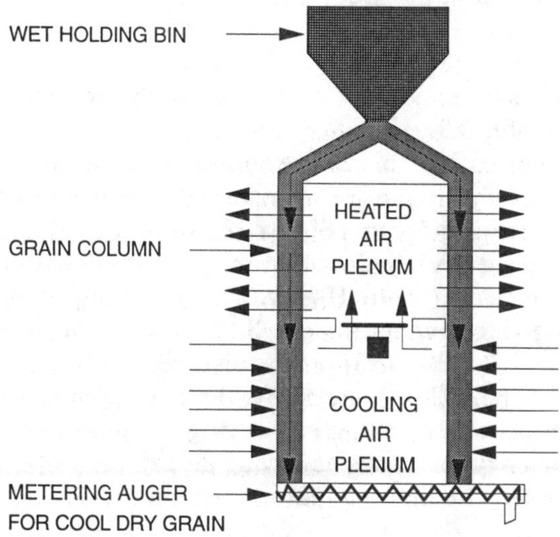

Figure 3.18–Continuous flow dryer with heat recovery.

stationary bed dryers. The grain mass does not move during the drying process, and the drying air is blown across the grain column. Drying air temperatures in portable batch dryers range from 160° to 200° F. Airflow ranges from 50 to 100 cfm/bu. Drying capacities for typical farm units normally range from 100 to 400 bushels/h for 10 points of moisture removal.

Continuous flow dryers may be categorized into three types:

1. Cross flow – drying air is blown across the grain column the same as in a portable batch.
2. Counter flow – drying air and the grain move in opposite directions.
3. Concurrent flow – drying air and grain move in the same direction.

Continuous flow farm units operate typically with drying temperatures ranging from 180° to 220° F and airflow rates from 75 to 125 cfm/bu. For on-farm units, generally drying capacities range from 125 to 600 bu/h for 10 points of moisture removal. Most continuous flow drying units used on the farm are the crossflow type, and the other types will not be discussed further.

An important consideration in the use of high speed grain dryers is drying air temperature. The maximum allowable drying air temperature depends on the following:

1. Final use of the grain.
2. Moisture content of the grain.
3. Type of grain.

Recommended safe maximum drying temperatures for shelled corn are presented in table 3.9.

Presented in figures 3.19 to 3.25 are simulated performance curves for a crossflow dryer drying shelled corn using representative operating temperatures ranging from 160° to 240° F and airflow of 50 to 125 cfm/bu. Heat recovery devices are not used and 10 points of moisture are removed (25 to 15% w.b.). As drying air temperature increases, drying rate increases (figs. 3.19 to 3.20), drying energy and cost decrease (figs. 3.21 to 3.23), and moisture spread across the grain column increases (fig. 3.24). Conversely, the drying energy increases as the airflow rate increases for a given drying temperature (fig. 3.21). In effect, dryer efficiency increases as drying air temperature increases for a given airflow rate. The limiting factor is the maximum allowable grain temperature.

Figures 3.20 and 3.25 present the drying rate and average grain temperature in the drying column, respectively, for the same ranges of temperature and airflow shown in figure 3.19 (drying rate is expressed in bu/h/ft^2 of dryer surface area assuming 1.25 ft of column thickness).

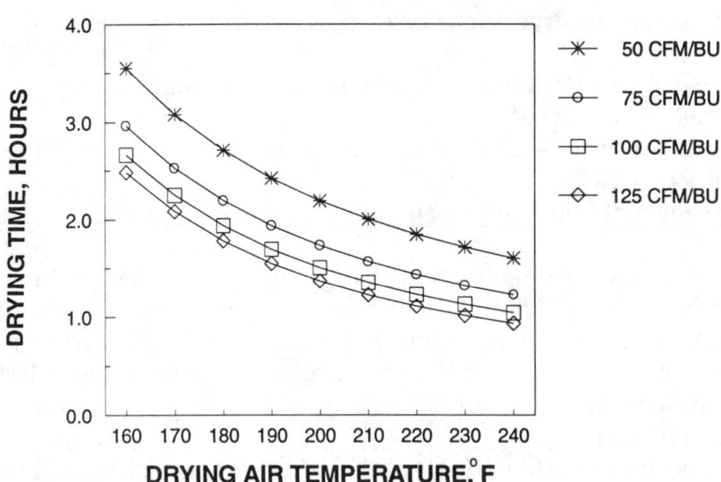

Figure 3.19–Drying time for corn in high temperature drying systems without heat recovery (Thompson et al., 1968).

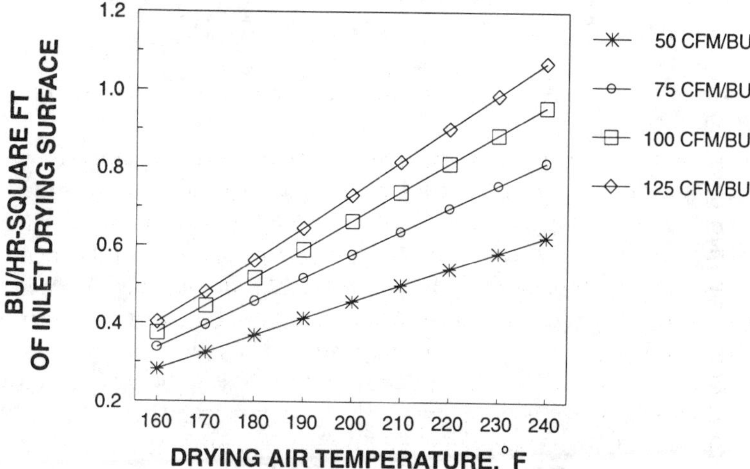

Figure 3.20–Drying rate for corn, bu/h/ft² of air inlet surface area without heat recovery (Thompson et al., 1968).

Note that both grain temperature and drying rate increase with drying temperature for constant airflow rates. For example, if the desired average grain temperature is limited to 180° F, the drying temperature

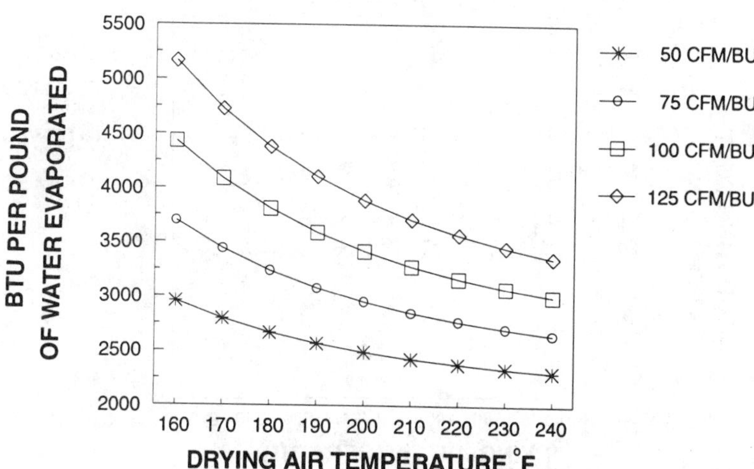

Figure 3.21–Energy required for drying corn, Btu/lb of water evaporated in high temperature drying systems without heat recovery (Thompson et al., 1968).

CORN

AMBIENT AIR AT 50 DEGREES F AND 55% RH
GRAIN DRIED FROM 25% TO 15%; LP GAS @ $0.80/GALLON

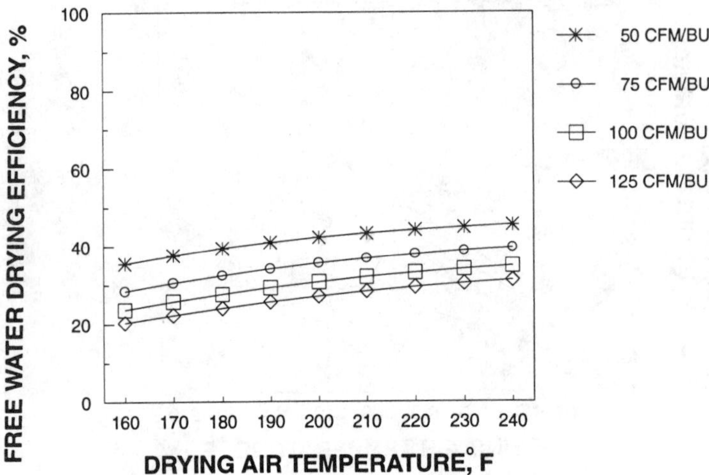

Figure 3.22–Corn drying efficiency compared to that of free water evaporation in high temperature drying systems without heat recovery (Thompson et al., 1968).

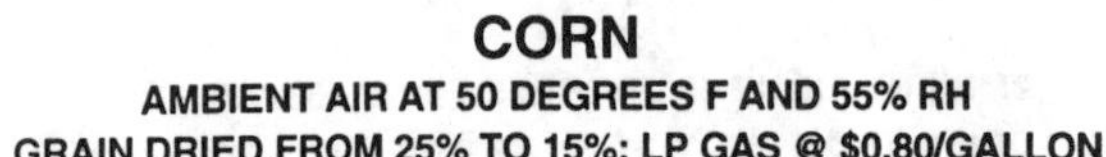

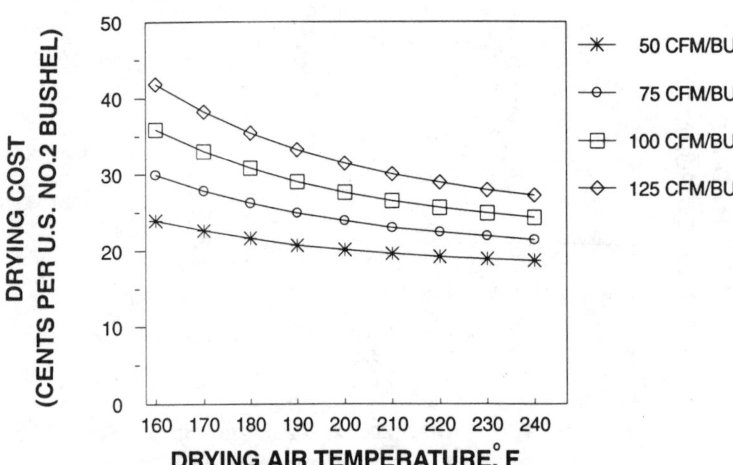

Figure 3.23–Cost of LP gas to dry a bushel of corn in high temperature drying systems without heat recovery (Thompson et al., 1968).

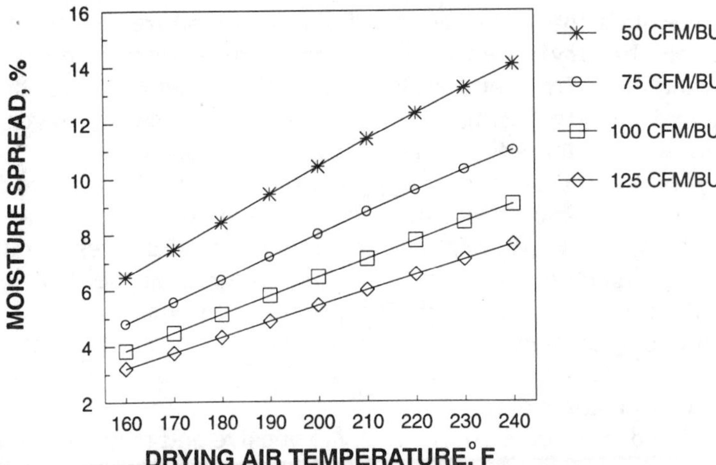

Figure 3.24–Moisture spread across the corn drying column in high temperature drying systems without heat recovery (Thompson et al., 1968).

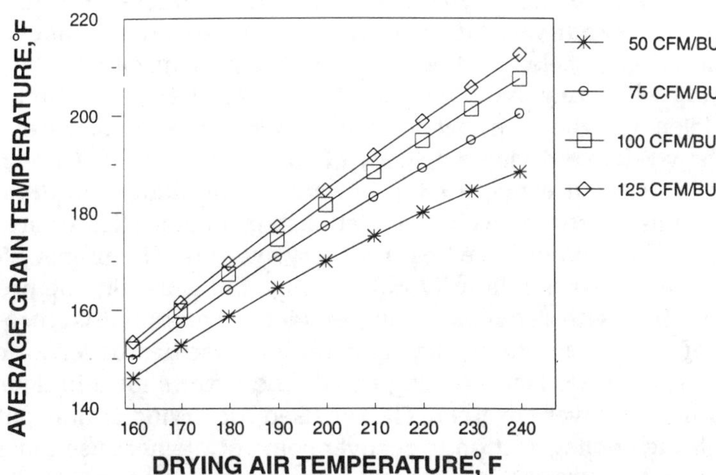

Figure 3.25–Average grain temperature across the corn drying column when corn has reached its final moisture content of 15% in high temperature drying systems without heat recovery (Thompson et al., 1968).

could range as high as 220° F depending on the airflow rate of the dryer. The expected moisture spread and drying energy for 220° F is approximately 6.3% and 2,300 Btu/lb of water removed, respectively. The trade-off is that although the drying temperature can be raised to decrease the drying energy, the grain quality may be reduced significantly if the temperature is too high. Grain temperature should be monitored closely during drying. If the grain exits the dryer "hot", dryeration procedures (see page 123) should be followed.

An easy method of monitoring grain temperature is to place a grain sample in a large-mouth thermos bottle containing a thermometer that reads up to 200° F. After filling the thermos, read the temperature as soon as the reading stabilizes. Leave the hot grain in the thermos until the next sample is to be taken. This reduces the heat drawn from the new sample by keeping the container warm.

Economic Considerations

Portable drying units are relatively expensive and may require a large capital outlay. Their advantage lies in drying large volumes of grain rapidly. These systems can be completely automated and will require the added expense of wet storage ahead of the dryer. When selecting or replacing these units, the following factors should be considered.

Drying Capacity. The dryer should be selected so that it is capable of drying the daily harvest in 24 h or less. The time span for dryer operation is an important factor in arriving at the necessary dryer capacity. This value will range normally from 16 to 18 h. However, these units may be selected and designed to operate for 24 h/day. This reduces the necessary dryer capacity and expense, but leaves few options if reserve capacity is needed to overcome breakdown, delays, etc. Future expansion of crop size or harvest rate must also be considered when selecting the appropriate dryer capacity.

Drying Efficiency. As LP gas and other fossil fuels increase in price, the efficiency of the dryer becomes even more important. Figures 3.21 to 3.23 show relative drying efficiency (Thompson et al., 1968) as related to Btu/lb of water evaporated, free water evaporation, and drying fuel cost per bushel, respectively. All relationships in these figures are based on dryers that do not have heat recovery systems. The primary feature of portable dryers is relatively rapid drying, but generally the producer pays for this with reduced drying efficiency in comparison to other types of dryers. In effect, drying rate decreases as dryer efficiency increases, all other factors being equal. The dryer shown in figure 3.18 uses a heat recovery system. The air used for drying is first pulled in through the cooling section to recover some of the heat used in drying. This air then moves into the drying section where the recovered heat is used. Depending upon dryer design and operating circumstances, energy requirements of a crossflow dryer may be reduced by as much as 35% by using a heat recovery system. Dryeration (discussed later in the chapter) provides an economic means for both increasing drying efficiency and capacity.

Summary

Portable batch and continuous flow drying are well-suited for on-farm systems where rapid drying is essential. Both dryer types require wet grain storage ahead of the dryer, and each may beautomated completely so that little or moderate supervision is necessary. When selecting a dryer of either type, the unit should provide a daily drying capacity no less than the daily harvest rate. These units are relatively expensive, but with proper selection and management they provide a viable high-speed drying alternative where rapid harvest rates are necessary. The ability to harvest rapidly and to add drying capacity by replacing an existing portable dryer with a larger unit often offsets the added energy and investment costs, especially for larger systems. A complete discussion of economic considerations is given in Chapter 9.

In-Bin Continuous Flow Drying

In-bin continuous flow drying most often utilizes the grain bin as a combination wet-holding and drying bin. Wet grain from the harvest is loaded directly into the drying bin. As the grain becomes dry, it is removed from the bottom of the bin by a tapered sweep auger. To some extent, this system is a "counterflow" dryer in that grain and air are moving in different directions. In-bin continuous flow drying systems are designed to operate continuously, drying one day's harvest in time for the next. Important considerations in evaluating performance of these systems include the daily harvest rate, the loading rate of wet grain into the drying bin, and the selection of the appropriate drying fan and bin combination that will provide the necessary drying capacity.

Equipment Requirements

Typical equipment for this type of dryer includes a fan and heater unit, a full perforated drying floor, a tapered sweep auger that removes dry grain from the bottom of the bin when drying, a vertical auger to transfer the dry grain from the bin to storage, and a grain spreader (fig. 3.26). The vertical auger usually is 4 to 6 in. in diameter with a capacity of 200 to 400 bu/h. The tapered sweep auger generally will remove a 4-in. grain layer and has an operating capacity of approximately 200 bu/h. In situations where very rapid drying is desired, the tapered sweep auger may be the limiting factor in drying capacity. In-bin continuous flow units may be equipped with a grain recirculation feature if it becomes necessary for the drying bin to also be used for storage. A high capacity unloading auger (independent of the drying system augers) may be installed if rapid unloading of the drying bin is required. The size of drying equipment will typically range from 10 to 20 hp for the drying fans and 18 to 36 ft for bin diameters. Multiple fans per bin may be used to obtain added drying capacity. Drying air temperatures will generally range from 140° to 200° F and require heater sizes of from 3 to 4 million Btu/h.

The operation of an in-bin continuous flow dryer is controlled by either a temperature sensing element located approximately 1 ft above the floor of the drying bin or a moisture sensing unit located in the vertical transfer auger. The drying system requires two thermostats, one to control the drying air temperature and another to control the operation of the tapered sweep auger that removes the dry grain. When the air temperature at the level of the sensing element reaches the desired setting, the tapered sweep auger is activated and rotates around the drying bin removing a layer of dry grain. As the moisture content of the incoming wet grain changes, the temperature setting may be varied to obtain the desired final moisture content.

Advantages and Disadvantages

In-bin continuous flow systems have several advantages over other in-bin drying systems. The use of higher drying temperatures increases the drying capacity without overdrying the bottom grain layers because the dried grain is removed upon reaching the desired final moisture content. The drying capacities of in-bin continuous flow units are similar to that of automatic batch-continuous flow dryers, but the in-bin systems have a higher drying efficiency if a minimum grain depth of 4 to

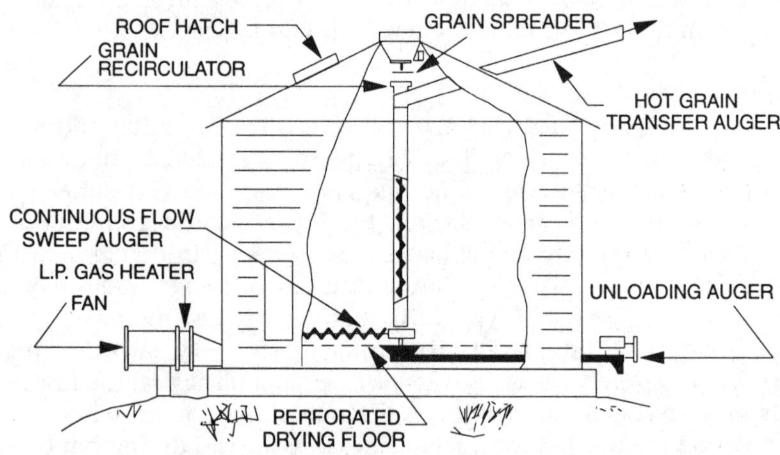

Figure 3.26–In-bin continuous flow drying equipment.

6 ft is maintained in the drying bin. As with portable drying units, higher drying temperatures allow the use of dryeration (discussion begins on page 123). Thus, many producers dry their grain to an intermediate moisture content, then use dryeration to condition the grain for final storage.

A disadvantage of continuous-in-bin systems relative to portable drying units is that one grain bin must be dedicated to drying (although the system may be used to "layer dry" the last grain to be dried, thus utilizing the drying bin to some extent). In addition, continuous operation of these units for three or four days may result in an accumulation of fines around the center of the drying bin because of repeated auger rotation. The accumulation of fine material may result in hot spots and uneven drying of the grain mass. Cleaning the bin may be required if the problem becomes significant.

Physical Performance

Several factors affect the drying performance of an in-bin continuous flow dryer: drying air temperature, grain depth in the drying bin, airflow delivered by the drying fan, drying bin diameter, and loading rate of wet grain into the drying bin. Drying capacity of in-bin continuous flow units is difficult to determine at any one time because of the changing grain depth in the drying bin associated with harvesting and drying. Simulated drying capacities and efficiencies for various grain depths and drying temperatures are shown in figure 3.27 for a 24-ft-diameter bin using a 13-hp fan. The curves indicate that for a given drying temperature, the drying capacity generally decreases as grain depth in the drying bin increases. For example, at 160° F the drying capacity is reduced from 170 bu/h to about 153 bu/h as the grain depth changes from 4 to 6 ft. Figure 3.27 also indicates that for a constant grain depth, increasing drying temperature increases drying capacity while reducing fuel consumption. For the example situation, increasing drying temperature from 120° to 200° F for a 4-ft grain depth increases the drying capacity from 115 bu/h to 220 bu/h while fuel use declines from 0.245 to 0.210 gal of LP gas/bushel. This represents a difference of $260 in the cost of drying 10,000 bushels with LP gas priced at $0.75/gal.

The capacities and efficiencies in figure 3.27 vary for different drying fans and bins and quantities of moisture removed. Maximum drying capacity and efficiency can most often be obtained when grain depths range from 4 to 6 ft, and drying temperatures exceed 160° F. Drying temperature is limited by ultimate use of the grain (table 3.9).

An important consideration is the selection of the drying fan and bin combination that provides adequate drying capacity for a given harvesting situation. Manufacturers' quoted drying rates, as well as those shown in figure 3.27, are for specified grain depths and do not consider the variation that may occur as wet grain is brought to the

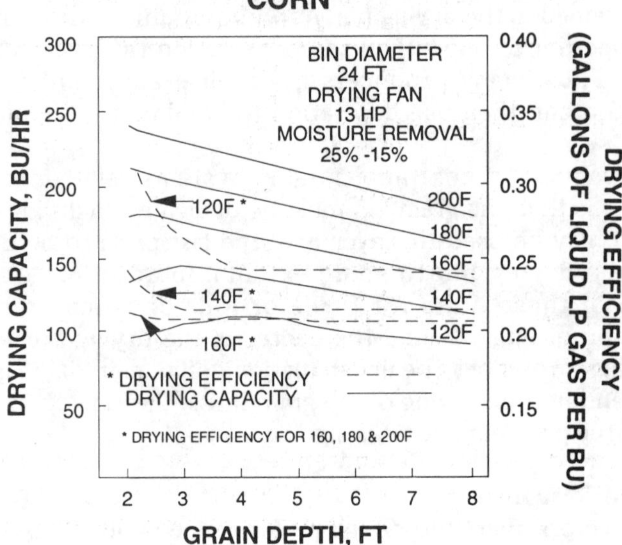

Figure 3.27–In-bin continuous flow performance data (Bridges et al., 1983).

drying facility over the harvest day. In-bin continuous flow units usually require a minimum depth of 1.5 ft of wet grain to start the drying process. If there is sufficient drying capacity over the drying day (24 h), no more than 1.5 ft of grain will remain for the start of the next day's operation. If these units do not "catch-up" in day-to-day operation, the producer eventually will cease harvesting operations because of inadequate drying capacity. This situation is illustrated by the following example.

Suppose a producer harvests 3,000 bushels of wet grain per day. The wet grain is delivered to the facility in 1-h intervals with each load containing 300 bu. The following drying conditions apply for each analysis (table 3.10): 10 points of moisture removal (25 to 15% w.b.), a drying temperature of 160° F, and an initial grain depth of 1.5 ft at the start of the drying day. The producer has selected a 13-hp drying fan but

Table 3.9. Maximum safe drying air temperatures of shelled corn and grain sorghum as categorized by use

Grain Use	Drying Temperature (° F)
Feed	180
Milling	140
Seed	110

Table 3.10. In-bin continuous flow drying example showing the effects of increasing drying bin diameter

Analysis No.	Drying Bin Diameter (Ft)*	Wet Grain Harvested (Bu)†	Wet Grain Dried (Bu)‡	Drying Time (h)	Drying Bin Depth After 24 h (Ft)
1	18	3,000	1,869	24	7.1
2	21	3,000	2,891	24	1.9
3	24	3,000	3,021	20	1.4

* An initial depth of 1.5 ft was assumed for each analysis

† A loading rate of ten 300-bu loads at 1 h intervals.

‡ For 10 points of moisture (25% to 15% w.b.), drying temperature of 160° F, 13-hp drying fan. (Bridges et al., 1983)

now needs to determine the appropriate drying bin diameter that will yield sufficient drying capacity to dry the daily harvest.

Simulated drying performances for three bin diameters (18, 21, and 24 ft) under the specified drying conditions are shown in table 3.10. The 18 ft bin can dry only 1,869 bushels leaving 7.1 ft of wet grain in the drying bin at the end of a 24-h period. Increasing the bin diameter to 21 ft (Analysis 2) increases daily drying capacity to 2,891 bushels with 1.9 ft of grain remaining to be dried at the start of the next harvest day. This system is much improved over that of Analysis 1, but is marginal if operated for many harvest days under the specified drying conditions. The third analysis, utilizing a 24-ft-diameter bin, attains the necessary drying capacity in approximately 20 h and provides sufficient drying capacity for future expansion.

Economic Considerations

When considering the purchase of high-speed dryers, the initial cost of an in-bin continuous flow system (including the drying bin) may be higher than a comparable portable drying counterpart (automatic batch or continuous flow) with a similar drying capacity. In-bin continuous flow units are capable of high drying rates and yield high quality grain. However, they are not as flexible as their portable counterparts in terms of expansion and in situations where the drying of multiple grain types is necessary.

One cost advantage of continuous in-bin systems over portable units is in modification of an existing system where a grain bin is already available for drying (i.e., batch-in-bin). In this situation, it is necessary to purchase only the tapered sweep auger, a central unloading auger and control equipment – the total cost at the margin usually being less than that of a portable drying unit. However, there are some additional cost considerations. Stresses will be placed on the bin walls and floor due to the auger rotation and weight of the grain. These stresses may require that additional wall stiffeners and floor supports be used. It may be necessary also to replace the existing drying fan to ensure sufficient drying capacity. Another consideration in cool climates is the use of bin wall liners to reduce condensation associated with high drying temperatures.

Summary

In-bin continuous flow dryers offer a viable, high-speed alternative to the portable drying units. These units, while somewhat more expensive than their portable counterparts, are popular on-farm units that have been shown to be reliable in operation and have obtained producer confidence.

Combination Drying

Combination drying is the drying method in which both high-temperature processes and in-bin drying (natural air, low-temperature or layer) procedures are combined. The high-temperature method is used to dry relatively wet grain to a sufficiently low moisture content so that in-bin drying can be used to complete successfully the drying process.

Equipment Requirements

Combination drying requires all the drying and handling equipment normally associated with both high temperature and in-bin drying (natural air, low temperature, layer). Combination drying equipment needs are (1) auger or bucket elevator with a pit or dump hopper for receiving grain, (2) wet holding bin, (3) high temperature dryer (typically a continuous flow or automatic batch dryer), (4) drying bin(s), each equipped with drying fan, heater (if low temperature or layer drying is to be used), perforated floor, sweep and unloading augers, and a grain spreader. Stirring devices may also be included.

Advantages and Disadvantages

The advantages of combination drying relate primarily to risk, drying fuel energy savings, facility expansion cost, and enhancement of grain quality as compared to using only high temperature drying. Combination drying allows the grain producer to begin harvesting whenever the grain is mature rather than waiting until sufficient field drying has occurred. Thus, there is a reduction of risk in that the grain may be dried with certainty to a sufficiently low moisture content so that an in-bin drying method may be used successfully.

Combination drying generally is more energy efficient than high temperature drying systems alone because in-bin drying is more energy efficient than high temperature drying. This is not to say that efficiency drops as drying air temperature alone increases. In fact the opposite is true. High temperature-high airflow drying systems generally use more energy for drying per pound of moisture removed than do typical in-bin drying systems because in-bin systems obtain a greater percentage of their drying potential from ambient air.

Natural air drying usually would be preferred to drying systems requiring the addition of heat if fuel costs (electricity and LP gas) for drying were the only economic considerations associated with grain production. An exception would be if the cost of LP or natural gas for increasing air temperature were significantly less than that of electricity for fan operation. However, field losses also represent an economic loss. As harvesting is delayed, field losses and the associated economic risks increase. Thus, fuel expenditures for drying are only part of the total economic picture, and differences between drying systems may not warrant the high-risk high-management aspects of in-bin drying. Combination drying offers something of a compromise between risk reduction and additional energy savings for drying.

It is difficult to compare the capital investment cost of combination drying to either in-bin or portable drying systems because the economic preference depends on the existing facility. As with any properly designed system, a combination drying system will not restrict the harvesting rate beyond acceptable limits. The high-temperature dryers will process the grain at an acceptable rate until the drying process can be completed successfully with in-bin drying. However, the reason that combination drying is selected is usually associated with drying and storage capacity expansion rather than energy savings or risk reduction. A typical situation is when a farmer begins with a single in-bin drying system. Eventually, the system includes several similar bins, all equipped with drying equipment. However, when further expansion of production occurs, the drying system becomes a bottleneck to the harvesting operation. At this point the farmer may decide to eliminate the problem by adding a high-temperature dryer. The question becomes one of sizing the unit. The most economical decision may be to buy a relatively small dryer to reduce the moisture of the grain to a level where existing in-bin drying can complete the process. Another option is to purchase a sufficiently large high temperature dryer to accommodate the present harvesting rate. The existing in-bin drying would then be used first as a "back-up" system and secondly as a means of absorbing future increases in drying capacity associated with expansion.

Suppose that the farmer in our example chooses to purchase a high temperature dryer that can dry all the grain to a safe storage moisture content that for the given geographical area is 13%. How can drying capacity be expanded as production increases without purchasing additional drying equipment? First, the dryeration process may be used to remove the last 2 points of moisture from the grain while eliminating the cooling portion of the dryer. By converting the cooling section to drying, an increase of approximately 50% can be obtained for a three-fan drying unit. If the grain were being dried normally from 23 to 13% (10 points removed), the 2-point reduction (after drying grain from 23% to 15% for 8 points removal) would further increase capacity by at least another 20% giving an approximate net increase of 70%. In addition, all this can be accomplished with existing in-bin drying facilities and limited additional capital investment. If this increase in capacity is not sufficient, the grain may be dried from 23 to 18% (rather than 15% for dryeration) and the in-bin dryer used to finish the drying. This further reduction in drying of 5 points removal, rather than the original 10 points, gives an added capacity of the combination drying system of approximately 150% (50% by removing the cooling section and at least 100% because 5 rather than 10 points of moisture are removed) over the original system in terms of bushels that may be received from the field each day. This results in a high quality product being dried with less energy and essentially no increase in capital investment.

Summary

Combination drying offers a low-risk method of utilizing in-bin drying. Drying efficiency is less than most in-bin drying methods but greater than high-temperature processes. Combination drying requires a relatively high level of capital investment and management if purchased as a unit because two complete drying systems are included. However, it may represent a relatively inexpensive way of adding drying capacity to an existing system.

Dryeration

Dryeration is the process by which high temperature grain, taken directly from a dryer, is tempered systematically and cooled in order to extract additional moisture from the grain without using any additional fossil fuel. A conceptualization of how heat is transferred in grain is needed to understand the dryeration process (Thompson and Foster, 1967; McKenzie et al., 1972; 1980).

Heat Transfer and the Dryeration Process

Heat is removed from grain primarily by two processes: sensible heat transfer and latent heat transfer. Sensible heat flow is based on temperature differences between the kernel and the surrounding air. If the air and the kernel are the same temperature, no sensible heat is transferred, and the temperature of both air and grain remain the same.

Latent heat transfer occurs when the excess heat stored in the grain is used to evaporate the moisture contained within the kernel. The forcing mechanism for this occurrence is the vapor pressure difference between the water vapor in the air and that in the grain. However, vapor pressure in the kernel is governed by temperature, so, in effect, the temperature of the air and the grain also influence the heat transfer process. As moisture within the kernel is vaporized and removed from the grain, heat is absorbed by the moisture and is removed also. This loss of heat from the grain results in a lowering of grain temperature.

There may be a distribution of moisture and temperature within the grain kernel; that is, not every point within the kernel is at the same moisture content and temperature. Materials such as grain expand and contract depending on temperature. Thus, when points within a single object, such as a grain kernel, are at different temperatures, stresses occur. If the stresses are great enough, "stress cracks" result. Stress cracks are not thought to occur in grain drying during the heating phase when the outside of the kernel is hotter than the kernel's center. They occur with cooling when the center of the kernel is at a higher temperature than its outside. While relatively rapid cooling results in more stress cracks, the moisture gradient is viewed as the primary cause.

Dryeration operates on the premise that (1) if the grain contains excess (or stored) heat, such as immediately after being exposed to the high temperatures involved in drying, and (2) if this stored heat can be removed using the latent rather than sensible heat transfer mechanism, then (3) grain quality can be maintained and energy efficiency increased at the same time. Each of these conditions will be examined.

Stored heat can be defined as the difference in the grain temperature and the outside air temperature, multiplied by the specific heat of the grain. Specific heat of grain is the ratio of the thermal capacity of grain as compared to that of water for a standard temperature and pressure. In equation form:

$$\underset{\text{(Btu)}}{\text{stored heat}} = (Tg - Ta) * Cp * (\text{lb of material}) \tag{3.4}$$

where

Stored heat	=	dryeration potential (Btu)
Tg	=	grain temperature (°F)
Ta	=	ambient air temperature (°F)
Cp	=	specific heat of grain, Btu/lb-°F, and equations for several types of grain are (Brooker et al., 1974):

$$= 0.325 + 0.00851 * MC \text{ (yellow dent corn)} \tag{3.5}$$

$$= 0.305 + 0.0078 * MC \text{ (oats, applicable in the 11.7 to 17.8\% moisture range)} \tag{3.6}$$

$= 0.265 + 0.0119*MC$ (rough rice, applicable in the 10.2 to 17.0 moisture content range) (3.7)

$= 0.283 + 0.00724*MC$ (hard white wheat, dry) (3.8)

$= 0.334 + 0.00977*MC$ (soft white wheat) (3.9)

MC = moisture content (% w.b.)

Note, the term "grain" includes some water in addition to the dry matter, so the specific heat of grain changes with grain moisture content. For example, suppose 1 bu of U.S. No. 2 corn (56 lb, 15.5% moisture content) is at 140° F, and that the specific heat of corn is 0.456 Btu/lb-°F. Then 0.456 Btu are stored in the grain for each pound of material and for each degree of temperature difference between 140° F and some base condition. If the ambient temperature is 40° F, the amount of stored thermal energy is:

stored thermal energy/bu = (140° F - 40° F) * (0.456 Btu/lb-°F) * 56 lb/bu

= 2,554 Btu/bu

Approximately 1,060 Btu of heat are required to evaporate 1 lb of water at 55° F. If all the stored heat in the above example were transferred as latent heat, 2.41 lb of water would be evaporated. The original 56.0 lb of material would now weigh 53.59 lb with the dry matter remaining constant. The new grain moisture content would be 11.7% [(0.155*56 lb – 2.41 lb water)/(56.0 lb grain – 2.41 lb water)] reflecting a reduction of 3.8 points of moisture. If the outside temperature were 90° F, the stored heat above the outside conditions (as computed by the above equation) would only be one-half of the original value for a moisture reduction potential of 1.9 points. Because ambient temperatures during the drying season usually lie between these extremes (40° to 90° F), a theoretical maximum moisture loss of approximately 3 points is expected when using the dryeration process. Potential changes in grain moisture content for a range of conditions are shown in figure 3.28.

In using the dryeration process, the moisture content of the grain exits the dryer approximately 2 points higher than the desired storage moisture content. Grain temperatures during drying tend to approach drying air temperatures as the moisture content of the grain decreases. In the initial stages of drying, most of the heat applied to the grain is used to evaporate moisture and is removed from the grain in the form of latent heat. Thus, the grain remains much cooler than the drying air

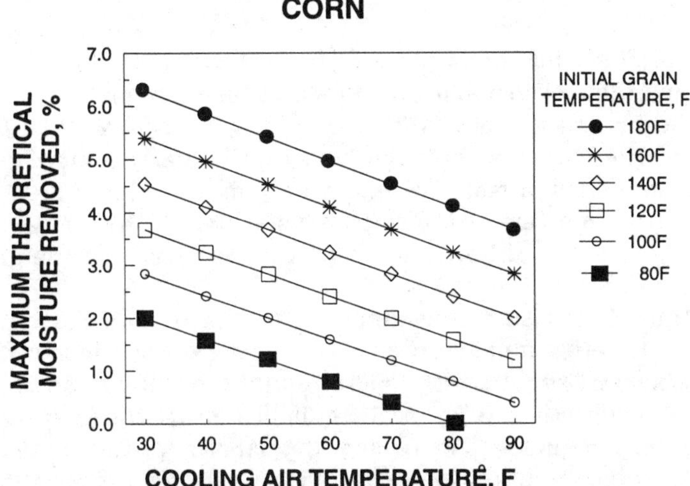

Figure 3.28–Potential moisture removal from dryeration.

earlier in the drying process. However, as more moisture is removed, more heat is transferred into the grain as sensible heat, and the temperature of the grain begins to increase. Drying efficiency increases as drying air temperature increases (other factors remaining the same). Thus, if drying is to be terminated at a higher grain moisture content, drying air temperatures may be increased while maintaining the same grain temperature to result in an even greater efficiency. The average grain temperature as a function of drying air temperature is shown in figure 3.25.

The objective of the dryeration process is to remove heat using the latent heat transfer mechanism; that is, by using the stored heat in the grain to vaporize moisture. If hot grain is placed into a bin and no air is passed through the grain, most of the heat in the grain will be used to vaporize water. The vapor pressure in the grain is higher than that of the air, forcing the moisture to transfer from inside the kernel to the air. Meanwhile, some of the heat will pass directly from the grain to the air using the sensible heat transfer mechanism. Gradually, the air and the grain will reach the same temperature and vapor pressure so that no additional moisture or heat transfer can occur. If air is blown through the grain, the new air is at ambient temperature and vapor pressure conditions so that moisture and heat are again transferred from the grain to the air. The transfer will continue until the grain is cooled to the same temperature as the air. The moisture actually removed from the grain is less than the theoretical values because some of the heat transfer results from the sensible method. Generally, about two thirds of the theoretical moisture removal potential is removed resulting in both higher quality grain and drying energy savings.

Tempering Time and Cooling Rate

The time that corn should be allowed to "temper" or "steep" before cooling begins is associated with the cooling airflow rate (Thompson and Foster, 1967) and is shown in figure 3.29. When averaged over all airflows, the points of moisture removed from the grain is greatest for corn when an 8-h tempering time is used. When averaging all the tempering times, an airflow rate of 0.5 cfm/bu results in the greatest loss.

Grain quality is also a consideration. Thompson and Foster (1967) examined the effects of tempering time and cooling airflow rate on sound kernels (without stress cracks) and the millability score of the grain. The results (figs. 3.30 and 3.31) indicate that the percentage of sound kernels continues to increase with tempering time over the range of values tested. Increases in cooling airflow rate tend to decrease the percentage of sound kernels initially while continuing to increase the millability score over the range of values tested.

The conclusions from the above study are for optimum moisture removal using dryeration, tempering time should be no less than 4 h or more than 10 h. The airflow cooling rate should range from 0.5 to 1.0 cfm/bu. The values are to be compared to the cooling cycle in a high temperature dryer where there is essentially no tempering time and the cooling airflow rate is in the range of 50 to 100 cfm/bu. Most of the heat removal in the high temperature situation utilizes the sensible means of heat transfer rather than the latent method so that most of the heat

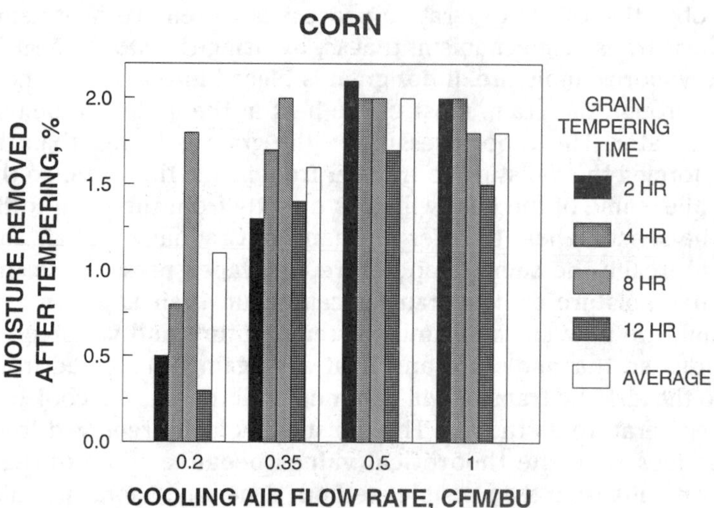

Figure 3.29–Effect of tempering time and cooling airflow rate on the percentage points of moisture removed during aeration of hot corn (Thompson and Foster, 1967).

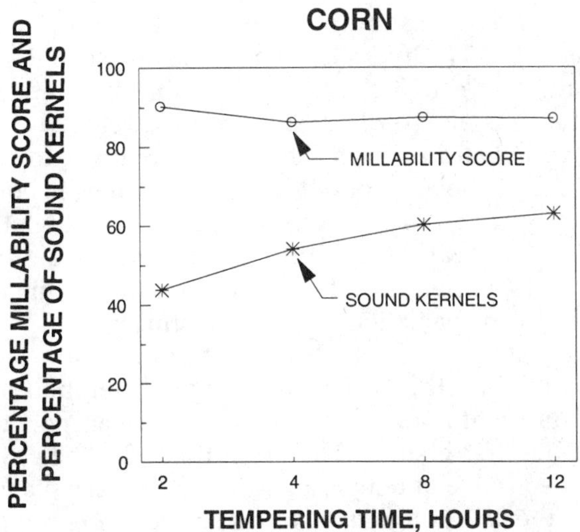

Figure 3.30–Effect of tempering time on corn quality (Thompson and Foster, 1967).

used in drying is lost insofar as additional moisture removal is concerned.

Equipment Requirements

The dryeration process requires that hot grain be conveyed to a holding or "tempering" bin where cooling will begin after a minimum of 4 h using an airflow rate of from 0.5 to 1.0 cfm/bu. Equipment

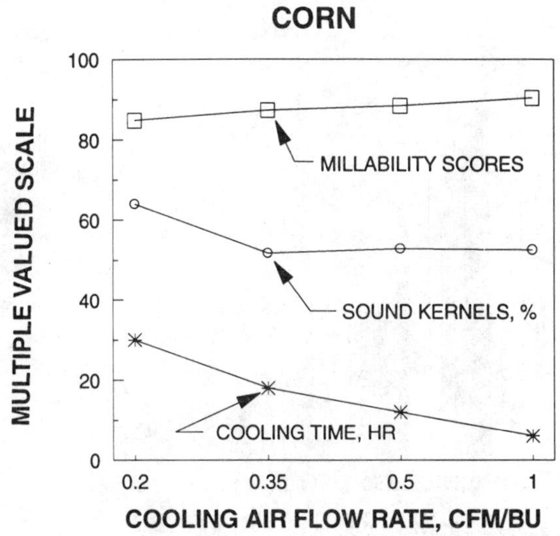

Figure 3.31–Effect of cooling rate on corn quality (Thompson and Foster, 1967).

requirements in the dryeration bin are like those used in natural air drying systems (fig. 3.1). Dryeration bin capacities should be sufficient to hold the daily harvest rate, and smaller bin diameters are usually acceptable because cooling airflow rates are relatively low.

Air is directed upward through the grain so there is no reheating or rewetting of grain first placed in the bin. After cooling is complete, the grain is transferred to storage. Unfortunately, the logistics of this process are somewhat more difficult to manage than with a conventional, high-temperature drying system. An examination of the system begins with an evaluation of the harvesting rate.

Every drying system should be designed to process the desired daily harvest as specified by the farmer. If a single dryeration bin is used, there must be sufficient time to temper, cool, and transport the grain to storage before the first grain arrives from the following day's harvest. The recommended range of tempering times is between 4 and 10 h. The time of day that the first hot grain will be delivered can be approximated as can the time required to transfer the dryeration bin's grain to storage. The allowable cooling time must be short enough to allow the cycle to complete itself in 24 h or less assuming that only one dryeration bin will be used. In equation form:

$$\begin{array}{c}\text{maximum time}\\ \text{allowable for}\\ \text{cooling in 1 bin}\\ \text{(h)}\end{array} = \begin{array}{l}\text{24 h - (tempering time, h) -}\\ \text{(daily harvest, bu) / (unloading rate of}\\ \text{dryeration bin, bu/h)}\end{array} \quad (3.10)$$

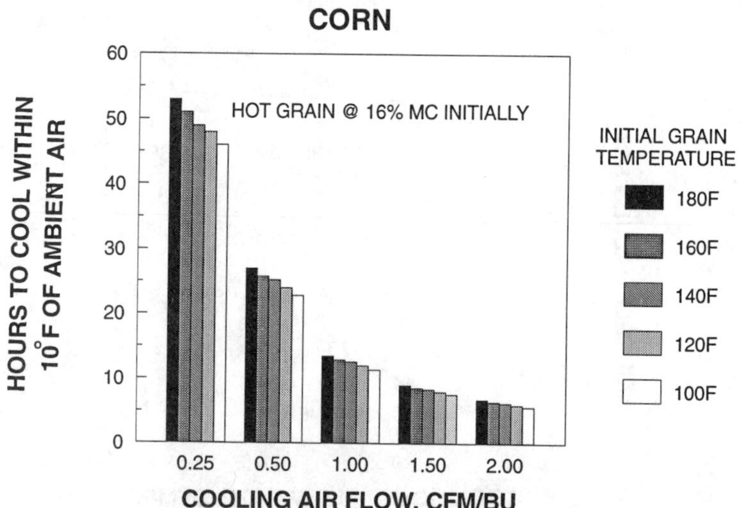

Figure 3.32–Cooling time for various temperatures of corn (Thompson, 1975a).

Cooling time is related directly to the airflow rate. Simulated cooling times are shown in figure 3.32 for several initial grain temperatures (Thompson, 1975a). Initial grain temperature has little influence on cooling time, and the average cooling time values across all grain temperatures are given in figure 3.33.

Airflow decreases greatly as the dryeration bin is filled. The influence of depth on airflow is shown in figures 3.34 to 3.36 for a three-bin diameters and three representative fans. Corresponding bin capacities are shown in figure 3.37. These figures may be used in combination with figure 3.33 to estimate cooling time as the bin fills. Maximum allowable time for cooling may be entered into figure 3.33 to obtain the minimum effective airflow rate. From figures 3.34 to 3.36, a fan and bin diameter combination that seems appropriate may be selected (remember that the bin must hold at least the number of bushels harvested each day).

There are several methods of estimating the average airflow over a day. A gross approximation is obtained by averaging the airflow rate when cooling begins with that of the bin when it is filled with the entire daily harvest. This approach overestimates the rate, and is equivalent to taking the midpoint ordinate value of a line drawn between the beginning and ending grain depths on the appropriate figure (figs. 3.34 to 3.36). The accuracy of this method improves as the depth - cfm/bu relationship becomes more linear.

An accurate estimate of average airflow/bu may be obtained from figures 3.34 through 3.36 by entering the average of the grain depths when cooling begins and when the bin is filled with the daily harvest.

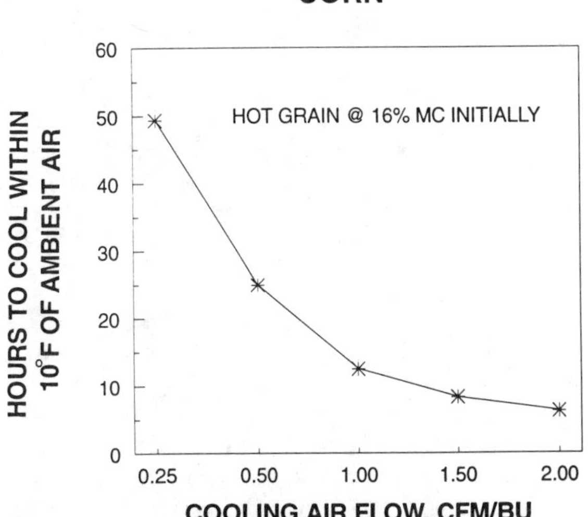

Figure 3.33–Average cooling time for corn (Thompson, 1975a).

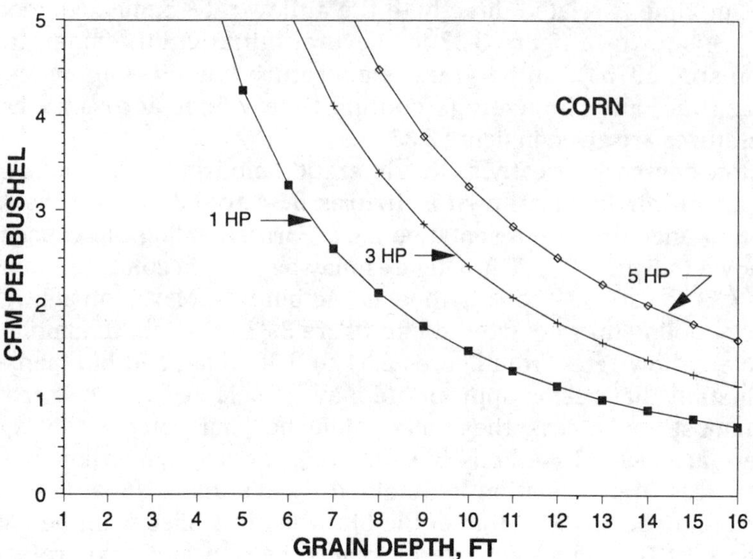

Figure 3.34–Fan performance in a 15-ft-diameter dryeration bin (Thompson, 1975b).

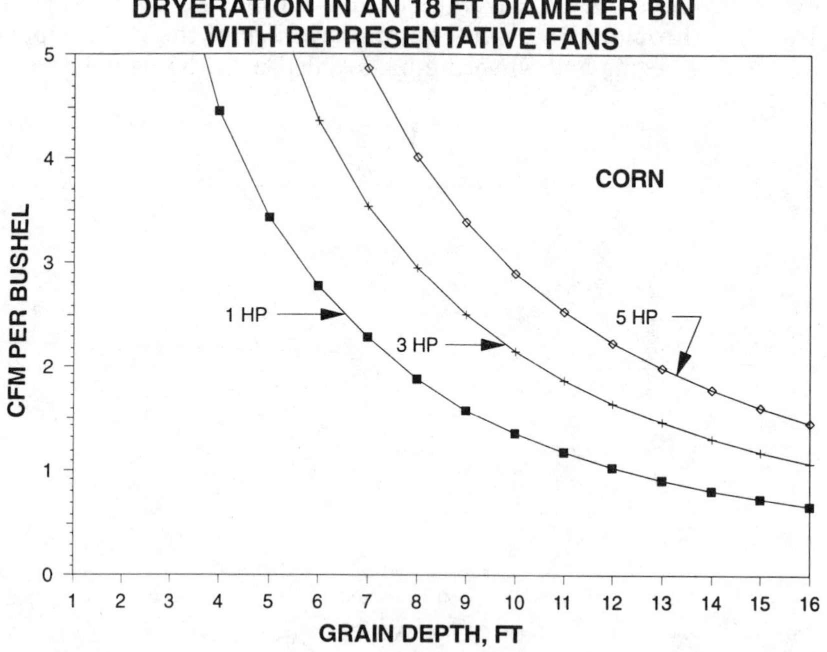

Figure 3.35–Fan performance in an 18-ft-diameter dryeration bin (Thompson, 1975b).

Accuracy of this method may be confirmed by recomputing the airflow with each dryer unloading to the bin. The depth of grain added to the bin with each batch dryer or surge bin unloading may be determined using the following equation:

$$\begin{array}{c}\text{depth change}\\ \text{in bin after}\\ \text{dumping, ft.}\end{array} = \frac{1.59 * (\text{dump size, bu})}{(\text{bin diameter, ft})^2} \tag{3.11}$$

By adding the depth change to the initial depth, the cfm/bu value may be determined using figures 3.34 through 3.36. Remember, the fan is not to operate until tempering is complete, so the initial grain depth is computed using the total bushels placed into the bin over this tempering period. The above process is repeated each time the bin receives grain over the day to be able to compute the average airflow rate.

Certain modifications are made to the above procedure if a continuous flow dryer is used and the grain is dumped directly into the dryeration bin. The bushels dried each hour represents the dump size value in equation 3.11, and the number of dumps equals the hours that the dryer operates.

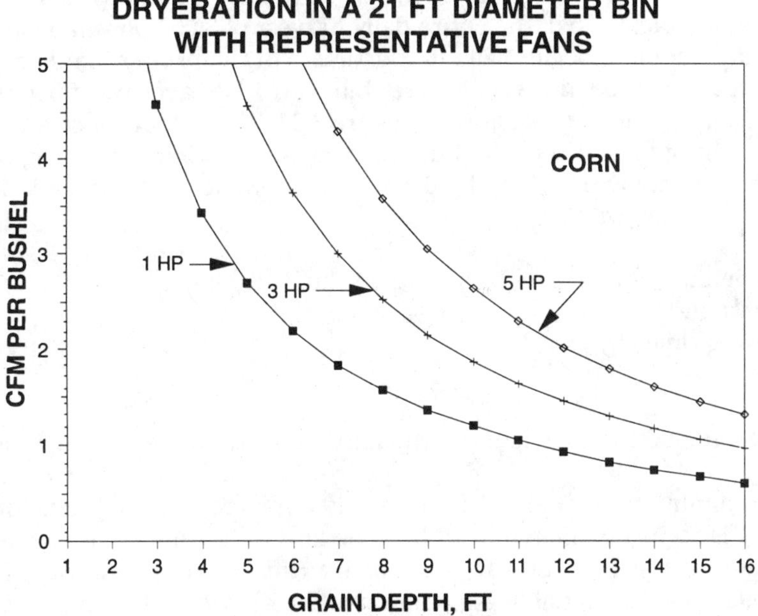

Figure 3.36–Fan performance in a 21-ft-diameter dryeration bin (Thompson, 1975b).

If the average airflow rate is between the allowable cooling time values obtained from equation 3.10 for 4 and 10 h tempering, the bin and fan combination will provide sufficient cooling. Otherwise, a new fan and bin combination should be selected and the process repeated. Ideally, the effective airflow rate should be between 0.5 and 1.0 cfm/bu. Continue to try to obtain this value while recognizing that not all possible bin and fan combinations are shown in figures 3.34 through 3.36, and it may be impossible to obtain the optimum cooling time for a particular cooling rate. If this is the case, a multiple-bin dryeration system might be selected, and equation 3.10 can be constructed in the following manner:

$$\begin{aligned}\text{maximum time allowable for cooling (h)} &= (\text{number of dryeration bins}) * 24\text{ h} - (\text{tempering time, h}) \\ &\quad - \frac{(\text{daily harvest, bu})}{\text{unloading rate of dryeration bin (bu/h)}} \qquad (3.12)\end{aligned}$$

To illustrate the above concepts, suppose that the daily harvest rate is 3,000 bu. An automatic batch dryer is used with a dump size and frequency of 300 bu once each hour. If the tempering time is 4 h, 1,200 bu will be in the dryeration bin when the fan is turned on. If a 10-h tempering time is used, the entire daily harvest of 3,000 bu will be in the bin before cooling begins. The first estimate of the proper bin diameter and fan size will be an 18-ft diameter bin with 16-ft eaves equipped with a 1-hp fan similar to that shown in figure 3.35. The unloading rate of the dryeration bin is 1,000 bu/h. If drying occurs over 10 h, tempering over 4 h and unloading over 3 h, the maximum allowable cooling time is 17 h as is shown below:

$$\begin{aligned}\text{maximum allowable cooling time (h)} &= 24\text{ h} - 4\text{ h} - (3{,}000\text{ bu}/1{,}000\text{ bu/h}) \\ &= 17\text{ h}\end{aligned}$$

If a 10-h tempering time is used, the maximum allowable cooling time is 11 h.

The minimum average airflow rate is approximately 0.8 cfm/bu if cooling is to be completed in 17 h, while an 11-h cooling time increases the required airflow rate to 1.2 cfm/bu (fig. 3.33). Assuming a flat-bottomed bin, the initial depth of grain in the bin when the fan is turned on (equation 3.11) is:

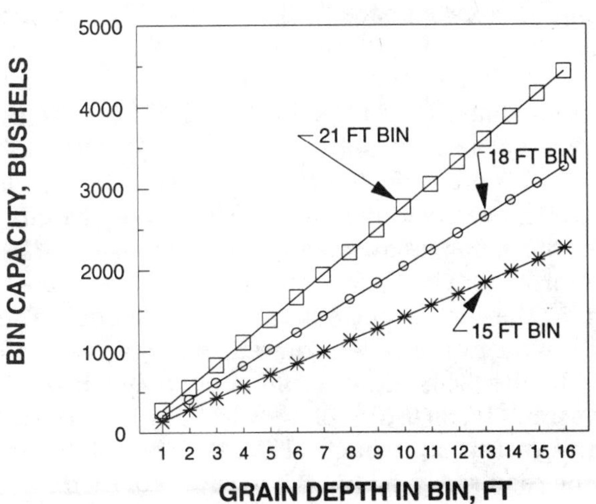

Figure 3.37–Typical flat-bottom dryeration bin capacities.

$$\text{initial depth (ft)} = 1.59 * 1{,}200 \text{ bu}/(18 \text{ ft})^2$$
$$= 5.89 \text{ ft}$$

The depth of the bin associated with the entire daily harvest of 3,000 bu is:

$$\text{final depth (ft)} = 1.59 * 3{,}000 \text{ bu}/(18 \text{ ft})^2$$
$$= 14.7 \text{ ft}$$

The average grain depth for the 17-h cooling time is 10.3 ft, requiring an average airflow rate of 1.3 cfm/bu as estimated from figure 3.35. If an 11-h cooling time is used, the average depth is the final depth and the associated airflow is 0.75 cfm/bu. In the 17-h case, the airflow rate exceeds the desired maximum of 1 cfm/bu for optimum moisture removal. In the 11-h case, the airflow is less than that required to cool the grain in time for unloading. In this situation, the tempering time should be adjusted to somewhere between 4 and 10 h. Rearranging equation 3.12 to determine this value:

$$\text{Tempering Time (h)} = 24 \text{ h} - \text{maximum time allowable for cooling in 1 bin (h)} - \frac{\text{(daily harvest rate, bu)}}{\text{(unloading rate of dryeration bin, bu/h)}} \quad (3.13)$$

Assuming that 1 cfm/bu is desired as the average cooling airflow rate, the cooling time is approximately 12.6 h (from fig. 3.33). Substituting into equation 3.13:

$$\text{Tempering Time (h)} = 24\text{ h} - 12.6\text{ h} - (3{,}000\text{ bu}/1{,}000\text{ bu/h}) = 8.4\text{ h}$$

For the example, the desired cooling airflow rate of 1 cfm/bu may be used in figure 3.35, giving an average grain depth of 12 ft. The maximum fill depth is approximately 15 ft. Thus, cooling would begin when the depth of grain in the bin reaches about 9 ft (1,800 bu). The tempering time will correspond to the time required to reach the 9 ft depth (6 h) rather than 8.4 h. The 8.4-h figure would apply only if it were possible to maintain a constant 1 cfm/bu cooling rate. It is important to note the iterative nature of this approach. That is, several "trial and error" attempts may be required to arrive at a feasible solution.

In-Bin Cooling

In-bin cooling is an alternative to conventional dryeration. In this process, the grain is cooled and stored in one bin so that extra handling is not required and logistical management is not needed for scheduling. As with conventional dryeration, air is blown upward through the grain with the hotter grain added on top of existing grain. The fan begins operation after the floor has been covered sufficiently. Fans operate continuously until the grain reaches the average daily temperature. Typically, this process is used with a system that had been employed previously for in-bin drying (natural air, low temperature or layer drying) so that little additional investment is needed. It offers another advantage in that any of these in-bin techniques may be used to further dry the grain, thus becoming a combination drying method.

Disadvantages of in-bin cooling, as compared to using an intermediate bin for dryeration, relate primarily to a lack of tempering time which results in reduced moisture removal, lower quality, and possible condensation along the bin walls. The reduced grain quality and moisture removal potential result from the relatively higher airflow rates associated with in-bin drying that reduce cooling time and the lack of tempering in some designs. In-bin cooling also results in condensation on the roof and side walls. Condensation may rewet the grain along the bin side wall and increase the risk of spoilage. The rewetting also occurs in the conventional dryeration bin. However, this wet grain is blended with the drier grain when it is conveyed to storage so that the moisture concentration problem is minimized.

Condensation is influenced by the bin wall temperature and the temperature and relative humidity of the air as it exits the grain, with higher relative humidity encountered when tempering in a dryeration bin. The bin wall or roof temperature below which condensation will occur (referred to as the dew point temperature) may be estimated from the Zone 1 conditions shown in figure 2.17 (from Chapter 2). The air exits the grain at near 100% relative humidity. If the bin surface is at or below the Point 3 temperature shown, water will condense on the wall or roof and may rewet the grain. Reduction of condensation may be accomplished by reducing or eliminating the tempering time so that the air will be less likely to be saturated; that is, the exit air relative humidity will be less. Unfortunately, the amount of moisture that may be removed is also reduced. Another option is to install a roof exhaust fan that pulls air across the grain surface rather than through the grain. In effect, this fan reduces the relative humidity and increases the exiting dryeration air temperature by blending in ambient air. This unit should be sized approximately 25% larger than the cooling fan with regard to airflow (McKenzie et al., 1980). However, the power requirements of the unit are less because this fan is not designed to pull air through the grain mass but across the grain.

Another method of reducing condensation is to cool only when the exiting air is above the dew point temperature. Effectively, this may mean that cooling cannot occur during the evening hours which may not be desirable. A compromise is to temper the grain only during the daytime when bin walls and roofs usually are warmer than the dew point temperature.

The problem of condensation worsens as the weather gets colder because the dryeration air tends to be more saturated, and dew point temperatures are more likely to be reached. In addition, more water may be removed from the dryeration air as dew point temperatures decrease. Thus, it may be necessary to alter the management strategy as winter approaches. One possibility is to use in-bin cooling only during the first part of the harvesting season and switch to a regular dryeration bin during the remaining harvest. Unfortunately, this strategy may create an even greater logistics problem in that potential drying capacity increases late in the harvest season as net moisture removal decreases. If drying through-put is the bottleneck in the system, even greater quantities of grain will have to be passed through the intermediate dryeration bin if harvesting capacity is not to be restricted.

If cooling bins are filled using bucket elevators, some of the drying air may exit the bin through the downspouting. This air is even more likely to reach the dew point temperature, thus adding to the condensation problem and increasing deterioration in the bucket elevator system. Again, a roof exhaust fan may help alleviate this problem by pulling some air down through the spouting. Another possibility is to restrict the movement of air upward through the spouting. Possible methods of doing this are shown in figure 3.38. It is very important that the manager know when downspouting may become blocked. Otherwise, the bucket elevator may be started inadvertently, clogging the leg and perhaps causing downspout failure.

Economic Considerations

The major cost of dryeration is in the increased capital investment associated with additional storage and handling equipment. Even this can be eliminated essentially with in-bin cooling. There are some added variable costs associated with fan operation, increased handling, and labor. However, the dryeration bins may be used also for storage at the end of harvest, and the operational costs are minor when compared to the savings of energy. For example, a point of moisture may be removed for approximately $0.02/bu when using LP gas priced at $1.00/gal. If a 5-hp fan is used to cool 9,000 bu in 10 h, the electricity cost would be $0.0003/bu assuming a price of electricity of $0.07/kW-h. Thus, energy savings totally dominate operation costs. Of course, the cost of the added storage and handling equipment must be considered also, but dryeration is the one process that will prove economical for practically all types of high temperature drying systems.

Summary

Dryeration offers a very economical method of maintaining grain quality and increasing drying rate and energy efficiency. The price of dryeration includes added facilities and management. Certainly, this

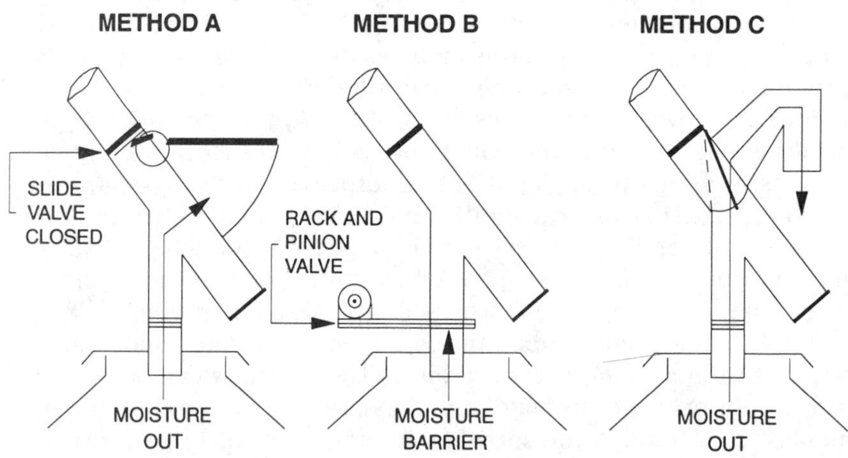

Figure 3.38–Alternative methods for blocking or bypassing gravity grain spouts to prevent vapor from traveling up the pipe.

process is worthy of consideration by any farmer using high temperature drying.

Stirring Devices for Grain Drying

Stirring devices are used in on-farm grain drying to enhance in-bin drying performance. These devices are suspended from the top of the bin with one or more augers extending into the grain mass. The augers are continually moved through the grain mass pulling grain kernels from the bottom of the bin and mixing them with kernels in the upper layers in order to constitute a mixing effect in the grain mass. Stirring devices may be used in conjunction with most in-bin drying systems (natural air, layer, and batch-in-bin) and may offer some advantages in terms of increasing drying capacity and maintaining grain quality as compared to the same drying situation without stirring.

Equipment Requirements

Stirring devices generally consist of a control arm that rotates around the bin with one or more vertical screw augers that extend into the grain mass (fig. 3.39). One end of the control arm is suspended from the bin roof in the center of the bin while the other end resides in a track that is placed around the circumference of the bin near the eaves. The number of screw augers suspended from the control arm varies from one to eight, depending upon the frequency of stirring and the diameter of the bin to be stirred. Most stirring devices are designed so that the individual augers have freedom to move back and forth along the control arm. This allows each auger to stir more than one path through the grain. The fewer the number of "down" augers, the more revolutions

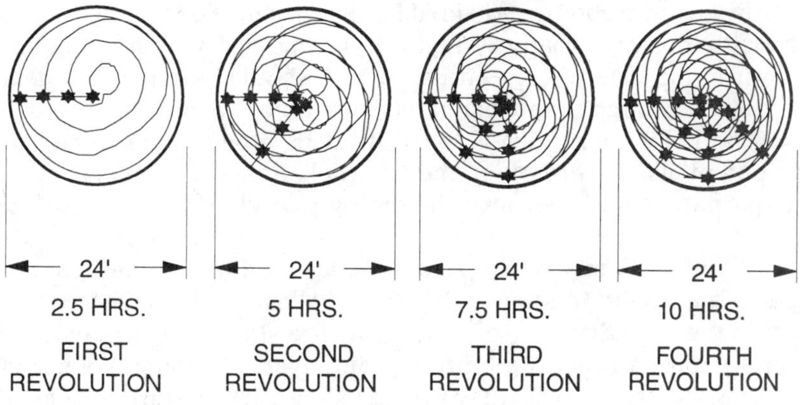

Figure 3.39–Stirring device patterns.

the device must make around the bin to stir the entire grain mass. The screw auger(s) are generally 2 in. in diameter and extend from the control arm to very near the bin floor. They are driven by 1- to 2-hp motors and rotate constantly, bringing grain from near the bottom of the bin to the top of the grain mass. The motor used to rotate the stirring device around the bin ranges from 5 to 10 hp depending on the number of down augers and the bin diameter.

Advantages and Disadvantages

Stirring devices offer some advantages when used with in-bin drying systems. The mixing effect of these devices allows reduction of the moisture gradient in the drying bin and practically alleviates overdrying of the bottom layers of grain. Stirring generally reduces the resistance to airflow in the bin which, in turn, provides a proportional increase in the drying rate. The use of stirring devices causes little mechanical damage to the grain itself and serves to break up problem areas in the grain mass where wet grain or trash may have become packed.

As to disadvantages, stirring devices are not economical in all drying situations (see Chapter 9). They occupy a portion of the bin for operational purposes, thus reducing bin storage capacity. Grain mass stirring causes fines to settle and concentrate near the bottom of the bin. In some situations, this may offset the expected increase in airflow and suggests that cleaning the grain is highly desirable prior to stirring.

The advantages and disadvantages of stirring devices are reflective of the controversy that has surrounded these items in the grain industry. Many claims have been made regarding their benefits which makes it necessary to evaluate carefully each drying situation before electing to use one.

Physical Performance

It is important that a stirring device mix the grain thoroughly relative to the drying rate and the potential for overdrying. For example, a batch-in-bin situation is a relatively rapid drying method with great potential for overdrying. Therefore, a stirring device should be capable of mixing the grain thoroughly approximately two times within the 16- to 18-h drying period if overdrying is to be avoided. In natural air drying (with little potential for overdrying), the bin would not require mixing as often (and perhaps never) because the drying period may extend for many days.

As the grain is mixed, the zone of influence for each vertical screw auger is parabolic in shape with the smallest width of the parabola situated near the drying floor of the bin. The stirring device must make several passes to mix the bottom grain layers thoroughly where the overdrying is taking place. Figure 3.39 shows the pattern of a stirring device rotating about the bin at 2.5-h intervals and covering most of the bin floor area in about 10 h. The outermost auger always remains near the bin wall. Stirring devices can be used with most in-bin drying methods. However, they are most beneficial in a high temperature in-bin drying system (layer, batch-in-bin) where the drying process can take advantage of the thorough mixing of the warmer grain layer with the cooler, more moist zones.

From the standpoint of the individual grain producer, the main concern with a stirring device is the effect that it will have on the individual's particular drying situation. For example, how much faster can the system dry and how much more wet grain may be dried at one time compared to the same system without stirring?

The increased drying capacity associated with adding a stirring device will vary among different drying situations and is difficult to predict. Figures 3.40 and 3.41 compare simulated stirred and unstirred average layer drying rates in bushels/day for a series of commercial drying fans (5, 7.5, 10, 15, and 20 hp) and a range of bin diameters (18 to 36 ft). The drying fans are the vane axial type and 42 in. in diameter except for the 5-hp fan which is 36 in. in diameter. Average drying rates for each fan and bin combination were calculated using total bin capacity at the eaves and the number of days to dry the bin. The

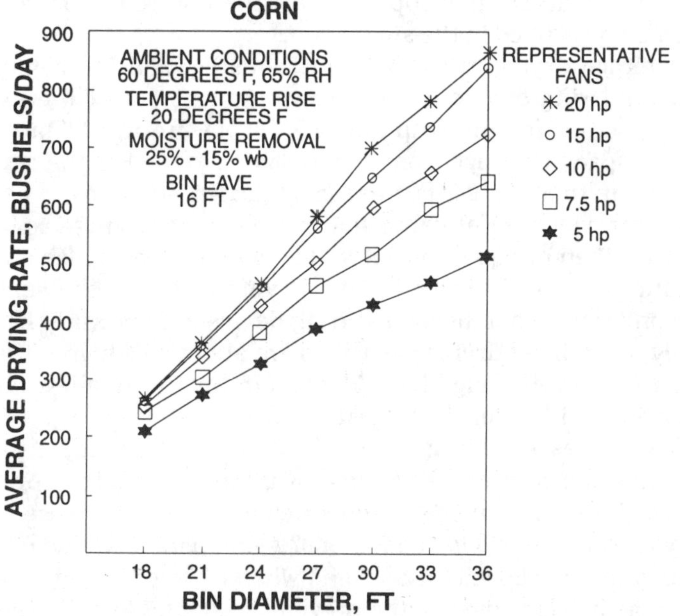

Figure 3.40–Layer drying situation without stirring (Bridges et al., 1982; 1984).

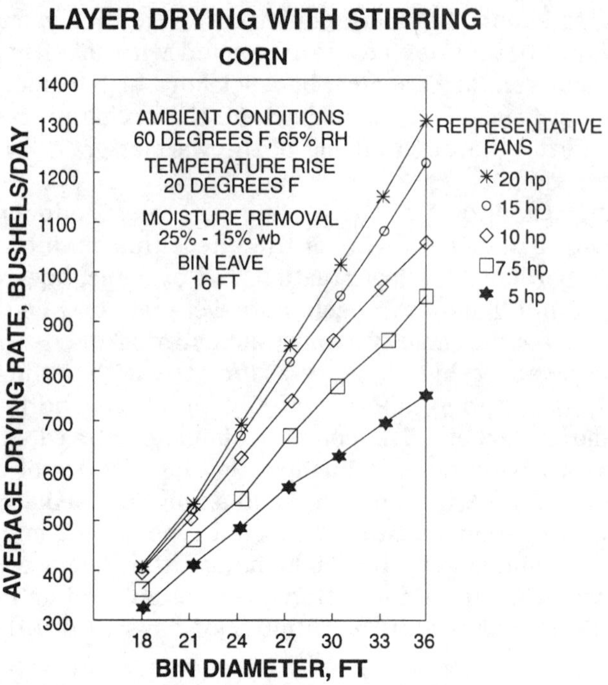

Figure 3.41–Layer drying situation with stirring (Bridges et al., 1982; 1984).

potential for aflatoxin development was the controlling factor in how fast each bin was filled in the simulation.

The average drying rates for the various fan and bin combinations without a stirring device are shown in figure 3.40. At the extremes, drying capacities for the 5-hp fan range from 216 to 511 bu/day. The 20-hp fan capacities range from 269 to 868 bu/day . For the smaller bin diameters (up to about 24 ft), increasing the fan horsepower for a given diameter bin yields relatively small increases in drying capacity. For bins greater than or equal to 24 ft in diameter, increasing fan size shows significant gains in drying capacity. Drying rates for the same set of drying conditions when using a stirring device are shown in figure 3.41. Generally, the stirred drying capacities are about 50% higher than those shown in figure 3.40 range from 329 to 775 for the smallest fan (5 hp) and from 407 to 1,329 for the largest (20 hp).

The capacities shown in figures 3.40 and 3.41 are for a particular drying situation and make the use of a stirring device appear very favorable. However, can the producer expect a 50% increase in drying capacity for all other drying conditions? A comparison of the percentage increase in drying capacity associated with stirring is given in table 3.11 for the same five fans used in the previous example. A 30-ft diameter bin and two sets of drying conditions are evaluated. The first set of drying

Table 3.11. Comparison of stirred and non-stirred situations (Bridges et al., 1984)

	Drying Situation No. 1* 60° F, 65% Rh, 20° F Rise 25 to 15%			Drying Situation No. 2* 40° F, 75% Rh, 10° F Rise 25 to 15%		
Drying Fan HP	Unstirred Drying Rate (bu/day)	Stirred Drying Rate (bu/day)	Percent Increase to Stirring (%)	Unstirred Drying Rate (bu/day)	Stirred Drying Rate (bu/day)	Percent Increase to Stirring (%)
5	432.9	646.3	49.3	178.0	202.3	13.6
7.5	523.0	793.7	51.8	202.9	242.6	19.6
10	591.4	878.4	48.6	218.8	270.1	23.4
15	641.7	972.9	51.6	235.8	302.1	28.1
20	696.0	1,028.2	47.7	240.2	318.4	32.2

* 30-ft-diameter drying bin.

conditions are the same as those used to generate figures 3.40 and 3.41 and show the same approximate 50% increase in drying capacity from stirring. The second set of conditions is for a cooler climate (40° F, 75% relative humidity) and a smaller drying air temperature increase above ambient (10° F). With the smaller temperature rise, the increase in drying capacity from stirring is considerably less for all fans as compared to the first situation. Further reductions in supplemental heat input will reduce the percentage increase to stirring even more. Thus, stirring will not increase drying capacity significantly unless there is a large potential for overdrying.

Stirring and Batch-in-Bin Drying

Stirring devices are often used in batch-in-bin drying to increase the daily drying capacity. Simulated stirred and unstirred daily drying capacities in bushels/day are shown in figure 3.42 for a batch-in-bin drying system for three different sets of drying circumstances. Common to each situation are the following: average ambient air conditions of 60° F and 65% relative humidity; 10 points of moisture removal (25 to 15% w.b.); a drying time of 17 h/day for the batch; and a 42-in.-diameter 15-hp vane axial fan. Bin diameters range from 18 to 36 ft.

The first example, shown in figure 3.42 and designated 140° F non-stirred, represents the daily drying capacities for each bin diameter with a drying temperature of 140° F and no stirring device. These drying capacities range from 1,171 bu/day for the 18-ft bin to 3,257 bu/day for the 36-ft bin. The second example (designated 140° F stirred) represents the drying capacities for the same drying conditions using a stirring device and shows an average increase of about 8.5% over the unstirred situation. The third example (designated 160° F stirred) shows the simulated drying capacities using a 160° F drying temperature and a

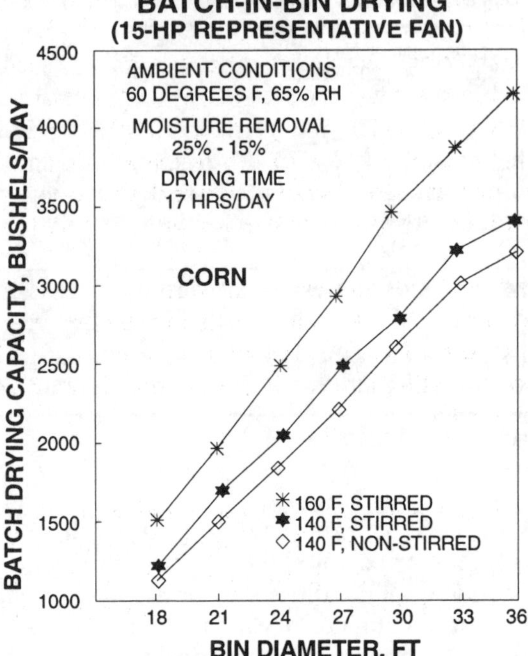

Figure 3.42–Stirring versus non-stirring comparison for batch-in-bin drying (Bridges et al., 1982; 1984).

stirring device. This curve represents an average increase in drying capacity of about 30% over the non-stirred 140° F example and is a viable alternative if more drying capacity is needed. This example shows that unless the drying air temperature is increased, the increase in drying capacity associated with a stirring device alone may not be significant when compared to the cost of the device. Drying temperatures above 140° F without stirring also increase drying capacity but usually are not recommended because of severe overdrying and a large moisture spread.

Economic Considerations

The economics of using stirring devices is dependent on many factors (cost of stirring device, fuel, and electricity, interest rate on the investment, tax bracket, individual drying situation, etc.). For most on-farm situations, a stirring device will pay for itself when there is great potential for overdrying (Loewer et al., 1984). This includes layer drying systems with a minimum 20° F temperature rise and batch-in-bin systems where drying temperatures may be raised above the recommended 140° F. In low-temperature (7° F or less), natural air, and batch-in-bin systems with no temperature increase over the non-stirred situation, the stirring device may not be economical even though some increase in drying capacity will occur. A detailed analysis of the economics of stirring is given in Chapter 9.

Another cost consideration is the use of a stirring device in an undersized drying system or one that is operating at 100% capacity. The addition of a stirring device may provide an attractive alternative when faced with the decision of how to best expand current drying capacity. Stirring, combined with the use of a higher drying temperature, may also allow the producer to harvest earlier and faster, thereby reducing harvest losses.

Stirring devices have numerous moving parts, and extensive use of these devices may require high maintenance cost over the life of the machine. The use of these devices may require bin walls to be supplemented with stiffeners to support the added stress caused by the rotation of the device.

Summary

Stirring devices are widely used for in-bin grain drying. Their major advantage is increased drying capacity for situations with overdrying potential. The major disadvantages are the investment cost of the equipment and whether the device will pay for itself over the usable life of the machine.

Special Considerations in Selected Grain Types

The principles of drying apply to all types of grain. Corn is used most frequently as an example because it is grown in most areas. However, each type and variety of grain is somewhat unique. The following discussion addresses some of the more common types of grain with regard to harvesting, drying, and storage. The harvesting information is supplemental to the grain drying and storage discussion and is presented as general rather than site-specific information.

White Corn

White corn provides a large portion of total corn utilized by the dry milling industry (White and Ross, 1972) while yellow corn is used primarily as animal feed. Accordingly, discoloration during the drying process is of greater concern when compared to yellow corn.

Drying. The primary objectives of a research effort by White and Ross (1972) were to (1) study color changes and stress-cracking in white corn dried with heated air as a function of drying air temperature and its cooling rate after drying, and (2) to compare these factors for white and yellow corn. Results indicated no apparent relationship between post-drying treatments and the values of a sample's color parameters, although some values were quite different from one year to the next. Slow cooling of both white and yellow corn after drying resulted in a dramatic reduction in the number of "checked" kernels (kernels with surface breaks), especially for drying air temperatures above 160° F. Color changes produced in white and yellow corn as a result of drying with heated air were similar. The number of checked kernels tended to be higher for the yellow corn but differences seldom exceeded 15 to 20%, even at the higher temperatures. Thus, from this research, it would appear that white and yellow corn are similar insofar as drying is concerned.

Rice

Rice quality is a primary factor in determining its price. Rice is graded using several classifications designed to characterize or quantify quality. Rough rice refers to grain that has not been milled. Rough rice is typically dried to approximately 13% moisture content for storage or immediate milling. The first unit operation of the milling process consists of removing the hull, leaving " brown rice". Further milling of the brown rice produces "white rice" and by-products (polish or bran). White rice is further characterized as being either "head rice" (¾ whole kernel or greater) or "brokens" (seconds, screenings, and "brewer's rice"). A primary measure of rice quality is the head rice yield which is greatly influenced by drying.

Drying. Methods for drying rice differ greatly between on-farm and commercial systems. In either situation, it is imperative that drying occur slowly if head rice yields are to be maintained during the milling process.

On-farm Drying. Typically, in-bin drying is used in on-farm systems. Natural air drying is a common technique in cooler, drier growing areas. Stirring devices may be used effectively in heated-air systems to reduce the stress induced by moisture gradients in the kernel. Rewetting the grain can reduc drasticallye head rice yields (Siebenmorgen and Jindal, 1986). Zones 2 and 3 conditions are preferable to those of Zone 1, and can be obtained only by limiting heat input or using natural air. Low temperature drying recommendations are well-suited to on-farm rice drying systems. Rice is said to be the most difficult grain to dry while preserving quality, and on-farm managers must be extremely diligent in controlling heat input if they are to be successful.

Commercial Drying. Commercial drying utilizes high temperature continuous flow, drying systems. Excessive grain damage is avoided by making multiple passes, limiting the moisture removal to 2 to 4% per pass (Steffe and Singh, 1980). The rice may be "tempered" for up to 24 h between passes to ensure that the grain moisture within the kernel has had time to equilibrate. In essence, commercial rice drying requires a form of dryeration be used. However, in rice drying, tempering is used primarily for preserving grain quality rather than for saving energy.

Grain Sorghum

Grain sorghum has certain harvesting, drying, and storage characteristics that are uncommon to other crops. With a few modifications, however, the equipment and machinery used for other grains can be adapted for use with sorghum.

Harvesting. *When to Harvest.* Grain sorghum should be harvested as early as possible to prevent heavy losses from birds, insects, molds, and adverse weather conditions. Humid conditions often prevail during the harvest season, delaying field drying and encouraging the development of mold. In addition, grain sorghum does not dry to moisture contents safe for storage when left in the field until it has been killed by a frost or by a chemical. Waiting for a frost in order to be able to harvest at low moisture contents increases the probability of excessive field losses. This risk may be reduced by early harvesting.

Grain sorghum is considered to be mature at about 30% moisture. However, at moistures higher than 25%, the seeds are too soft to withstand adequate threshing action, thereby requiring delay of harvest until the moisture content of the grain is 25% or less. Because of the advantages associated with harvesting at higher moisture contents (up to 25%), it is important that the producer have either drying equipment available or plan to store the grain sorghum as high moisture grain.

Operating the Combine. Grain sorghum can be combined with a regular grain header. The stalk should be cut as high as possible without excessive loss of heads. The combine must be adjusted properly and operated at an appropriate speed if clean grain is to be harvested. Because sorghum seeds crack easily, the cylinder speed should be reduced below that used for wheat, but the rest of the machine should run at a normal speed. It may be necessary to remove some of the concave bars to prevent damage to the seed. Thrashing action should be only enough to detach the seeds from the heads.

Stalks of grain sorghum plants are normally much wetter at harvest than corn stalks, and they are more likely to be chopped up and delivered into the grain tank. As little material as possible should be returned to the cylinder because a portion of the recycled matter eventually could be broken into pieces small enough to go into the grain. It is normally advisable to keep the chafer extension closed enough to prevent excessive return even at the expense of a slight loss of grain. This may be accomplished in part by using a piece of sheet metal over the chafer extension.

The sieve should be watched closely for indications of matting since this will lead to excessive grain loss. The sieves must be adjusted properly and must not be overloaded.

The operator's manual for the combine gives recommended speeds and adjustments for the cylinder, concaves, chafer sieves, and fan. It also contains information on lubrication and maintenance. Careful adjustment and operation of the combine will help reduce the amount of cracked grain and foreign material.

Holding Limitations. Holding wet grain sorghum is somewhat more difficult than holding corn at the same moisture content and temperature. Grain sorghum has a higher rate of respiration than corn, resulting in temperatures that could cause germination or spoilage more rapidly.

Drying. *Cleaning.* Cleaning grain sorghum before drying is desirable because of the large amount of trash normally contained in the grain after harvesting. Cleaning wet grain is difficult because of juices coating the cleaning screens. However, drying with direct-fired crop dryers could result in a fire hazard if the trash content of the sorghum is high and if trash from the area surrounding the dryer is sucked into the fan and ignited by the burner. This hazard is increased with sorghum because the trash in the grain will sometimes accumulate in "pockets" and dry excessively because of long exposure to the high temperature and high velocity drying air. A fire could result if trash which enters the blower is ignited by the burner and ignites the very dry trash in the grain. Therefore, precautions must be taken to clean the grain before drying.

Airflow Characteristics. The small round sorghum kernel offers a higher resistance to airflow than larger grain kernels such as corn. This results in larger horsepower requirements for fans to obtain the same drying capacity as for corn. In the case of systems that have been designed for corn, a shallower depth of sorghum must be used to assure adequate drying without overloading the system. This depth is $\frac{2}{3}$ to $\frac{3}{4}$ the depth normally used for corn. For example, if a 4 ft depth of corn can be dried in a grain bin, the sorghum depth should be reduced to approximately 3 ft, assuming both grains to be at the same moisture content. A system to be used primarily for drying sorghum should be designed for higher static pressures than a similar system used for corn. Chapter 4 addresses airflow considerations in detail.

Dryer Operation. Grain sorghum dries at approximately the same rate as corn provided that the same amount of air and heat are supplied to each bushel. Using shallower depths of grain sorghum in layer and batch-in-bin systems that were designed for drying corn allows the airflow per bushel of each grain to be approximately the same. Again, a guideline for the depth of sorghum layer is two-thirds to three-quarters the depth of corn for the same moisture content.

A longer drying time is required for grain sorghum than for corn. The rate of discharge should be decreased to allow the grain sorghum to be exposed a longer time to the drying air when using a continuous flow dryer. The ground around the fan intake should be kept clean of trash to reduce the fire hazard in grain sorghum drying. It may be necessary to cover the nearby ground to prevent trash from entering the fan.

Likewise, it may be desirable to build an intake for the fan which extends 8 to 10 ft above the ground to prevent pulling trash through the burner. However, care should be taken to keep the trash prevention mechanisms from reducing the airflow efficiency through the fan.

Temperature. When drying, the maximum recommended sorghum kernel temperatures are the same as for corn and depend on the intended use of the grain. Recommended upper temperature limits are the same as for shelled corn (see table 3.10)

Deep layer drying (in-storage layer) requires limiting the increase in air temperatures to 10° F to 20° F above the outside air temperature to keep from overdrying the lower layers of grain. Shallow layer drying (batch-in-bin) with grain depths up to 4 ft and airflow between 10 and 25 cfm/bu allows a maximum air temperature of 140° F (110° F for seed). Column drying (batch or continuous flow) grain sorghum requires slightly lower air temperatures than for corn because of longer exposure. For these dryers, the drying air temperature should be 160° F to 200° F, depending on the amount of trash in the sorghum. Measuring the grain temperature as it leaves the dryer may be desirable to avoid high temperature damage. The reduced drying capacity of systems with grain sorghum as compared to corn may encourage either the use of higher temperatures or failure to dry the grain to a safe moisture content for storage. Drying with higher temperatures increases the danger of fire, especially when high trash contents are present. Accordingly, precautions should be taken at all times. An especially dangerous situation is the operation of fully automatic processes for long time periods without observation and supervision.

Storage. Storage considerations in grain sorghum (moisture content, aeration, insect control) are much like other grain, and are discussed in Chapter 4. However, grain cleanliness is more critical when drying grain sorghum, especially when high temperatures are used. Excessive trash should be removed from grain sorghum before it is stored. Otherwise, the accumulation of trash in small pockets in the grain bin can cause heating which, in turn, induces insect and mold problems. Normally, cleaning sorghum requires the removal of particles larger than the grain kernels instead of the removal of fines.

Soybeans

It has been estimated that an average operator will leave from 2½ to 4 bu of soybeans per acre in the field. Certain precautions can be taken to reduce field losses and maintain the quality of soybeans.

Harvesting. *When to Harvest.* The moisture of the bean and the hull determine the best time for harvesting soybeans. The range for harvesting soybeans is from 9 to 20% moisture content (w.b.). There is less seed damage at higher moisture contents. However, the soybean is considered to be fully mature well before the bean can be harvested with present field equipment. There is considerably less visible damage when it is harvested at moisture contents above 13.5%. Damage increases more rapidly when moisture content falls below 13.5%, and there is excessive cracking if harvesting is delayed until seed moisture content is less than 11.5%. It has been observed that soybeans that have field dried to a moisture content below 13.5%, and then subjected to rain, tend to have a higher percentage of cracked seedcoats when harvested.

Operating the Combine. Among the most important items to consider when operating the combine are ground speed, reel speed, cutterbar height, and separation of soybeans from pods. The combine should be operated at an average speed of from 2½ to 4 mph. Fast ground speed can result in high losses from the header. The ground speed can be estimated by dividing the number of 3-ft steps covered by the combine in 20 s by a factor of 10. This will give an estimated miles per hour ground speed.

Ground speed should also be matched to the combine reel speed. The best speed ratio is a reel speed 1.25 times the combine ground speed. A low reel speed reduces shattering.

Soybeans should be cut as close to the ground as possible. For best results, a cutting height of about 2½ in. above the ground is recommended. Normall this willy require special equipment such as a floating cutterbar or an automatic header control. Field losses are increased about 10% when the beans are cut 3.5 to 4 in. above the ground. Cylinder speed is quite important in reducing the amount of seed cracking. Cylinder speeds of 700 and 900 rpm result in higher seed cracking than a speed of 500 rpm. It may be necessary to use a cylinder speed slower than 500 rpm if excessive seed cracking continues. The operator's manual should be followed in determining the initial settings for the cylinder speed and concave spacing. It may be necessary to change the cylinder speed as the harvest progresses and the moisture content changes. Soybeans at high moisture contents require more energy to remove the bean from the pod. Therefore, the cylinder speed has to be higher and the concaves set closer for high moisture conditions. As the beans dry out, the cylinder speed can be slowed and the concaves adjusted open. The combine blower fan may have to be slowed down to prevent beans from being carried over the back of the combine as the moisture content drops.

The reel axle should be 6 to 12 in. ahead of the cutterbar and the reel-bats should leave the plants just as they are cut. The reel depth should be just enough to control the beans. A six-bat-reel normally will give more uniform feeding. Measuring harvest losses is quite important in determining combine adjustments. An average of four beans/ft^2 is equivalent to about 1 bu/acre loss.

Drying. Soybeans can be dried with conventional drying equipment. However, the percentage of cracked beans increases rapidly with the drying temperature. Research at the University of Kentucky (White et al., 1976, 1980) indicates that little cracking occurs if the temperature of the drying air is 70° F or below, but there are significant differences among the varieties with regard to cracking when higher drying temperatures are used. For 100° F, the percentage of the beans that had skin cracks ranged from 10 to 60%. For 130° F, the range was 50 to 90%; and for 160° F, 80 to 100%. For 100° F drying temperatures, the beans that cracked below the skin ranged from 5 to 20%. For 130° F, this range was 20 to 70%; and for 160° F, 30 to 80% of the beans cracked.

Problems

(NOTE: The same fan is used throughout the problem set.)

3.1. Table 3.1 and figures 3.3 to 3.4 were developed using the software package STIRDRY. Using the conditions listed at the top of table 3.1 as input for STIRDRY, calculate the average drying rate in bushels per day for an initial corn moisture content of 24%. What is the new drying rate if ambient conditions are changed to 55° F and 65% Rh and the temperature of the drying air is increased to 3° F above ambient?

3.2. An 18-ft-diameter bin is filled with 22% moisture content shelled corn to a depth of 5 ft. The grain is to be dried to a moisture content of 12%. If a drying fan supplies 3 cfm/bu of 70° F Tdb and 55% Rh air, how long will the drying process take ?

3.3. A 30-ft-diameter bin is filled to a depth of 16 ft with grain. Using figure 3.2, determine the airflow in cfm that a typical 10-hp axial fan would deliver.

3.4. A 21-ft-diameter bin is filled to a depth of 16 ft with shelled corn that is to be dried to a final moisture content of 13% using a low temperature drying system. The expected average daily ambient conditions are 70° F and 60% Rh. If the allowable temperature rise to keep the equilibrium moisture content below 13% is 2° F and if a 15-hp axial fan is used:

(a) What quantity of energy would be required for a thermal system?

(b) What quantity of energy would be required for an electrical system?

3.5. Corn is to be dried using a low-temperature drying system. If the expected daily average weather conditions are 60° F and 70% Rh, what is the desired temperature of the drying air if the corn is to be dried to a moisture content of 12%?

3.6. Using the conditions listed at the top of table 3.2 as input for STIRDRY, produce a drying schedule for an initial grain moisture content of 24%. Compare the total drying days for layer 11 using ambient conditions of 65° F and 55% Rh and a drying temperature of 75° F.

3.7. Using the conditions listed at the top of table 3.3 as input for STIRDRY, calculate the average drying rate in feet per day for an initial grain moisture content of 24%. Determine the new drying rate for average ambient conditions of 38° F and 70% Rh and a drying temperature of 46° F.

3.8. Using the conditions listed at the top of table 3.4 as input for STIRDRY, calculate a drying schedule for an initial grain moisture content of 24%, and compare the drying days per layer output to the values shown in table 3.4. Recompute the drying days per layer for a situation where bin diameter and eave height are 18 ft and 16 ft, respectively, and drying air temperature is 70° F.

3.9. Using the conditions listed at the top of table 3.5 as input for STIRDRY, calculate the average drying rate in bushels per day for an initial moisture content of 24% and compare the output to the values shown in table 3.5. Recompute this average drying rate for a situation where the drying temperature is 62° F and the bin diameter and eave height are 21 ft and 16 ft, respectively.

3.10. Table 3.8 was developed using the software package CROSSFLOW.
   (a) Using the conditions listed on table 3.8 as input for CROSSFLOW, calculate the drying time for a 4-ft grain depth and 120° F drying air temperature. (Assume the price of LP gas is $1.00/gal).
   (b) Use CROSSFLOW to calculate the drying time for the same conditions listed at the bottom of table 3.8 and a drying air temperature of 128° F and a 3-ft grain depth.

3.11. Estimate the size of a wet holding tank needed for an operation harvesting 10,000 bushels of corn per day. The dryer is operated 10 h/day and the corn is dried at an effective rate of 800 bu/h.

3.12. Use the software package NATAIR to calculate the amount of moisture removed from corn with an initial moisture content of 25% and an initial grain temperature of 100° F, assuming the grain is cooled with air with a dry bulb temperature of 80° F and relative humidity equal to 55%. It is desired to dry the grain to a final moisture content of 13%. The fan delivers 3 cfm/bu of corn. Compare these results to the maximum theoretical moisture that can be removed as presented in figure 3.28.

3.13. A bin contains corn at a moisture content of 15.5% and a temperature of 120° F. The specific heat of the grain is 0.49 Btu/lb° F, and the 14-ft diameter bin is filled to a depth of 10 ft with grain.
   (a) How much energy is stored in the grain if the outside temperature is 80° F?
   (b) What will the final moisture content be after dryeration is completed?

3.14. What is the maximum allowable time for cooling a dryeration bin that is being filled at a rate of 2,000 bu/day? The corn must be tempered for 6 h and the dryeration bin is unloaded at a rate of 750 bu/h.

3.15. If a tank dumps 200 bu of wheat into a 15-ft diameter dryeration bin, what is the change in grain depth?

3.16. A dryeration system uses 2 bins that each can be unloaded at a rate of 100 bu/h. What is the maximum time allowable for cooling if the grain is tempered for 8 h and the daily harvest is 1,000 bu?

3.17. Using the conditions listed in table 3.12 for DRYING SITUATION NO. 1 as input for STIRDRY, calculate the non-stirred and stirred drying rates for a 10-hp drying fan (from Chapter 4 - table 4.2, Fan 8A) with thermostatic control. Recompute these drying rates for a 24-ft diameter bin.

References for all chapters begin on page 541.

4

Fan Performance and Selection

The design of grain drying facilities is greatly influenced by fan performance and selection. Essential to design is an understanding of the engineering principles associated with air flowing through grain and the differences in performance among different types of fans.

Principles of Airflow Through Grain

When air is forced through grain, it encounters a certain resistance to flow. This resistance, that is often referred to as pressure drop, is a result of friction in the grain mass. Its magnitude is dependent on: properties of the drying air; airflow rate; surface and shape characteristics of the grain kernels and associated size, shape and configuration of the grain void spaces; amount, size, and distribution of any broken kernels and fine materials contained in the grain; and overall depth of the grain mass.

Most pressure-drop data for airflow through grain are based on experimental results that have been determined for loose-fill (unpacked) clean grain and unheated ambient air. In practice these values must be adjusted to account for variations from these test conditions (Stephens and Foster, 1976; 1978). Figure 4.1 is a graphical presentation of the relationship between airflow rate (cfm/ft^2) and static pressure drop expressed in inches of water per foot of grain (ASAE Standards, 1985, based on research by Shedd, see Appendix B page 535).

Static Pressure

Static pressure (or static head) can be expressed in several ways. The most commonly used unit of pressure is pounds per square inch (psi). However, in ventilation and crop drying applications, pressure drop is usually less than 1 psi. Therefore, it is more convenient to express these pressures in terms of the number of inches of water (in. H_2O) that a fan can displace in the vertical direction. The weight of a 27.7 in. column of water that is one square inch in cross-section is 1 lb. Therefore, the column will produce a pressure of 1 psi at its base. Pressure expressed as psi can be converted to equivalent inches of water and vice versa. The static pressures normally encountered in crop drying installations are on the order of 1 to 8 in. of water.

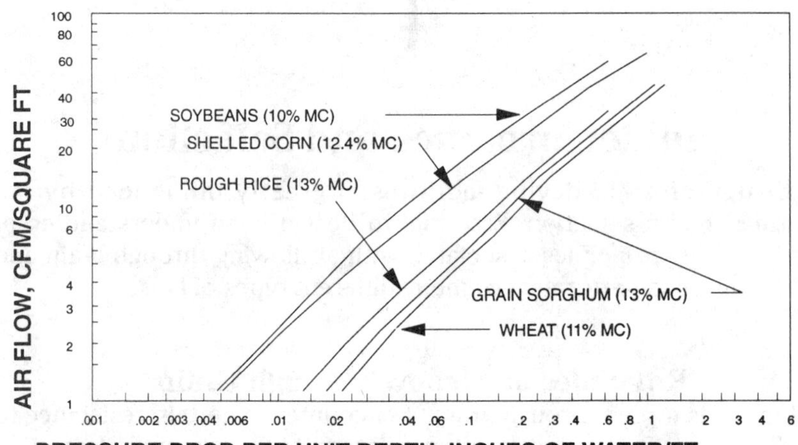

Figure 4.1–Shedd's data for airflow (ASAE Standards, 1985).

Static pressure may have either a positive or negative value. When static pressure is measured in the air plenum, a "positive" static pressure results when the fan is "pushing" air through the grain. Conversely, a "negative" static pressure system exists when the fan "pulls" air through the grain mass.

The static pressure produced by a drying fan across a depth of grain can be measured with a pressure gauge. However, it is most often measured directly in inches of water by using a manometer as shown in figure 4.2. A manometer consists of a glass or plastic U-tube containing a quantity of water. When one side of the manometer is connected to the outlet duct of a fan or to the air plenum of a grain bin (the space beneath the drying floor), the pressure in the duct (or plenum) will either push or pull against the water (depending on whether it's a positive or negative system), thus creating a height differential in the tube. The difference in the water level of the two columns (measured in inches) is the static pressure generated by the fan. The manometer shown in figure 4.2 is measuring positive static pressure because the fan is pushing air through the grain.

In figure 4.1, airflow rates are expressed as cubic feet per minute per square foot of floor area (cfm/ft^2). This is the bulk (or apparent) air velocity through the grain and can be computed by dividing the total airflow rate (cfm) by the cross-sectional area of the bin (ft^2). When the airflow rate in cubic feet per minute per bushel (cfm/bu) is known, the

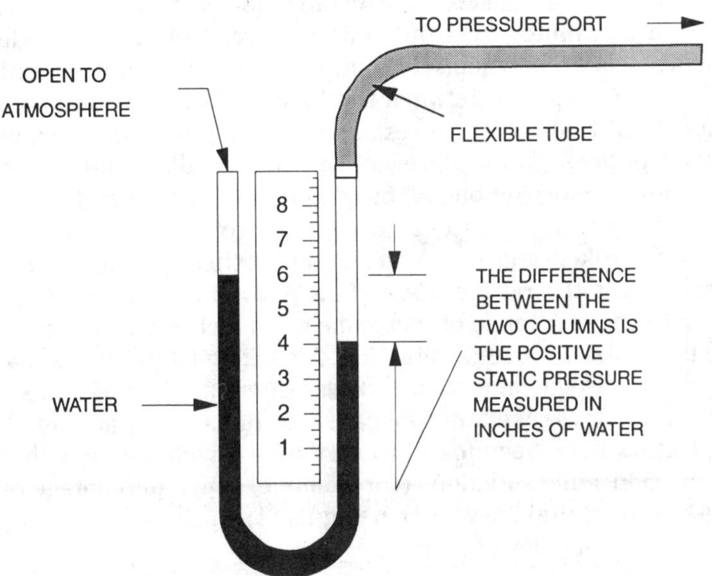

Figure 4.2–Manometer for measuring static pressure.

cfm/ft^2 in a bin-type dryer can be computed directly by the following equation:

$$cfm/ft^2 = (\text{grain depth, ft}) * (cfm/bu)/1.25 \quad (4.1)$$

Resistance to airflow is highly dependent on grain type and airflow rate (fig. 4.1). The static pressure drop per foot of grain depth increases at an increasing rate with air velocity; that is, doubling air velocity more than doubles the pressure head required to force air through a given grain depth. For example, doubling of airflow from 20 to 40 cfm/ft^2 increases the required static head per foot of shelled corn nearly three times from 0.18 to 0.525 in. of water.

Influence of Grain Type

The difference in grain type greatly influences the resistance to airflow. Reasons for these differences are directly related to the shapes of the different types of kernels, some of which fit more tightlytogether, thus altering the pathway that air must follow through the grain.

The curves in figure 4.1 represent the airflow resistance of clean grain of several grain types under loose fill conditions. Loose fill occurs when grain is introduced into a bin at a very low velocity (without significant dropping height) and allowed to flow naturally into a cone-shaped pile. These conditions seldom occur in actual drying bins because most grain contains some broken and fine material ("fines"). In addition, grain usually is dropped a considerable distance into the drying bin, or it is thrown by a distributor over the grain surface. As grain is distributed, the grain kernels orient themselves into compacted positions, and fines sift into the open spaces between the kernels. This situation produces "compacted fill" that increases resistance sharply to airflow. An uneven distribution of fines also will prevent the uniform distribution of drying air throughout the drying bin. A "packing factor" is applied to the figure 4.1 values for purposes of design in order to compensate for grain compaction, broken and fine material, and other system losses. This factor represents the ratio of the airflow resistance of a given grain lot and depth to the resistance of that same grain if cleaned and introduced into the bin under a loose fill situation. A packing factor of 1.5 is a good representative design value to use with most grains. This assumes a 50% increase in airflow resistance because of fines and packing. Actual packing factors have been found to vary over a wide range with values as high as 2.5 in some situations (for example, a high percentage of fines in the grain and a grain spreader being used) to as low as 1.0 in clean grain when a stirring device is used.

Types of Drying Fans

There are two types of fans normally used in grain drying and aeration: axial flow and centrifugal (figs. 4.3 to 4.4).

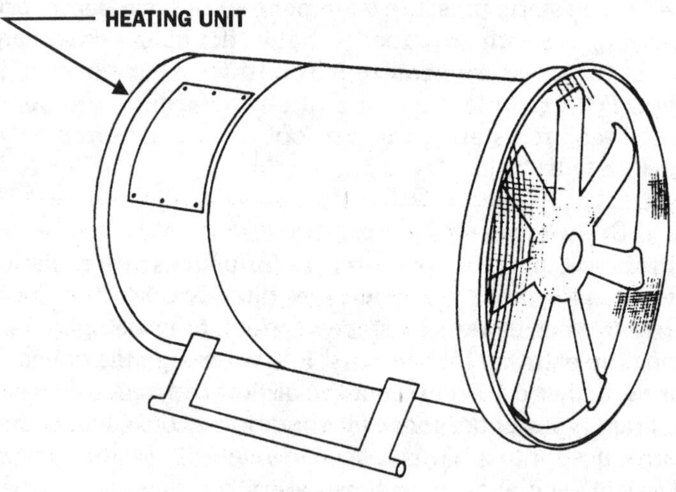

Figure 4.3–Axial flow fan.

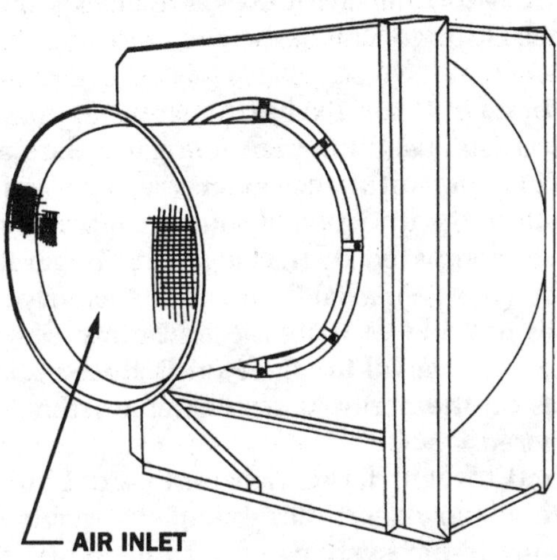

Figure 4.4–Centrifugal fan.

Axial-flow Fans

In axial-flow fans, the air passes through the fan in a direction nearly parallel to the fan axis. The fan impeller consists of a number of wide blades attached to a large fan hub. The fan impeller is enclosed in a cylindrical housing, and this assembly may be classified as either a tube-axial or vane-axial fan depending on its construction. The tube-axial fan does not have guide vanes while the vane-axial fan has a number of fixed guide vanes located either just upstream or downstream from the fan impeller. Vanes reduce air turbulence and improve fan performance. Vane-axial fans can be used effectively at higher pressure than tube-axial fans (up to 5 in. of H_2O as compared to 4 in. for the tube-axial fan). Thus, they are often a good choice when aerating or drying grain in deep storage bins.

Centrifugal Fans

Centrifugal fans have a blower wheel located in the center of a spiral housing. This wheel has air moving blades (either forward or backward curved) mounted along its outer edge. As the fan turns, the resulting centrifugal force causes the pressure at the center of the blower wheel to decrease. At the same time, the pressure in the spiral housing outside the blower wheel increases. The result is that air flows axially into the center of the blower wheel and then out radially through the impeller blades on its periphery. Fan performance characteristics depend on the fan diameter, speed and type of fan blades (forward or backward curved). Centrifugal fans are quieter operating but are generally more expensive than axial-flow fans. However, they can work against higher static pressure heads (up to 8 in. of H_2O).

Airflow Comparisons in Grain Drying

Most centrifugal fans used in grain drying are the backward curved type. In these fans, the rotor blades are relatively deep and curve backwards away from the direction of rotation. If performance curves of axial-flow fans are compared to that of a centrifugal fan of the same horsepower rating (fig. 4.5), axial-flow fans will usually supply more air at static pressures of 2.5 in. of H_2O or less. The centrifugal fan generally will perform better above 4.0 in. of water. Between 2.5 and 4.0 in., fan selection depends on the performance characteristics of the individual fans being considered.

When compared to centrifugal fans, vane-axial fans generally are better adapted to shallow-depth bin drying systems, such as batch-in-bin and continuous in-bin systems, and to deep-bed drying systems requiring low airflow rates (1 cfm/bu or less). They are generally lower in initial cost but operate at higher noise levels than centrifugal fans.

Centrifugal fans are selected over vane-axial fans when the application requires a relatively high air volume (2 cfm/bu or more) for deep-bed drying systems. These fans are also preferred when a low noise level is an important consideration.

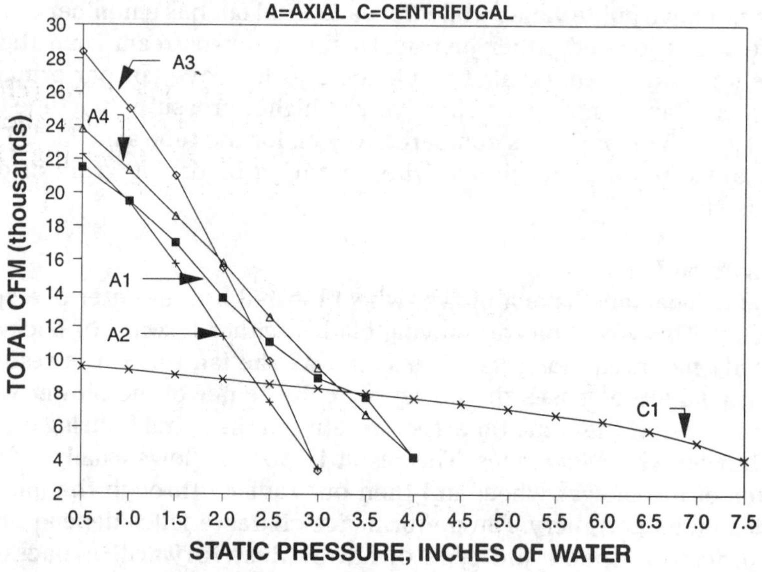

Figure 4.5–Fan performance for different fans of the same horsepower (based on Thompson, 1975).

Selection of Drying Fans

Pressure Considerations

When selecting a fan for drying or aerating grain, primary considerations are the quantity of air to be moved per unit of grain (cfm/bu), the static pressure head required to provide this quantity, and the horsepower of the fan motor. Representative airflow rates for aeration, dryeration, and drying systems are listed in table 4.1. These values are typical for the conditions listed but can be varied if drying or cooling rates need to be increased or decreased. Once the desired cfm/bu has been selected, the total design air delivery rate of the fan can be calculated by multiplying the total number of bushels to be cooled or dried by the cfm/bu rate. For round bins the bushel capacity for various grain depths can be obtained from table 3.6 from Chapter 3. The static pressure required to force air through the specified depth of grain at the desired rate can be computed using the procedures discussed earlier in the chapter. For example, assume 1 cfm/bu is to be moved through a 15 ft depth of shelled corn. Using equation 4.1:

$$\text{cfm/ft}^2 = 15 \text{ ft} * 1 \text{ cfm/bu} / 1.25 = 12.0$$

Using figure 4.1 for shelled corn, the pressure drop per foot of grain is 0.09 in. of water. Assuming a packing factor of 1.5, the total static pressure required to provide drying air at the indicated rate is:

$$P_s = 0.09 \text{ in. } H_2O * 15 \text{ ft} * 1.5 = 2.03 \text{ in. } H_2O.$$

Airflow Considerations

A fan must be selected that is rated to provide the total airflow (cfm) needed at a given static pressure, essentially 2.0 in. of H_2O for the above example. Selection of a fan is accomplished by reviewing the performance ratings of the various crop drying fans that are

Table 4.1. Representative airflow rates for grain drying and storage systems

Description	Range (CFM/Bu)
Aeration	0.1 - 0.5
Dryeration	0.5 - 1.0
Natural Air Drying	1 - 3
Low Temperature Drying	1 - 3
Layer Drying	1 - 3
Batch-in-bin Drying	5 - 15
In-bin Continuous Flow Drying	1 - 20
Automatic Batch Drying	50 - 100
Continuous Flow Drying	75 - 125

commercially available and meet the criteria for static pressure and total airflow.

Table 4.2 contains fan performance data for several fans including those used for purposes of illustration in this chapter. Assume that a fan is to be selected for a 36-ft-diameter bin filled with corn to a height of 15 ft above the plenum chamber for a total capacity of 12,215 bushels. Given the 1 cfm/bu airflow rate used in the previous example, the total airflow to be supplied is 12,215 cfm. The next step is to select a fan that will provide at least 12,215 cfm at a static pressure of 2.0 in. of water. In reviewing table 4.2, Fan ID's 3 and 5-13 satisfy the above criteria. If the bin diameter were increased to 48 ft, the total airflow required increases to 21,715 cfm, and only Fan ID's 10-13 would meet the design criteria.

If several fans meet the indicated performance criteria, then select the fan that supplies the highest cfm per unit of input power. This selection criterion provides a fan that is more energy efficient and economical to operate.

Again, axial-flow fans generally are the best choice at relatively low static pressures and high airflow rates. Centrifugal fans are better suited to high pressure, high airflow operating conditions. For heated air drying, the fan performance data used to select the drying fan should consider the flow resistance produced by the burner and its associated accessories. When used for natural air drying, a fan will supply significantly more air at a given static pressure than it will when combined with heating equipment. Most manufacturers provide rating tables for their fans for both natural and heated air applications. Fans

Table 4.2. Example of fan performance information for selected fans

Fan ID and type*	Horse power	Static Pressure (In. of H_2O) 0.5	1.0	2.0	3.0	4.0	5.0	6.0
1.A	5	19,020	15,500	5,500	–	–	–	–
2.A	5	16,800	14,720	9,800	5,600	–	–	–
3.A	7.5	21,500	19,500	13,700	8,900	4,200	–	–
4.A	7.5	22,275	19,400	11,400	3,550	–	–	–
5.A	7.5	28,400	25,000	15,500	3,400	–	–	–
6.A	7.5	23,800	21,350	15,800	9,530	4,200	–	–
7.A	10	24,530	22,420	16,200	11,000	6,400	–	–
8.A	10	28,400	25,900	19,900	13,730	7,800	–	–
9.A	10	30,500	27,030	17,600	6,000	–	–	–
10.A	15	34,800	31,850	23,650	14,650	4,750	–	–
11.A	15	34,500	32,150	26,750	19,140	13,000	–	–
12.A	20	40,100	37,100	29,200	18,500	9,800	–	–
13.A	20	39,700	37,212	31,700	23,000	16,100	7,075	6,350
14.C	7.5	9,600	9,375	8,825	6,300	7,700	9,425	8,325
15.C	10	12,875	12,550	11,850	11,150	10,400	11,675	10,375
16.C	15	15,975	15,575	14,700	13,800	12,800	14,700	13,300
17.C	20	19,750	19,200	18,150	17,120	16,100		

* A = axial fan; C = centrifugal fan.

purchased as part of a crop dryer will be rated with the equipment included in the drying system. The efficiency of crop-drying and aeration fans typically will range from a high of 60 to 65% to as low as 40%. The actual installed efficiency will depend on the environmental conditions, the quality and type of fan, and the static pressure of the drying system in relation to the fan curve.

Horsepower Considerations

Fan size often is defined in terms of the horsepower of the fan motor. This is often confusing because airflow rates differ among fans and fan types of the same horsepower (fig 4.5 and table 4.2). The best method of selecting a fan is to compare performance data such as that given in table 4.2. In addition, the fan motor should be large enough to handle the indicated load with some margin of safety.

The actual horsepower required to drive a fan at the specified operating point can be calculated if the fan efficiency is known at that point. Horsepower is a measure of the rate at which work is being done when the fan is delivering a specified airflow against an existing static pressure. A convenient equation for determining the horsepower requirement for a drying fan is:

$$hp = (cfm * Ps)/(63.46 * efficiency) \qquad (4.2)$$

where

cfm = airflow in cfm
Ps = static pressure in inches of water
efficiency = fan efficiency (%)

If the actual fan efficiency is not known, 50% efficiency can be assumed for design purposes. The fan motor horsepower rating should equal or exceed the computed horsepower or the fan motor may overload. Fan horsepower may be limited by the local electric power supplier to prevent relatively high starting currents from lowering line voltage. In considering the installation of any drying system one should check with the power company to determine the availability and types of electric service, the size fan motors allowed by the company, the electric rate schedule, and the magnitude of any demand charges.

A consideration when designing a bin drying system is that decreasing the depth of grain results in less static pressure for the same airflow rate. Therefore, less horsepower is required to deliver a given airflow through the grain mass. Decreasing maximum grain depth can help maintain reasonable levels of fan horsepower at high airflows. This can be accomplished by using larger diameter bins to dry any specified batch size whenever it is practical to do so.

Performance Curves

Fans of differing design have different performance characteristics. Fan performance information in figure 4.6 shows the broad range of the airflow - static pressure relationships for a typical set of fans that were used in the following examples. In grain drying, fan performance also differs with grain type, grain depth, and bin diameter although the general shapes of the curves are similar. Measures of fan performance for a given diameter of bin and depth of grain include total airflow (cfm), airflow per unit volume of grain (cfm/bu), air velocity (feet per minute, fpm), static pressure, and percentage efficiency. These factors are shown in figures 4.7-4.8 for 10-hp axial and centrifugal, respectively. In the following examples, only cfm/bu will be used as a basis for comparison. Corn will be the grain type unless stated otherwise. A computer model developed by Thompson (1975) was used to calculate fan performance.

Bin Diameter. In figure 4.9, fan performance curves for the same axial fan are shown with different bin diameters. As diameter increases, cfm per bushel decreases for the same grain depth. However, for the same number of bushels, cfm per bushel increases as diameter increases because airflow is affected more by grain depth than grain volume. This may be confirmed by using table 4.1 and figure 4.9 in combination. An 18-ft diameter bin holds 2,036 bu when filled to a depth of 10 ft. A 36-ft-diameter bin contains 2,036 bu when filled to a depth of 2.5 ft. For the

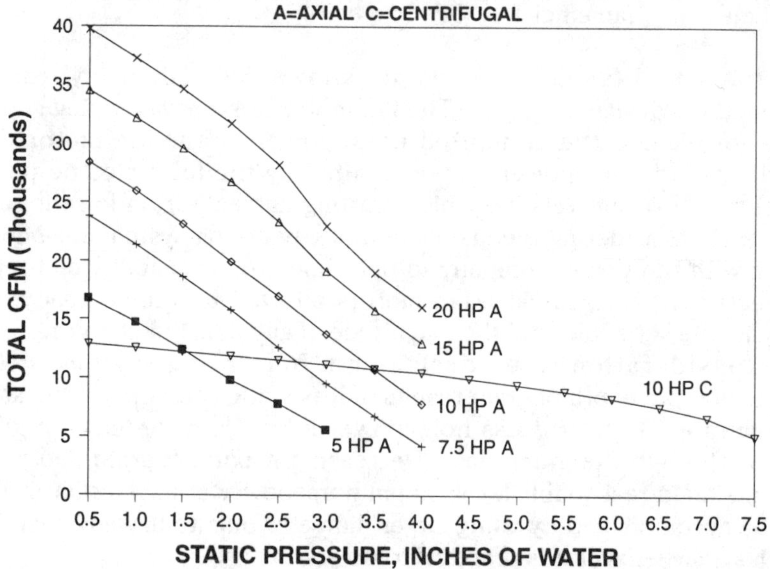

Figure 4.6–Airflow - static pressure relationships for different fans (based on Thompson, 1975).

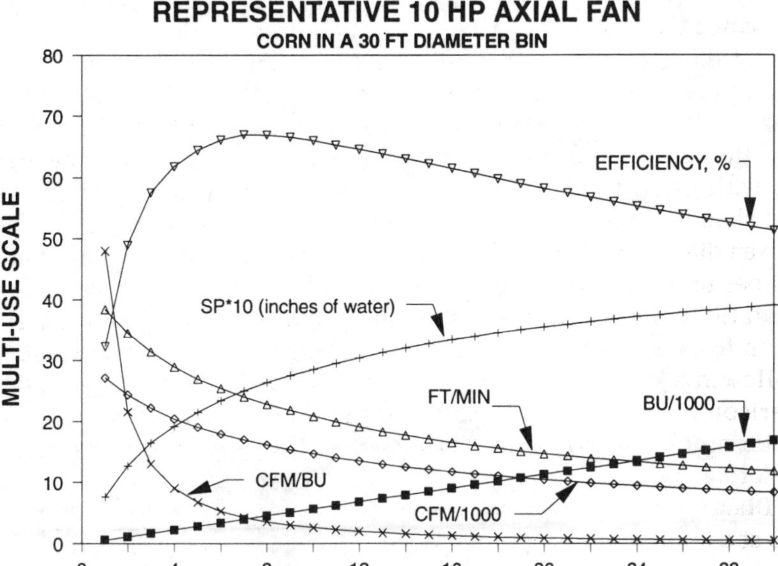

Figure 4.7–Axial fan performance (based on Thompson, 1975).

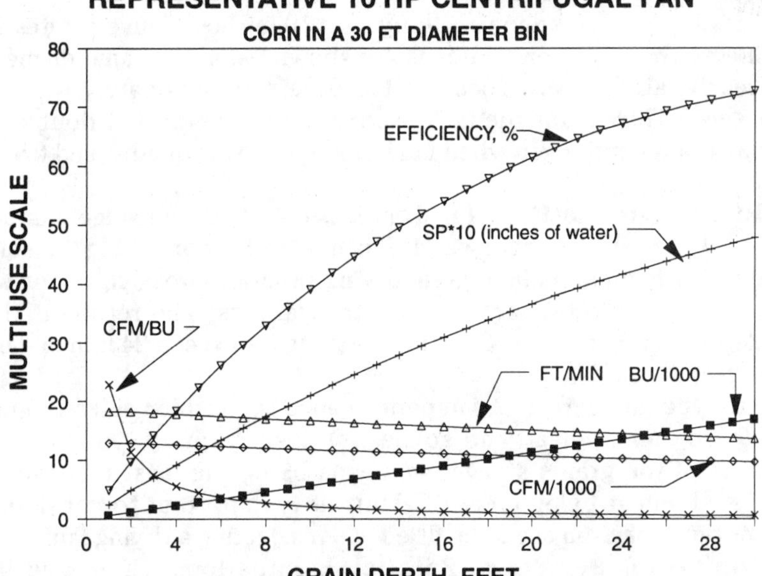

Figure 4.8–Centrifugal fan performance (based on Thompson, 1975).

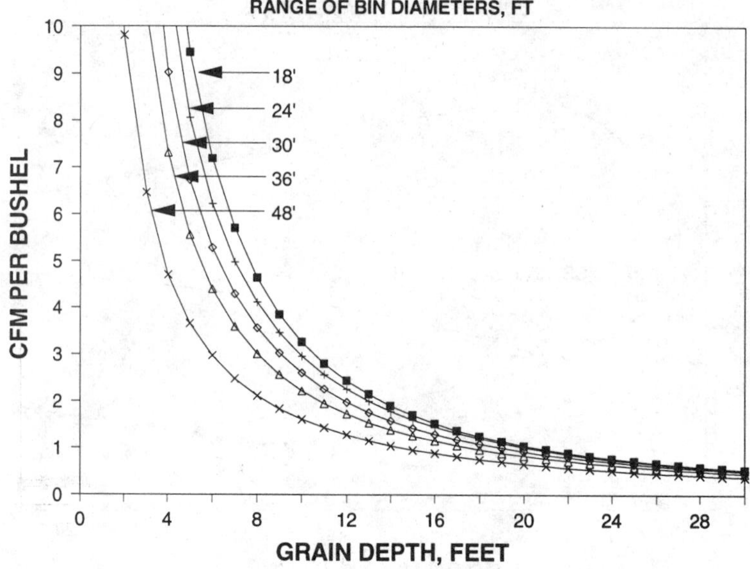

Figure 4.9–Influence of bin diameter on airflow (based on Thompson, 1975).

fans shown in figure 4.9, the airflow rate for the 36-ft bin is approximately 10 cfm/bu, while the same fan delivers about 1.7 cfm/bu in the 18-ft bin. Increases in grain depth disproportionately reduce the airflow rate for all bin diameters.

Increasing Fan Horsepower. In figure 4.10, airflow delivery rates for five different horsepower axial fans are shown using the same diameter bin. Clearly, airflow rate does not increase proportionately with fan horsepower. That is, for the fans used, airflow rate did not double for the same grain depth even when fan horsepower was quadrupled from 5 hp to 20 hp.

Axial Compared to Centrifugal. In figure 4.11, a centrifugal fan and axial fan of the same horsepower are compared for corn using the same diameter bin. In most on-farm grain drying situations, axial-flow fans will deliver a higher airflow rate than centrifugal fans. The reason is that grain depths rarely exceed 16 ft and almost never exceed 24 ft in on-farm storage facilities.

Grain Type. In figure 4.12, fan performance is given for different grain types (corn, milo, wheat, and soybeans). Clearly, grain type makes a difference of the grains shown, soybeans being the least resistant to airflow and wheat being greatest. Thus, it is important to consider if different types of grain are to be dried when selecting a drying fan.

Multiple Fans Per Bin in Parallel. In situations where a utility company limits the fan motor size that can be started on its system, the

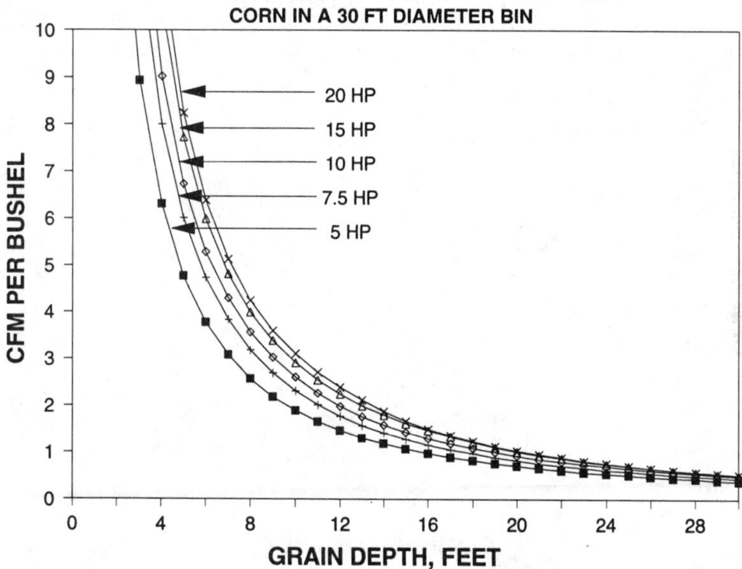

Figure 4.10–Airflow for different horsepower fans on the same bin (based on Thompson, 1975).

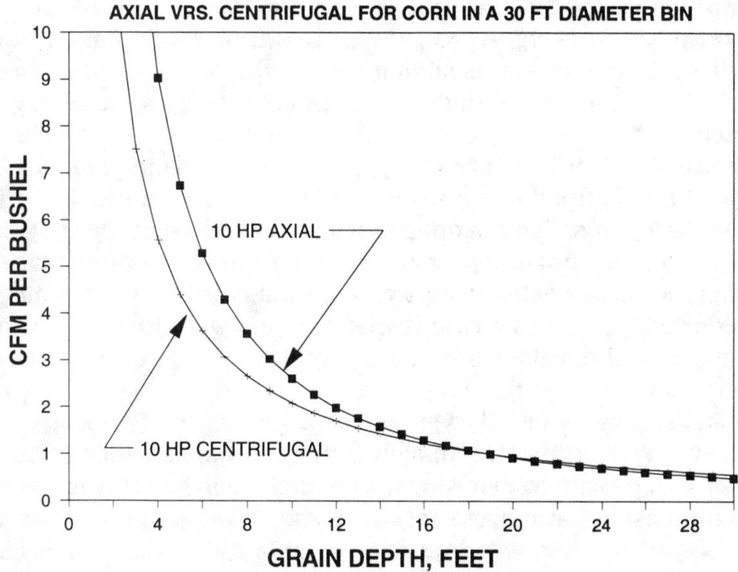

Figure 4.11–Airflow for axial and centrifugal fans that have the same horsepower (based on Thompson, 1975).

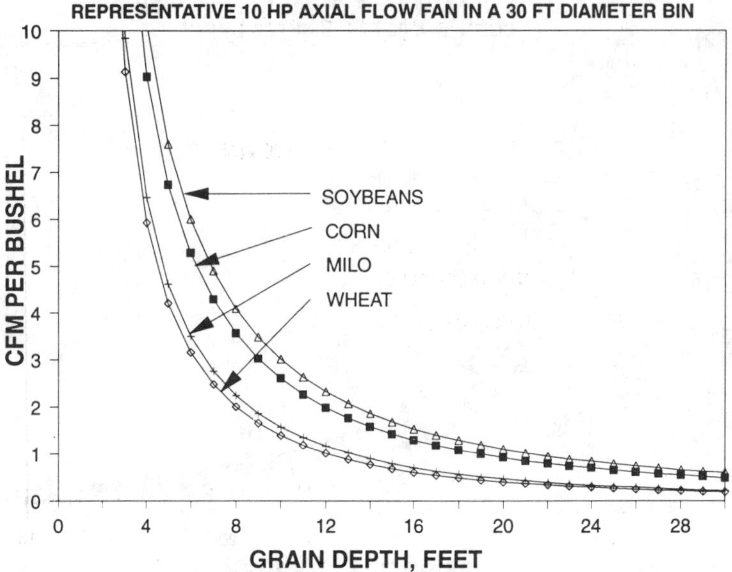

Figure 4.12–Airflow as influenced by type of grain (based on Thompson, 1975).

needed airflow can sometimes be obtained by installing two identical fans on the same drying bin with the individual fan motors falling within the power company's specified limit. This arrangement produces less starting current than one motor alone having the same total horsepower rating provided the two fan motors are not started at the same time. Controls must be designed to prevent this from happening. Likewise, devices such as louvers or sliding valves should be used to prevent reverse airflow through the second fan, thus causing problems in getting it started.

The expected performance curve for fans operating in parallel may be generated by multiplying the airflow delivery at each static pressure by the number of fans. For example, in figure 4.13 two fans were placed in "parallel" to dry corn. The performance curve was generated by doubling the airflow delivery for the same static pressure, assuming that the performance curves of the respective fans were identical. Again, a computer model developed by Thompson (1975) was used to compute fan performance. The two-fan arrangement did not provide as much additional airflow for grain drying as one might expect. Performance was enhanced only at the more shallow grain depths, and only about 20% more air was provided than when using one fan alone. This is the result of the increased static pressure head that develops across the grain depth as the airflow rate is increased when both fans are operating on the same system. The efficiency of energy utilization can also be expected to decrease, because two fans will probably be operating at a

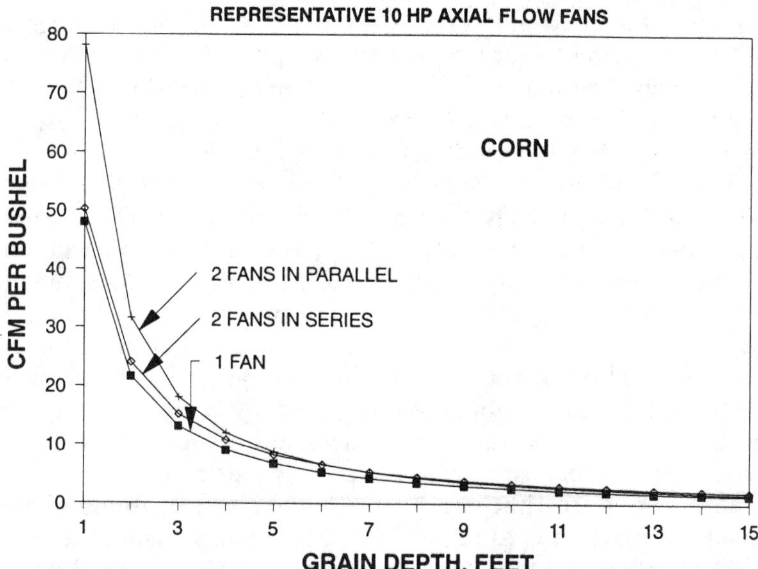

Figure 4.13–Airflow for fans in series and parallel (based on Thompson, 1975).

somewhat less efficient point on their respective performance curves. In addition, two smaller fans usually will cost more than one fan for the same total horsepower.

Multiple Fans Per Bin in Series. Fans may be placed in series (end-to-end) in situations where air is to be delivered against a relatively high static pressure. Performance for fans operating in series may be determined for similar fans by multiplying the static pressure readings for the same airflow as given in the fan performance data by the number of fans. The limited net effect of placing two similar fans in series is shown in figure 4.13. Series placement is not often used in grain drying because there is little additional airflow to be gained at typical on-farm grain depths.

Air Distribution Considerations

One goal of any drying or aeration system is to provide the same airflow rate (cfm/bu) to all parts of the grain being dried or conditioned. This can be achieved only if the geometry of the air distribution system is such that all airflow paths are of equal length, and each provides the same resistance to air movement. In bin drying and aeration systems, the best method of obtaining equal flow paths is to use a full perforated floor system. The fan supplies air to a plenum chamber located beneath the floor, and all flow paths will be equal provided the grain has been leveled to a constant depth. In portable batch and continuous-flow dryers, constant width drying columns are used to promote uniform airflow paths. Non-uniform depths cause more air to pass through the shallow depths rather than through the deeper portions of the grain mass. Uneven air distribution creates uneven drying or cooling.

Non-uniform air distribution may also be caused by uneven resistance to airflow in different regions of a drying or aeration bin. This can be caused by the accumulation of fines and other trash in the center (or other areas) of a bin as it is loaded. This region may provide such a high resistance to airflow that the majority of the drying air will move through the less dense material surrounding this region. Consequently, this zone may not dry properly and, in the case of in-bin dryers, can provide an excellent environment for mold and insect activity. This potential problem may be reduced by using a grain spreader that evenly distributes the fines throughout the grain mass. Even distribution of fine material increases the resistance to airflow in other areas of the bin, but it provides for more uniform drying and cooling patterns.

Another procedure that can help alleviate the problem of non-uniform air distribution is to remove some of the material in the center of the bin using a center unloading auger, and then spreading this material uniformly over the top surface of the grain. (If the material is too "trashy", it may have to be disposed of by other methods such as feeding to animals). Although this procedure produces a top layer of grain and fine material with a high resistance to airflow, it also provides for more uniform airflow over the bin cross-section. Of course, an even better procedure is to clean all the grain before it is put in the bin or dryer in order to prevent the problem from occurring.

Sometimes the location of an underfloor unloading auger or the supports for the perforated floor can create situations that result in uneven air distribution. The placement of such obstructions and/or the placement of the drying fan should be such that airflow is restricted as little as possible.

While most bin-type dryers utilize a perforated floor for air distribution, many aeration systems employ less expensive subfloor duct systems. These produce non-uniform flow paths (with resulting non-uniform cooling patterns); however, if properly designed, their performance generally is satisfactory. Subfloor systems may be of several types with the " X", " Y", and " I" patterns (discussed in detail in Chapter 5) typical of those employed in round storage bins. The air leaving the ducts will spread out in a semi-cylindrical pattern around the duct until it reaches the wall or a point half-way to an adjacent duct. At this point, the air will turn and move vertically. Airflow patterns in duct systems tend to create regions of low velocity between ducts and along the walls of the storage. Thus, these areas will require a longer time to cool when the grain is being aerated.

Problems

4.1. A bin is filled to a depth of 8 ft with 12.4% moisture content shelled corn. Assuming a packing factor of 1.3, what static pressure is required to provide drying air at a rate of 5 cfm/bu?

4.2. An airflow of 20,000 cfm is to be delivered to a drying bin at a static pressure of 1.5 psi. If the fan is 47% efficient, what is the horsepower required of the motor driving the fan?

4.3. From figure 4.6, what is the efficiency of the 7.5-hp axial fan when operating against a static pressure of 2 psi?

4.4. From figure 4.7, for corn in a 30-ft-diameter bin equipped with a 10-hp axial-flow fan and filled to a grain depth of 12 ft:
(a) What is the capacity of the bin in bushels?
(b) What is the static pressure?
(c) What is the efficiency of the fan?

4.5. From figure 4.7, for corn in a 30-ft-diameter bin equipped with a 10-hp axial-flow fan, what is the maximum grain depth that can be dried if an airflow of 10 cfm/bu is required?

4.6. From figure 4.8 for corn in a 30-ft-diameter bin equipped with a 10-hp centrifugal fan and filled to a grain depth of 12 ft:
(a) What is the static pressure?
(b) What is the efficiency of the fan?

4.7. A 24-ft-diameter bin is filled with corn. The bin is equipped with a 10-hp axial-flow fan. Using figure 4.9, if an airflow of 6 cfm/bu is required, what is the maximum depth of grain that can be dried in the bin?

4.8. A 30-ft-diameter bin is filled to a depth of 6 ft with corn. If an airflow of 6 cfm/bu is required, what horsepower motor is needed to drive the axial flow fan?

4.9. From figure 4.11, if an airflow of 8 cfm/bu is required:
(a) What is the maximum grain depth the 10-hp axial-flow fan can be used to aerate?
(b) What is the maximum depth the 10-hp centrifugal fan can be used to aerate?

4.10. From figure 4.12, for a grain depth of 10 ft, how many cfm/bu will be provided by the 10-hp axial-flow fan for:
(a) Soybeans?
(b) Wheat?
(c) Corn?
(d) Milo?

4.11. From figure 4.13, for a 5-ft grain depth, what quantity of airflow will be provided by:
(a) 1 fan?
(b) 2 fans in series?
(c) 2 fans in parallel?

References for all chapters begin on page 541.

5

On-Farm Storage

Once grain has been dried, it must be stored in an environmentally proper manner in order to avoid spoilage. The key to successfully storing grain is to develop an understanding of the principles of grain storage and to use these principles in aeration, inspection, and sampling (Ross et al., 1973).

Principles of Grain Storage

Grain can be stored several years, under proper conditions, with little or no detectable loss of quality. Under improper conditions, however, grain can begin to spoil within a few hours. What constitutes spoilage or loss of quality in stored grain is relative. Drying certain seed grain with heated air at temperatures above 80° F may reduce germination significantly. A maximum drying temperature of 100° F is recommended for corn and soybean seed. When using corn to produce grits or alcohol, drying with air temperatures that raise the grain temperature above 140° F has adverse effects. However, if the grain is to be fed to animals, there is little evidence to indicate that drying temperature affects quality significantly.

Grain spoilage is the result of microorganisms (bacteria, yeast, and fungi) using grain nutrients for growth and reproductive processes. Microorganisms also produce heat during growth that can increase the temperature of stored grain. The result can be "heat damage" that sometimes renders grain unfit for feed. Such conditions also have been known to cause fires and dust explosions in storage structures.

Under the proper environmental conditions, certain microorganisms can produce toxins or other products that can cause serious illness and even death when consumed by livestock or humans. However, not all microbial action in grain and feed materials is considered detrimental. Microorganisms are used in the manufacture of feeds such as ensiled whole plant corn, hay, and high moisture shelled corn. This action is called fermentation rather than spoilage. The process is the same with the difference being that the action is controlled and considered beneficial.

To store grain successfully, grain and the atmosphere in which it is stored must be maintained under conditions that discourage or prevent the growth of microorganisms that cause spoilage. The major influences on the growth and reproduction of microorganisms in grain include:

1. moisture
2. temperature
3. oxygen supply
4. pH
5. condition or soundness of the grain
6. storage time
7. initial infestation
8. amount of foreign material present

Moisture

Moisture content is the most important factor affecting the growth of microorganisms in stored grain. If moisture can be maintained at a sufficiently low level, the other factors affecting growth will not greatly influence spoilage of the grain. Moisture content can be expressed in several ways, but most often a wet basis percentage is used. Mathematically, the wet basis percentage is:

$$\begin{matrix}\text{moisture} \\ \text{content} \\ \%\end{matrix} = \frac{\text{weight of water in sample} * 100}{\text{total weight of sample}} \qquad (5.1)$$

Total weight of the sample includes both the weight of the dry matter and the weight of the water, hence the term "wet basis".

The moisture content of the grain and the relative humidity of the surrounding air both affect microbial growth and spoilage. Relative humidity is a term describing the relative amount of water contained in the air. A relative humidity of 100% indicates that the air contains all the water it can hold normally at that temperature, whereas a relative humidity of 0% indicates that there is no water in the air; that is, the air is completely dry.

If a sample of grain is placed in a closed jar, water will move both from the grain into the surrounding air and from the air into the grain. If the grain and air are maintained at a constant temperature, a condition will be established in which the rate of water movement from the grain will equal exactly the rate of water movement into the grain. The net amount of moisture in the grain and in the air will be constant, establishing an equilibrium condition. The grain moisture content in this condition is known as the equilibrium moisture content, and the relative humidity of the air is known as the equilibrium relative humidity. The relationship between these is shown for shelled corn in figure 5.1. At a given moisture content, a higher temperature will result in a higher

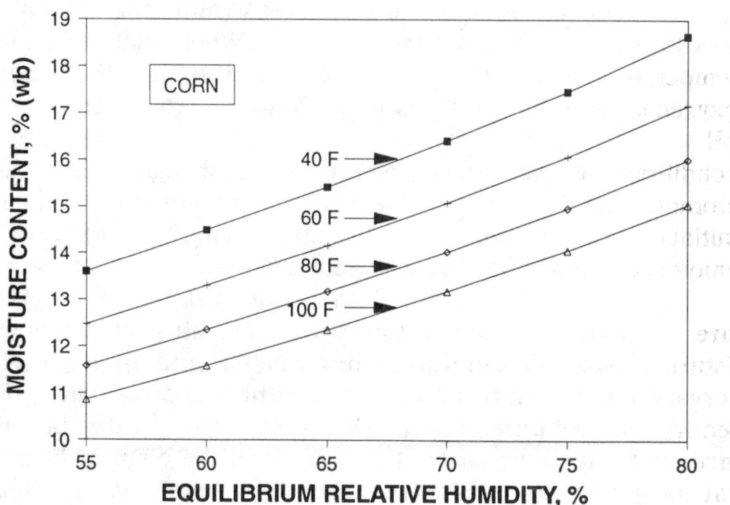

Figure 5.1–Equilibrium moisture contents for corn over a range of temperatures and relative humidities (Ross et al., 1973).

equilibrium relative humidity while a lower temperature will result in a lower equilibrium relative humidity.

For example, if a relatively wet sample of shelled corn with a moisture content of 25% is placed in an environment being maintained at 80° F and 65% relative humidity, it will dry to the moisture equilibrium corresponding to 65% relative humidity, about 13% moisture content. Likewise, if a dry sample is placed in this environment, it will absorb moisture until it reaches the 13% equilibrium moisture content. Grain is dried by forcing air with low relative humidities through the grain. The grain dries as it attempts to come into equilibrium with the low relative humidity air.

Conversely, if a large quantity of grain is placed in a tight storage structure, the air between the particles in the grain mass will come into moisture equilibrium with the grain. For example, if shelled corn at a moisture content of 13% and temperature of 80° F is placed in a bin where the air has a relative humidity of 40%, the grain will lose moisture, and the relative humidity of the air will increase to approximately 65% (fig. 5.1). The moisture content of the grain decreases slightly during the equalization period. However, in view of the normally large amount of grain and small amount of closed air, this loss of moisture is insignificant and will affect the grain's moisture content only slightly.

Microorganisms respond to their environment in somewhat the same way as grain. They absorb water in which the nutrients required for their growth and reproduction are dissolved (nutrients can enter the microorganism cell only in a dissolved state). When relative humidity is sufficiently high, microorganisms absorb moisture. If the relative humidity drops below a critical level, however, they cannot absorb water causing their growth and reproduction to cease. Figure 5.2 indicates the relationships between growth and relative humidity for fungi and bacteria. Note that at 62% relative humidity, growth is minimal for fungi. Bacteria generally require relative humidities of 90% or more for growth. Fungi have been recognized as a major cause of spoilage in grain. If the grain moisture content is maintained at such a level that the equilibrium relative humidity of the surrounding air is at or below 62%, microbial growth will be minimal or arrested, and spoilage will not occur. For shelled corn, this moisture content is approximately 13% for normal summer temperatures in the central United States but will be different in the northern or southern areas of the country because of differences in average temperature. Thus, it is generally recommended for the central United States that shelled corn be dried to 13% moisture content if it is to be stored through summer months. However, a higher or lower moisture content may be recommended in colder or warmer areas of the country, respectively.

Temperature affects the equilibrium relative humidity of grain. At higher temperatures, the equilibrium relative humidity increases for a

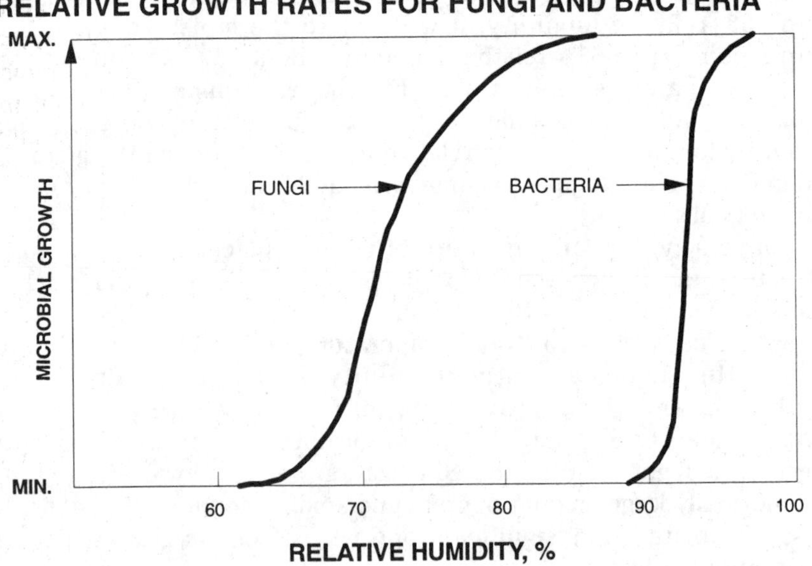

Figure 5.2–Growth of fungi and bacteria as influenced by relative humidity (Ross et al., 1973).

given grain moisture content (fig. 5.1). Grain that has a moisture content safe for storage at 75° F (for shelled corn, this corresponds to a moisture content of 13% and equilibrium relative humidity of approximately 65%) may not be safe for storage at 95° F because of the increased equilibrium relative humidity. This is the reason (along with other factors to be discussed later) that in warmer climates, the recommended long-term safe storage moisture content for shelled corn is lower than 13% while in colder climates it is higher.

Average Moisture Contents. It is not sufficient to have a "safe" average moisture content if that moisture content is determined by averaging a very wet lot of grain with a very dry lot. If the wet lot is placed in the bin in one location (not mixed with the dry grain), microbial growth can take place in the wet lot and cause problems in the entire bin. Also, if moisture migrates to a certain area within the storage bin, that area may become wet enough to support microbial growth that will result in spoilage. When thoroughly mixed, wet and dry grain will equilibrate to a moisture content at some point between the original moisture contents.

Research has shown that well-mixed white corn lots of 8, 20, and 25% moisture content will equilibrate to a point where the final moisture content of the high-moisture portion is 1.4 to 2.6% above that of the low-moisture fraction (White et al., 1972). The higher the temperature, the smaller the difference. Also, the lower the temperature, the slower the rate of equalization. Again, if the wettest kernels of the mixed lots are not below a safe storage moisture content, they are subject to microbial growth.

Moisture Migration. Moisture often accumulates in the top layers of stored grain even though initially the grain is stored at a safe moisture content in weather-tight bins. The accumulation is a result of moisture migration caused by temperature differences within the grain mass. Grain that is harvested and placed in storage during the warm months of late summer or early fall slowly loses heat as the weather gets colder. Grain near the surface and next to the walls cools first, while grain in the center of the bin remains warm. This temperature difference creates slowly moving air currents as illustrated in figure 5.3. Cool air near the walls moves downward, forcing warm air upward through the center of the grain mass. When the warm air reaches the cold grain near the top surface, condensation may occur in the same way moisture condenses on the exterior of a glass of ice water. Although the moisture migrates slowly, it continues as long as temperature differences exist in the grain. If allowed to continue for months or even a few weeks, the accumulated moisture may promote insect activity, microbial growth, and spoilage in the upper layers of the stored grain, particularly in large bins. The process reverses itself when the weather changes from winter to

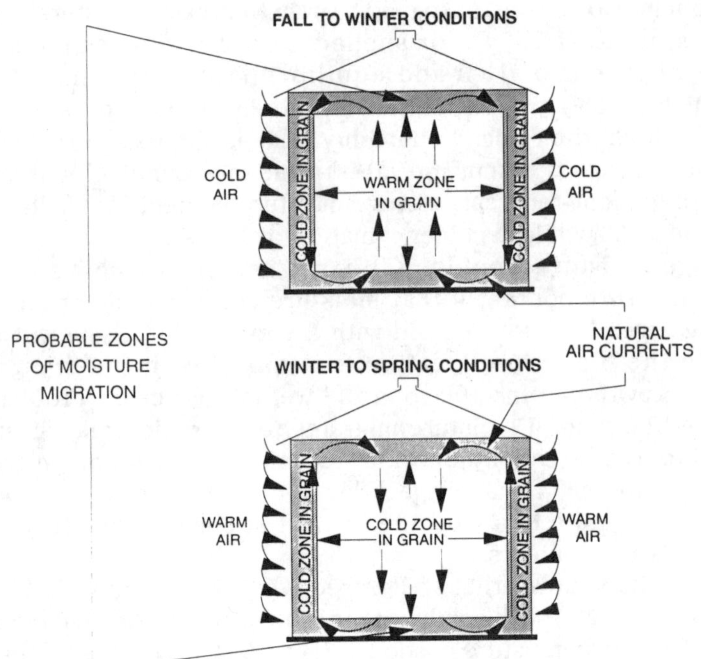

Figure 5.3–Natural air currents in a bin for summer and winter conditions.

springtime conditions with potential moisture condensation occurring near the bottom center of the grain mass.

Moisture migration can be controlled by equalizing the temperature throughout the grain. An effective method of doing this is to move small quantities (1/10 cfm/bu) of air through the grain more or less continuously, or to move larger quantities of air (greater than 0.5 cfm/bu or as provided by a drying fan) through the grain periodically. Ventilating grain to control moisture migration is referred to as "grain aeration" or "grain cooling". Proper operation of aeration equipment can keep dry-stored grain close to the average air temperature throughout the fall and winter. Aeration is a practical way to help maintain stored grain quality by providing better storage conditions and is discussed in detail later in this chapter.

Temperature

The effect of temperature on the growth of fungi is illustrated in figure 5.4 for two general groups of microorganisms — the thermophiles whose growth is optimum at higher temperatures and the mesophiles whose growth is optimum at normal atmospheric temperatures. The growth curve for a third group, the psychrophiles, whose growth is

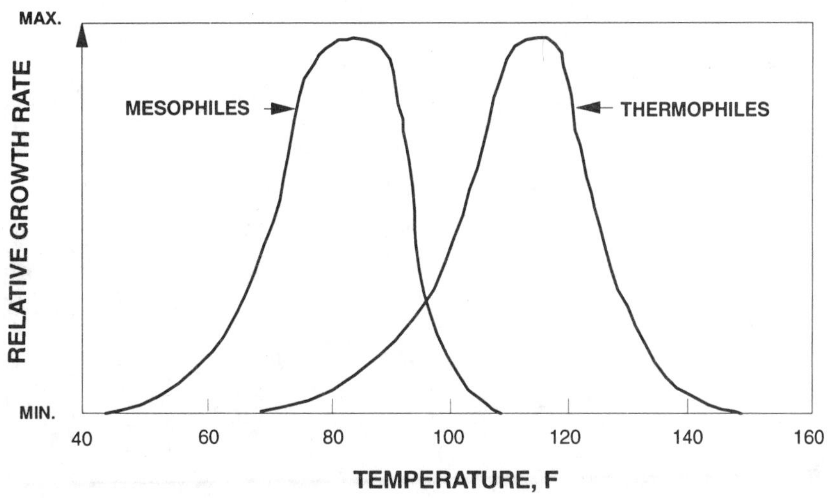

Figure 5.4–Growth of fungi as influenced by temperature (Ross et al., 1973).

optimum at low temperatures, is not shown. Psychrophiles are not a major factor in grain spoilage.

Common storage fungi grow most rapidly at temperatures of 85 to 90° F. Below these temperatures, fungi growth rates decrease and reach a minimum at 35 to 40° F. This is why stored grain should be cooled by aeration to about 40° F whenever possible. Cooling below 40° F is not necessary because storage fungi activity is already at a minimum. Cooling to below 32° F may freeze the grain. If the grain is frozen, the fan(s) must run longer to warm the grain and condensation may occur.

Equilibrium moisture content (EMC) varies with temperature. With corn for example, the following relationships exist for a constant equilibrium relative humidity of 65%:

1. At 40° F, the EMC is 15.4%.
2. At 60° F, the EMC is 14.0%.
3. At 80° F, the EMC is 13.1%.
4. At 100° F, the EMC is 12.2%.

The relative humidity of 65% was chosen for purposes of example because it is near the moisture level at which microbial growth is minimal and spoilage is essentially stopped (fig. 5.2). EMC data for corn can be determined from figure 5.5 for a broad range of temperature/ relative humidity combinations.

Safe storage moisture content recommendations may vary with geographical location because of differences in seasonal temperatures.

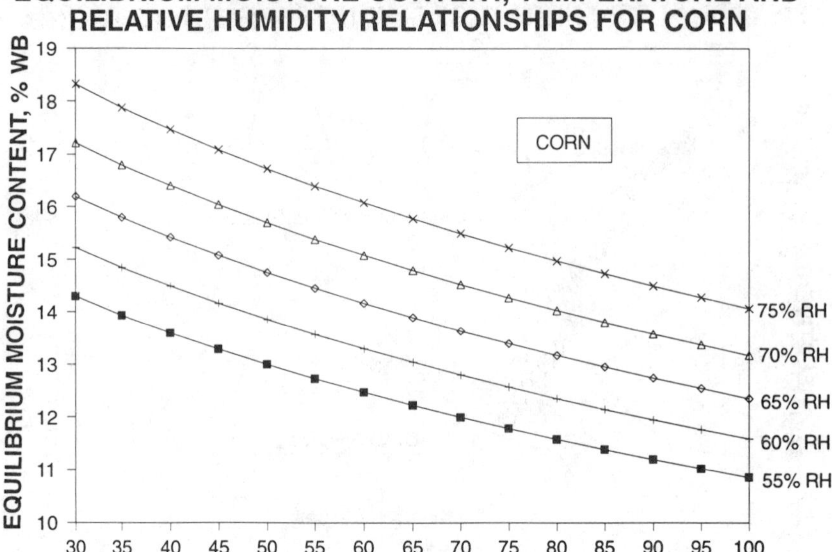

Figure 5.5–Equilibrium moisture content, temperature, and relative humidity relationships for corn (Ross et al., 1973).

Consider the mean temperatures during the fall, winter, and early spring for three U.S. locations displaced by approximately 250 miles of latitude as shown in the graph of figure 5.6. The locations are Birmingham, Alabama, Owensboro, Kentucky, and Chicago, Illinois. The approximate harvest periods for corn at these three locations are indicated on the graph. Figure 5.7 shows how the equilibrium moisture content varies with air and grain temperature for four different levels of equilibrium relative humidity. Also shown are lines indicating the mean temperature during the harvest period for the three locations. Two lines are shown for Chicago, one for the average mid-harvest period and another for an average late-harvest period.

Descriptive adjectives have been used to indicate the degree of safety associated with storing shelled corn at various equilibrium relative humidities. The term "safe" is used for the 65% equilibrium relative humidity line since, as stated earlier, microorganisms have minimal growth at or near this relative humidity. The term "more safe" and "very safe" are used for the 60 and 55% relative humidity lines, respectively, since less microbial growth would be expected for grain stored at these relative humidities. The terms "some risk" and "more risk" are used for the 70 and 75% relative humidity lines, respectively, to indicate that as the equilibrium relative humidity increases, there is some additional risk involved in storage. No statistical or quantitative significance can be

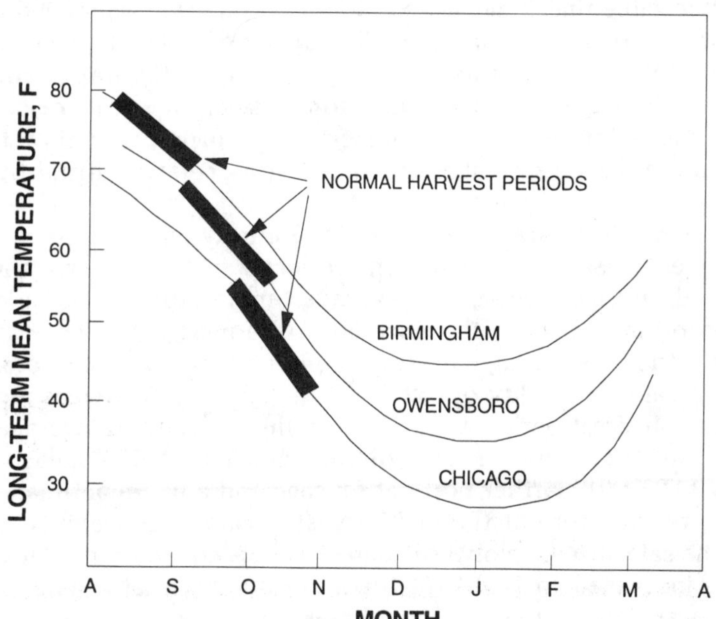

Figure 5.6–Harvest time data for example problem (Ross et al., 1973).

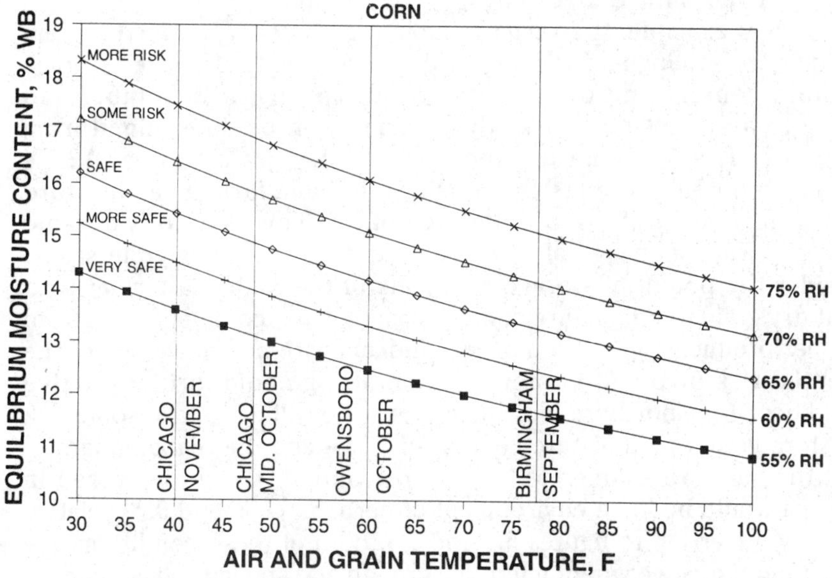

Figure 5.7–EMC data and storage risk - example problem (Ross et al., 1973).

attached to these terms; they are intended only to describe trends. It is entirely conceivable that grain stored in an environment maintained at either 55 or 60% relative humidity could support the development of toxin-producing microorganisms. The chances of this happening increase as the relative humidity of the storage environment increases. Again, nothing is intended by the descriptive words other than to indicate increased risk of storage spoiling as relative humidity increases.

The question of why grain producers in Kentucky might store corn safely at higher moisture contents than producers in Alabama will now be addressed. In Birmingham, the average temperature at the mid-harvest period is indicated in figure 5.6 to be near 77° F. For this temperature, the safe storage moisture content corresponding to an equilibrium relative humidity of 65% is 13.5% (fig. 5.7). In Owensboro during the mid-harvest period a month later, the temperature averages near 60° F and the safe storage moisture content at 65% relative humidity is 14.2%. Still farther north at Chicago and a half month later, the average temperature at the mid-harvest period is approximately 48° F, and the safe storage moisture content at 65% equilibrium relative humidity is increased to 14.8%. At the more northern latitudes, the harvest dates are later in the fall. The decrease in temperature with later harvest dates plus the average decline in temperature with the more northern latitudes combine to give lower relative temperatures at harvest time. As indicated in figure 5.7, the decrease in temperature allows grain to be stored safely at higher moisture contents in the more northern areas of the United States.

As a final example, the range in moisture contents between a grain producer in Birmingham wishing to dry corn to a "more safe" storage moisture content and one in Chicago harvesting in November and willing to assume "more risk" than normal can be determined from figure 5.7. The Alabama producer would dry grain to near 12.5% moisture content whereas the northern Indiana farmer might store shelled corn at 17.5% — a difference of 5%! This is a considerable difference when a change of 0.2% can be critical to storing grain safely. Consider the potential storage problems of the Alabama farmer who might dry grain to a moisture content based on a recommendation from a farmer producing grain in northern Indiana with no knowledge of the potential risk involved in storing grain at high equilibrium relative humidities. If the northern Indiana farmer were willing to take some risk and store grain at a moisture content of 17.5% and the Alabama farmer used the same recommended moisture content, the grain stored in Alabama would be in an environment of perhaps 77° F and 95% relative humidity. Referring to figures 5.2 and 5.4, both of these conditions are well within the range where microbial growth and spoilage will occur.

The foregoing examples were based on monthly temperatures estimated from long-term averages. In practice, the actual seasonal temperatures during the harvest and storage period are the ones of real importance and the ones on which the producer bases management decisions. If the season is cold, it may not be necessary to dry to as low a moisture level as when the season is warm. However, it is impossible to predict the weather accurately. Also, the examples given were based on the temperature at the time of harvest. Temperatures will generally decrease after harvest so that storage conditions would be safer for higher moisture grain from this standpoint. However, grain not used by the time the temperature begins to increase in the spring will need to be dried for safe storage during the warmer months.

Oxygen Supply

One or more groups of microorganisms will grow and reproduce at any oxygen concentration (fig. 5.8). Microorganisms are classified as anaerobic, aerobic, facultatively anaerobic, or microaerophilic, according to their oxygen requirement. Anaerobes grow without free oxygen while aerobes require free oxygen. Facultatively anaerobic microorganisms grow either in the presence or absence of free oxygen, and microaerophilic microorganisms grow in the presence of minute quantities of free oxygen. Bacteria are represented in all of these categories. Yeasts are moderate aerobes, and molds are strong aerobes.

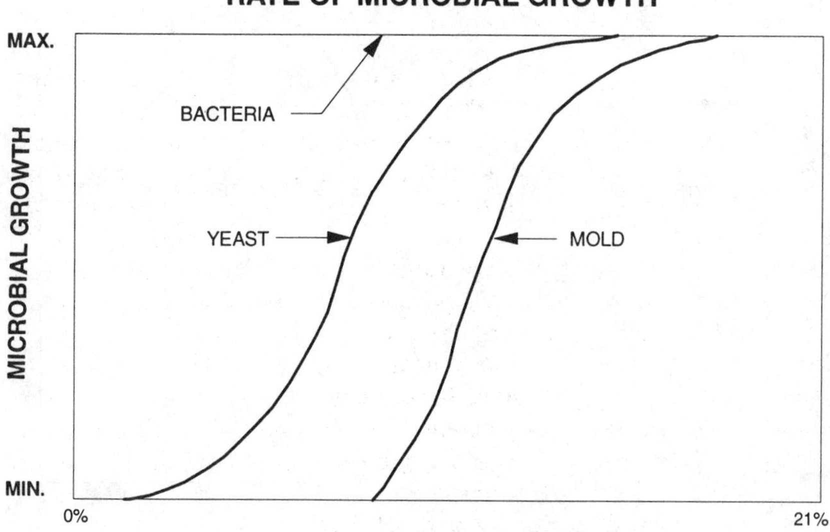

Figure 5.8–Yeast, mold, and bacteria growth as influenced by oxygen (Ross et al., 1973).

Managing oxygen concentrations in storage should help control some bacteria and most yeast and molds. Attempts have been made to control oxygen concentration in large grain storage where grain has been dried to near safe moisture contents. This is not a widespread practice since it is difficult to maintain the oxygen concentration at a low enough level to be effective.

Acidity

Acidity is measured by pH which represents the hydrogen ion concentration in a solution. Pure water pH values range from 1 to 14 with neutral near 7.0. When acids are placed in solutions, they will increase the hydrogen ion concentration and the acidity of the solution, thus lowering the pH value.

The effect of pH on microbial growth for fungi and bacteria is illustrated in figure 5.9. The pH is lowered and the oxygen concentration controlled in sealed storage for ensiling grain.

The pH of grain may be lowered by the addition of propionic, acetic or formic acids directly to the grain while filling the bin. However, this procedure results in increased deterioration of bin walls and grain handling equipment, and prevents the marketing of grain through normal channels.

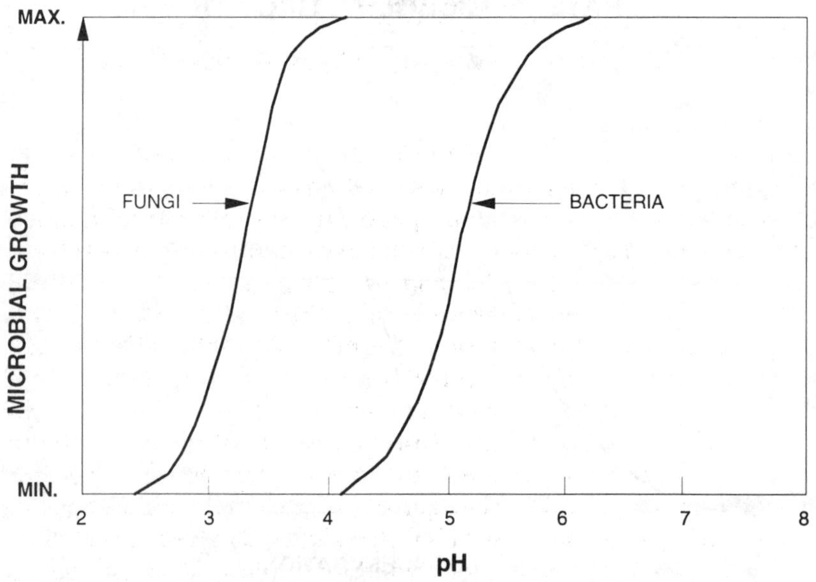

Figure 5.9–Fungi and bacteria growth as influenced by pH (Ross et al., 1973).

Grain Condition

Grain condition refers to its quality. An increase in the amount of cracked, damaged kernels in stored grain can increase spoilage and thereby reduce grain quality. This is one of the reasons that it is impossible to establish an absolute maximum safe storage moisture for grain. Although the reasons why cracked or damaged kernels affect spoilage are not completely known, their presence certainly increases the ability of microorganisms to invade the grain kernel. Grain that is in poor condition must be dried to a lower moisture content than grain in good condition if it is to be stored safely over long periods. The relationship between condition and safe-storage moisture content has not yet been quantified.

Storage Time

The length of the storage period influences the amount of spoilage in grain. Microbial growth and reproduction can occur even under conditions considered safe for storage, although the growth would be difficult to detect. Over long storage periods, this effect accumulates and can be detected eventually. In general, the longer the storage period, the lower the moisture content should be to ensure safe storage.

Initial Infestation

The degree of grain contamination by microorganisms is also important. Suppose high-moisture grain in poor condition is mixed with sound grain at a low-moisture content in order to bring the mixed lot to an average safe-storage moisture. The chance for spoilage in the mixed lot is greatly enhanced because the mixed lot would have a heavy infestation of microorganisms provided by the partially spoiled grain.

High-Moisture Storage

High-moisture grain (above 23%) can be stored under conditions where oxygen is excluded from the stored mass. If high-moisture grain is placed in a hermetically sealed space, the aerobic microorganisms present in the grain will use the existing oxygen. However, anaerobic (or facultatively anaerobic) bacteria grow under oxygen-free conditions (fig. 5.8). These microorganisms produce chemical compounds, as by-products of their growth, that lower the pH of the grain. When the pH is lowered to about 4 to 4.5, bacterial growth will be minimal — the bacteria are essentially stopped by their own wastes (fig. 5.9). Since fungi do not grow under oxygen-free (anaerobic) conditions, all microbial growth will be minimal, and the grain will not heat. Thus, grain can be stored successfully. Management of high-moisture systems can be very critical, and careful consideration should be given to their use.

Organic acids (propionic, acetic and formic) may be used to prevent microbial growth in high-moisture grain so that the grain can be stored under normal atmospheric conditions (i.e., the storage is not hermetically sealed). When sprayed on the grain, these acids lower the pH to approximately 4, thus reducing microbial growth (fig. 5.9). In addition, propionic acid appears to have fungicidal properties. Although the concept of using this preservation method dates back to 1918, it has only been in the last few years that it has found commercial use.

Summary

There are several mechanisms that may be used in storing grain successfully. The most common technique is to control grain moisture content. The level to which grain must be stored for long-term storage is determined primarily by temperature. Because temperature is a function in large part of latitude, recommendations vary with geographic location.

Aeration

A farmer may have the returns from much of a year's work stored in the form of grain for future feeding or sale. The grain must be aerated if the risk of storage losses is to be minimized, and a regular and accurate method of inspection and sampling must be followed. Potential problems exist when: (1) damaged and/or high moisture grain is stored,(2) the aeration system is inadequate, or (3) the grain bin is incorrectly filled or unloaded.

Many storage problems result from inadequate aeration. Maintaining grain quality, especially over long storage periods, is dependent on a functional aeration system coupled with an adequate management strategy. Adequate management is based on an understanding of heat and mass transfer within a grain mass.

Natural Air Movement

Grain is a good insulator. Heat loss from grain is relatively low in comparison to most materials. For this reason, when grain is placed in a bin in the fall, the portion of the grain near the center tends to remain near the temperature at which it came from the dryer or field. The grain near the bin wall tends to cool to near the average outside temperature. As the weather grows colder, the difference in temperature between the grain in the center of the bin and that near the bin wall produces natural air currents inside the grain. The air near the bin wall becomes cool and more dense and tends to fall, forcing the warmer air up through the center of the grain mass (fig. 5.3). As the relatively cooler air passes through the grain mass, it warms and absorbs moisture. As it nears the top center surface of the grain, the air cools to a point at which it can no longer hold the moisture it has absorbed. The excess moisture condenses on the surface of the grain, increasing the moisture content of the grain and creating an environment that enhances mold or insect growth. This can occur even though the average grain moisture content is at or below recommended levels. The reverse situation occurs during the summer months (fig. 5.3). In this case, the moisture condenses near the bottom center of the grain mass.

Management Guidelines

Generally, the problem of natural air currents developing within a bin may be minimized by keeping the temperature of grain in the center of the bin within 10° F of the average grain temperature near the bin wall. This may be accomplished in most farm structures by using aeration fans that pull air down through the grain at airflow rates of at least 0.1 cfm (cubic ft/min) for each bushel of grain in the bin until the temperature of the grain mass is within 10° F of the average monthly temperature. (A representative system is shown in figure 5.10.) A slightly lower airflow rate may be used in very large farm or commercial structures, and a higher rate of 0.2 cfm/bu is suggested for flat storage structures. However, it is not necessary to cool the grain mass below 35 to 40° F because the activity of important storage fungi is very low below these temperatures. In addition, colder grain requires a longer warming time. Condensation may also occur when air warmer than 32° F comes in contact with frozen grain. Condensation may be adsorbed by

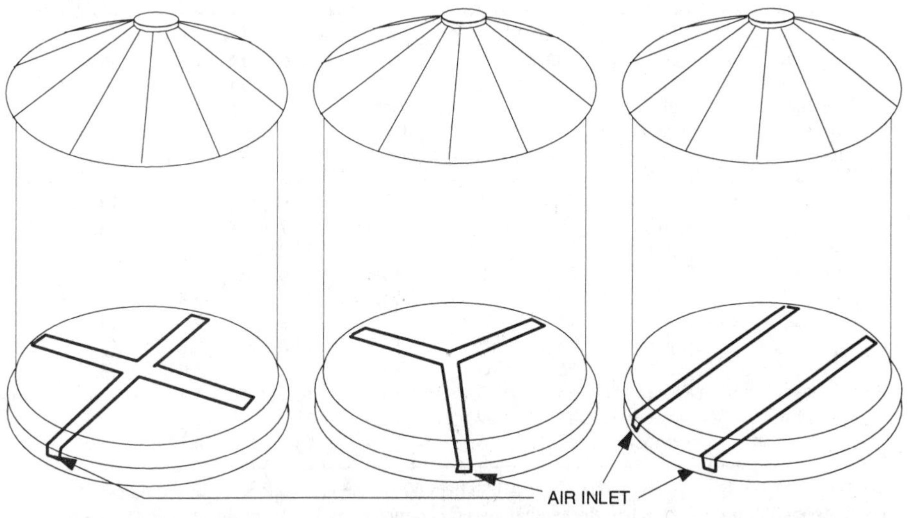

Figure 5.10–Representative aeration duct arrangements (Loewer et al., 1979).

the grain, thereby raising its moisture content and making it more susceptible to spoilage.

The aeration system should not be used to raise the grain temperature above 60° F because mold and insect growth occur at a much faster rate above this temperature. Unfortunately, these recommendations may be in conflict during the times of the year when temperatures are at the extremes. An example of the types of compromises that must be made are shown in table 5.1 for one particular location. Each geographic area would be somewhat different depending on climate.

Positive and Negative Pressure Systems

Aeration fans may be used as either positive or negative pressure systems. If air is "pushed" through the grain, the system produces a positive pressure in the grain mass. Likewise, if the air is "pulled" through the grain, the system produces a negative pressure. Either system is acceptable and has its advantages.

A consideration when choosing between these alternatives is that positive pressure systems add heat to the air that enters the grain mass because of the inefficiency of the fan. The additional heat results in a temperature increase that is usually less than 2° F. A positive pressure system should be selected if adding heat is considered advantageous, as is the case with most in-bin drying systems. A negative pressure system

Table 5.1. Aeration recommendations for Owensboro, Kentucky

Average Monthly Temperature °F	Month of Storage	Grain Temperature °F*		Maximum Allowable Recommended Storage Moisture Content			
		Min	Max	Corn	Soybeans	Wheat	Sorghum
34.9	January	35	45	15.7	14.3	14.2	14.3
37.9	February	35	45	15.5	14.0	14.1	14.2
46.0	March	41	51	14.9	13.7	13.7	14.1
57.9	April	53	63	14.3	12.2	13.3	13.8
66.8	May	55	65	13.8	11.6	13.1	13.7
75.3	June	55	65	13.3	11.1	12.8	13.5
78.2	July	55	65	13.2	10.9	12.7	13.4
76.8	August	55	65	13.3	11.0	12.8	13.5
70.2	September	55	65	13.6	11.4	13.0	13.6
59.5	October	55	65	14.2	12.1	13.3	13.8
46.6	November	42	52	14.9	13.2	13.7	14.1
37.3	December	35	45	15.5	14.0	14.1	14.2

* When the average monthly temperature is less than 40° F, the grain temperature should range between 35° and 45° F. When the average monthly temperature is greater than 60° F, the grain temperature should range between 55° and 65° F. For average monthly temperatures between 40° and 60° F, the grain temperature should be in the range of ± 5° F of the monthly average temperature.

is preferable if dust control at the surface of the grain is considered more important.

A negative system operation prevents condensation on the roof as the grain is cooled and allows use of solar heat in the roof when warming the grain. One disadvantage of this method is that the bottom layers of grain are the last to cool, and it may be difficult to know when cooling is complete.

Positive aeration systems allow the addition of grain layers without rewarming or cooling the grain that is already there. It is also easier to determine when the grain is in proper condition because the top layer is the last to cool. There is some evidence that positive systems provide better air distribution than negative systems. Based on the maximum air velocity recommendations (2,500 ft/min vs. 1,500 ft/min, table 5.2) the positive systems will generally require smaller duct diameters.

Airflow Rates and Direction

Temperature and moisture equilibrium conditions of the grain and air determine the potential for cooling or warming. Given these conditions, the rate at which a warming or cooling front moves through the grain is directly proportional to airflow.

Table 5.2. Allowable velocities and surface areas for ducts placed in flat storage and upright silos

Storage Type	Duct Length (if applicable)	Recommended Maximum Velocities within Duct	Recommended Maximum Surface Air Velocities
Flat Storage Systems	Up to 25 ft of duct length	1,500 to 2,000 fpm * †	20 fpm* 25 to 30 fpm†
Based on Duct Length	25 ft of duct length and above	1,000 to 1,500 fpm * †	20 fpm* 25 to 30 fpm †
Flat Storage Systems	negative pressure systems	1,500 fpm‡	25 fpm‡
Based on Type of Pressure†	positive pressure systems	2,500 fpm‡	25 fpm‡
Upright Storage (Silos)*			25 to 30 fpm * †

* Noyes, 1967.
† Frus, 1967.
‡ Bridges et al., 1988.

An airflow rate of 0.1 cfm/bu is often considered as satisfactory for aeration, especially for commercial facilities (Kline and Converse, 1961; Holman, 1960). However, small fans that are installed on farm-size bins usually deliver considerably more than this rate (fig. 5.11). A typical cooling time for winter conditions is 160 h of fan operation using 0.1 cfm/bu. This time would be reduced to 32 h if the airflow rate were increased to 0.5 cfm/bu (Thompson, 1975a).

Typically, crop drying fans are able to push a drying front through grain in a circular bin in a few hours. This provides a tremendous management advantage in that fan operation can be restricted to times when environmental conditions are satisfactory. When airflow rates are low, the time required to move a cooling or warming front may be several times greater than expected because the environment is changing constantly both during the day and over the season. Higher aeration rates provide the manager greater reserve capacity to preserve grain that is going out of condition (Converse, 1973). This excess capacity can be detrimental if used indiscriminately in that excess moisture may be added or removed if the system is not monitored and managed correctly. The manager must always be aware that a safe storage environment must be present everywhere in the bin rather than just on average conditions for the bin as a whole. Slow aeration rates tend to create a series of fronts within the mass, thus making sampling

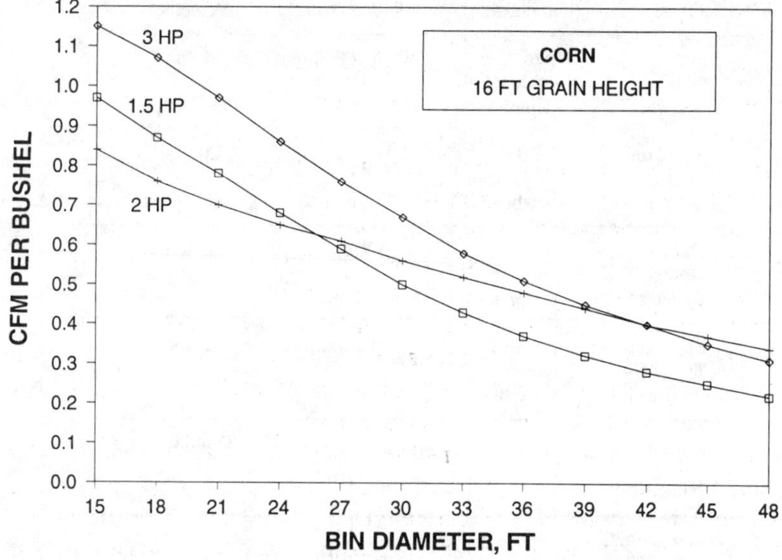

Figure 5.11–Air flow capacities for selected aeration fans (based on Thompson, 1975b).

and inspection an even more important part of grain storage management.

Generally, the direction of airflow during aeration is downward (although choice of direction is not considered critical in most farm systems). Downward movement is preferred because the center of the grain mass is usually warmer than the outside air during fall and winter conditions when most grain is placed in storage. The bin roof and the grain surface exposed to outside air are therefore likely to be cooler than the grain at the bottom center of the bin. Under these conditions and when the airflow rates are relatively slow (as with 0.1 cfm/bu), there is more likelihood for condensation to occur at the top of the grain mass than at the center (fig. 5.3). The downward movement of air through the center of the bin tends to reverse the naturally occurring air currents during the fall and winter. The reverse situation occurs during the spring and summer. However, the usual practice is to have sold or fed the grain before the summer season. Regardless, air direction makes little difference when large quantities of air are passed through the grain (as when using drying fans) because the front moves rapidly in comparison.

Grain Cooling and Heating

A difference in airflow only alters the rate at which the cooling or warming front moves through the grain so long as the grain is in moisture equilibrium with the air. If this is not the case (and it rarely is), cooling or warming time may be altered depending on the extent to which drying and rewetting occur. The influence of air and grain conditions on warming and cooling is shown in figures 2.16 to2.23 and is discussed in detail in Chapter 2.

When air passes through grain, heat is exchanged until both are at the same temperature given sufficient time. It is interesting to note that cooling and warming times are not influenced significantly by the temperature difference between the grain and the air. This is because the temperature gain (or loss) of the air is equal to the temperature loss (or gain) of the grain. Heat may be exchanged in either the sensible or latent form. Sensible heat exchange involves transfer of heat because the temperatures of the grain and air are different. Latent heat transfer occurs when there is an exchange of water between the grain and air.

For purposes of illustration, assume that summer conditions refer to when the average temperature of the air is greater than the temperature of the grain. Similarly, winter conditions refer to when the average temperature of the grain is greater than the temperature of the air. Thus, summer conditions are comparable to those in Zones 1, 2, 3, and 7, while winter conditions are, on average, reflective of Zones 4, 5, and 6 [fig. 2.16 (Chapter 2)]. However, aeration differs from drying in that the incoming air varies directly with ambient conditions in both temperature and relative humidity. Thus, any of the zone conditions discussed in Chapter 2 can and probably will occur sometime over the storage period somewhere in the bin. As airflow rates become larger, the probability increases that the grain will be maintained at somewhat uniform conditions. Overall, under summer conditions (as defined above), there will be, on average, a warming of the grain. Similarly, winter conditions result in a long-term cooling of the grain. However, the extent and rate at which grain cools, warms, dries, and rewets varies with moisture exchange and the many fronts that may exist simultaneously within a grain mass. Fortunately, there is usually little moisture that is added or removed through aeration.

Predicting the time to cool or warm grain becomes difficult as the grain and air temperatures move toward equilibrium with each other and as ambient air conditions change. These factors have been incorporated into a computer model (Thompson, 1975a), and time estimates are shown in figures 3.32 to 3.33 (Chapter 3).

Automatic Controls

Aeration fans may be controlled automatically using thermostats and/or humidistats to operate only during times when acceptable temperature and relative humidity conditions exist. Unfortunately, insofar as the control of aeration systems is concerned, temperature and relative humidity tend to change in opposite directions. That is, an increase in ambient air temperature is usually accompanied by a decrease in relative humidity and vice versa. For example, suppose under winter conditions that corn is to be maintained at 13% moisture, and no additional drying is desired. A humidistat setting of 55% relative humidity represents the equilibrium moisture content of corn to be stored at 13% moisture and 50° F. An upper thermostat setting of 60° F would limit fan operation to prevent overdrying if temperatures exceed this level. Thus, the aeration fan would not be operated when either the relative humidity is above 55% (in order to prevent rewetting) or when the temperature of the air is at least 10° F warmer than the grain. What may happen is that high relative humidities may limit fan operation in the morning while the higher air temperature may restrict operation in the evening, thus severely limiting aeration, especially with low airflow rates.

Automatic Aeration Controllers

There are several companies that provide computerized aeration controllers that decide automatically when the aeration fans should be turned on and off. The advantages of using these units over manual controls are generally threefold. The first advantage is energy savings. There is a tendency by grain producers to let the fans run too long when controlling manually. With the automatic controllers the fans can be shut off immediately when the desired grain temperature is reached. A second advantage is the ability to minimize the moisture content differential within the bin. More precise control of the fan run-time prevents overdrying of the bottom layers of grain, reduces the moisture spread within the bin, and provides a better quality product. The third advantage is the reduced worry and stress on the grain system manager. Automatic control of the fan units is a more reliable mechanism for maintaining the condition of the stored product and alleviates some of the decision making process necessary with manual controls.

Generally the controller units contain a microprocessor unit that allows them to be programmed to monitor the grain condition within the bin. The controller unit runs the fans automatically using temperature and humidity values for the specific crop and time of year. These control values may be entered and/or modified by the producer. The basic controller units generally are capable of monitoring two to four bins at a time with expansion units available for control of up to 12 bins. These units are capable of aerating bins with different crops, and most have a control feature for running the fans manually.

An additional feature of the controller units is their capability to monitor grain temperatures within each bin. Usually, these units can interface with existing temperature probes to determine the occurrence of any "hot" spots within a bin. They can provide a daily record of several statistics including the average grain temperature in each bin, the average ambient air temperature and relative humidity, the highest temperature reached at each bin sensor, and the total hours of fan operation. Some units come with battery backup which is a desirable option in case of a power failure. Other desirable features include the control of the number of fans running at a single time and lock-out capability to prevent all fans from starting at the same time.

These units are designed to supplement the conventional drying systems rather than to replace them. The key in using these devices is a good understanding of the proper aeration management practices and guidelines. The actual payback for these automatic controllers is difficult to define, but preventing the loss of a single bin of grain in a five-year period would make them well worth the investment. These units are a welcome addition for any size operation, but in general the larger the operation, the more economical the controllers become.

Duct Types Used in Aeration Systems

Two types of above-floor ducts, metal and plastic, generally are available. The metal ducts will be found in both round and half-round diameters ranging from 6 to 36 in. The number of diameters for the plastic ducts is smaller ranging from 8 to 24 in.

There are several trade-offs that must be considered in selection. First and foremost is the amount of open area. Generally, the metal ducts have perforations equal to about 10% of their duct surface area. In contrast, plastic drainage pipe normally will have a maximum perforated area in the range of 3 to 4%. Caution should be taken in using ducts with less than this amount of open area because the effective cubic feet per minute delivered to the grain under normal design conditions may be reduced drastically.

Other considerations in duct selection include structural and cost factors. Structurally, the metal is stronger, but the plastic is lighter and easier to install and remove. Initially, the plastic is cheaper, but depending on how long the facility is to be used, the metal ducts might be a better investment. Some plastic pipe is sold with a fabric "sock" to keep the kernels away from the perforations. When using this type of duct, a positive pressure system should be used in order to prevent fines and foreign material from clogging the duct.

Aeration Systems for Round Storage Bins

Most modern grain storage bins are circular and equipped with either subfloor aeration ducts or perforated floors. Certainly, above-floor ducts may be used but tend to hamper the unloading and clean-out operation when emptying the bin. Examples of typical aeration duct arrangements are given in figure 5.10. The choice of arrangement is based on air distribution and how the unloading auger is to be incorporated into the storage bin. The unloading auger outlet should be sealed when it is not in use in order to restrict the flow of natural air currents and prevent the aeration fan air from bypassing the grain and escaping.

Sizing Aeration Ducts. The design of aeration ducts within round grain storage structures is based on obtaining reasonably uniform air distribution with acceptable friction losses. A level grain surface is assumed generally. Two factors are important in duct design:

1. Cross-sectional area.
2. Surface area exposed to the grain.

Cross-sectional area may be determined by using the following relationships:

$$\begin{array}{c}\text{minimum cross-}\\\text{sectional area}\\(\text{ft}^2)\end{array} = \frac{\text{total air volume (cfm)}}{\begin{array}{c}\text{allowable air velocity}\\\text{within the duct (fpm)}\end{array}} \qquad (5.2)$$

Total air volume depends on the particular fan, bin, and grain type combination. Allowable air velocities within a duct are given in table 5.1.

There must be sufficient surface area also to allow the air to pass from the duct into the grain mass. As with duct diameter, there is a maximum exit velocity for the escaping air. The following relationship is used:

$$\text{minimum total surface area}\left(\text{ft}^2\right) = \frac{\text{total air volume (cfm)}}{\text{maximum allowable velocity of duct exit air (fpm)}} \qquad (5.3)$$

Again, the recommended maximum surface air velocities are given in table 5.1.

There are several configurations of ducts as shown in figure 5.10. Some ducts are perforated while other are positioned above the floor with spacer blocks. Perforated ducts should have an open area equivalent to 7 to 10% of the total duct surface area (Noyes, 1967). Cross sectional and surface areas for several duct types are shown in tables 5.3 to 5.5. The flush floor ducts include only the exposed perforated surface as the true areas. Also, the exposed area resulting from the spacer blocks (duct top and two sides) is the surface area for the non-perforated ducts.

Once the total surface area is known, the minimum associated duct length may be computed using the following relationship:

$$\text{minimum duct length (ft)} = \frac{\text{minimum total surface area}\left(\text{ft}^2\right)}{\text{duct surface area}\left(\text{ft}^2\right)\text{ per ft of length}} \qquad (5.4)$$

The minimum duct length does not address the problem of air distribution. Similarly, the air delivery of the fan is very difficult to predict when the grain surface is irregular as often is the case with flat storage systems.

Duct Design Example. Suppose corn is to be stored in a 24-ft diameter bin to a height of 16 ft above the floor and a minimum airflow rate of 0.25 cfm/bu is specified. A duct system placed on the floor of the structure is to be used to aerate the bin. A minimum airflow rate of 1,448 cfm would be needed based on a bin capacity of 5,791 bushels and a minimum specified aeration design rate of 0.25 cfm/bu. For this example, the maximum allowable airflow velocity within the duct is set at 2,000 fpm (table 5.2). From equation 5.2:

$$\text{minimum cross-sectional area}\left(\text{ft}^2\right) = \frac{1{,}448 \text{ cfm}}{2{,}000 \text{ fpm}} = 0.724 \text{ ft}^2$$

For this example, a 12-in. diameter circular duct satisfies the cross-sectional area requirement (table 5.3). The maximum allowable velocity of duct exit air is set at 30 fpm (table 5.2). Using equation 5.3:

$$\text{minimum total surface area}\left(ft^2\right) = \frac{1{,}448 \text{ cfm}}{30 \text{ fpm}} = 48.26 \text{ ft}^2$$

Table 5.3. Cross-sectional and surface areas for circular and semicircular ducts

	Circular		Semicircular	
Duct Diameter (in.)	Cross-Sectional Area (ft^2)	Surface Area Per Unit Length* (ft^2/ft)	Cross-Sectional Area (ft^2)	Surface Area Per Unit Length (ft^2/ft)
4	0.087	0.84	0.044	0.52
5	0.136	1.05	0.068	0.65
6	0.196	1.26	0.098	0.79
7	0.267	1.47	0.134	0.92
8	0.349	1.68	0.175	1.05
9	0.442	1.88	0.221	1.18
10	0.545	2.09	0.273	1.31
11	0.660	2.30	0.330	1.44
12	0.785	2.51	0.393	1.57
13	0.922	2.72	0.461	1.70
14	1.069	2.93	0.535	1.83
15	1.227	3.14	0.614	1.96
16	1.396	3.35	0.698	2.09
17	1.576	3.56	0.788	2.23
18	1.767	3.77	0.884	2.36
19	1.969	3.98	0.984	2.49
20	2.182	4.19	1.091	2.62
21	2.405	4.40	1.203	2.75
22	2.640	4.61	1.320	2.88
23	2.885	4.82	1.443	3.01
24	3.142	5.03	1.571	3.14
25	3.409	5.24	1.704	3.27
26	3.687	5.45	1.844	3.40
27	3.976	5.65	1.988	3.53
28	4.276	5.86	2.138	3.67
29	4.587	6.07	2.293	3.80
30	4.909	6.28	2.454	3.93
31	5.241	6.49	2.621	4.06
32	5.585	6.70	2.793	4.19
33	5.940	6.91	2.970	4.32
34	6.305	7.12	3.153	4.45
35	6.681	7.33	3.341	4.58
36	7.069	7.54	3.534	4.71

* Assumes that 20% of the duct is lying on the floor resulting in an 80% efficiency (Noyes, 1967).

If a 12-in.-diameter circular duct is used, the surface area per foot of length is 2.51 ft (table 5.4). From equation 5.4:

$$\text{minimum duct length (ft)} = \frac{48.26 \text{ ft}}{2.51 \text{ ft/ft}} = 19.23 \text{ ft}$$

The bin is 24 ft in diameter. Thus, the minimum length required for a 12-in.-diameter circular duct meets the design requirements.

Suppose a flush-floor rectangular duct system is to be used that is 24 in. wide with a surface area of 2.0 ft^2/ft long (table 5.5). Again, using equation 5.4:

$$\text{minimum duct length (ft)} = \frac{48.26 \text{ ft}}{2.0 \text{ ft/ft}} = 24.13 \text{ ft}$$

Essentially the minimum duct length satisfies the surface area requirement associated with placement in a 24-ft-diameter bin. A 5 in. depth would provide 0.83 ft^2 of cross-sectional area (table 5.4) which

Table 5.4. Cross-sectional area for rectangular ducts

	Cross-sectional Area (ft^2)							
Depth of Duct (in.)	Width of Duct (in.)							
	9	12	15	18	21	24	27	30
4	0.25	0.33	0.42	0.50	0.58	0.67	0.75	0.83
5	0.31	0.42	0.52	0.63	0.73	0.83	0.94	1.04
6	0.38	0.50	0.63	0.75	0.88	1.00	1.13	1.25
7	0.44	0.58	0.73	0.88	1.02	1.17	1.31	1.46
8	0.50	0.67	0.83	1.00	1.17	1.33	1.50	1.67
9	0.56	0.75	0.94	1.13	1.31	1.50	1.69	1.88
10	0.63	0.83	1.04	1.25	1.46	1.67	1.88	2.08
11	0.69	0.92	1.15	1.38	1.60	1.83	2.06	2.29
12	0.75	1.00	1.25	1.50	1.75	2.00	2.25	2.50
13	0.81	1.08	1.35	1.63	1.90	2.17	2.44	2.71
14	0.88	1.17	1.46	1.75	2.04	2.33	2.63	2.92
15	0.94	1.25	1.56	1.88	2.19	2.50	2.81	3.13
16	1.00	1.33	1.67	2.00	2.33	2.67	3.00	3.33
17	1.06	1.42	1.77	2.13	2.48	2.83	3.19	3.54
18	1.13	1.50	1.88	2.25	2.63	3.00	3.38	3.75
19	1.19	1.58	1.98	2.38	2.77	3.17	3.56	3.96
20	1.25	1.67	2.08	2.50	2.92	3.33	3.75	4.17
21	1.31	1.75	2.19	2.63	3.06	3.50	3.94	4.38
22	1.38	1.83	2.29	2.75	3.21	3.67	4.13	4.58
23	1.44	1.92	2.40	2.88	3.35	3.83	4.31	4.79
24	1.50	2.00	2.50	3.00	3.50	4.00	4.50	5.00

exceeds the minimum requirement of 0.724 ft^2 as previously computed. However, there are other factors to consider if the flush-floor aeration system is used. Depending on the design, it must be determined if an unloading auger can fit inside the duct. In addition, duct air velocity may be affected adversely if the duct size is large enough to contain an unloading auger and if the space occupied by the auger restricts airflow in the cross-section.

For the example problem, assume a 6-in. tube-type unloading auger with an additional 2 in. of support structure to be placed inside the duct. This auger has a cross-sectional area of 0.2 ft^2. The duct design has a top width of 24 in. A duct depth of approximately 8 in. will be required to contain the unloading auger and its support structure. The 8-in. duct has a cross-sectional area of 1.33 ft^2 (table 5.4). The net minimum cross-sectional area needed is 0.724 ft^2. Thus, the total cross-sectional area of the duct would have to be 0.924 ft^2 (0.724 design minimum + 0.2 from the auger) which is less than the 1.33 ft^2 provided by the 8-in. duct depth. Therefore, in this situation, the addition of an unloading auger increased the depth of the duct required to physically contain the auger

Table 5.5. Surface areas for the sides and top of rectangular ducts

	Surface Area (ft^2)							
Duct Depth (in.)	Duct Width (in.)							
	9	12	15	18	21	24	27	30
0	0.75	1.00	1.25	1.50	1.75	2.00	2.25	2.50
4	1.42	1.67	1.92	2.17	2.42	2.67	2.92	3.17
5	1.58	1.83	2.08	2.33	2.58	2.83	3.08	3.33
6	1.75	2.00	2.25	2.50	2.75	3.00	3.25	3.50
7	1.92	2.17	2.42	2.67	2.92	3.17	3.42	3.67
8	2.08	2.33	2.58	2.83	3.08	3.33	3.58	3.83
9	2.25	2.50	2.75	3.00	3.25	3.50	3.75	4.00
10	2.42	2.67	2.92	3.17	3.42	3.67	3.92	4.17
11	2.58	2.83	3.08	3.33	3.58	3.83	4.08	4.33
12	2.75	3.00	3.25	3.50	3.75	4.00	4.25	4.50
13	2.92	3.17	3.42	3.67	3.92	4.17	4.42	4.67
14	3.08	3.33	3.58	3.83	4.08	4.33	4.58	4.83
15	3.25	3.50	3.75	4.00	4.25	4.50	4.75	5.00
16	3.42	3.67	3.92	4.17	4.42	4.67	4.92	5.17
17	3.58	3.83	4.08	4.33	4.58	4.83	5.08	5.33
18	3.75	4.00	4.25	4.50	4.75	5.00	5.25	5.50
19	3.92	4.17	4.42	4.67	4.92	5.17	5.42	5.67
20	4.08	4.33	4.58	4.83	5.08	5.33	5.58	5.83
21	4.25	4.50	4.75	5.00	5.25	5.50	5.75	6.00
22	4.42	4.67	4.92	5.17	5.42	5.67	5.92	6.17
23	4.58	4.83	5.08	5.33	5.58	5.83	6.08	6.33
24	4.75	5.00	5.25	5.50	5.75	6.00	6.25	6.50

but did not require a greater depth to satisfy the maximum velocity allowed.

Aeration Systems for Flat Storage Structures

Traditionally, most grain is stored in circular-type bins that provide a convenient means for grain handling and management. However, rectangular structures and covered piles may also be used to store grain (a discussion of flat storage is given in Chapter 7). The management practices and design principles used with round storage are still required for flat storage, but the geometry of rectangular structures may make implementation of these practices more difficult. Thus, design of aeration systems in flat storage facilities may be more critical in terms of maintaining a quality product.

While the reasons for grain aeration are the same regardless of structural type, the problems associated with the design and use of aeration systems in rectangular buildings require different guidelines than those used in round grain bins.

Air Distribution in Flat Storage. Air distribution is a key factor in the effectiveness of any grain aeration system. Ideally, the airflow should be distributed uniformly throughout the grain mass. However, there is the potential for "dead zones" in any type of storage resulting in air movement that is less than desired. These zones are more prevalent in flat storage systems (fig. 5.12). Some airflow movement will be experienced in the dead zones, but it may be less than one-half of the design rate. To help offset this problem in flat storage structures, it is recommended that 0.2 cfm/bushel be used as the minimum design airflow rate. This rate is twice the minimum airflow rate generally recommended for circular storage.

Another consideration in providing a more uniform airflow pattern relates to the grain peak that usually occurs in flat storage structures (fig. 5.12). Since the air will flow toward the path of least resistance, the more level the grain surface the more evenly the airflow will be distributed.

Fines may also cause more airflow distribution problems in flat storage than in circular bins. Circular storage bins are often filled using grain spreaders which tend to distribute the fines more evenly throughout the grain mass. This method of filling is not as practical in rectangular structures, so the fines concentrate under the filling auger causing uneven airflow at these points in the grain mass.

Aeration System Design for Flat Storage. Rectangular storage structures usually result in less uniform airflow than their round counterparts. The aeration design criteria have evolved from general rules of thumb. For any given rectangular storage, the amount of aeration required depends on the type of air distribution pattern, how fast the grain is to be cooled, and, to some extent, the investment the

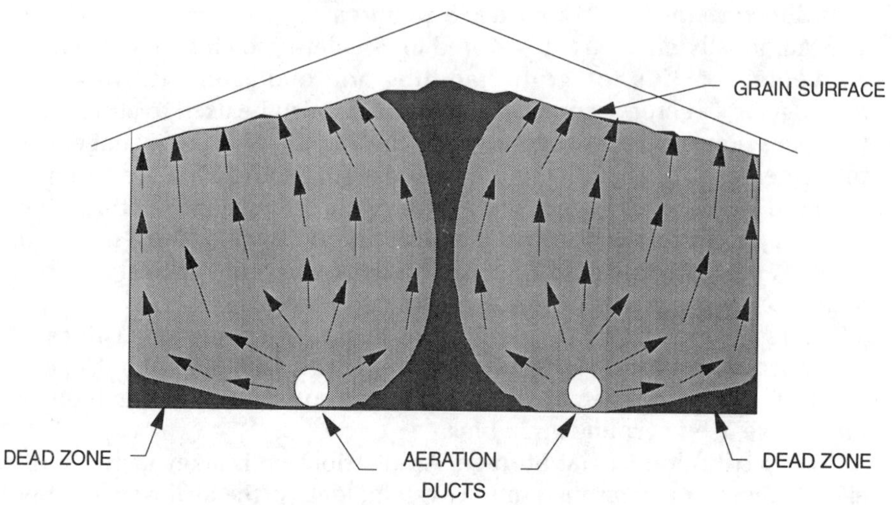

Figure 5.12–Air distribution in a flat storage system.

producer is willing to spend for this practice. Design of such systems includes determination of the number of ducts, duct spacing, duct sizing, duct type, fan selection discussed and system operation. In general, the design recommendations set forth in the following sections will pertain to on-floor duct systems.

Duct Spacing and Number for Flat Storage. The number of ducts is generally determined by the spacing required for uniform airflow. Two general rules of thumb (illustrated in fig. 5.13) are used for spacing ducts in a flat storage:

1. The spacing between ducts should be less than or equal to the maximum grain depth. Using the dimensions shown in figure 5.13, "D_1" should always be less than or equal to "H_1".
2. The longest air path served by a given duct should be less than or equal to 1.5 times the shortest path. From figure 5.13, "$X_1 + Y_1$" should be less than or equal to 1.5 times "S_1" for any duct in the system. As can be seen from the figure, the longest air path is a function of the slope of the grain surface.

If the grain in the flat storage structure is level, the required number of ducts can be determined by dividing the appropriate storage dimension by the grain depth and then rounding up to the next whole number. This will ensure that the duct spacing is less than the grain depth. Care should be taken in using duct systems designed for level

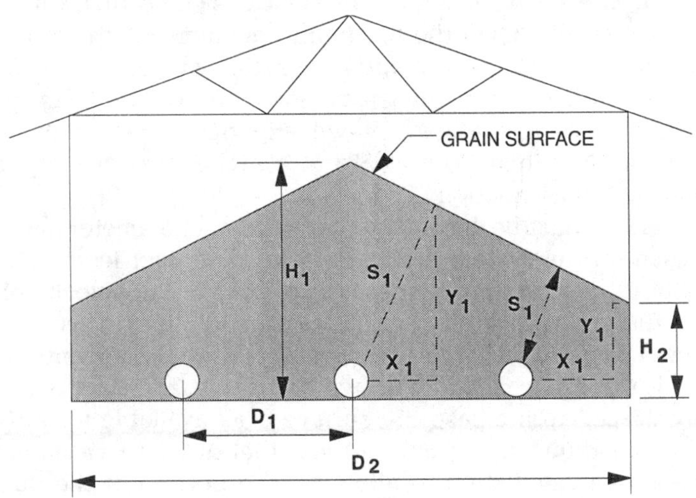

Figure 5.13–Illustration of the general spacing rules for aeration ducts in a flat storage system (Bridges et al., 1988).

storage conditions in situations where the grain surface has a peak. These systems are seldom adequate if the surface of the grain is sloped more than 10%. Remember, a key element in system design is uniform air distribution throughout the grain mass.

Another rule of thumb used to determine the duct spacing is that the length of the non-perforated duct section at the fan should be approximately the same as the distance from the first duct to the side of the structure ("X_1" in fig. 5.13). This should also be the distance from the end of the duct farthest from the fan to the end wall of the structure opposite the fans.

Duct Sizing Consideration in Flat Storage. Once the number of ducts and the spacing have been determined, the individual duct diameters may be chosen. Generally, this will be a function of the number of bushels each duct is to aerate, the design airflow rate, and the design velocity. If the ducts are chosen to run perpendicular to the peak, each duct will be required to aerate the same amount of grain, and all ducts will be the same size. This will also be true regardless of duct direction, if the storage has a level surface. If the ducts are chosen to run parallel with the grain peak, the center ducts generally will aerate more bushels and need to be larger.

It is important to size the ducts properly so that a uniform airflow and pressure will be maintained throughout the tube. Typically, the grain farthest from the fan will be the last to cool. Improper sizing of either the duct diameter or the fan may result in excessive friction losses and uneven airflow in the duct, thereby requiring more time for the fan to cool the grain. In order to facilitate the duct design so that a uniform pressure drop is maintained, the maximum recommended duct design air velocity is 2,500 fpm for a positive aeration system (duct under pressure) and 1,500 fpm for a negative aeration system (duct under suction). Positive systems, in general, will experience a static pressure regain which allows them to operate at a higher velocity and still provide uniform air distribution.

The duct length must be considered once the duct diameter has been selected based on the system design velocity. The duct length should provide sufficient perforated surface area so that the exit velocity of the air entering the grain from the duct is less than 25 fpm. This value is recommended to prevent large pressure drops as the air enters the grain mass. If the exit velocity is above this value, the duct diameter should be enlarged to increase the surface area available for airflow. Generally for round ducts, only 80% of the actual surface area should be used in the above calculation to allow for the portion of the duct in contact with the floor.

Table 5.3 presents cross-sectional areas and surface area per foot of length for common duct diameters. Table 5.6 gives approximate design static pressures for different grain depths and types (Hellevang,1984). The following design example uses values from these tables to demonstrate the principles discussed above.

Design Example: Size the center duct in figure 5.13 if it is to be used in aerating 10,000 bushels. Assume a round duct is to be placed in a positive pressure system.

Required cfm
for the duct = 10,000 bu * 0.2 cfm/bu
= 2,000 cfm

Duct
cross-sectional
area = 2,000 cfm/2,500 fpm
= 0.8 ft^2

From table 5.3, the cross-sectional area of 0.8 ft^2 requires a duct diameter of 13 in. because the 12-in. size would require an air velocity larger than the 2,500 fpm allowed for a positive system.

Suppose that the designer wishes to build in an added margin of safety and selects a 14-in. duct rather than a 13-in. duct. What is the air velocity using the 14-in. duct (cross-sectional area is 1.069 ft^2)?

Air velocity = 2,000 cfm / 1.069 ft^2
= 1,871 fpm

Now determine the minimum perforated surface area to maintain an exit velocity from the duct of 25 fpm or less.

Minimum
surface area = 1,871 cfm / 25 fpm
= 74.8 ft^2

From table 5.3, the surface area per ft of length for a 14-in. duct is 2.93 ft^2 (this includes the factor of 0.8 for contact with the floor). Duct length may be computed as follows:

length = 74.8 ft^2 / 2.93 ft^2 per ft of duct length
= 25.5 ft

Table 5.6. Approximate design static pressures for different grain types (Hellevang, 1984)

Crop Type	Crop Depth (ft)	Design Air Flow Rates in CFM/Bu 0.10	0.20	0.50
		Static pressure (in. of H_2O)*		
Wheat	10	0.74	0.98	1.70
	15	1.02	1.58	3.31
	20	1.45	2.39	5.90
	25	2.00	3.50	9.13
	30	2.66	4.91	12.60
Barley	10	0.62	0.76	1.19
Oats	15	0.79	1.10	2.08
Sunflower	20	1.01	1.61	3.50
	25	1.31	2.21	5.56
	30	1.69	3.07	7.70
Corn	10	0.56	0.61	0.83
Soybeans	15	0.62	0.77	1.31
Edible Beans	20	0.73	1.01	2.06
	25	0.86	1.31	3.09
	30	1.04	1.72	4.40

* Static pressure based on Shedd's data with a packing factor of 1.5 plus 0.5 in. of H_2O for entrance and duct loss.

The above calculation indicates that if a 14-in. round duct is to be used, the duct must have at least 25.5 ft of perforated surface in length. If the duct length is less than 25.5, the diameter must be increased to get the exit velocity below the maximum value allowed. If a half-round duct is to be used in this same situation, an 18-in. duct would be required, and the minimum perforated length should be 38.3 ft to satisfy all airflow requirements.

Generally, the duct length should be limited to a maximum of 100 ft because of severe pressure drops that may be encountered beyond this distance. One alternative for buildings over 100 ft in length is to branch from the center with two ducts or use ducts extending from both ends of the building treating each half as a separate bin.

Other Considerations

Aeration systems are designed to provide uniform airflow distribution to the grain mass. Any design, however, should also consider the grain handling and unloading system used in the structure. Above-floor duct systems incorporate unique problems in handling that are not common to round storages. The producer is encouraged to develop designs that will provide easy access to the building and facilitate handling in the structure while still providing adequate airflow. The general design rules stated above allow flexibility in these systems, and producers should use these in creating a workable design for their individual structure. A computer model (Bridges et al., 1988) has been developed to provide aeration design information for rectangular storages. Generally, this model considers level or peaked grain masses in rectangular storages and allows the user several options in changing duct sizes and spacing for input designs.

Cost of Aeration

Aeration should be viewed as a necessary part of grain storage, the added cost is for capital investment (aeration fan, ducts, perforated floor, controls, etc.) and the operation of the fan. Fan operation cost is the power used by the fan multiplied by the hours of fan operation and the cost of electricity. The following equation may be used:

$$\begin{array}{c}\text{total}\\ \text{cost of}\\ \text{aeration}\\ (\$)\end{array} = \begin{array}{c}0.7457 * \text{fan hp} * \text{hours of operation}\\ * \text{ electricity cost } (\$ / \text{kW-h})\end{array} \qquad (5.5)$$

For purposes of illustration, suppose that shelled corn is to be stored to a depth of 16 ft in a 36-ft diameter bin (13,029 bu). The selected airflow rate is 0.5 cfm/bu, and so a 3-hp fan is required (fig. 5.14). A

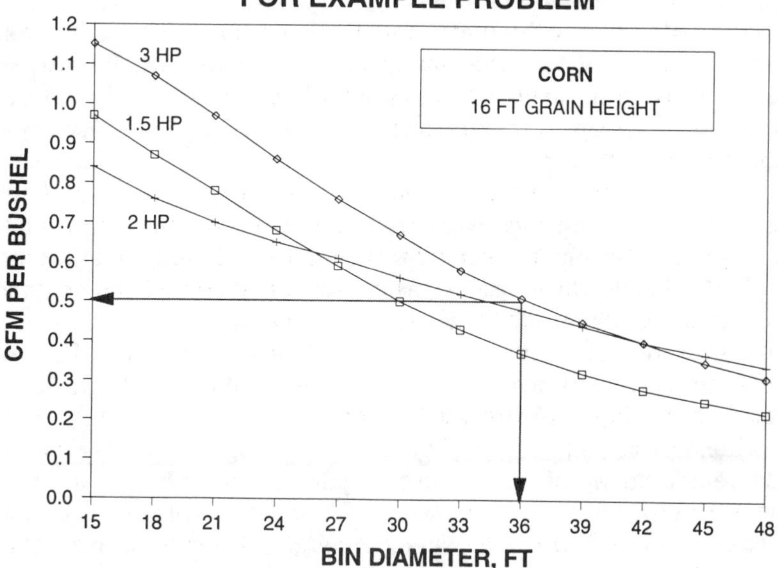

Figure 5.14–Airflow capacity - example problem.

cooling front can pass through the grain in approximately 24 h as estimated from figures 3.32 to 3.33 (Chapter 3). If the grain is to be stored for 32 weeks and the aeration fan to be operated 24 h every two weeks, the total fan operation time is 384 h (16 days * 24 h/day). If the cost of electricity is \$0.06/kW-h:

total aeration cost (\$) = 0.7457 * 3 hp * 384 h * \$0.06/kW-h

= \$51.54 (\$0.004/bu)

The cost of fan operation is low when compared to the risk of losing an entire bin filled with grain because of spoilage and faulty management.

Summary

Aeration is a very critical part of grain storage management and is a major contributor to the prevention of grain spoilage. Rewetting or overdrying occurs to some extent in all aeration systems but usually offset each other and are not major concerns if the fans are operated only as long as needed to maintain proper grain temperature.

Stored Grain Sampling and Inspection

Sampling and inspecting stored grain are essential if quality and economic value are to be maintained. Generally, spoilage occurs in particular areas of the storage structure rather than uniformly throughout the grain mass. Hence, successful sampling and inspection techniques are dependent upon recognizing and identifying the probable source and location of grain spoilage.

Inspection Through Surface Observation

One method of grain inspection is through surface observation, both outside and inside the storage structure. For example, leaking grain and/or budging bin walls may indicate that the grain is expanding because of water absorption. The water may be from rain, condensation or biological activity (fungus, insects, etc.). Regardless, adding moisture to grain eventually will lead to spoilage.

Another tell-tale sign of grain spoilage is melting snow on the bin roof during very cold weather. This may again indicate abnormally warm temperatures within the grain mass associated with biological activity. Likewise, standing water and weeds around a storage bin provide an ideal environment for insects, and finding fresh grain away from the bin is sign of rodent activity.

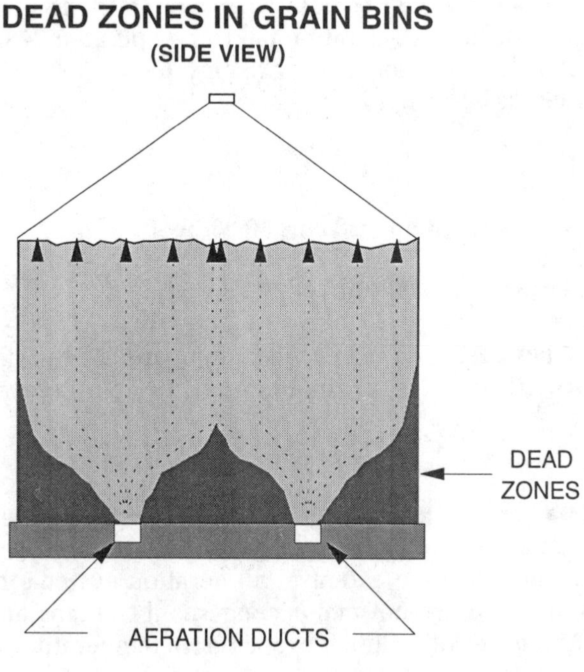

Figure 5.15–Dead space zones in grain bins (side view).

Aeration Systems and Spoilage

The key to locating potential spoilage areas within aerated storage structures is to recognize that air flows along the path of least resistance. Hence, there may be "dead space" areas through which very little air passes when using a duct-type aeration system (figs. 5.15 to 5.16). Likewise, overfilling a bin may create "dead space" zones (fig. 5.17). Therefore, when inspecting a bin for possible trouble spots, be sure to probe into the "dead space" zones if at all possible.

The best method of distributing air evenly through the grain is to use a perforated floor (fig. 5.18). However, if aeration is not managed properly, possible trouble zones could still occur, especially in the top and bottom centers of the bin (fig. 5.19). Likewise, overfilling may present the same problem for bins equipped with perforated floors as for those with duct systems (fig. 5.17). Note also that trash and fine material tend to collect under a perforated floor. This material provides an excellent breeding place for insects, so treating and cleaning may be required. The same material collects under the subfloor aeration ducts. However, cleaning and treating is much easier when compared to the effort required to remove a perforated floor.

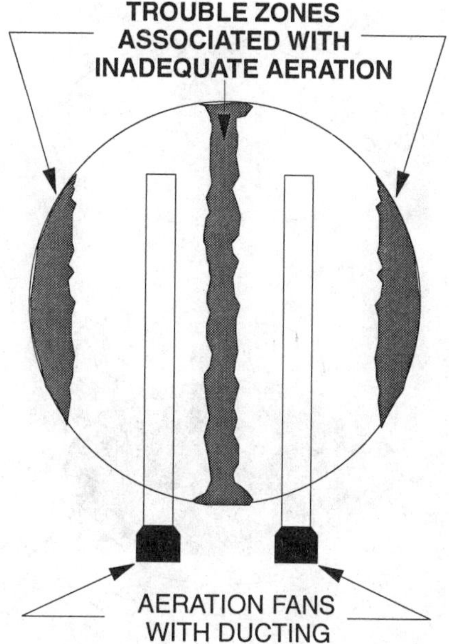

Figure 5.16–Dead space zones in grain bins (top view).

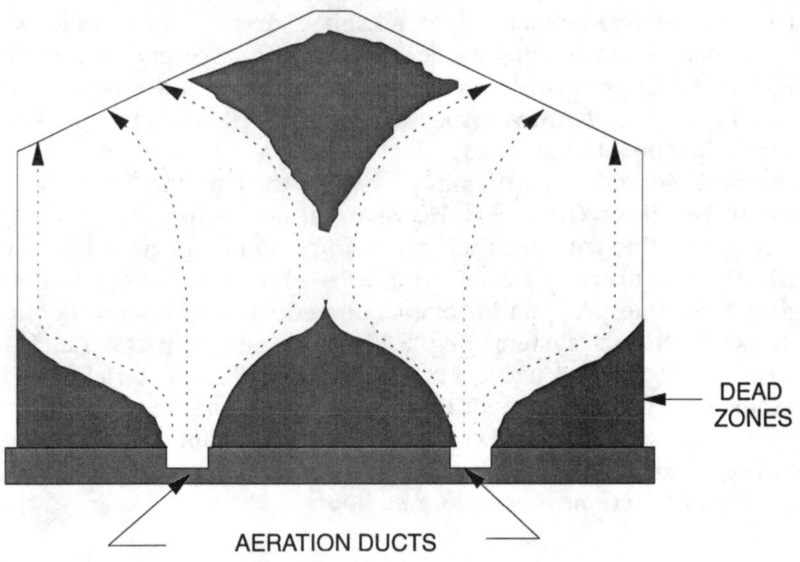

Figure 5.17–Dead space zones from overfilling.

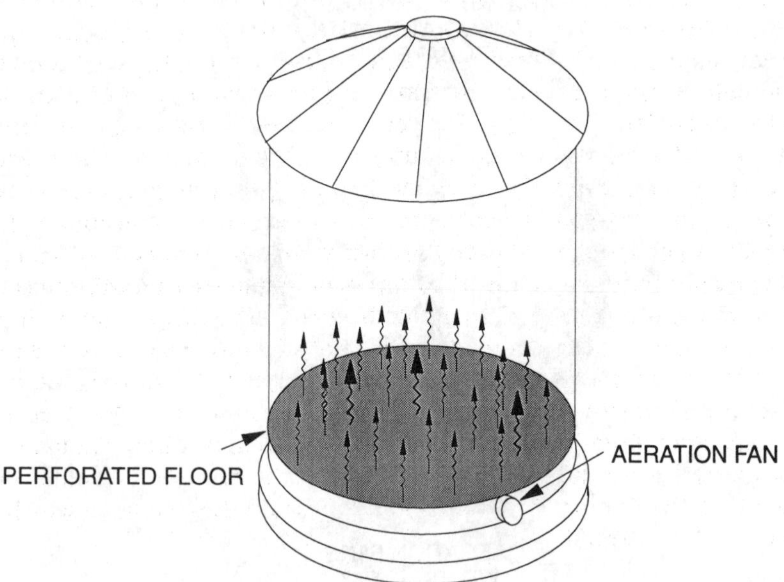

Figure 5.18–Perforated floor systems.

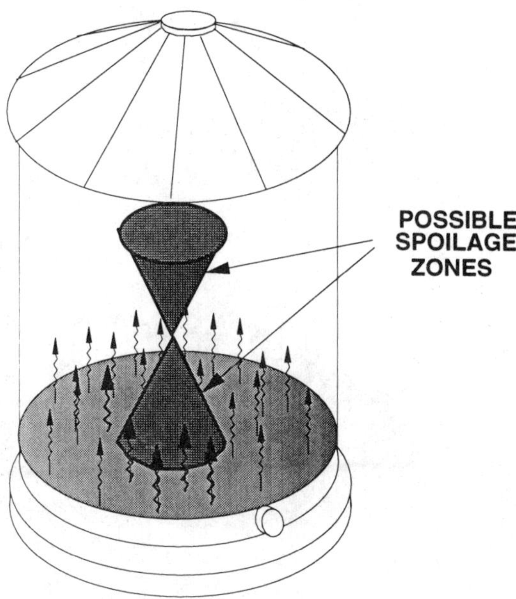

Figure 5.19–Trouble zones using perforated floor systems.

Filling and Unloading Grain Bins

Storage problems may result from factors other than inadequate aeration. For example, when grain bins are filled, foreign and light material (such as trash, weed seed, and broken parts of kernels) tend to accumulate in the center of the bin and may form a core of material from top to bottom (fig. 5.20). The "core" may be so tightly packed that aeration or drying air will go around it, passing through the looser cleaner grain. Consequently, this zone may not properly dry. In the case of in-bin drying systems, the core provides an excellent environment for mold and insect growth. Potential problems may be reduced by using a grain spreader that distributes the fines evenly. Another possibility is to remove the center material by unloading the bin with a center draw unloading auger and then spreading this material uniformly over the top surface of the grain after leveling. This procedure may involve some risk to workers (Loewer and Loewer, 1975; Loewer et al., 1979) so be extremely careful that no one is inside the bin when it is being unloaded. The preferred procedure is to clean the grain before it is placed in the bin. Probing the center of the bin should indicate the extent to which a center core has formed.

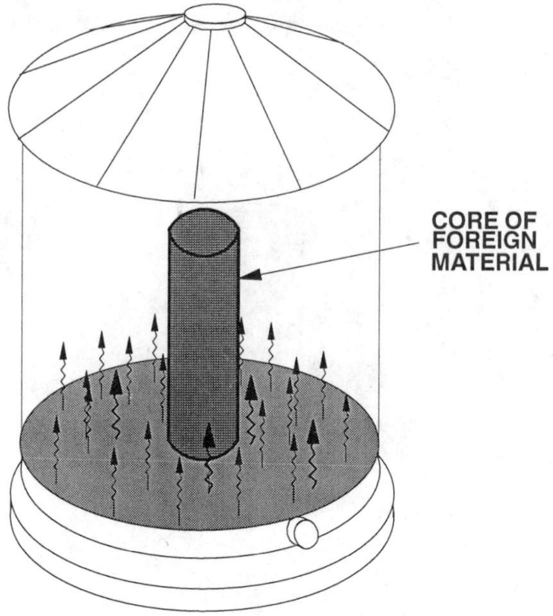

Figure 5.20–Core of foreign material.

"Hot spots" may be found in any part of the grain mass. These trouble zones usually result from accumulations of trash or foreign material. However, if a load of relatively wet grain is placed into a bin of dry grain, the wet grain may begin to spoil regardless of the average moisture content of the entire mass of grain (fig. 5.21). When probing a bin, investigate points where the probe has relative difficulty in penetrating. Generally, wet grain or trash offers more resistance to probe penetration than does dry grain. Again, the safety aspects associated with entering a bin of grain are important in that the grain may have bridged across the bin.

When a typical farm grain bin is unloaded, grain from the top portion of the bin is removed first. The grain will continue to flow until it reaches a natural angle with the bin floor called the angle of repose (fig. 5.22). The angle of repose usually ranges from 25 to 35° depending in part on its moisture content. A bin may continue to be filled and unloaded without ever removing that portion of the grain in the stagnant areas (fig. 5.22), as would be the case of a wet holding bin for a dryer. This "stagnant" grain should be examined carefully because it may be at a higher moisture content or contain different levels of foreign material than the rest of the grain.

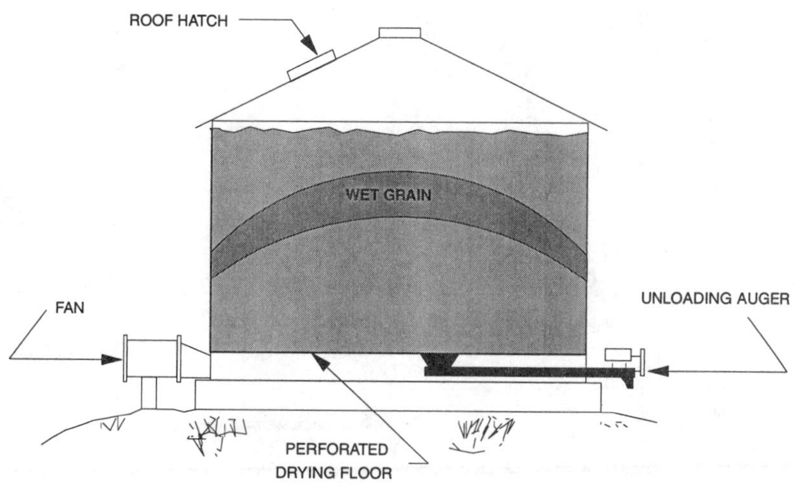

Figure 5.21–Layer of wet grain inside a bin.

When grain is drying in a bin, a drying front moves in the direction of the airflow, usually from the bottom to the top of bin (fig. 5.23). The grain above the drying front remains essentially at the same moisture content as when it entered the bin. The grain below the drying front will

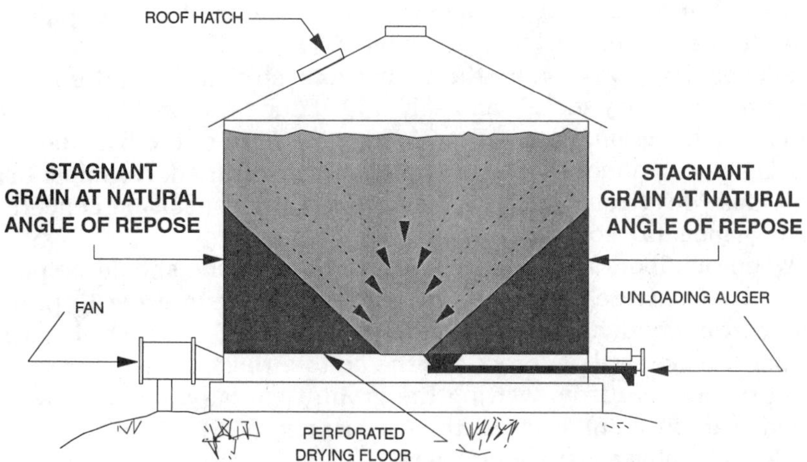

Figure 5.22–Unloading grain using the angle of repose.

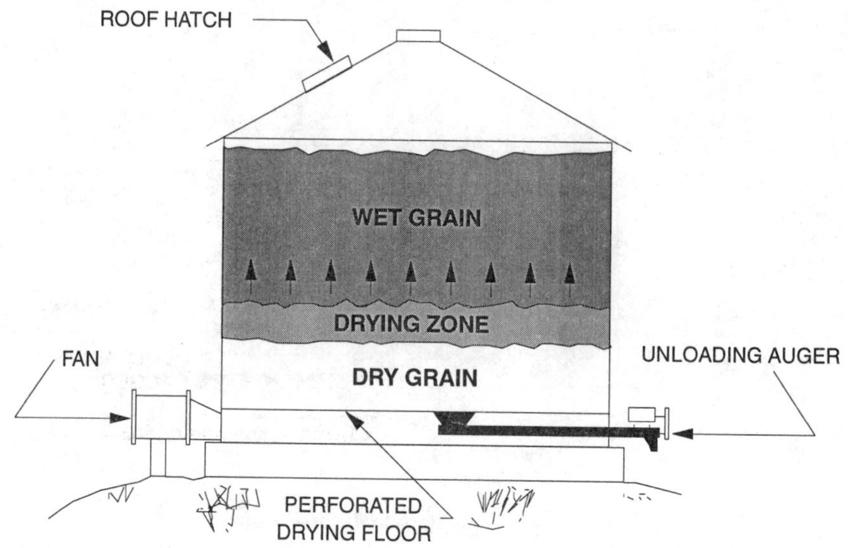

Figure 5.23–Drying front movement.

have a lower moisture content and be in equilibrium with the drying air. The level of the drying front may be found by locating the point where the temperature of the grain increases significantly. This level also will be where a probe begins to move more freely through the grain indicating a drier portion of the grain mass has been reached. When examining grain dried in a bin, compare the moisture content of the grain a few inches from the surface with that a few feet lower to determine if the system was allowed to operate until all the grain was dried to the desired final moisture content (fig. 5.24).

Stirring devices break up the drying zone so that the entire mass of grain tends to dry more uniformly and reduce the problem near the surface of the grain. However, grain near the floor of the bin where the stirring auger cannot reach may be severely overdried. A sample taken from this "dead zone" area may lead to a faulty conclusion concerning average moisture content (fig. 5.25).

When bin floors are constructed, a layer of plastic should be placed under the concrete floor to serve as vapor barrier in order to prevent water from condensing on the floor and wetting the grain. Likewise, sealing the side wall and roof and the bottom ring on the concrete slab will prevent rain from wetting the grain. Generally, spoilage will be found near any point where wetting occurs (fig. 5.26).

The unloading auger should be cleaned before grain is placed in a bin or after a partial unloading. Otherwise, when the grain is unloaded, the

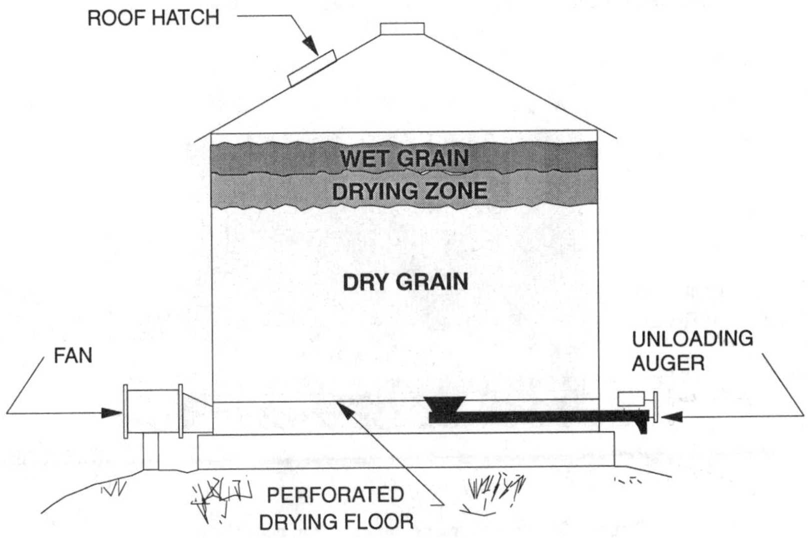

Figure 5.24–Undried grain at surface.

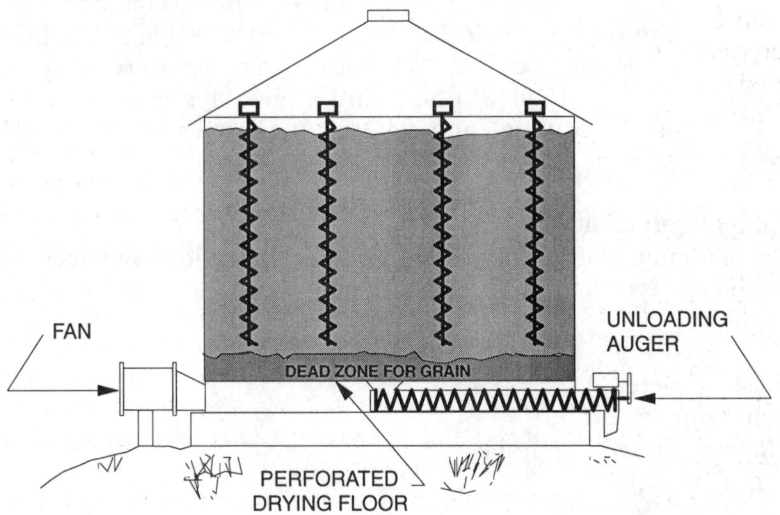

Figure 5.25–"Dead zone" associated with stirring devices.

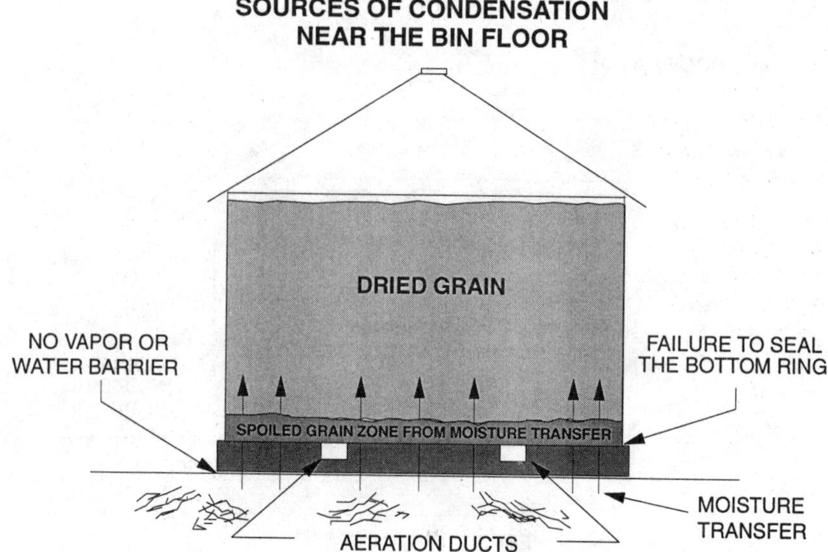

Figure 5.26– Prevention of condensation near the floor.

sample taken from the truck may indicate contamination by mold or insects at some level higher than that actually present inside the bin (fig. 5.27). Also, water may collect inside an auger and wet the grain remaining from a previous unloading.

When grain is first removed from a dryer, moisture meters generally indicate a lower grain moisture content than actually exists. Once the moisture in the grain has reached equilibrium within the kernel, a more accurate moisture reading may be obtained. When examining a grain sample, be sure that the grain has been cooled and that it has been a few days since drying occurred. Otherwise, the moisture reading might be somewhat lower than will be found a few days later. Generally, aeration does not produce any significant changes in the moisture content of the grain.

Sampling Equipment

The minimum equipment needed for checking a grain bin accurately for problems associated with moisture content and insects includes the following:

moisture meter
grain sampling probe
temperature sensing probe
measuring tape
grain dividers
weighing scale for determining foreign material

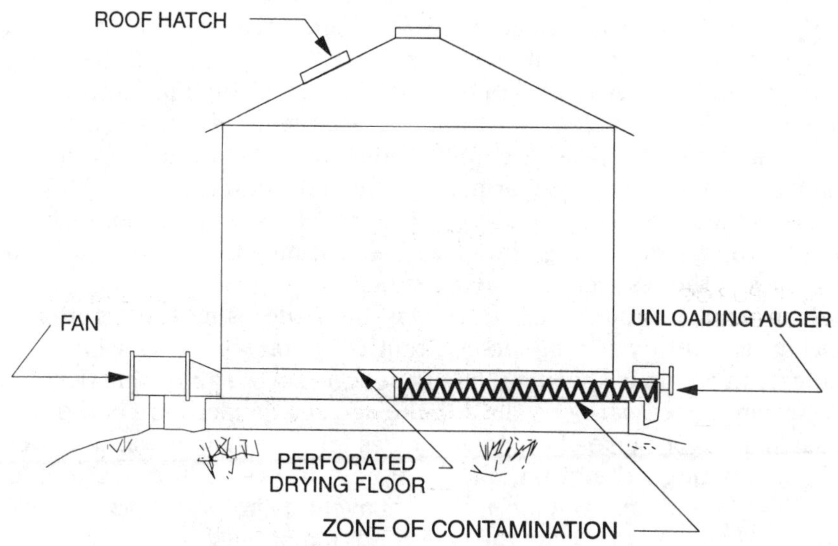

Figure 5.27–Possible grain contamination in auger.

bushel test weight stand
grain sieves
black light

Sampling Procedure

When sampling a bin, know the past history of filling and unloading, and take the necessary safety precautions. Be familiar with the suffocation hazards in grain bins (Loewer and Loewer, 1975), in particular those associated with the way grain bins unload and what can happen when probing through bridged grain.

The sampling procedure may be divided into three separate tests, each designed to locate and determine possible trouble areas.

Average Moisture and Temperature Test. Probing the grain for representative samples is required for determining average moisture and temperature. Accuracy increases with greater numbers of representative samples. Under some circumstances, such as the bin being too full, it may be impossible to obtain all the data needed for a complete test. When the number of probes is limited, sampling should occur in a north-south orientation in order to obtain the greatest potential difference in grain temperatures. The following procedure is suggested when at all possible. The bin should be divided into five equal volumes. The sampling distances from the center of the bin associated with these volumes are shown in table 5.7. The surface of the grain should be divided into five sections or "pies". A probe test should be made on the boundary line separating each "pie" for one of the equal volume diameters given in table 5.7. The probe should record temperature and moisture contents for depths that are 6 in. below the surface and for 5-ft depth increments below the surface until the floor is reached. Again, it may be impossible to force a sampling probe deep enough for the complete bin sampling. However, as many samples as possible should be taken. Sampling points are shown in figure 5.28. A bin history and bin sample form are presented in tables 5.8 and 5.9.

The temperature and moisture content of each sampling point should be recorded. The average temperature and moisture content are the averages of the respective sample readings.

Fan Test. An additional check may be made using the aeration or drying fan and by paying close attention to musty odors which may indicate the grain is spoiling. The temperature of the air exiting the grain bin may not necessarily be close to the average temperature of the grain mass. This is because of the time it takes for a cooling or warming front to pass through the grain, as mentioned in the earlier sections on aeration. In addition, the air may circumvent a "hot spot", as shown in figure 5.20, and not indicate the true condition of the grain.

Spot Test. Once the probe and fan tests have been conducted, the bin should be examined for possible trouble areas or hot spots that indicate rapid mold or insect development. Several random probes for temperature and moisture should be made in the center of the bin and near the grain walls and floor. The bin door may have an opening for a probe in order to allow sampling near the floor. Suspicious areas should be examined closely. In addition to temperature and moisture content

Table 5.7. Sampling distances for equal volumes of grain

Bin Diameter (ft)	Distance from Center of Bin to Point of Sampling (ft) — Multiplier* Bin Diameter for Distance				
	0.1580	0.2740	0.3535	0.4185	0.4745
15	2.37	4.11	5.31	6.28	7.12
18	2.85	4.93	6.37	7.54	8.54
21	3.32	5.76	7.43	8.79	9.97
24	3.79	6.58	8.49	10.05	11.39
27	4.27	7.40	9.55	11.30	12.81
30	4.74	8.22	10.61	12.56	14.24
33	5.22	9.04	11.67	13.81	15.66
36	5.69	9.86	12.73	15.07	17.08
39	6.16	10.69	13.79	16.32	18.51
42	6.64	11.51	14.85	17.58	19.93
48	7.58	13.15	16.97	20.09	22.78
60	9.48	16.44	21.21	25.11	28.47

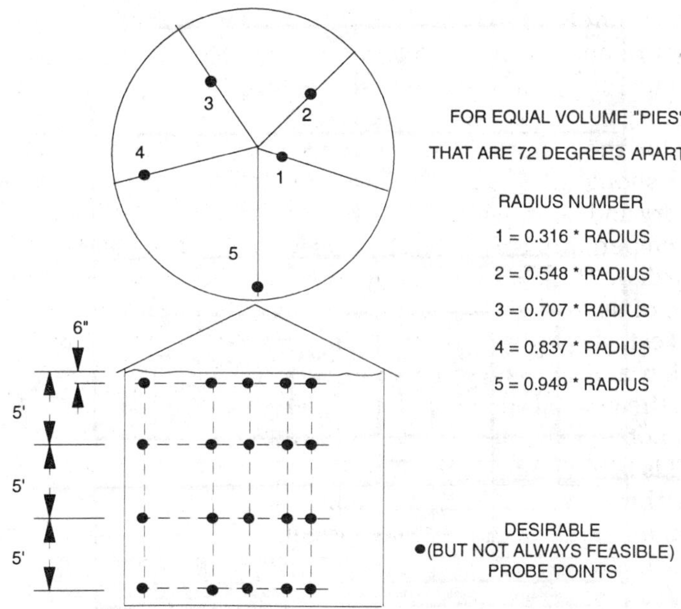

Figure 5.28–Probe points and depths of sampling (Loewer et al., 1979).

Table 5.8. Example form for recording bin history

Name: Address						
Bin ID:			Bin Capacity:			
Aeration or Drying System Information:						
OBSER-VATION NO.	DATE	BUSHELS	ACTION TAKEN C= CHECKED F= FILLED U= UN-LOADED	% MC	GRAIN TEMP.	COMMENTS

Table 5.9. Example form for recording bin grain samples

Name: Address					Date: Phone Number:	
Bin ID: Bin Capacity:					Average Moisture Content: Average Grain Temperature:	
Recommendation:						
SAMPLE NO.	DEGREES RIGHT FROM CENTER LINE *	FEET FROM BIN CENTER POINT	FEET DOWN FROM SURFACE	% MC	GRAIN TEMP.	COMMENTS

* Center line is a reference extending from the top center of the bin to the bin wall.

readings, these same areas should be examined for foreign material and the presence of insects and aflatoxin. A black light will give some indication of the presence of aflatoxin by giving the grain a green-yellowish appearance. However, a positive black light reading does not necessarily mean that aflatoxin is present, but does indicate that a more sophisticated test should be performed by a lab that conducts this test on a regular basis.

Corrective Action

Generally, search for trouble spots by investigating potential problem zones. Once the tests have been completed, a course of action should be taken. Certainly, if insects are present, a fumigation procedure should be followed (Gregory, 1973). If aflatoxin is present in quantities exceeding minimum market standards, the grain may not be sold through the normal channels. However, the grain may be fed to livestock under some circumstances involving relatively low levels of contamination (Spruill, 1977) or proper treatment of the grain by ammonia (Brekke et al., 1975; Hammond, 1977). Usually, most of the problems associated with farm storage are with excess grain moisture and temperature. The information shown in table 5.10 and figure 5.29 should be a guide to safe storage under most conditions if the grain has near uniform temperature and moisture conditions. Additional drying

Table 5.10. Safe storage conditions for various grain types

Temperature (°F)	Safe Grain Moisture Content (% w.b.) in Equilibrium with Air at 65% Rh							
	Barley	Corn (yd)	Rice (Rough)	Sorghum	Soybeans	Wheat (Duram)	Wheat (Hard)	Wheat (Soft)
30	13.1	16.2	14.9	14.4	12.7	15.2	15.9	14.4
35	13.0	15.8	14.6	14.3	12.5	15.0	15.7	14.2
40	13.0	15.4	14.4	14.2	12.3	14.7	15.4	14.0
45	12.9	15.1	14.1	14.1	12.1	14.5	15.1	13.8
50	12.8	14.8	13.9	14.0	12.0	14.3	14.9	13.6
55	12.7	14.2	13.7	13.9	11.8	14.2	14.7	13.4
60	12.7	14.2	13.5	13.8	11.6	14.0	14.5	13.3
65	12.6	13.9	13.3	13.7	11.5	13.8	14.2	13.1
70	12.5	13.6	13.1	13.6	11.3	13.6	14.1	13.0
75	12.5	13.4	12.9	13.5	11.2	13.5	13.9	12.8
80	12.4	13.2	12.7	13.4	11.0	13.3	13.7	12.7
85	12.3	13.0	12.6	13.3	10.9	13.2	13.5	12.6
90	12.3	12.7	12.4	13.2	10.8	13.0	13.4	12.5
95	12.2	12.6	12.3	13.1	10.6	12.9	13.2	12.3
100	12.1	12.4	12.2	13.0	10.5	12.8	13.0	12.2

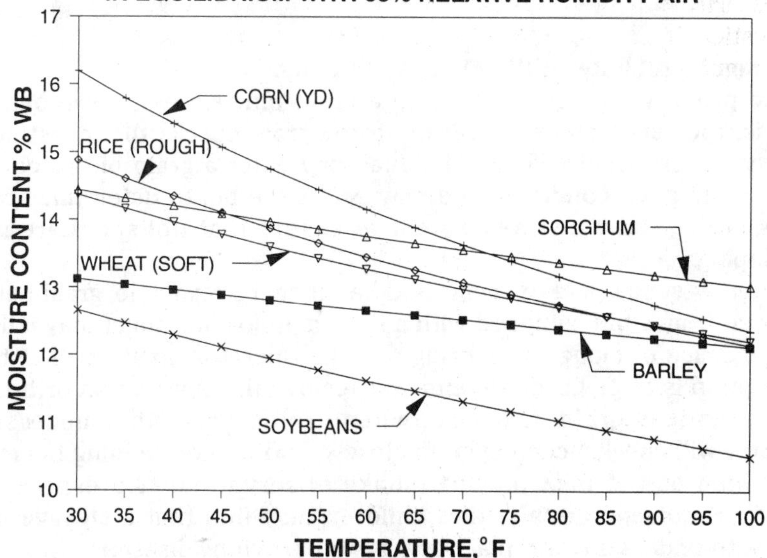

Figure 5.29–Safe storage conditions for various grain types (Loewer et al., 1979).

may be required if the grain is at too high a moisture content, and the grain should be aerated if temperature limits are exceeded.

When sampling for the average grain conditions, if it is found that any particular probe sample exceeds the minimum specifications, then corrective action should be taken regardless of the "average" conditions of the grain. This usually involves removing the grain from the bin before additional spoilage occurs, and drying it to a safe moisture content as quickly as possible.

Again, take all the safety precautions when sampling. The value of the sampler is greater than the value of the sample.

Suffocation Hazards in Grain Bins

Suffocations in grain storage systems are a major concern (Loewer and Loewer, 1975; McKenzie ; Willsey, 1972). The hazard may be increasing for the following reasons:

1. Increase in on-farm grain drying and handling systems.
2. Grain bins on the farm are getting bigger.
3. Grain handling rates are increasing.
4. More operators are working alone because of increased mechanization.
5. Most operators are not aware of how grain flows from bins and therefore do not understand the dangers involved.

Don't make the mistake of your life. Be aware of the dangers of flowing grain.

Why You Might Enter A Bin Filled With Grain

The primary reason for entering a grain bin relates to economics. That is, the successful manager of stored grain checks the investment closely and frequently. The individual may enter a grain bin to check visually the grain condition, and may probe the bin to determine grain temperature and moisture content to ensure that hot-spots are not developing.

Other reasons relate to drying and handling. For example, grain being removed from a bin equipped with a bottom unloading auger may fail to flow because of clogging or bridging. The operator may feel that the only option is to go inside the bin and remove the obstruction or break up the bridged grain. Likewise, when drying grain, the successful operator will check incoming grain closely, and the wet holding bin may be viewed as the best place to make observations. An especially important consideration is that children may find that a storage bin filled with grain is an ideal place to play, thus inviting disaster.

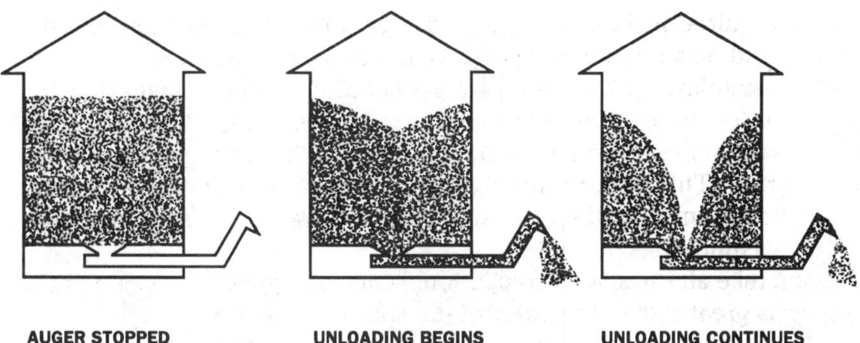

Figure 5.30–When bins unload, the grain at the top of the bin is removed first.

Reasons Why You May Not Come Out Alive

Why is flowing grain so dangerous? To better comprehend the hazard, the way in which most farm storage bins unload must be understood. Grain storage structures should be, and usually are, unloaded from the center. When a valve is opened in the center or a bottom unloading auger is started, grain flows from the top surface down a center core to the unloading port or auger. This is called "enveloping flow" and is illustrated in figure 5.30. The grain across the bottom and around the sides of the bin does not flow (fig. 5.30). The rate at which the grain is removed is what makes the enveloping flow so dangerous. A typical rate for a bin unloading auger is 1,000 bu/h. This is equivalent to 1,250 ft^3/h or approximately 21 ft^3/min. A man 6-ft tall displaces about 7.5 ft^3, assuming an average body diameter of 15 in. This means that the entire body could be submerged in the envelope of grain in approximately 22 s. Even more importantly, you could be up to your knees in grain and totally helpless to free yourself in less than 5 s (fig. 5.31). In fact, it requires up to 2,000 pounds of force to pull a totally submerged man up through grain (Schwab et al., 1982).

You must remember that flowing grain is like water in that it will exert pressure over the entire area of any object that is submerged in it. However, the amount of force required to pull someone up through grain is much greater than required in water because grain exerts no buoyant force and has much greater internal friction. People who have helped pull partially submerged children from grain have commented on how hard they had to pull and, frequently, that shoes were pulled off in the grain. This may mean that rescue efforts will fail unless grain movement is stopped.

Grain that bridges across a bin can be another hazard. Bridging grain may create air spaces in a partially unloaded bin (fig. 5.32). This situation presents several dangers. The first is that the person may break through the surface and be trapped instantly in the flowing grain (fig. 5.33). Another danger is that a large void may be created under the

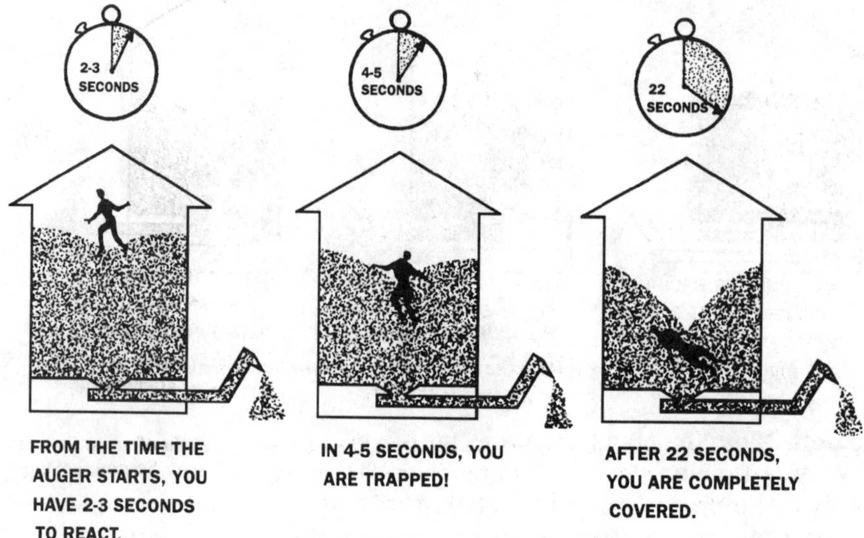
2-3 SECONDS
4-5 SECONDS
22 SECONDS
FROM THE TIME THE AUGER STARTS, YOU HAVE 2-3 SECONDS TO REACT.
IN 4-5 SECONDS, YOU ARE TRAPPED!
AFTER 22 SECONDS, YOU ARE COMPLETELY COVERED.

Figure 5.31–Twenty-two seconds to suffocation (Loewer and Loewer, 1975).

bridged grain by previous unloading so that a person who breaks through the crust may be buried under the grain and suffocate even though the unloading auger may not be in operation at the time (fig. 5.34). A third hazard is that, if the grain is wet enough to mold and bridge across a bin, there may be little oxygen present in the cavity because of microbial action. Therefore, a person falling into this void may be forced to breathe toxic gases and microbial spores even if the individual's head stays above the level of the surrounding grain.

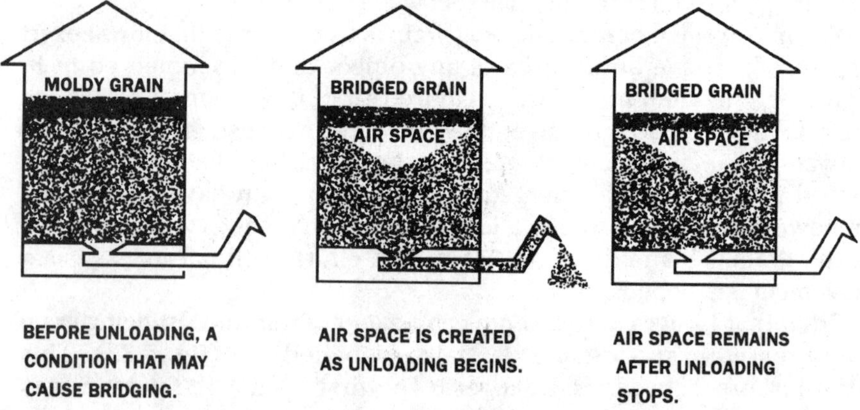
MOLDY GRAIN
BRIDGED GRAIN
AIR SPACE
BRIDGED GRAIN
AIR SPACE
BEFORE UNLOADING, A CONDITION THAT MAY CAUSE BRIDGING.
AIR SPACE IS CREATED AS UNLOADING BEGINS.
AIR SPACE REMAINS AFTER UNLOADING STOPS.

Figure 5.32–Potential hazard created by bridging. Note also that when the air space becomes large enough, the bin walls may buckle.

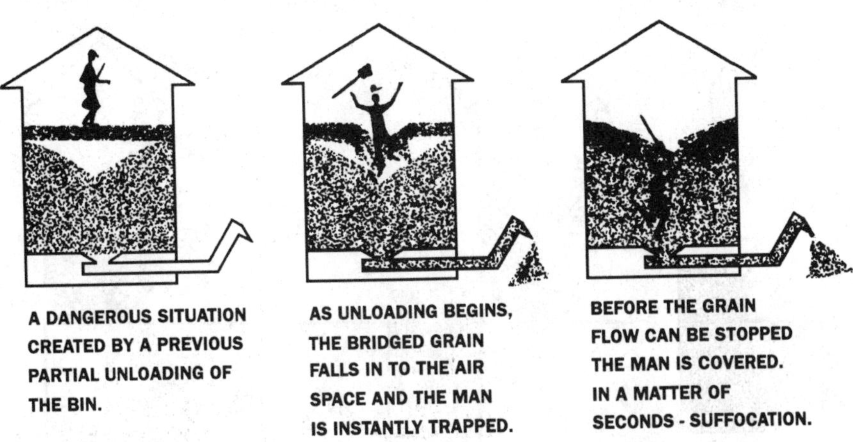

A DANGEROUS SITUATION CREATED BY A PREVIOUS PARTIAL UNLOADING OF THE BIN.
AS UNLOADING BEGINS, THE BRIDGED GRAIN FALLS IN TO THE AIR SPACE AND THE MAN IS INSTANTLY TRAPPED.
BEFORE THE GRAIN FLOW CAN BE STOPPED THE MAN IS COVERED. IN A MATTER OF SECONDS - SUFFOCATION.

Figure 5.33–Two basic principles were violated. First, the man entered a bin of grain that was out of condition without seriously considering its previous unloading history. Second, he didn't ensure that unloading could not occur while he was inside.

Safety hazards in grain bins are not limited to those with bottom unloading augers. Gravity unloaded bins may present a similar danger through bridging or unloading. A definite danger exists with wet holding bins that feed automatic-batch grain dryers. When the dryer completes its drying cycle and reloads, a person in the wet holding bin can be drawn below the surface of the grain in a matter of seconds (fig. 5.35).

Flowing grain hazards, in addition to mold and dust health hazards, exist when working with grain that has gone out of condition or has built up in a tall pile. A wall of grain may look perfectly safe but one scoopful could pry out the "foundation" and start an avalanche or "cave-off" (fig. 5.36). Grain is deceptively heavy. For example, a 6-ft tall man, prone and covered by 1 ft of corn, will be under about 300 lb of grain. People

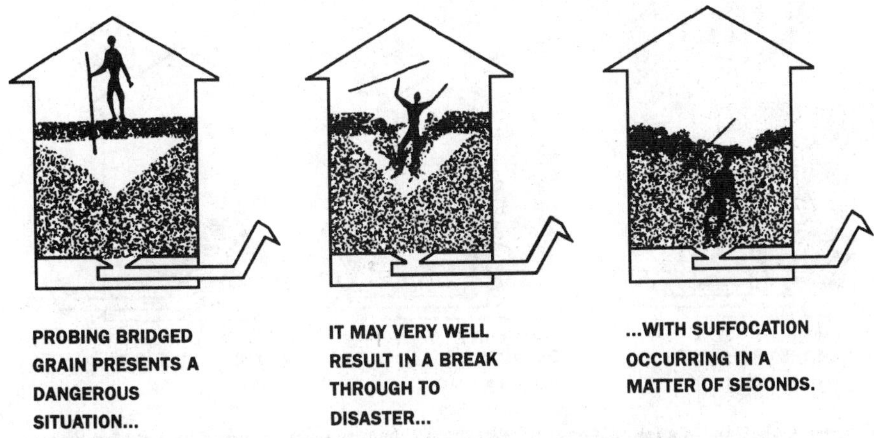

PROBING BRIDGED GRAIN PRESENTS A DANGEROUS SITUATION...
IT MAY VERY WELL RESULT IN A BREAK THROUGH TO DISASTER...
...WITH SUFFOCATION OCCURRING IN A MATTER OF SECONDS.

Figure 5.34–Bridged grain presents a danger, even when the bin is not being unloaded.

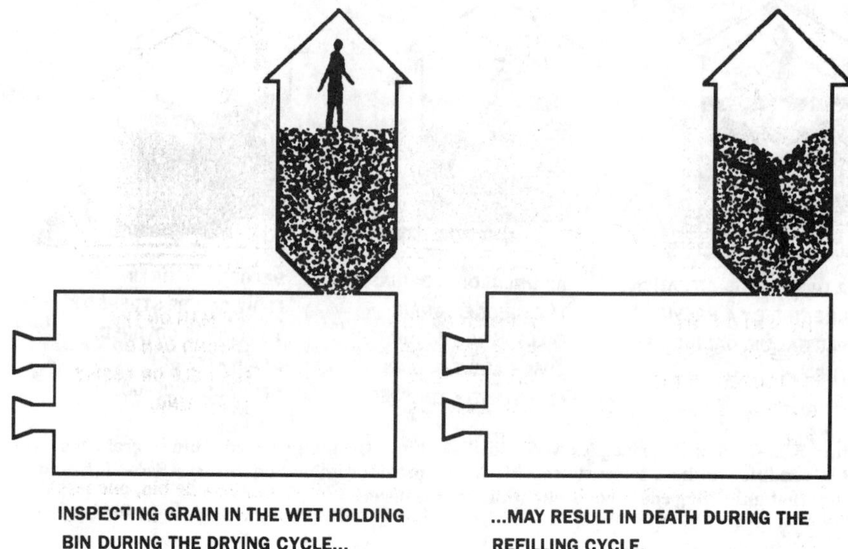

INSPECTING GRAIN IN THE WET HOLDING
BIN DURING THE DRYING CYCLE...
...MAY RESULT IN DEATH DURING THE
REFILLING CYCLE.

Figure 5.35–In a modern grain facility, bins may load or unload automatically adding to the suffocation hazard.

who hear of suffocations are often surprised to learn that the victim was under only a shallow pile.

How To Reduce The Risk

Rule 1. A man entering a grain bin should be fastened to a safety rope or harness that is tied to a point outside the structure. Two additional people should be involved - a second person who can see the man inside the bin and a third person on the ground who can (1) assist in lifting the inside person to safety, (2) quickly go for aid without the

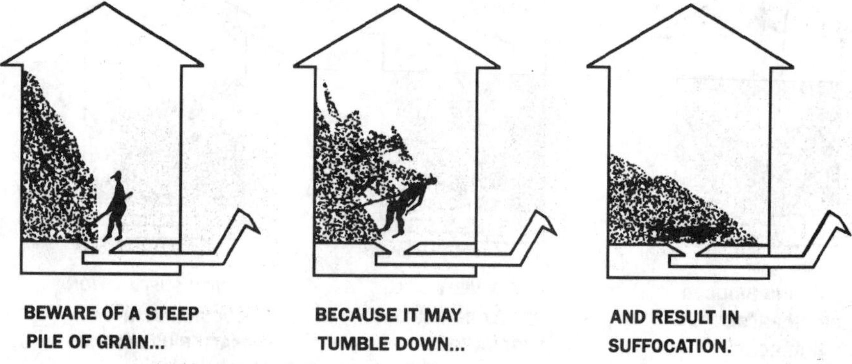

BEWARE OF A STEEP
PILE OF GRAIN...
BECAUSE IT MAY
TUMBLE DOWN...
AND RESULT IN
SUFFOCATION.

Figure 5.36–Beware of a tall pile of grain. A person lying prone and covered by 1 ft of grain will be subjected to a force of over 300 lb.

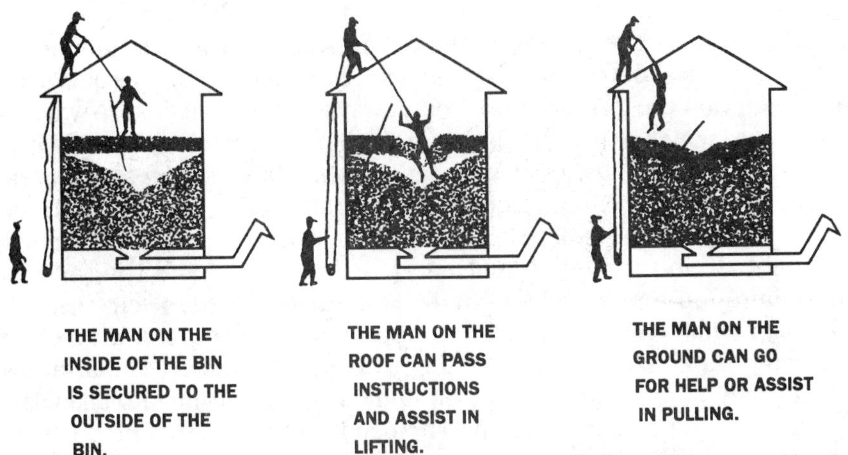

THE MAN ON THE
INSIDE OF THE BIN
IS SECURED TO THE
OUTSIDE OF THE
BIN.
THE MAN ON THE
ROOF CAN PASS
INSTRUCTIONS
AND ASSIST IN
LIFTING.
THE MAN ON THE
GROUND CAN GO
FOR HELP OR ASSIST
IN PULLING.

Figure 5.37–You should use three individuals when investigating a questionable bin, one inside, one in direct communication with the individual inside, and one on the outside to go for help or assist in lifting.

danger of falling off the bin in a panic to climb down, and (3) ensure that no one starts the unloading equipment (fig. 5.37). Don't depend on being able to communicate from the inside to the outside of the bin. It is difficult to hear under any circumstances, especially when unloading equipment or drying fans are in operation. The use of prearranged arm and hand signals is suggested under these conditions.

Rule 2. Never enter a bin of flowing grain. If you drop a grain probe or shovel, first stop the flow of grain, take the precautions given in Rule 1, then retrieve the lost item. Remember, no piece of equipment is worth a human life (fig. 5.38).

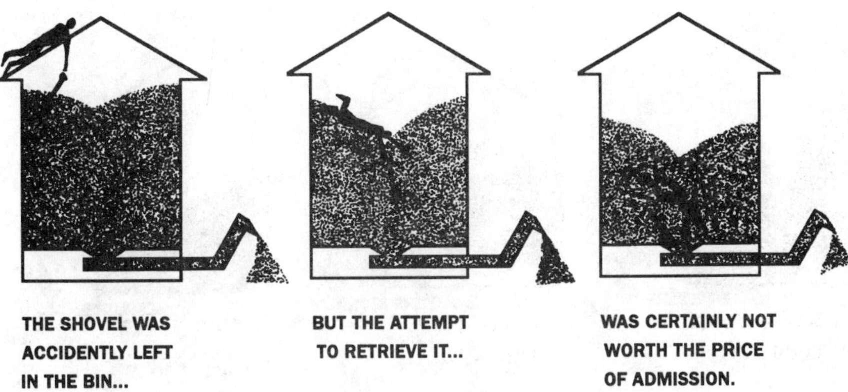

THE SHOVEL WAS
ACCIDENTLY LEFT
IN THE BIN...
BUT THE ATTEMPT
TO RETRIEVE IT...
WAS CERTAINLY NOT
WORTH THE PRICE
OF ADMISSION.

Figure 5.38–Never enter a bin of flowing grain for any reason. Always stop the flow of grain and "lock-out" the unloading system before entering.

Rule 3. Don't enter a bin without knowing its previous unloading history. This is especially true if the surface appears crusty because that may mean that the grain has bridged. Always be cautious before walking on any surface crust. If the bin has been out of condition, be sure it is well ventilated and enter slowly because of the danger from toxic gases, microbial spores and a reduced oxygen content. For this situation, be sure to follow the procedure suggested in Rule 1.

Rule 4. If you feel you must enter the bin alone and the bin has unloading equipment, you should lock out the control circuit, tell someone what you are doing, and post a sign on the control switch informing other workers that you are in the bin. Otherwise, a fellow worker may start the unloading equipment with you inside. Likewise, check each bin before you begin to unload it to be sure that no one is in the bin. For bins that unload by gravity flow, lock out the control gate and follow the same general procedure as with bins that have unloading equipment.

Rule 5. Be careful in any rescue attempt to avoid being pulled into the flowing grain and becoming a second accident (fig. 5.39). Likewise, be especially cautious when attempting to rescue someone who has been overcome by toxic gases or by breathing air with a reduced oxygen content. In these circumstances it will probably be impossible for you to enter the bin and pull the individual to safety without being overcome in the same way. To avoid placing yourself in this situation, it is imperative that the bin be well ventilated, that you enter cautiously, and that you follow the instructions given in Rule 1. Running the aeration system would probably help ventilate the bin. However, air movement through the grain mass may be severely limited if the grain is out of condition.

Rule 6. Safety measures should include the installation of ladders and ropes on the inside of the bin. Note that possibly you can "walk down" a bin if you stay near the outside of the bin wall and keep moving although walking in the soft grain will be very difficult. However, the

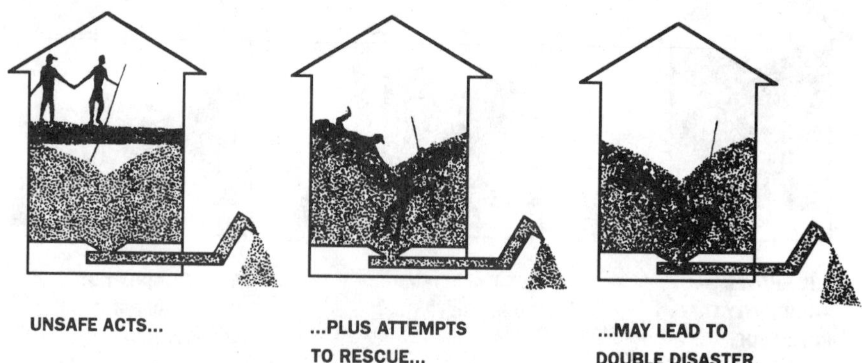
UNSAFE ACTS...
...PLUS ATTEMPTS TO RESCUE...
...MAY LEAD TO DOUBLE DISASTER.

Figure 5.39–Take the proper precautions the first time. You may not have a second chance.

best preventive measure is to avoid being caught in a potentially dangerous situation by practicing the rules of safety when working with grain.

Please - Before It's Too Late

Discuss the safety hazards of flowing grain with your family, employees, or fellow workers. It is the responsibility of each of us to keep informed of possible unsafe situations and take the necessary precautions to prevent their occurrence. The dangers associated with suffocation in flowing grain are no exception.

Problems

5.1 1,200 cfm of air is to be moved through a duct at a velocity of 1,000 fpm. What is the minimum cross-sectional area required?

5.2 If the maximum acceptable exit velocity from a duct carrying an airflow of 1,200 cfm is 25 fpm, what is the minimum surface area?

5.3 A 15-in. square duct used to supply air for a drying system requires a minimum total surface area of 100 ft^2. What is the minimum duct length?

5.4 Size the side duct in figure 5.13 if it is to be used in a system aerating 17,000 bushels. Assume a round duct in a positive pressure system carrying 0.2 cfm/bushel of airflow.

5.5 Shelled corn is to be stored at a depth of 18 ft in a 30-ft diameter bin. An airflow rate of 0.6 cfm/bu is required to aerate the grain. The grain is to be stored for 24 weeks and the system is operated 24 h every 2 weeks. The cost of electricity is $0.06/kW-h.
(a) What is the total cost of operating the system?
(b) What is the cost per bushel to operate the system?

References for all chapters begin on page 541.

6

On-Farm Materials Handling

Grain handling systems are composed of a series of materials handling mechanisms connected by surge containers. Each of these mechanisms has its place in on-farm materials handling. Proper selection and sizing of each conveyor and storage container requires an understanding of the principles of materials handling and their influence on engineering design.

Principles of Materials Handling

The principles of materials handling involve an understanding of (1) the basic principles of physics in order to determine power requirements for transporting material, and (2) the basic principles of flow in serial and parallel transport modes. The objectives of this chapter are to relate each of these areas to the flow of granular material.

Energy

Energy may be defined as the capacity or ability to do work. Energy may be found in several forms including electrical and chemical. However, mechanical energy will be the form addressed here. Mechanical energy may be defined further as being either the energy of motion (called kinetic energy) or the energy of position (called potential energy). Energy cannot be created or destroyed. However, it can change form, and some forms of energy, such as heat losses from friction, may not be usable in conveying granular materials.

Work

Work is defined as a force acting over a distance. In equation form:

$$\text{work} = (\text{force}) * (\text{distance in direction of force}) \tag{6.1}$$

Note that work is defined as being in the same direction as the movement of the object. The total force being applied to an object may be much greater than its portion being directed along the path that the object is moving. However, only the “in-line motion” portion of the force is used to determine work. The “in-line” portion of the force may be determined by multiplying the total force by the cosine of the angle

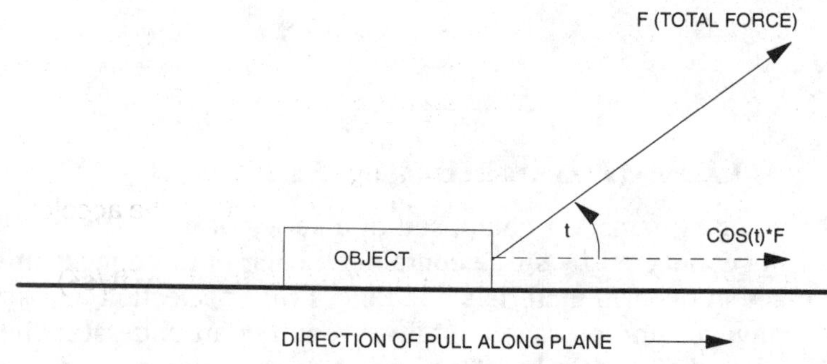

Figure 6.1–Force and direction of movement

between the direction of movement (fig. 6.1). The general form of the "work" equation is:

$$\text{work} = \text{force} * \text{distance} * \text{COS}(t) \tag{6.2}$$

where t represents the angle between the direction of the force and the direction of the movement. Again, if the object does not move, work is not being done regardless of the force being applied.

Force

A force is a "push" or "pull" exerted on a body. It has both a magnitude and a direction. Newton's Laws of Motion state the following:

1. A body will maintain either at a state of rest or of constant motion along a straight line unless acted upon by some unbalanced force to change that state.
2. An unbalanced force acting on a body produces in it an acceleration that is in the direction of the force, in direct proportion to the force, and inversely proportional to the mass of the body.
3. To every action (or force), there is an equal and opposite reaction (or force).

In terms of materials handling, Newton's Laws of Motion state that forces will be required to accelerate or decelerate the flow of material or to change the direction of flow. Likewise, any force that is applied will be countered by an equal and opposite force. In essence a "pull" in one direction generates a "push" of equal magnitude but in the opposite direction. The forces do not cancel each other, however, because they are applied to separate bodies. For example, suppose a man drags an object using a rope. The rope "pulls" the object in the direction of the man while at the same time the man "pushes" in the opposite direction against the earth.

Force is defined as:

$$\text{force} = \text{mass} * \text{acceleration} \quad (6.3)$$

where for purposes of materials handling:

mass = weight of the object (lb) divided by the acceleration associated with gravity (32 ft/s^2)

acceleration = rate at which an object is accelerating (ft/s^2)

Work and Force

Work results from a force acting over a distance. The types of forces associated with work include (Merkel, 1974):

1. Work against frictional forces.
2. Work against gravitational forces.
3. Work against elastic forces.
4. Work against inertial force.

Frictional Forces. The resistance to an object sliding across a surface is referred to as friction. Frictional forces attempt to impede the motion of the object and convert the kinetic energy of the object into heat energy. This can be verified easily by placing one's hand along the bottom of a downspout through which grain is being discharged. In fact, if the frictional forces are great enough, the moving object will be stopped completely. In the case of "falling" through air or other fluids, an object will reach a "terminal" velocity when the frictional forces associated with moving through the air equals the gravitational forces.

The mathematical equation for determining frictional force is:

$$F = M * N \quad (6.4)$$

where

F = frictional force (lb)

M = coefficient of friction (dimensionless)

N = normal (perpendicular) force of the object against the plane

The normal (or perpendicular) force of the object against the plane may be defined as:

$$N = (\text{weight of the object, lb}) * \text{COS}\ (j) \quad (6.5)$$

where j represents the angle between the direction of travel for the object and a horizontal plane. [Technically in equation 6.5, the object's weight is equal to its mass times the acceleration associated with gravity (32 ft/s^2). For purposes of grain flow, "at earth level" weight will be used for simplicity.]

In the case of flowing grain, frictional forces exist all along the grain's contact path with a relatively stationary plane. This results in the relatively large frictional forces associated with augers because granular materials are in contact with the flighting for the entire length of the auger.

Gravitational Forces. Work is required to move an object against gravitational forces to a higher plane:

$$\text{work} = (\text{weight of the object, lb}) * (\text{vertical distance, ft}) \quad (6.6)$$

The work required to move grain vertically is the weight of the grain times the vertical distance that the grain is to be transported.

Elastic Forces. Work may be accomplished by elastic forces. Elasticity refers to a characteristic of a material that allows it to return to its original shape after an external force is removed. A spring is the most common example. It is assumed that force is directly proportional to the amount of deflection of the spring. In equation form:

$$\text{force} = k * (\text{deflection of spring, ft}) \quad (6.7)$$

where

k = spring constant and is a measure of the stiffness of the spring (lb/ft)

The work by or against a spring is:

$$\text{work} = (\text{average force applied, lb}) * (\text{displacement, ft}) = k * X^2/2 \quad (6.8)$$

where X = displacement (ft)

Generally, the use of springs has little application to materials handling systems other than in some forms of scales or as shock absorbers and dampers.

Pressure Forces. Pressure is defined as force per unit area. Water pressure, for example, is expressed in pounds per square inch. Force is the product of area and pressure. In equation form:

$$\text{force (lb)} = (\text{pressure, lb/in.}^2) * (\text{area, in.}^2) \quad (6.9)$$

where area is the total contact area of the fluid.

By substituting force into the equation for work:

$$\text{work (ft-lb)} = (\text{pressure, lb/in.}^2) * (\text{area, in.}^2) * (\text{displacement, ft}) \qquad (6.10)$$

where

area = the cross-section of the fluid in the direction of travel, usually the cross-sectional area of a hydraulic cylinder

displacement = distance that the "piston" travels in the cylinder

With materials handling, the work against pressure forces is associated with hydraulic pumps and pneumatic conveyors.

Inertial Forces. Inertial forces refer to the forces required to alter the motion of a body. From Newton's Laws of Motion, a body at rest will remain at rest, and a body in motion will remain in motion unless altered by outside forces. With materials handling, force will be required to either increase or decrease the velocity of flowing grain. In equation form:

$$\text{inertial work (ft-lb)} = \frac{(\text{weight of object, lb}) * (V_f^2 - V_i^2)}{(64\ \text{ft/s}^2)} \qquad (6.11)$$

where

V_f = final velocity of object (ft/s)

V_i = initial velocity of object (ft/s)

For example, work will be required to increase the velocity of grain falling on a conveyor belt.

Power

The term of greatest interest to designers of materials handling systems is "power". Power refers to the rate at which work is being done. In equation form:

$$\text{power} = \text{work/time} \qquad (6.12)$$

The most common measure of power in the United States, insofar as materials handling is concerned, is horsepower. By definition:

$$1 \text{ horsepower} = 33{,}000 \text{ ft-lb/min}$$
or
$$= 550 \text{ ft-lb/s}$$

One horsepower is required to lift a 550 lb object to a height of 1 ft in 1 s, or to lift a 1 lb object 550 ft in 1 s. In designing materials handling systems, the rate of flow translates directly to the power requirements for obtaining the flow.

Summary

The conveying of granular materials requires that forces act on the material over distances. The forces encountered relate to friction, gravity, elasticity, pressure and inertia. The work required to move granular material is the sum of the product of each of these forces times the distance that each force is applied. Power requirements depend directly on how fast the work is to be accomplished.

Examples of Horsepower Calculations

Practically every form of materials handling employs all the forces previously mentioned. The following examples relate to application of these concepts recognizing that many simplistic assumptions have to be made concerning the handling method.

Example 1: Frictional Losses. Grain is flowing through a downspout at the rate of 1,000 bu (60 lb/bu)/h. Each bushel occupies 1.25 ft^3. The downspout is 8 in. in diameter, 20 ft in length, and is at a 40° angle to the ground. One half of the circumference is in direct contact with the grain. The coefficient of friction between the pipe and the grain is 0.4. What is the horsepower equivalent to the friction losses associated with the flowing grain?

The first step in solving this problem is to determine the force of the grain against the downspout.

The volume of the downspout is:

$$\begin{aligned} \text{downspout volume (ft}^3\text{)} &= \text{pi} * (\text{radius})^2 * 20 \text{ ft} \\ &= \text{pi} * (4 \text{ in.}/12 \text{ in./ft})^2 * 20 \text{ ft} \\ &= 6.9813 \text{ ft}^3 \end{aligned}$$

where pi = 3.1416

If the downspout is half filled, then it contains 3.4907 ft^3 of grain or 167.55 lb of grain at any one time. However, the force of the grain against the downspouting is at a 40° angle. From equation 6.5:

$$\begin{aligned} N &= 167.55 \text{ lb} * \text{COS } 40° \\ &= 128.35 \text{ lb} \end{aligned}$$

From equation 6.4:

$$\text{force} = M * N = 0.4 * 128.35 \text{ lb} = 51.34 \text{ lb}$$

The work against friction for grain traveling the 20 ft of downspouting is:

$$\text{work} = 51.34 \text{ lb} * 20 \text{ ft} = 1{,}026.8 \text{ ft-lb}$$

The horsepower equivalent of the frictional force is dependent on the rate at which work is done. The rate for this example may be found by calculating how long the grain remains in the 20-ft length of downspouting:

$$\text{effective time in downspouting (min)} = \frac{\text{effective volume of downspouting}}{\text{flow rate of grain}}$$

$$= \frac{3.49 \text{ ft}^3}{(1{,}000 \text{ bu/h} * 1.25 \text{ ft}^3\text{/bu}) / 60 \text{ min / h}}$$

$$= 0.16752 \text{ min}$$

The horsepower equivalent for friction may now be computed:

$$\text{horsepower (friction)} = \frac{\text{work/minute}}{33{,}000 \text{ ft-lb/min-hp}}$$

$$= \frac{1{,}026.8 \text{ ft-lb/0.16752 min}}{33{,}000 \text{ ft-lb/min-hp}}$$

$$= 0.19 \text{ hp}$$

Example 2: Gravitational Forces. The design rate for a bucket elevator is 2,000 bu/h (56 lb/bu). How much power is required to overcome the effects of gravitational pull if the grain is to be elevated 100 ft?

$$\text{work} = 2{,}000 \text{ bu} * 56 \text{ lb/bu} * 100 \text{ ft} = 11{,}200{,}000 \text{ ft-lb}$$

This is to be done in 60 min. The horsepower requirement is:

$$\text{horsepower (gravitational pull)} = \frac{11{,}200{,}000 \text{ ft-lb}/60 \text{ min}}{33{,}000 \text{ ft-lb min}}$$

$$= 5.66 \text{ hp}$$

Example 3: Elastic Forces. A spring is to be used to determine the weight of sacks of grain. A calibration is made using a 100-lb weight with the spring compressed a total of 4 in. What is the spring constant if it is linearly proportional?

From equation 6.7:

$$\begin{aligned} k &= \text{force/deflection} \\ &= 100 \text{ lb}/4 \text{ in.} \\ &= 25 \text{ lb/in.} \end{aligned}$$

Example 4: Pressure Forces. Two hydraulic cylinders, each 8 in. in diameter (4 in. radius), are to be used in lifting a platform for the dumping of trucks. Maximum loaded truck weight is 90,000 lb. What horsepower is needed to lift the largest loaded truck to a height of 1 ft in 5 s if the cylinder fluid pressure is 50 lb/in.^{2}?

From equation 6.10:

$$\begin{aligned} \text{work} &= 50 \text{ lb/in.}^2 * 2 \text{ cylinders} * (4 \text{ in.})^2 * \text{pi} * 1 \text{ ft} \\ &= 5{,}026.5 \text{ ft-lb} \end{aligned}$$

This is to be accomplished in 5 s. Therefore, the horsepower requirement is:

$$\text{horsepower (pressure)} = \frac{5{,}026.5 \text{ ft-lb}/5 \text{ s}}{550 \text{ ft-lb s}}$$

$$= 1.83 \text{ hp}$$

Example 5: Inertial Forces. Grain is to be placed on a belt traveling 200 ft/min at a rate of 60,000 lb in 30 min. What horsepower requirements will be needed to overcome the forces of inertia?

From equation 6.11:

$$\text{work} = \frac{60{,}000 \text{ lb} * \left(\left[\left\{200 \text{ ft/min}\right\} / 60 \text{ s/min}\right]^2 - 0^2\right)}{64 \text{ ft-s}^2}$$

$$= 10{,}417 \text{ ft-lb}$$

This work is to be accomplished in 30 min.

$$\begin{array}{c}\text{horsepower}\\ \text{(inertial forces)}\end{array} = \frac{10{,}417 \text{ ft-lb/30 min}}{33{,}000 \text{ ft-lb/min}}$$

$$= 0.01 \text{ hp}$$

As can be seen from this example, inertial forces are generally small when compared to the gravitational forces needed to convey grain.

Principles of Sizing Components

The principles of sizing materials handling components are:

1. At least one component of the materials handling system will always be the limiting factor in terms of capacity.
2. The limiting component should have sufficient capacity to handle the specified design flow rates of the system.
3. Each materials handling component must be able to handle the flow of material into it from all other equipment components that operate simultaneously and feed it directly.
4. There should be safeguards against accidental overloading of handling components that are located "downstream" from larger capacity equipment.
5. Selection of a component should be based on its ability to handle the material at the least possible cost per unit of material processed.

The first step in designing a materials handling system is to specify the desired processing rate between delivery and receiving stations. The designer must recognize that at least one component in the materials handling system will always be limiting. Otherwise, the system would process material at an infinitely high rate. The key to optimizing the design is to select the materials handling component with the smallest capacity that is sufficient to meet or exceed the design specifications. When optimizing material flow, it is desirable that all

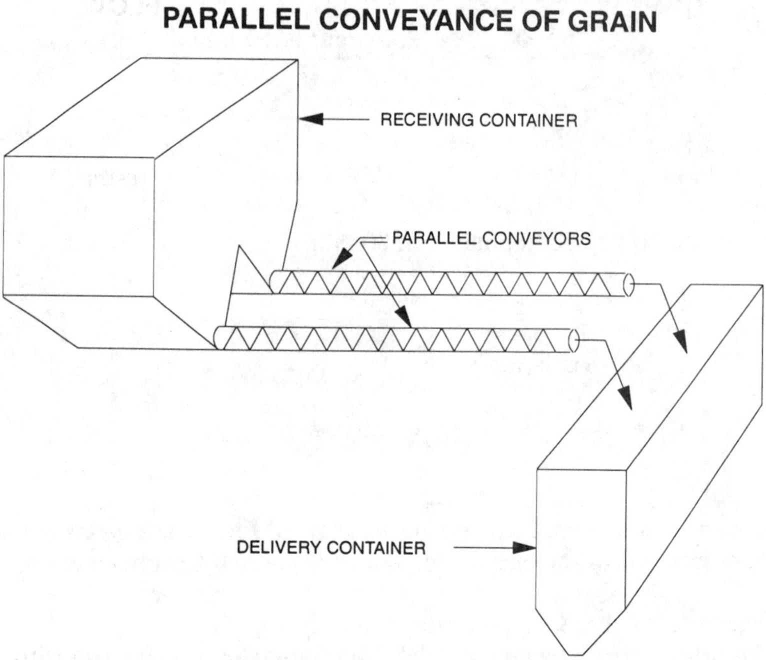

PARALLEL CONVEYANCE OF GRAIN

RECEIVING CONTAINER

PARALLEL CONVEYORS

DELIVERY CONTAINER

Figure 6.2–Parallel conveyance of grain.

equipment components operate at their maximum capacities but none are to exceed the design capacity.

Materials handling equipment may be placed in series or parallel (figs. 6.2 and 6.3). A certain amount of "surge" capacity exists at each junction between conveyors or between conveyors and receiving or delivery points (fig. 6.3). The sizing of these surge capacities is primarily dependent on economics. Engineering considerations (which translate directly to economic considerations) include having space for surge capacity and ensuring that downstream conveyors do not limit capacity, especially if flow stoppages may occur.

It is preferable to size the downstream conveyor in serial systems slightly larger than the upstream conveyor in order to minimize the surge capacity between the two conveyors. Theoretically, surge is not needed in serial systems if the flow rates from one conveyor to another are the same or if all conveyors are operating at less than maximum capacity. In the later case, the excess capacity serves to alleviate surges in the system flow.

One method of regulating granular materials flow through the handling system is to place control valves or gates at the exit point of each surge capacity. The set points for these devices must ensure that the downstream conveyor will not be overloaded, while, at the same

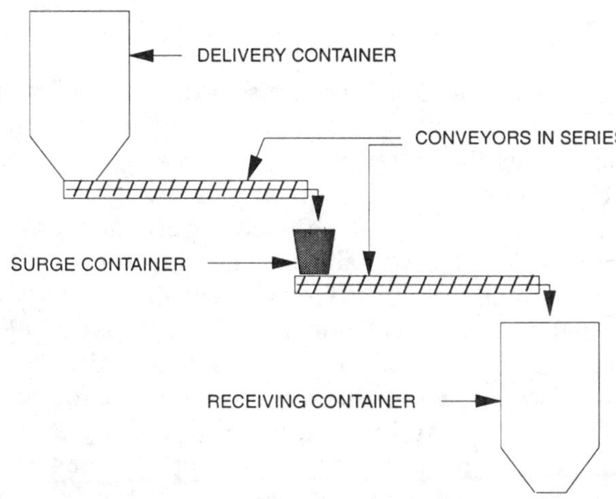

Figure 6.3–Surge protection in serial grain flow.

time, keeping the surge capacity from being exceeded and causing an eventual stoppage of the upstream conveyor.

The designer should consider that some materials handling components, most notably augers, will have different capacities depending on the type and condition of the material being handled. For example, the same auger will convey different quantities of grain depending on the grain type (i.e., corn or soybeans), moisture content (higher capacities at lower grain moistures) and degree of cleanliness (lower capacities with trashy grain). Conveyor design in series should be based on the maximum expected capacity and include control valves for further flow regulation.

Another consideration is the starting of the materials handling system. The order of starting should begin with the most downstream conveyor, working back to the receiving point. This ensures that the entire pathway is clean and able to convey material without overloading the surge capacities or, in some cases, exceeding the starting current needed for electrical motors.

Conveyors have multiple uses in most grain handling systems, especially those at the farm level. For example, the same bucket elevator may be used to receive wet grain and convey it to the wet holding bin, receive dry grain from the dryers and convey it to storage, and receive grain from storage and convey it to trucks for delivery to market. Although it may be oversized for some of its applications, the bucket elevator should be of sufficient capacity to satisfy each of the situations. The design procedure is to trace each flow pattern to ensure each conveyor in series is of sufficient capacity to handle the incoming material.

Summary

Ultimately, economic considerations in their broadest sense govern the design. The decision maker must recognize that many different systems will convey the material and may have a similar purchase cost. However, the economics of materials damage, maintenance, future expansion and utilization, convenience, safety, and processing speed must all be considered, especially as related to the laws of diminishing returns. For example, a bucket elevator with a capacity of 1,600 bu/h will load an 800-bu trailer in 30 min. A 3,200-bu/h bucket elevator will fill the same truck in 15 min. Will the 15 min saved justify the added cost of the elevator? The answer to this question will depend in large part on how many trucks will be loaded, both on a daily and yearly basis. Clearly, the design of materials handling systems requires much subjective as well as objective judgement. In the following sections, different characteristics of typical grain handling equipment will be addressed.

Augers

Augers are used extensively for materials handling, especially in on-farm situations. Farm applications include (1) portable augers for loading and emptying grain carried by delivery vehicles, (2) augers attached directly to trucks and wagons for unloading, (3) permanent augers placed under grain bins for unloading, (4) flexible augers for conveying feed, (5) sweep augers for pulling grain to a center unloading point, and (6) screw augers used with stirring devices.

Auger Components

The auger system is composed of several parts (fig. 6.4). The screw conveyor is composed of a pipe with a welded steel strip that is formed into a continuous helix. The helix is referred to as the flighting. The distance along the pipe from one point on the flighting to the next similar point is called the "pitch". Couplings and shafts refer to the mechanisms by which two screw conveyors are joined. Hangers are used to provide support and maintain alignment of the screw conveyor. The screw conveyor may be housed in a "tube" or "trough". The tube is a hollow cylinder while the trough has a "U" shape, hence the term "U-trough" augers.

Design Capacity and Power Requirements

The theoretical capacity of a full screw conveyor is:

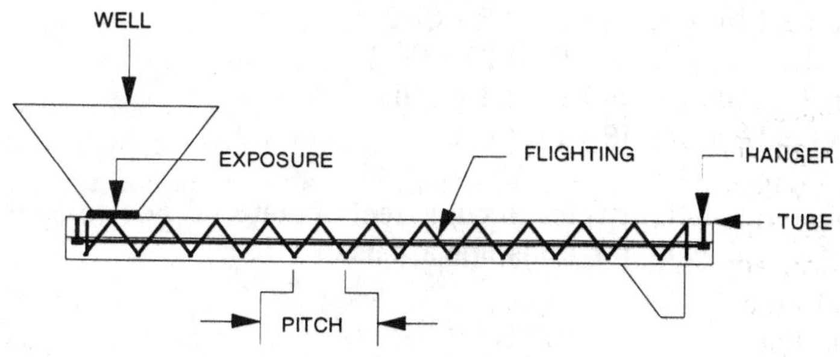

Figure 6.4—Auger components.

$$CFTHR = \frac{(D^2 - d^2) * P * RPM}{36.6} \tag{6.13}$$

where

CFTHR = ft^3/h of material being moved by a full auger
D = diameter of screw (in.)
d = diameter of shaft (in.)
P = pitch of the auger, in. (usually the pitch is equal to D)
RPM = revolutions per minute (rpm) of the shaft

The actual capacity of the screw conveyor may be one third to one half of the theoretical capacity because of material characteristics, screw-housing clearance, and the degree of elevation.

Horsepower requirements are difficult to determine because of the diversity of different augers and materials. The following equation approximates the horsepower required for an auger operating in the horizontal position (Henderson and Perry, 1966):

$$CHP = \frac{C * L * W * F}{33{,}000} \tag{6.14}$$

where

CHP = computed horsepower
C = conveyor capacity (ft^3/min)
L = conveyor length (ft)
W = bulk weight of material (lb/ft^3)
F = material factor (see table 6.1)

The horsepower in equation 6.14 must be adjusted under the following conditions (HP = horsepower of horizontal auger):

a. If CHP < 1, set HP = 2.0 * CHP
b. If 1 ≤ CHP < 2, set HP = 1.5 * CHP
c. If 2 ≤ CHP < 4, set HP = 1.25 * CHP
d. If 4 ≤ CHP < 5, set HP = 1.1 * CHP
e. If CHP ≥ 5, set HP = 1.0 * CHP

The above equations refer to cubic feet of material. For conversion to bushels, the following relationships are used:

$$1 \text{ bushel} = 1.25 \text{ ft}^3$$
$$1 \text{ ft}^3 = 0.8 \text{ bu}$$

A more thorough engineering approach to screw conveyor design is beyond the scope of this book. For most farm applications, manufacturer's literature may be used. It should be noted that augers become considerably less efficient when they are used to convey material vertically. Estimates of capacity and horsepower requirements are given in tables 6.2-6.8 for representative farm situations (Kentucky Agricultural Engineering Handbook; Midwest Plan Service, 1968). (For a summary of related data, see Pierce and McKenzie, 1984).

Table 6.1. Horsepower factors for screw conveyors and bulk densities of grain (Henderson and Perry, 1966)

Material	Bulk Weight (lb/ft³)	Horsepower Material Factor (F)
Barley	38	0.4
Beans	48	0.4
Beans, caster	36	0.5
Beans, soy	45-50	0.5
Bran	16	0.4
Clover seed	48	0.4
Corn, shelled	45	0.4
Cornmeal	40	0.4
Cotton seed (dry)	25	0.9
Cotton seed hulls	12	0.9
Lime, ground	60	0.6
Milk, dried	36	1.0
Oats	26	0.4
Peanuts, unshelled	15-20	0.7
Rice, rough	36	0.4
Rye	44	0.4
Timothy seed	36	0.7
Wheat	48	0.4

Tube and U-trough Augers

Tube and U-trough augers are the two most common conveyors of agricultural materials, especially grain. Generally, U-trough augers operate at a lower speed than tube augers. Their screw diameters are usually larger giving them greater capacity per revolution. Because of its lower speed, the U-trough auger generally is considered to cause less grain damage than tube augers. However, tube augers are less expensive and meet the needs of most on-farm materials handling situations.

Portable Augers

Portable augers are used extensively for transferring grain into and out of storage and are most often tube-type augers. Diameters typically range from 6 to 12 in. with maximum capacities near 4,500 bu/h. Portable augers range in length from about 27 ft to 81 ft. Augering grain with this type of auger at angles exceeding 45° is considered impractical.

One of the important considerations when filling bins is to be able to reach the bin's center opening (unless a roof auger is used). The auger must be able to reach the center of the bin while being able to clear the bin's eaves (fig. 6.5). At the same time, it is desirable to keep the auger

Table 6.2. Capacities and horsepower requirements (at the drive shaft) for representative screw conveyors and selected types of grain. (Based on a 6-in. auger handling dry shelled corn*†‡)

Items		Angle of Elevation of Screw									
Auger SP.	Intake Exposure	0°		22.5°		45°		67.5°		90°	
RPM	(in.)	(bu/h)	(hp/10 ft)	(bu/h)	(hp/10 ft)	(bu/h)	(hp/10 ft)	(bu/h)	(hp/10 ft)	(bu/h)	(hp/10 ft)
200	6	590	0.20	520	0.30	370	0.33	280	0.31	220	0.25
	12	590	0.38	550	0.41	500	0.44	400	0.44	280	0.32
	18	620	0.32	570	0.43	510	0.47	430	0.45	310	0.35
	24	630	0.44	590	0.50	550	0.55	470	0.54	350	0.40
400	6	970	0.35	850	0.52	650	0.60	480	0.57	380	0.46
	12	1,090	0.56	1,010	0.82	850	0.88	690	0.83	520	0.70
	18	1,170	0.74	1,070	0.92	940	1.02	720	0.92	560	0.80
	24	1,190	0.97	1,110	1.13	1,010	1.18	830	1.07	660	0.92
600	6	1,210	0.49	1,050	0.72	820	0.82	590	0.77	490	0.64
	12	1,510	0.84	1,400	1.22	1,160	1.28	910	1.16	740	1.05
	18	1,650	1.17	1,500	1.42	1,270	1.52	1,010	1.42	800	1.23
	24	1,700	1.47	1,570	1.74	1,440	1.80	1,140	1.60	920	1.40
800	6	1,320	0.58	1,100	0.86	890	0.95	640	0.92	540	0.77
	12	1,760	1.07	1,660	1.54	1,370	1.62	1,080	1.46	890	1.32
	18	1,990	1.57	1,790	1.96	1,510	2.08	1,220	1.94	1,000	1.64
	24	2,140	1.95	1,910	2.32	1,740	2.39	1,360	2.12	1,100	1.89

* 54-56 lb/bu @ 14.5% w.b.
† Add 10% to table values to determine total hp requirements.
‡ Kentucky Agricultural Engineering Handbook.

Table 6.3. Capacities and horsepower requirements (at the drive shaft) for representative screw conveyors and selected types of grain. (Based on a 6-in. auger handling wet shelled corn*†‡)

Items		Angle of Elevation of Screw									
Auger		0°		22.5°		45°		67.5°		90°	
RPM	MC (%)	(bu/ h)	(hp/ 10 ft)	(bu/ h)	(hp/ 10 ft)	(bu/ h)	(hp/ 10 ft)	(bu/ h)	(hp/ 10 ft)	(bu/ h)	(hp/ 10 ft)
200	14%	590	0.28	550	0.41	500	0.44	400	0.44	280	0.32
	25%	370	1.37	320	1.40	280	1.31	200	0.97	160	0.31
400	14%	1,090	0.56	1,010	0.82	850	0.88	690	0.83	520	0.70
	25%	700	1.84	620	1.89	510	1.78	400	1.45	300	0.70
600	14%	1,510	0.84	1,400	1.22	1,160	1.28	910	1.16	740	1.05
	25%	950	2.32	820	2.34	680	2.27	520	1.92	410	2.09
800	14%	1,760	1.07	1,660	1.54	1,370	1.62	1,080	1.46	890	1.32
	25%	1,100	2.80	950	2.85	770	2.75	580	2.44	470	1.55

* All with 12-in. exposure but at either 14% or 25% w.b.
† Add 10% to table values to determine total hp requirements.
‡ Kentucky Agricultural Engineering Handbook.

Table 6.4. Capacities and horsepower requirements (at the drive shaft) for representative screw conveyors and selected types of grain. (Based on a 6-in. auger handling dry soybeans *†‡)

Items		Angle of Elevation of Screw									
Auger SP.	Intake Exposure	0°		22.5°		45°		67.5°		90°	
RPM	(in.)	(bu/ h)	(hp/ 10 ft)	(bu/ h)	(hp/ 10 ft)	(bu/ h)	(hp/ 10 ft)	(bu/ h)	(hp/ 10 ft)	(bu/ h)	(hp/ 10 ft)
200	6	490	0.30	410	0.41	320	0.41	240	0.38	180	0.34
	12	500	0.40	430	0.53	360	0.57	290	0.50	220	0.40
	18	520	0.50	500	0.60	440	0.66	360	0.60	240	0.45
	24	540	0.60	520	0.67	470	0.68	390	0.64	290	0.52
400	6	880	0.52	710	0.71	570	0.77	400	0.70	310	0.60
	12	990	0.84	830	1.14	690	1.20	540	1.04	390	0.79
	18	1,110	0.98	1,030	1.18	880	1.29	740	1.23	460	0.95
	24	1,180	1.36	1,040	1.62	900	1.63	800	1.54	560	1.14
600	6	1,080	0.68	890	0.96	700	1.07	510	1.00	390	0.87
	12	1,350	1.20	1,130	1.61	930	1.71	710	1.48	500	1.10
	18	1,620	1.45	1,510	1.74	1,280	1.94	1,050	1.88	660	1.47
	24	1,690	2.13	1,520	2.52	1,320	2.51	1,100	2.32	790	1.76
800	6	1,180	0.78	960	1.12	740	1.28	550	1.22	420	1.10
	12	1,610	1.51	1,310	1.98	1,080	2.10	820	1.84	640	1.50
	18	1,980	1.93	1,840	2.29	1,530	2.54	1,230	2.44	810	1.98
	24	2,020	2.93	1,850	3.43	1,640	3.48	1,320	3.24	1,000	2.56

* 54-56 lb/bu @ 11-12% w.b.
† Add 10% to table values to determine total hp requirements.
‡ Kentucky Agricultural Engineering Handbook.

Table 6.5. Capacities and horsepower requirements (at the drive shaft) for representative screw conveyors and selected types of grain. (Based on an 8-in. U-trough auger handling dry shelled corn*†‡)

Items		Angle of Elevation of Screw							
Auger Speed	Intake Exposure	0°		15°		25°		35°	
RPM	(in.)	(bu/h)	(hp/10 ft)	(bu/h)	(hp/10 ft)	(bu/h)	(hp/10 ft)	(bu/h)	(hp/10 ft)
100	14	980	0.61	875	0.78	850	0.79	755	0.80
200	14	1,880	1.65	1,730	1.60	1,620	1.45	1,500	1.68
300	14	2,800	2.01	2,550	2.15	2,370	2.26	2,140	2.27
400	14	3,525	2.53	3,250	2.56	2,950	2.68	2,575	2.75

* 55 lb/bu @ 14% w.b.
* Add 10% to table values to determine total hp requirements.
‡ Kentucky Agricultural Engineering Handbook.

as near horizontal as possible in order to maximize capacity for the same power requirement. In addition, the location of the auger wheels may determine the angle required to reach the bin center. Minimum auger lengths needed for representative bin loading are given in figure 6.6. The portable auger length for a given bin can be determined from the following equations.

For auger angles greater than the bin roof slope:

$$L = \frac{BEHGT}{SIN(t)} + \frac{(0.5 * BND)}{COS(t)} \tag{6.15}$$

Table 6.6. Capacities and horsepower requirements (at the drive shaft) for representative screw conveyors and selected types of grain. (Based on an 8-in. U-trough auger handling dirty, chaffy wet shelled corn*†‡)

Items		Angle of Elevation of Screw							
Auger Speed	Intake Exposure	0°		15°		25°		35°	
RPM	(in.)	(bu/h)	(hp/10 ft)	(bu/h)	(hp/10 ft)	(bu/h)	(hp/10 ft)	(bu/h)	(hp/10 ft)
100	7.5	702	1.15	664	0.99	613	0.98	502	0.92
	15.0	712	1.17	682	1.19	668	1.23	640	1.30
	22.5	722	1.12	685	1.15	672	1.33	687	1.37
200	7.5	1,515	1.96	1,318	2.00	1,078	1.94	832	1.72
	15.0	1,450	2.27	1,432	2.65	1,342	2.60	1,205	2.55
	22.5	1,515	2.38	1,438	2.70	1,360	2.78	1,328	2.77

* 53 lb/bu @ 27.5 to 28.5% w.b.
† Add 10% to table values to determine total hp requirements.
‡ Kentucky Agricultural Engineering Handbook.

Table 6.7. Capacities and horsepower requirements (at the drive shaft) for representative screw conveyors and selected types of grain. (Based on an 8-in. U-trough auger handling wet shelled corn*†‡)

Items		Angle of Elevation of Screw							
Auger Speed	Intake Exposure	0°		15°		25°		35°	
RPM	(in.)	(bu/h)	(hp/10 ft)	(bu/h)	(hp/10 ft)	(bu/h)	(hp/10 ft)	(bu/h)	(hp/10 ft)
100	7.5	774	1.52	760	1.45	745	1.45	533	1.10
	15.0	769	1.57	758	1.66	705	1.61	663	1.41
	22.5	775	1.52	783	1.53	734	1.67	706	1.46
200	7.5	1,574	2.22	1,312	1.98	1,130	1.95	950	1.70
	15.0	1,540	2.54	1,476	2.56	1,418	2.72	1,278	2.38
	22.5	1,608	2.61	1,520	2.85	1,424	2.82	1,406	2.72

* 52.5 lb/bu @ 25.0 to 26.5% w.b.
† Add 10% to table values to determine total hp requirements.
‡ Kentucky Agricultural Engineering Handbook.

For auger angles less than or equal to the bin roof slope:

$$L = \frac{BEHGT + 0.5 * BND * TAN(r)}{SIN(t)} \tag{6.16}$$

Table 6.8. Capacities and horsepower requirements (at the drive shaft) for representative screw conveyors and selected types of grain. (Based on an 8-in. tube auger handling wet shelled corn*†‡)

Items		Angle of Elevation of Screw							
Auger Speed	Intake Exposure	0°		15°		25°		35°	
RPM	(in.)	(bu/h)	(hp/10 ft)	(bu/h)	(hp/10 ft)	(bu/h)	(hp/10 ft)	(bu/h)	(hp/10 ft)
100	7.5	690	0.68	660	0.72	610	0.78	575	0.82
	15.0	715	0.78	700	0.86	685	0.95	650	1.04
200	7.5	1,200	1.20	980	1.34	930	1.42	856	1.50
	15.0	1,200	1.80	1,170	1.90	1,130	2.05	1,070	2.20
300	7.5	1,590	1.60	1,240	1.80	1,200	1.90	1,010	2.00
	15.0	1,760	2.62	1,650	2.85	1,570	3.00	1,440	3.20
400	7.5	1,710	1.93	1,380	2.20	1,285	2.40	1,100	2.50
	15.0	–	–	2,000	3.70	1,800	3.82	1,640	4.00

* 51 to 55 lb/bu @ 23.5 to 26.5% w.b.
† Add 10% to table values to determine total hp requirements.
‡ Kentucky Agricultural Engineering Handbook.

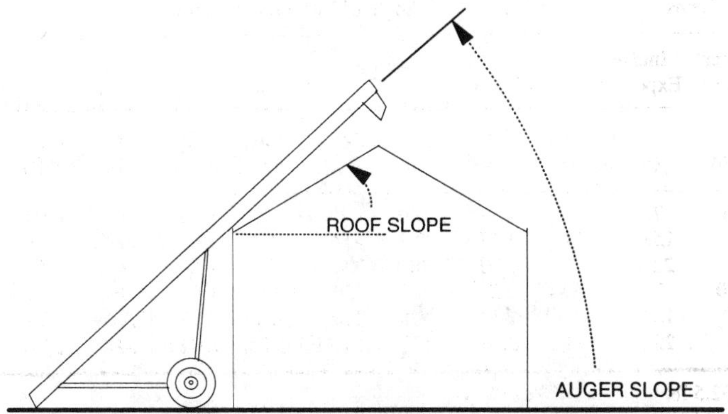

Figure 6.5–Bin eave height influence on auger discharge height.

where

L	=	length of auger to reach the center of the bin (assuming wheels and related assembly do not interfere) (ft)
BEHGT	=	eave height of the bin (ft)
BND	=	diameter of bin (ft)
t	=	angle of auger relative to the ground (°)
r	=	slope of bin roof (°)

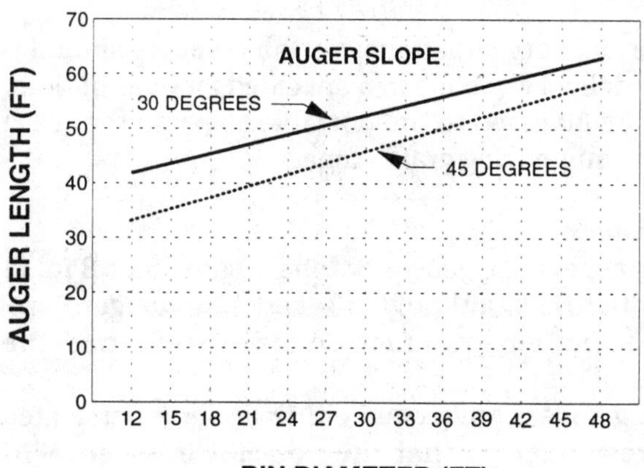

Figure 6.6–Minimum auger lengths needed to unload bins with a 17.5-ft eave height.

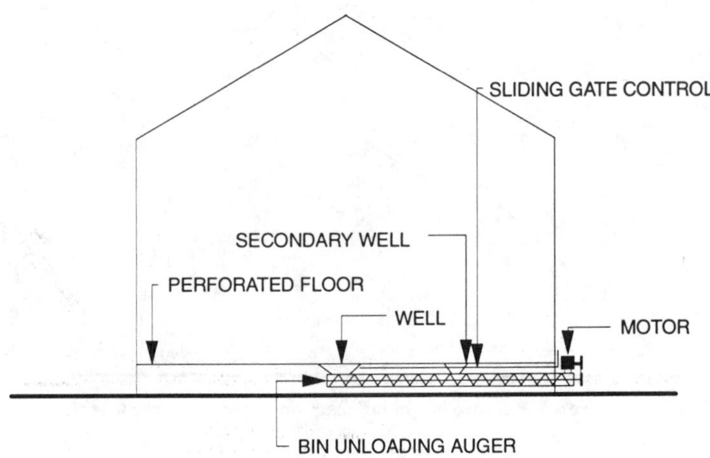

Figure 6.7–Bin unloading auger with secondary well.

From the above equations the greater the angle of the portable auger, the shorter the required auger length may be up to 45° above which the required auger length begins to increase again. Thus, if an existing auger at an angle of 45° won't reach the center of the bin, further changes in its angle will not help. The roof slope of most bins is 30°. Therefore, auger angles greater than 30° will "pivot" about the eave height of the bin. Again, the portable auger wheel assembly may result in auger lengths greater than those calculated using the above equations. With materials handling, the more important question is whether the reduction in augering capacity at increased angles justifies the savings in auger length.

In receiving or delivery situations, portable augers should be sized to load or unload a delivery vehicle in a specified amount of time. Incoming conveyors must be matched to the portable auger so that all run at near full capacity, thus minimizing grain damage.

Bin Unloading Augers

Bin unloading augers are generally tube augers, 6 or 8 in. in diameter for most farm situations, although 10- and 12-in. augers are available (fig. 6.7). Unloading augers must be sized to unload the bin in a specified amount of time.

Most unloading augers are placed either under a perforated bin floor or into a concrete duct so that they do not interfere with surface unloading. Bins should be unloaded from the center to prevent excess

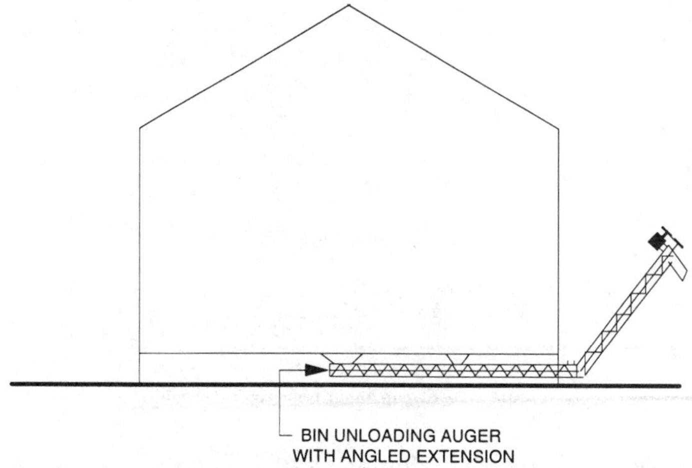

Figure 6.8–Bin unloading auger with angled extension.

stress on the bin walls. However, additional grain inlets should be located along the auger in case the center well becomes clogged.

Unloading augers generally deposit the grain directly into another

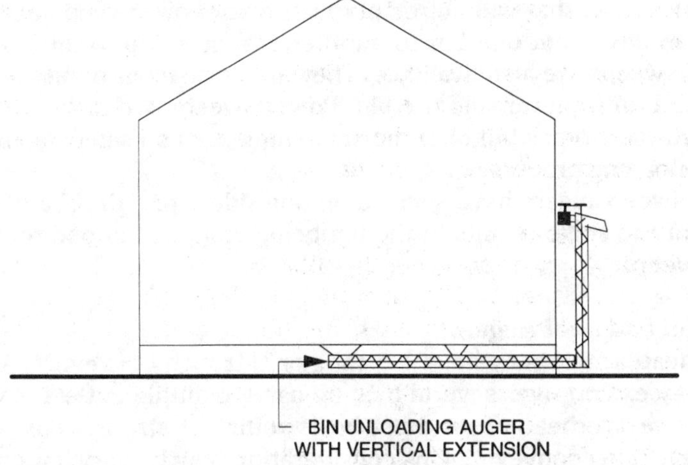

Figure 6.9–Bin unloading auger with vertical extension.

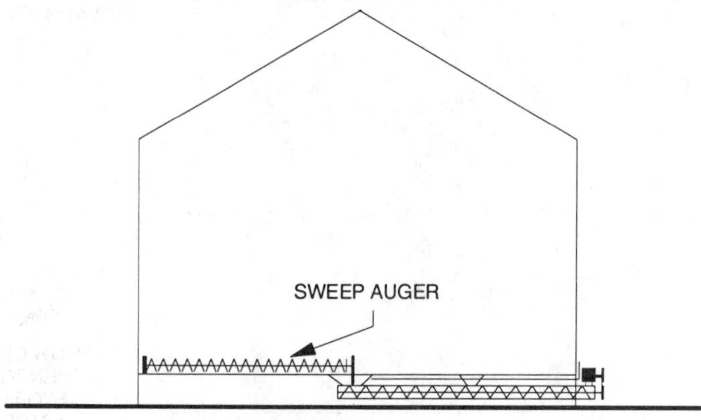

Figure 6.10–Sweep auger powered externally.

conveyor or into a receiving hopper. If the auger is too low for direct grain deposition, an angled extension may be added to elevate the grain to the desired height (fig. 6.8). Vertical auger kits may also be attached so that delivery vehicles may receive grain directly (fig. 6.9). Generally, it is more economical to direct the flow of grain to a central loading conveyor rather than duplicate a vertical unloading system for every bin. The auger maintenance increases greatly when there are changes in elevation of flow within the auger such as with angled or vertical auger attachments. Some attachments are moved from bin-to-bin as required, thus reducing the investment expense associated with multiple purchases.

Sweep augers (fig. 6.10) pivot about the center well in the bin to unload the grain that will not empty by gravity flow. Sweep augers that may be moved from one bin to another are called "portable" sweeps. "Power" sweeps are also available. They are permanent bin fixtures and are turned on from outside the bin. Power sweeps are more expensive than portable sweeps but offer more convenience and safety advantages in the unloading process.

Most sweep augers have a shield on one side. This "drag" against the motion of the auger results in the bin being unloaded in one revolution of the sweep.

Overhead and Roof Augers

Overhead augers may be either tube or U-trough conveyors. A single auger may extend over several bins by using multiple outlets along the trough. The overhead auger must be of sufficient size to convey grain from the feeding conveyor. This consideration may be very important in the case of bucket elevators with relatively large capacities. A safe guard

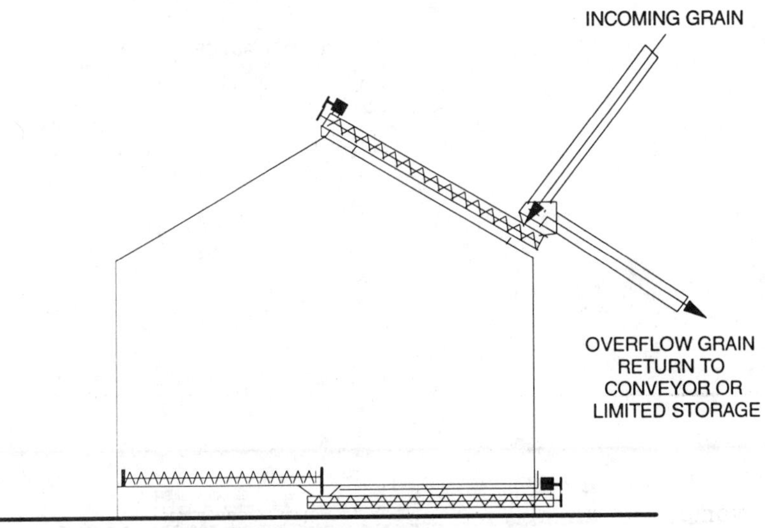

Figure 6.11–Overflow return on bin roof auger.

may be constructed at the inlet of the auger to redirect excess grain rather than clog the incoming downspout (fig. 6.11).

Roof augers (figs. 6.12 to 6.13) provide a means of reaching tall or "out-of-the-way" bins with existing materials handling equipment. In

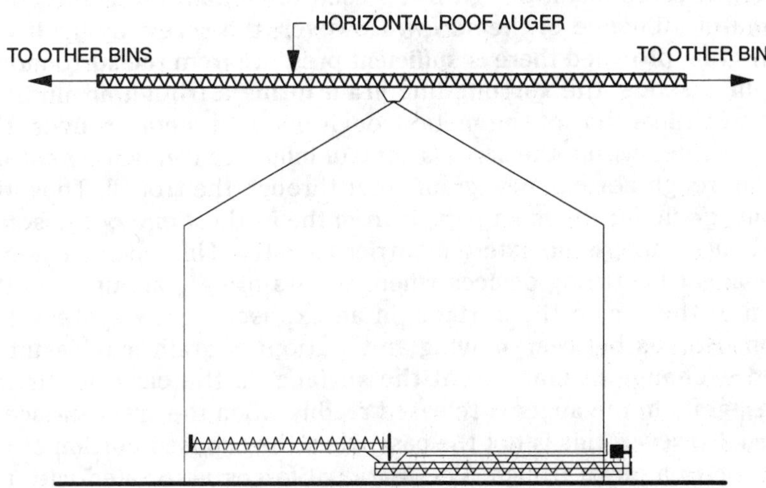

Table 6.12–Horizontal roof auger.

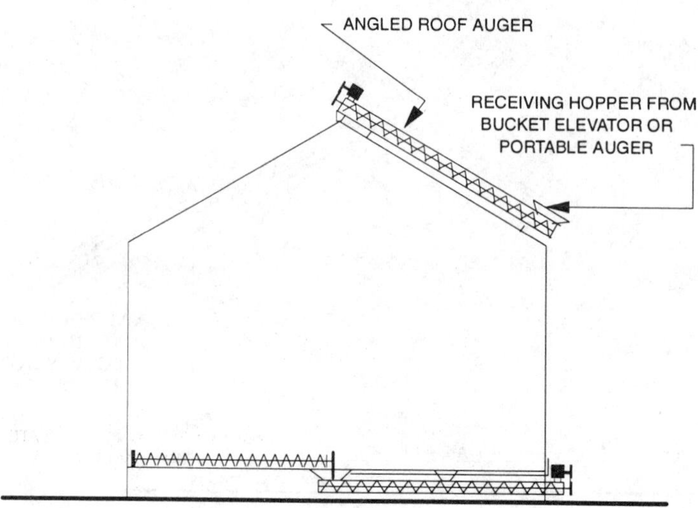

Figure 6.13–Angled roof auger.

particular, the roof auger receiving point may be reached directly by a bucket elevator while bin's center point is not accessible. Similarly, expansion bins in circular arrangements may receive grain directly from portable augers by using roof augers.

Screw Movement in Grain

Most often, grain enters the auger at one point and is conveyed to another, with the metal trough being used to contain the grain within the area of influence of the screw. However, the screw by itself will convey grain provided there is sufficient pressure from the surrounding grain. In essence, the surrounding grain forms a trough around the screw much like that of the metal trough. As with metal troughs, the grain contained within the screw's zone of influence can neither escape past the trough nor can new grain enter through the trough. Thus, the incoming grain, for the most part, is from the furthest end of the screw even though there is no external barrier to entry. This phenomenon is most evident in stirring devices where screws bring grain up from the bottom of the bin to the surface. In an exposed screw system, the frictional forces between moving and stationary grain may result in limited exchange of material at the surface. In the case of stirring devices, grain in the auger is released readily when the grain surface is reached. However, this is not the case when the exposed portion of the screw enters a metal trough. The frictional forces associated with the exposed portion extend the zone of influence of the screw beyond its

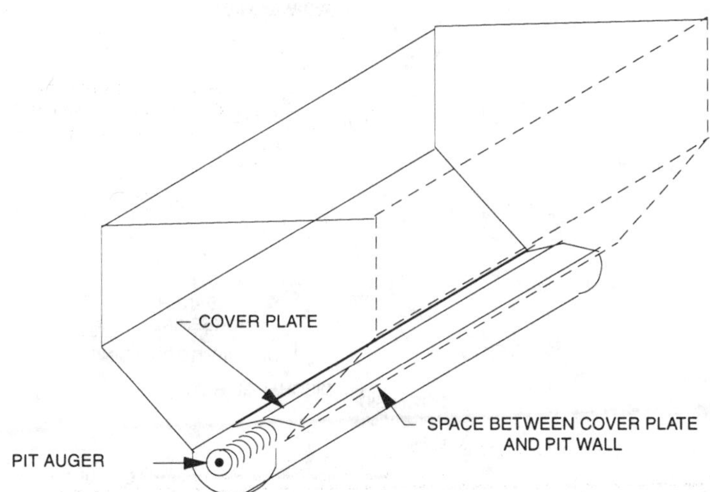

Figure 6.14–Pit unloading auger.

diameter. The greater the exposure, the more efficiently the screw is filled resulting in greater capacity as can be seen in tables 6.2-6.8. However, greater filling efficiency results in greater back pressure at the interface where the exposed portion of the screw enters the metal trough. If the exposed length is long enough, the back pressure may keep the screw from turning. This may be the case in auger-unloaded pits. To solve this problem, an auger cover plate is used (fig. 6.14). The cover plate reduces the grain pressure over the top and along the length of the auger. It may be raised, if necessary, to increase capacity. However, different materials will not flow the same given the same cover plate spacing.

Summary

Augers are used extensively to convey grain and similar products. Power requirements increase and throughput capacity decreases with increases in elevation. Exact sizing of augers requires manufacturer's engineering literature concerning performance of a particular type of conveyor.

For on-farm applications, U-trough augers are preferred over tube augers when high capacities are needed at relatively low auger speeds. However, U-trough augers generally are more expensive for the same design capacity.

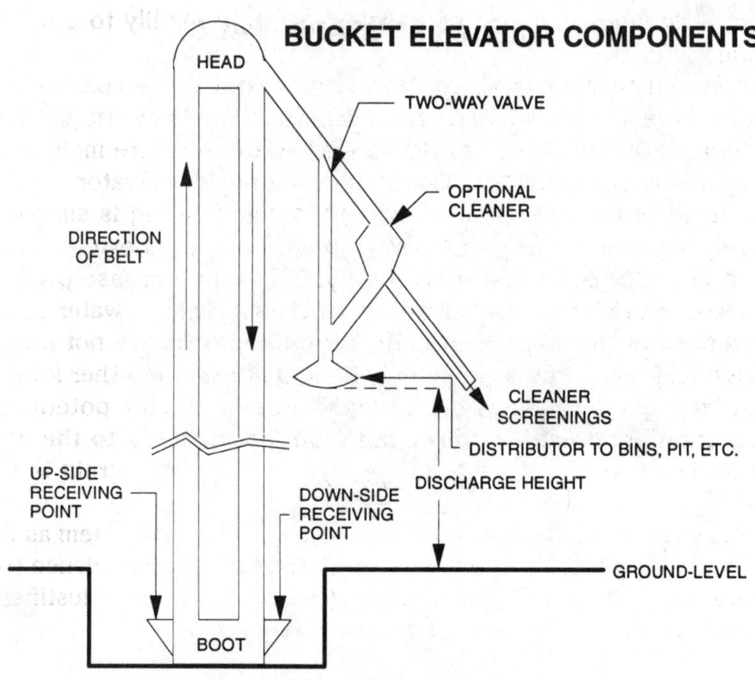

Figure 6.15–Bucket elevator components.

Bucket Elevators

Bucket elevators are one of the more commonly used conveyors of grain. The typical bucket elevator (often referred to as a "leg") conveys grain vertically using buckets attached along a belt (fig. 6.15). The grain (or other type of material) exits the bucket elevator by centrifugal force as the buckets begin their downward movement. From the top of the bucket elevator, grain is directed through a "downspout" and "distributor" to either a holding container or another conveyor such as an auger.

Advantages and Disadvantages

Bucket elevators are very efficient to operate because grain does not slide along exposed surfaces, thus reducing frictional losses significantly. Buckets occupy relatively little horizontal area, and can be purchased in a wide range of capacities to meet practically any grain handling need. In addition, bucket elevators can be operated at less than full capacity without damaging the grain nearly as much as most other conveyors. Bucket elevators are relatively quiet grain conveyors that are relatively free of maintenance and have long productive lives. They require little labor in terms of transferring grain readily to a different location.

The primary disadvantage of bucket elevators is the comparatively high purchase and installation costs, especially for farm installations of less than 20,000 bu. Other considerations relate to where maintenance and repair will be conducted. The motor for a bucket elevator is located in the head of the conveyor. Similarly, downspouting is suspended relatively high above the ground. Most bucket elevators extend into the ground in order to be fed more easily. This can increase problems associated with clean-out of the conveyor, especially, if water collects around the elevator boot. Certainly, bucket elevators are not portable although additional downspouts may be added to serve other locations within "reaching" distance of the leg. There is greater potential for mixing grain as a result of directing grain accidentally to the wrong location because one does not directly see where the grain is being transferred.

Individuals who have bucket elevators usually describe them as being "very convenient". Generally, they believe the added convenience (to be translated as increases in total system efficiency) more than justifies the additional cost.

Design Considerations

Bucket elevator design begins with determining the through-put capacity and discharge height followed by the selection of auxiliary components to the system.

Through-put Capacity. The bucket elevator is usually a multiple-use conveyor, especially in on-farm systems. Possible uses include:

1. Transporting wet grain from a receiving location to a wet-holding bin.
2. Transporting dry grain from a receiving pit to a storage bin.
3. Conveying dry grain from a grain dryer or dry grain surge bin to a storage bin, load-out bin, or transport vehicle.
4. Conveying dry grain from storage into a load-out bin or transport vehicle.
5. Transport of feed ingredients into a feed processing facility.
6. Conveying of processed feed from storage into a load-out bin or transport vehicle.

The design capacity should be based on the largest demand of the possible uses. For example, a bucket elevator may be fed directly by conveyors or gravity flow from a pit, load-out bin, surge bin or wet-holding bins if they are close enough. Similarly, the bucket elevator may unload directly into an auger so long as the leg capacity does not exceed that of the receiving auger.

Handling capacity is most important when receiving wet grain during harvest. It is highly desirable that transport vehicles be able to unload and return to the field in sufficient time to avoid causing harvesting delays. The allowable time that a vehicle has to complete unloading must be specified. The bucket elevator, in conjunction with the receiving pit, must be able to receive all the grain from the largest transport vehicle within this allocated time.

The limit to receiving grain may be associated with the rate at which grain can exit the transport vehicle, especially if mechanical unloading is not available. There is also the trade-off between pit size and bucket elevator capacity. A relatively large pit allows the vehicle to unload at its maximum rate with no restrictions. The bucket elevator must be able to empty the pit sufficiently in order to accommodate the next vehicle and not impede the dryer. For example, assume a 300-bu capacity truck is to be unloaded in 3 min with 20 min between truck deliveries. Thus, the minimum average effective unloading rate is 100 bu/min. If the receiving pit is very small, i.e. 1 bu, the bucket elevator must process the grain at the rate of 6,000 bu/h, very high for a farm conveyor. Conversely if the pit were 300 bu in capacity, the truck could be unloaded instantly, and the bucket elevator would have at least 20 min to empty the pit giving a minimum average effective design rate of 900 bu/h. Typically, grain is gravity fed into the wet-holding bin from the leg so that the bucket elevator has no exit restrictions.

The same basic design criteria exists when the bucket elevator is transporting dry grain to storage from a delivery vehicle. The exception occurs when downspouting is connected to an auger-type conveyor for a bin located too far from the leg to be reached by gravity flow. In this case, the safest design precaution is to size the auger slightly larger than the leg capacity to ensure that the downspouting does not restrict and stop the leg eventually from functioning. With increases in capacity, augers increase in cost at a faster rate than do bucket elevators. Therefore, sizing the leg at lower capacities offers some cost advantages for augers as well.

The unloading rate from batch type dryers can be fairly high, in the order of several thousand bushels per hour. Conversely, continuous flow dryers unload at the drying rate and do not exceed 600 bu/h for most farm situations. For either dryer, it may be advantageous to add a surge bin that receives grain from the dryer. This bin can be unloaded periodically, and its exit rate regulated to match the receiving conveyor.

Conveying grain from storage into a load-out bin or transport vehicle is a somewhat unique situation in that more than one bin may be unloaded at once so long as the bucket elevator has sufficient capacity to handle the grain. The load-out bin allows for very rapid unloading and utilization of a smaller leg. However, the designer should weigh carefully the costs associated with rapid load-out. Rarely are the monetary savings associated with a 15-min decrease in the time for loading worth the added cost.

The rates at which feed ingredients are delivered (or processed-feed removed) are usually not major factors for farm-type systems. This is because the capacity of the bucket elevator for other grain transport functions is usually acceptable for feed ingredients. Another consideration is that stored grain usually will be the primary feed ingredient. Regardless, the design capacity should be based on the greatest capacity demand.

Discharge Height. The discharge height is determined by identifying the elevation above ground where grain (or a similar product) is to be delivered as referenced by the location of the bucket elevator. The angle of the downspouting must be sufficiently steep so that the material will flow easily to the desired delivery point. The following minimum angles, referenced to ground level, are recommended for downspouting common grains within "normal" ranges of moisture:

Dry Grain – 37°
Wet Grain – 45°
Feed Material – 60°

The equation for discharge height is:

$$\text{discharge height (ft)} = D + \text{TAN}(t) * B \tag{6.17}$$

where

D = height of point where material is to be delivered (ft)
B = horizontal distance between the delivery point and the bucket elevator discharge point (ft)

The values of the tangent of "t" at 37°, 45°, and 60° are, respectively, 0.754, 1.000, and 1.732. Equation 6.17 can then be rewritten as:

$$\text{dry grain discharge height (ft)} = D + 0.754 * B \tag{6.18}$$

$$\text{wet grain discharge height (ft)} = D + B \tag{6.19}$$

$$\text{feed material discharge height (ft)} = D + 1.732 * B \tag{6.20}$$

The designer calculates the discharge height associated with the delivery point based on the intended flow (wet grain, dry grain, or feed material). The design discharge height is the maximum height calculated. In effect, only one delivery point determines discharge height, and all others are at angles that exceed the minimum design angles.

The discharge height is referenced to ground level and is not the same as the total length of the bucket elevator (fig. 6.15). The leg may extend below the ground to allow for feeding from other conveyors and pits. In-line cleaners must be located above the discharge point for grain handling systems. The longer the leg, the greater its cost, so increased pit depth and bin height also increases bucket elevator costs.

Auxiliary Components. Bucket elevators have many auxiliary components. Some are essential while others are only needed for special applications.

Distributors. Grain is usually discharged from the bucket elevator into a distributor. The distributor is a device that allows someone on the ground to direct grain to different locations. The distributor contains an internal downspout (referred to as an outlet selector) that may be moved to lock into an external downspout going to the desired delivery point. The number of outlets ranges normally from 4 to 12. The outlet selector is controlled manually by a pipe or cable connection, with pipe being the preference for systems less than 100 ft tall. Electrically controlled distributors are also available. The internal angle of the outlet selector may be either 45° or 60°, depending on whether the distributor is to be used for feed-type materials.

Downspouts. Downspouts are normally made of 12 to 14 gauge metal but heavier spouting is available for more intensive applications. Common diameters are 6, 8, 10, and 12 in. in both hot rolled and galvanized form. Recommended maximum flow capacities are (Midwest Plan Service, 1968):

6 in. – 1,500 bu/h
8 in. – 3,000 bu/h
10 in. – 5,000 bu/h
12 in. – over 5,000 bu/h

Downspout length may be determined using the Pythagorean Theorem:

$$C = (A^2 + B^2)^{0.5} \tag{6.21}$$

where
C = downspout length (ft)
A = horizontal distance from the delivery point to the outlet on the distributor
B = vertical distance from the delivery point to the outlet of the distributor

Truss Kits for Downspouting. Trusses are available to support downspouts. Generally, they are not used for downspout lengths of less than 40 ft, but become more important as the length of the downspouting increases.

Grain Retarders and Cushion Boxes. Grain retarders are placed in downspouts to slow the rate of descent of the grain. A device is suggested for each 40 ft of grain travel.

Cushion boxes perform a similar function by acting to "cushion" the grain at the end of the downspouting. The basic idea behind this device is that grain striking against grain is less damaging to seed structure than when grain strikes metal. The slowing of the grain also allows it to fall straight down into the bin and be distributed more evenly by a grain spreader.

In-Line Cleaners. Grain cleaners may be installed as an integral part of the bucket elevator system (fig. 6.15). The usual configuration is to have a two-way valve at the grain discharge point with the option of sending the grain through the cleaner or directly into the distributor.

In-line grain cleaners generally have screens for removing only the fine or course material but not both during a single pass. Screenings are directed to a collection point while the clean grain passes on through the distributor. In-line cleaners provide a convenient way of cleaning grain before drying, thereby decreasing the drying energy required because part of the wet material is removed, and resistance to airflow is lessoned. Likewise, grain that is cleaned before being placed into storage reduces the chance of problems associated with collection of fine materials. The cost of cleaners includes both the purchases of the unit and the added bucket elevator height so that grain discharge point is sufficiently high. As with all cleaning operations, removal of fine materials reduces the quantity of material to be sold or fed while not necessarily increasing its grade.

Platforms and Ladders. The bucket elevator is serviced, in part, at its top so that some means of accessing this area must be provided. Usually, ladders (preferably caged) extend from the bottom to the elevator top with intermediate platforms being located to serve bucket elevator components, such as cleaners and distributors. Ladders may be hinged at the bottom to prevent unauthorized individuals (most notably children) from climbing the conveyor. Safety harness equipment is also available.

Backstop. Should there be a power failure when a bucket elevator is operating, the loaded buckets will cause the belt to reverse directions. The grain will then be deposited in the boot of the leg and may prevent restarting until cleanup is completed. A "backstop" serves as a brake against the belt reversal, thus preventing a potential stoppage problem.

Feeding Direction. Bucket elevators may be "fed" from either the "up" or "down" side, with "up" or "down" referring to the direction that the belt is moving. Grain may be "force-fed" by an auger or "gravity fed" by a receiving pit or hopper. The most common practice is either to "force feed" the elevator on the "down" side or "gravity feed" the leg on the "up" side. The main reason for these practices is that "force feeding" may be conducted at a lower point on the down side of the bucket elevator, thereby decreasing pit depth and the associated length of the leg. The cups are filled nearer to capacity with less grain damage because belt and grain directions are more nearly the same with some surge capacity in the boot of the leg. Conversely, gravity discharge into the "up" side can be done anywhere along the leg because there is less restriction of the belt from the entering grain.

Other considerations include the location of the leg's discharge point relative to the receiving point and the required angle for the downspouting. Most bucket elevators will receive material from both the "up" and "down" sides. The "force-fed" system is directly attached to the leg and prohibits the entry of grain except through the forcing auger. Conversely, gravity-feeding allows for deposition of material from many sources into a common receiving pit or hopper attached to the bucket elevator.

Power Requirement. Bucket elevators are very efficient handlers of grain. The following equation may be used to determine power requirements:

$$\text{horsepower (bucket elevator)} = \frac{H * BPH * WPBU}{EFF * (19{,}800)} \qquad (6.22)$$

where

H = total height of the bucket elevator (top to bottom, ft)
BPH = bushel per hour capacity
WPBU = weight per bushel (lb)
EFF = efficiency of bucket elevator, (% approximately 80% for design purposes, Norder and Weiss, 1984)

Horsepowers for a range of bucket elevator capacities and heights for soybeans, one of the heavier grains in terms of bulk density, are given in figure 6.16.

Belt and Cup Specifications. Many combinations of bucket spacing and belt speed will deliver the desired throughput of material. Specifications are provided by the manufacturer. Their recommendations are based on the principle that momentum of the grain should be sufficient at the discharge point to project the grain and clean out the bucket elevator cups while not letting the material fall through the "down" side of the leg. The following equations (Henderson and Perry, 1966) may be used to determine cup spacing and belt speed:

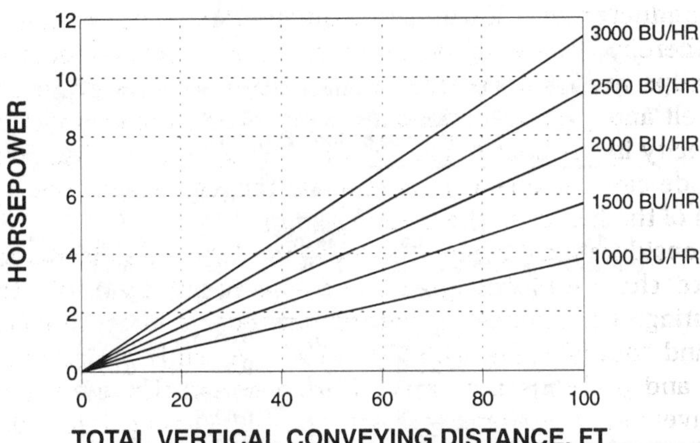

Figure 6.16–Horsepower requirements for bucket elevators.

$$NRPMHP = 54.19 / RADHP^{0.5} \tag{6.23}$$

where

- NRPMHP = revolutions per minute of the head pulley
- RADHP = effective radius of the head pulley (in.)

The effective radius is approximately equal to one-half the diameter of the head pulley plus one-half the projected width of the cup perpendicular to the belt. The discharge from the cup is neither instantaneous nor uniform because the true radius relative to individual grain kernels extends from one edge of the cup to the other. In equation form:

$$RADHP = (DHPUL + CUPP) / 2 \tag{6.24}$$

where

- RADHP = approximate effective radius of the head pulley (in.)
- DHPUL = diameter of the head pulley (in.)
- CUPP = perpendicular distance from the tip of cup to belt (in.)

The rotation of the pulley is used to compute belt speed in the following manner:

$$BELTFM = (pi * DHPUL * NRPMHP) / 12 \tag{6.25}$$

Table 6.9. Dimensions and capacities for a typical type of centrifugal discharge elevator bucket (Salem made by Link Belt, 1968)

Cup Length (in.)	Cup Projection (in.)	Cup Depth (in.)	Angle From Lip to Back (°)	Capacity When Totally Filled (ft³)	Horizontal Capacity to Lip (ft³)
2.5	2 5/16	2 3/8	70	0.0040	0.0028
3	2 5/16	2 3/8	70	0.0049	0.0035
3	2 3/4	2 7/8	67	0.0070	0.0047
3.5	2 5/16	2 3/8	70	0.0058	0.0041
3.5	2 3/4	2 7/8	67	0.0083	0.0057
4	2 3/4	2 7/8	67	0.0097	0.0067
4	3 1/4	3 3/8	68	0.0134	0.0094
4.5	2 3/4	2 7/8	67	0.0111	0.0076
4.5	3 1/4	3 3/8	68	0.0154	0.0108
5	3 1/4	3 3/8	68	0.0173	0.0122
5	3 3/4	3 7/8	64	0.0219	0.0140
5.5	3 3/4	3 7/8	64	0.0244	0.0157
6	3 3/4	3 7/8	64	0.0269	0.0173
7	4 1/8	4 3/8	66	0.0407	0.0282
8	4 1/2	5 1/8	62	0.0570	0.0343
9	4 1/2	5 1/8	62	0.0648	0.0391
9	5 3/8	6	62	0.0915	0.0576
10	4 1/2	5 1/8	62	0.0727	0.0440
10	5 1/8	5 5/8	66	0.0946	0.0651
10	5 3/8	6	62	0.1027	0.0648
10	6 3/8	6 3/4	69	0.1455	0.1098
11	5 3/8	6	62	0.1140	0.0721
11	6 3/8	6 3/4	69	0.1615	0.1220
12	5 3/8	6	62	0.1252	0.0793
12	6 3/8	6 3/4	69	0.1775	0.1343
12	7 1/4	7 1/2	67	0.2217	0.1537
13	5 3/8	6	62	0.1364	0.0865
13	6 3/8	6 3/4	69	0.1935	0.1465
14	5 3/8	6	62	0.1476	0.0937
14	6 3/8	6 3/4	69	0.2095	0.1588
14	7 1/4	7 1/2	67	0.2621	0.1822
15	5 3/8	6	62	0.1588	0.1010
15	6 3/8	6 3/4	69	0.2255	0.1711
16	5 3/8	6	62	0.1700	0.1082
16	6 3/8	6 3/4	69	0.2414	0.1833
16	7 1/4	7 1/2	67	0.3024	0.2107
18	5 3/8	6	62	0.1924	0.1226
18	6 3/8	6 3/4	69	0.2734	0.2078
18	7 1/4	7 1/2	67	0.3428	0.2392
20	6 3/8	6 3/4	69	0.3054	0.2323
20	7 1/4	7 1/2	67	0.3830	0.2677
22	7 1/4	7 1/2	67	0.4234	0.2962
24	7 1/4	7 1/2	67	0.4638	0.3247

where

BELTFM = belt speed (ft/min)
pi = 3.1416
DHPUL = diameter of the head pulley (in.)
NRPMHP = revolutions per minute of the head pulley

Note that the belt speed is not influenced by cup size or spacing. Generally, cup size is determined by the size of the leg opening. The cup width may extend to the width of the leg, and the cup depth must be small enough to fit the opening. Dimensions and capacities of various cup sizes are given in table 6.9. The number and spacing of a given cup size may be determined by the following sequence of equations:

$$BELTL = 2 * BEAXLE + 0.5 * pi * \left[\left(DHPUL + BPULD\right) / 12\right] \quad (6.26)$$

where

BELTL = length of belt (ft)
BEAXLE = distance between the head and boot pulley axles (ft)
DHPUL = diameter of head pulley (in.)
BPULD = diameter of boot pulley (in.)

Belt revolutions:

$$BLTRPM = BELTFM / BELTL \quad (6.27)$$

where

BLTRPM = rpm of the belt

Bushels per revolution:

$$BUPREV = BECAP / \left(60 * BLTRPM\right) \quad (6.28)$$

where

BUPREV = bushels of grain per revolution of the belt
BECAP = bucket elevator capacity (bu/h)

Weight per cup:

$$CUPWT = \left(CCAP * 0.8 * WPBU\right) / 1{,}728 \quad (6.29)$$

where

CUPWT = weight of material conveyed by each cup (lb)
CCAP = capacity of each cup when filled ($in.^3$)
WPBU = weight per bushel of grain (lb)

Number of cups:

$$NCUPS = (BUPREV * WPBU) / CUPWT \quad (6.30)$$

where

NCUPS = number of cups required (round to next largest whole number)

Spacing between cups:

$$CUPSPA = (BELTL * 12) / NCUPS \quad (6.31)$$

where

CUPSPA = spacing between cups (in.)

Placement and Construction. The bucket elevator should be placed in position to be accessible by as many present and future incoming conveyors as possible. Similarly, it should be able to service as many points as may be needed in present and future expansions. Excess discharge height in the initial design may result in substantial savings in the future. Do not block access to the bucket elevator by locating a structure in the path of future expansion. The relatively small savings in auger length in the short term may prove to be very expensive when expansion occurs. Remember, good design allows for future growth. Most farm facilities were built with the idea that no expansion would occur. The end result is unnecessary costs of future modifications and additions.

Bucket elevators usually are assembled on the ground and lifted into position by a portable crane. Once the bucket elevator is positioned, it must be anchored into place by guy wires extending from several points on the leg to at least three anchor points on the ground. Do not tie the leg onto bins or other structures because their failure might also cause failure of the bucket elevator. The manufacturer's anchoring and guy wire specifications should be followed exactly.

Economic Considerations

The bucket elevator has a relatively large investment cost when compared to most portable conveyors. Portable augers are generally less expensive than bucket elevators in annual costs. However, the absolute difference is relatively small for facilities greater than 30,000 bu in capacity. Likewise, the cost differential is very little when based on a per-bushel-processed measure. The value of a bucket elevator increases considerably when labor is in short supply and the "convenience value" is considered. Similarly, the cost of adding a new delivery or receiving point is relatively low when compared to other types of conveyors.

Summary

Bucket elevators are relatively expensive long-term investments that provide a very efficient means of conveying grain. Even though the initial cost is high, most individuals will replace an obsolete bucket elevator with another bucket elevator rather than with portable conveyors, thus confirming the satisfaction most users have with this type of conveyor.

Receiving Pits and Hoppers

Receiving pits and hoppers are used as surge storage for incoming grain or similar material. A pit generally refers to a permanent structure while a hopper is generally portable. Conceptually, a pit is larger than a receiving hopper, but functionally both serve the same purpose. For that reason, future references to a "pit" generally will apply to a "hopper" as well unless otherwise noted.

Advantages and Disadvantages

Every receiving conveyor has some surge capacity at its inlet unless it is force-fed by another conveyor. Thus, a receiving pit or hopper is part of any transfer of grain from field to storage. The relative advantages and disadvantages of pits relate primarily to size and unloading rate.

Pit Design

There are two basic types of pits as described by the method of unloading. Gravity unloaded pits feed the receiving conveyor through one or more openings in the bottom of the pit. A conveyor-unloaded pit feeds grain onto a conveyor extending across all or part of the bottom of the pit. Many of the criterion for design is the same for both types of pits. Key design considerations are:

1. The pit, in combination with the receiving conveyor, must be able to process the incoming vehicles in a specified amount of time.
2. The width of the pit (pit width refers to its measurement perpendicular to the direction of vehicular travel) must be able to accommodate the unloading width of the delivery vehicle plus an

additional allowance for flow of the material to either side if build-up in the pit occurs. The length of the pit should be sufficient to keep grain from overflowing.

3. The angle of all the pit walls should be sufficiently steep so that incoming material will gravity-feed to the bottom of the pit. For grain, the minimum angle is 45°. For poor flowing material such as ground or powdered meals, the minimum angle is 60°.

Any one of the above three factors may determine pit size regardless of the other two. Allowable space and ground conditions may force the choice of one type of pit over another. Ideally, both the gravity and conveyor pits should be designed and cost compared before actual selection is made.

Pit Capacity. Pit capacity depends on the inflow and outflow rates of grain. The first step in design is to determine the maximum volume of grain to be unloaded at any one time and the maximum time to complete unloading. The equation for pit capacity is:

$$\text{PITCAP} = \text{M} * (\text{VUR} - \text{GRRP}) \tag{6.32}$$

where

PITCAP = capacity of pit, not to be less than zero (bu)
M = maximum time to unload a delivery vehicle (min)
VUR = maximum delivery vehicle unloading rate (bu/min)
GRRP = simultaneous grain removal rate from the pit (or receiving hopper) (bu/min)

It can be seen from the above equation that if the grain is not to be removed from the pit at the time of unloading, the pit must be able to hold all the grain from the delivery vehicle. Likewise, a grain removal rate from the pit in excess of the vehicle unloading rate removes the need for pit grain storage except to collect the grain.

Top Surface Dimensions. The width of the pit refers to the distance perpendicular to the path of the delivery vehicle, and its length is the distance in the direction of vehicle travel. Both dimensions should be sufficient to keep grain within the pit structure rather than spilling over and having to be placed back manually into the pit after every unloading (fig. 6.17). The following equations apply:

$$\text{MNPW} = \text{MXBUW} + 2*\text{UHBD} / (\text{TAN}(b)) \tag{6.33}$$

$$\text{MNLP} = 2*\text{UHBD} / (\text{TAN}(b)) \tag{6.34}$$

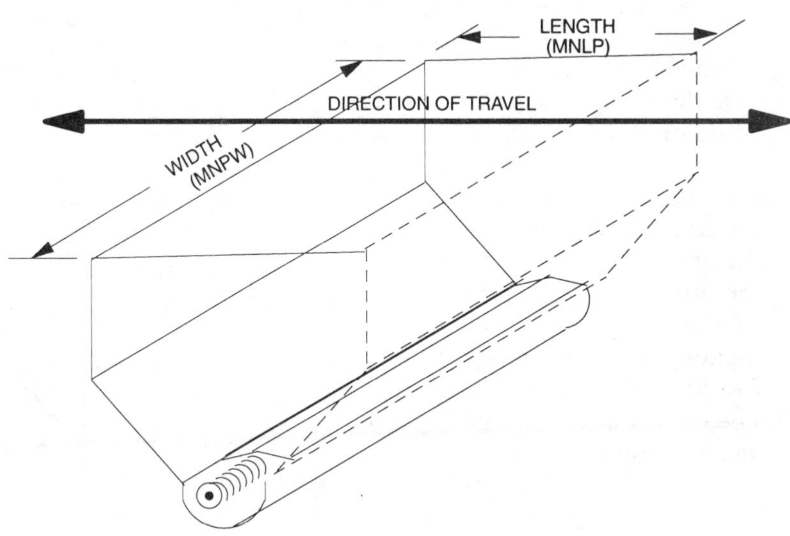

Figure 6.17–Pit length and width.

where

MNPW	=	minimum pit width (ft)
MNLP	=	minimum pit length (ft)
MXBUW	=	maximum bed unloading width (ft)
UHBD	=	unloading height for the end of the truck (or other delivery vehicle) bed during discharge
b	=	least angle of repose for the types of material being unloaded (°)

If the pit capacity is sized correctly, the width will need only to be slightly larger because overflow will not occur. Similarly, the length would have to be only slightly larger than the projection of the grain striking the surface. However, relatively smaller top surface dimensions will require a deeper pit to meet capacity specifications. Deeper pits generally cost more and are harder to keep clean and dry. In addition, grain has to be elevated further to reach ground level, thus increasing power and length requirements of bucket elevator or auger conveyors.

Space limitations may restrict pit widths, lengths, and capacities. Pit structures that are purchased commercially generally come in 8, 10, and 12 ft widths with 6 ft being a common length. Side unloading vehicles may require special considerations.

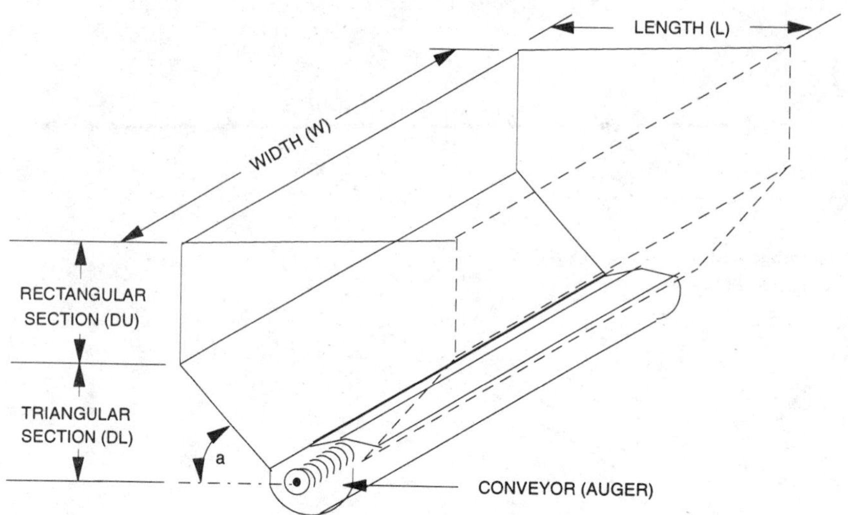

Figure 6.18–Auger-unloaded pit components.

Depth Dimensions. The depth of the pit must be sufficient to allow for self-cleaning while holding the design capacity. In some situations, the required angles of the pit walls will result in a pit of greater volume than the design capacity, in which case the design is complete. If more volume is needed to reach the design capacity, the pit will have a rectangular cross-section with vertical walls.

The depth dimensions are influenced greatly by whether the pit is a gravity flow or conveyor unloaded pit. The most common pit types are shown in figures 6.18 and 6.19.

Lower section of conveyor unloaded pits. In conveyor unloaded pits (fig. 6.18) the conveyor usually extends along the entire width of the pit. In this case, pit wall slopes are vertical along the length of pit. The following equation applies:

$$DL = \left(TAN\,(a)\right) * L / 2 \tag{6.35}$$

where

DL = depth of the lower portion of the pit (ft)
L = pit length (ft)
a = minimum angle of repose of incoming materials (°)

For situations when the conveyor extends along the entire width, the formula for capacity of the lower section is:

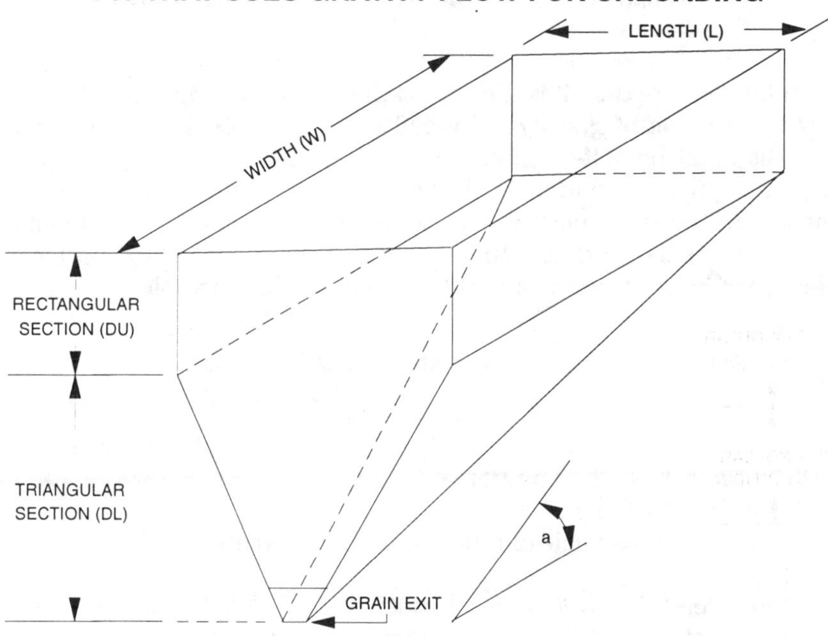

Figure 6.19–Gravity-flow pit components.

$$CAPL = DL*L*W/2 \tag{6.36}$$

where

CAPL = capacity of the lower section of the pit only (ft^3) (multiply by 0.8 to obtain bushels)

L = length of pit (ft)

W = width of pit (ft)

If the capacity of the lower portion of the pit is equal to or greater than the design capacity, the design is complete. Otherwise, the following equation is used, which applies to both gravity unloaded and conveyor unloaded pits, to determine the depth of the upper portion:

$$DU = \frac{(DCAP\text{-}CAPL)}{W*L} \tag{6.37}$$

where

DU = depth of the upper portion of the pit (ft)

DCAP = design capacity of the total pit (ft^3) (multiply by 0.8 to obtain bushels)

The total depth of the pit is the sum of the lower and upper depths.

Lower section of gravity unloaded pits. In gravity unloaded pits, all pit walls must be self-cleaning. The least depth for this type of pit is when the pit is unloaded in the center of its length along the wall adjacent to the receiving conveyor. The pit length in this situation may be up to twice its width and the pit still be self-cleaning. If the pit length is less than twice the pit width, the following equation applies:

$$DL = \left(TAN\,(a)\right) * W \tag{6.38}$$

where
DL = lower depth of pit (ft)
W = pit width (ft)
a = maximum angle of repose of incoming materials (°)

If the pit length is over twice its width, substitute pit length (L) for pit width in equation 6.38. The lower pit capacity is found by:

$$CAPL = \left(DL * L * W\right) / 3 \tag{6.39}$$

where
CAPL = capacity of the lower portion of the pit (multiply by 0.8 to obtain bushels) (ft^3)

If the lower section capacity is greater than the design capacity, the design is complete. Otherwise, a rectangular section must be added as discussed in the section on conveyor unloaded pits.

Other Conveyor Pit Considerations

Conveyor pits may be unloaded by augers, flight conveyors, belt conveyor or vibrators with U-trough augers being the usual mechanism. The unloading mechanism should be locked in electrically with the primary conveyor, i.e., bucket elevator, so that the pit will not become empty when the receiving conveyor is not functioning. The conveying mechanism will need a cover plate for reasons discussed in the section on augers. The conveying mechanism may extend past the pit so that it may receive grain from storage bins opposite the bucket elevator. The pit conveyor generally needs a means of regulating incoming grain in order to be sure of matching flows. Variable spacing of the cover plate or a variable speed belt or auger driving mechanism may be used.

Other Gravity Pit Considerations

Gravity pits generally require greater pit depths. Flow is controlled by variable spacing of the grain outlet. This allows for relatively easy control of flow but can cause major restarting problems of the bucket elevator if there is a power failure.

General Design Considerations

The pit grate should be made of pipe or angle-iron so that it will be as self-cleaning as possible. It should be removed easily to enter the pit for cleaning or repair, and it must be able to support the weight of the loaded delivery vehicle.

The bottom of the pit should be accessible for purposes of clean-up and repair. There should be sufficient room for two individuals to maneuver with items as shovels, brooms, and vacuum cleaners. The bucket elevator also requires space for repair and clean-up.

A primary consideration is the dryness of the area. The pit support area should be well drained externally, and include provision for a sump pump if required. The sump pump is mandatory for pits that are not sheltered.

If the pit dumping area is to be sheltered, the building must be of sufficient height to allow for a truck to raise its bed and pass through the facility. Otherwise, a driver will someday forget to lower the dump bed and damage the truck, the building, and perhaps individuals working in the area.

Wet Holding Bins

The wet holding bin acts as a surge tank for wet grain prior to drying. The wet holding bin serves two primary roles:

1. It allows for more rapid unloading of delivery vehicles so that harvesting is not delayed.
2. It allows for immediate filling of the dryer, thus increasing through-put capacity.

Sizing Wet Holding Bins

A wet holding bin is used for short-term storage of grain. If it is to be used to hold wet grain for more than one day, provisions for aeration and/or drying should be considered depending on ambient temperatures during harvesting (see Chapter 2). The size of the wet holding bin is influenced by the type of dryer.

Continuous-Flow Drying Systems. In continuous-flow drying systems, grain is being removed in a continuous and uniform manner. The design capacity is:

$$\text{WHCAP} = (\text{HR} - \text{DR}) * \text{HTIME} - \text{PCAP} \tag{6.40}$$

where

WHCAP	=	wet holding capacity for a continuous flow dryer (bu)
HR	=	maximum harvesting rate for the system (bu/h)
DR	=	potential drying rate for the dryer if there are no restrictions in the system (bu/h)
HTIME	=	time that the harvesting occurs (h)
PCAP	=	capacity of pit (bu)

Equation 6.40 is based on the premise that the wet holding bin will be empty at the beginning of the harvest day. Otherwise, grain will continue to accumulate over the harvest season except on those days when harvesting cannot occur because of adverse weather. The accumulation of ever increasing quantities of wet grain over more than one day should be avoided. In no instance should the total wet holding capacity, to include the wet holding bin and the pit, be smaller than the capacity of the largest delivery vehicle.

Batch Drying Systems. Batch dryers have a filling and unloading time in addition to a drying time. In essence, these dryers operate in distinct stages so that grain removal from the wet holding bin is relatively rapid as compared to continuous-flow dryers. The wet holding capacity is computed using the same equation as for continuous-flow dryers (equation 6.40). However, for batch drying systems, the minimum wet holding capacity (bin and pit) should not be less than the capacity of the largest vehicle plus the size of the dryer batch.

Filling and Unloading Considerations

The wet holding bin is a link in the materials handling chain. Its filling rate is a function of the incoming conveyor, usually a portable auger or bucket elevator. If the inflow rates are less than the drying rates, only minimal wet holding capacity is needed. This situation may occur when a relatively large pit is used in conjunction with relatively low capacity conveying to wet holding. The pit auger and receiving conveyor should be locked-in electrically with pressure switches in the wet holding bin to avoid blockage associated with over filling (fig. 6.20).

Wet holding bins are much like pits in that they may be unloaded by conveyors or gravity flow. Gravity flow wet holding bins must be placed over the dryer, thus requiring more expense for the structure needed to support the bin. However, they take up less ground space than auger unloaded bins in that they occupy much of the same ground area as the dryer (fig. 6.21). Gravity flow wet holding bins are especially advantageous in filling continuous flow dryers because the flow rates are relatively low and no matching of conveyors in series is required.

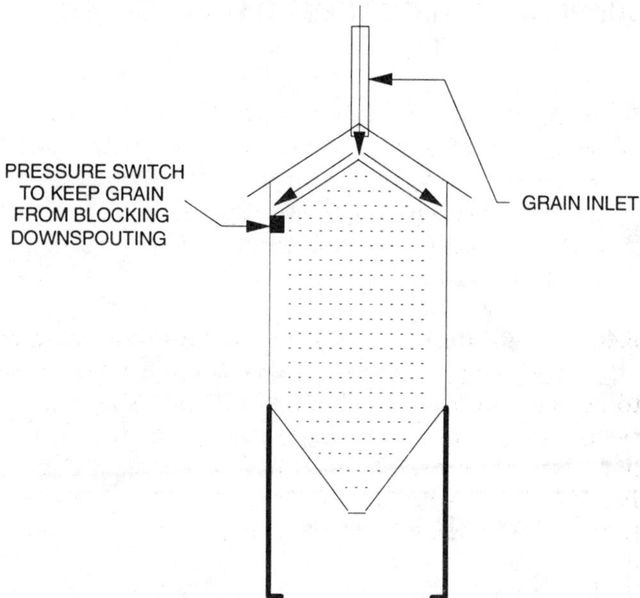

Figure 6.20–Pressure switch in a wet holding bin.

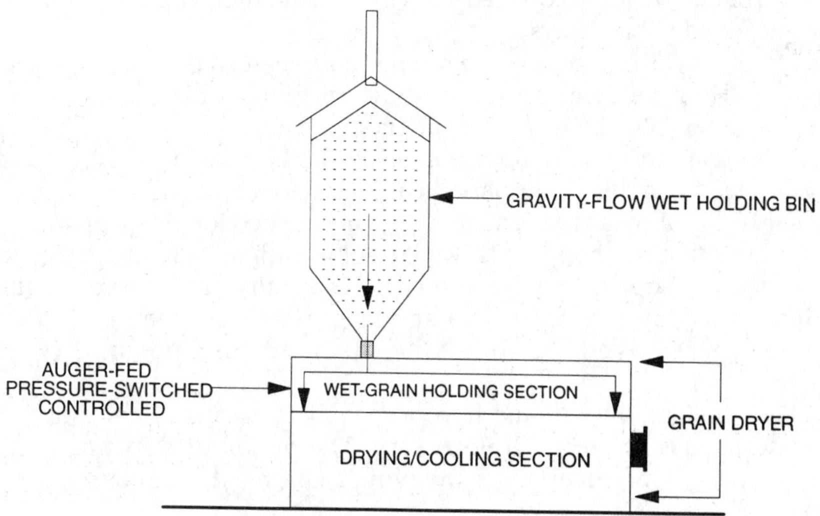

Figure 6.21–Gravity-flow wet holding bin.

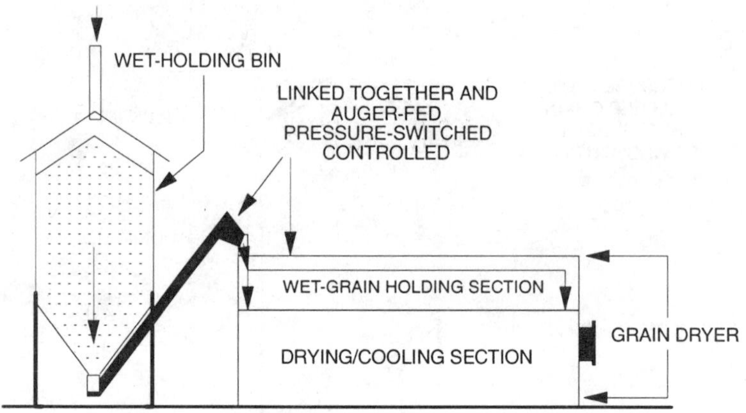

Figure 6.22–Auger-unloaded wet holding bin.

Wet holding bins equipped with unloading conveyors (fig. 6.22) must be sized and matched electrically to the dryer. Typically, portable dryers have some excess grain capacity above the drying columns. In essence, this is also a gravity flow wet holding bin in that the primary wet holding bin unloading conveyor does not have to operate all the time, even with a continuous flow dryer. The wet holding bin unloading conveyor should have sufficient capacity to fill either the active batch drying volume or its associated internal wet holding capacity without causing added filling delays.

Wet holding bins generally are hopper-bottomed so that they are self-cleaning. However, flat-bottomed bins with appropriate unloading equipment may also be used. One method of using flat bottomed bins is to let the grain form its natural hopper in the bottom of the bin, being sure to clean out this grain periodically, preferably at the end of each day (fig. 6.23). A means of reducing the time needed for cleaning out the grain in the "natural hopper" is to fill the bin with sufficiently dry grain until a natural hopper may be formed. Because dry grain serves as the hopper, it is less susceptible to spoilage. However, even this grain should be removed periodically to avoid any possibility of contamination.

All wet holding bins should have a secondary means of unloading in order to bypass the dryer if necessary. The option of directing grain back to the pit or receiving conveyor is the most frequently used alternative (fig. 6.24).

FLAT-BOTTOMED WET HOLDING BIN SUPPLYING A HIGH TEMPERATURE DRYER

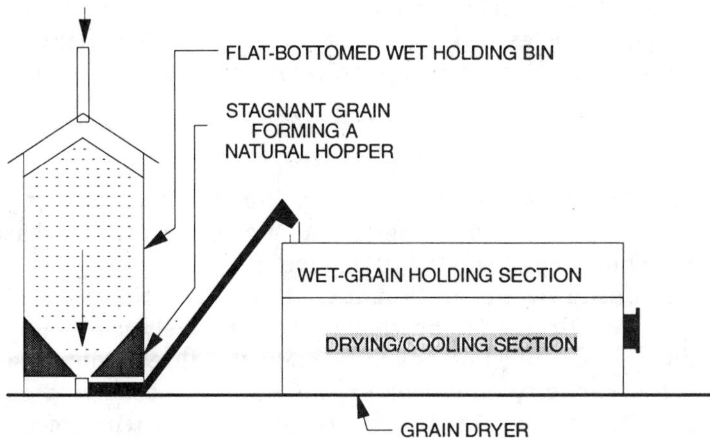

Figure 6.23–Flat-bottomed wet holding bin using grain as a hopper.

GRAVITY FLOW WET HOLDING BIN WITH EMERGENCY UNLOADING

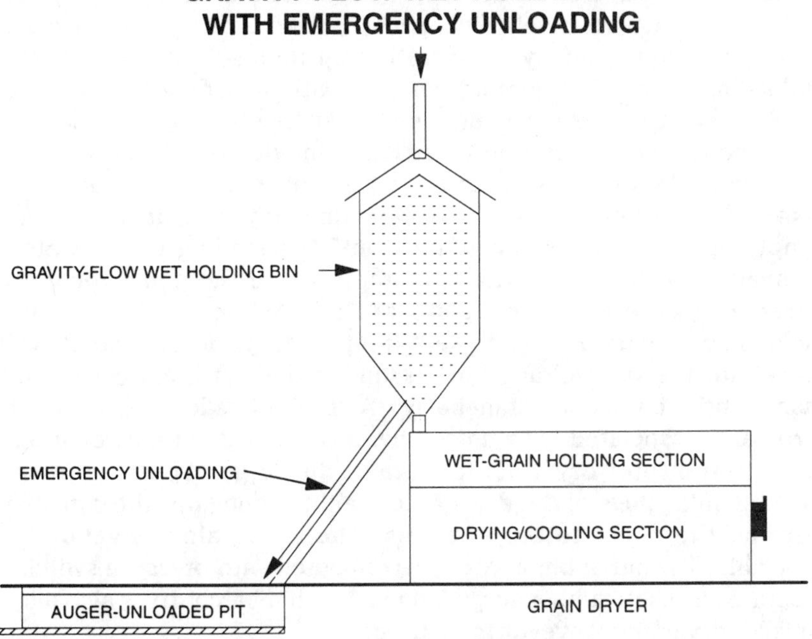

Figure 6.24–Gravity-flow wet holding bin with emergency unloading.

Summary

All grain drying systems have some form of wet holding capacity. Possibilities include conventional wet holding bins, pits, delivery vehicles, and combine harvesters. The conventional wet holding bin is usually the least costly. The most expensive means of holding wet grain is to place it "on wheels", that is, with portable containers such as combines and vehicles. The economic loss usually is much greater when undersizing rather than oversizing conventional wet holding bins.

Surge Bins

A surge bin collects dry grain directly from the dryer and empties its contents periodically into storage through a conveyor. The reasons for using surge bins depend on the drying method.

Continuous-Flow Dryer Considerations. The flow of grain from continuous-flow dryers is constant and of relatively low volume. If no surge bin is used, this grain has to be conveyed directly into storage by either a low-capacity conveyor or the wet grain receiving conveyor (with a redirected grain flow pattern). The addition of a second conveyor for transporting grain to storage may be relatively expensive, and normally a low-capacity bucket elevator is used. If the existing wet grain conveyor is used, the dryer will have to be stopped each time a vehicle unloads because of the simultaneous demands for use.

Batch Dryer Considerations. When using automatic-batch dryers, the unloading rate is relatively high, perhaps even greater than the capacity of the receiving conveyor. As with continuous-flow drying systems, unloading the batch dryer may compete with unloading a vehicle for use of a receiving conveyor. Again, one alternative is to use a separate dry grain conveyor that is tied electrically in with the dryer so that the conveyor operates only when the dryer unloads. Another option is to use the wet grain conveyor in the same way. In this situation, the emptying of the dryer must "lock out" the feeding of the wet grain conveyor by the pit or receiving hopper, and the conveyor must be directed to the appropriate storage bin. When a delivery truck is unloading, the dryer must be kept from dumping or its contents will go back into the wet holding bin (assuming the wet grain conveyor can handle both inputs simultaneously). A surge bin alleviates many of the problems associated with unloading. If the wet grain conveyor also is used to transport dry grain to storage, the surge bin serves to collect several unloadings of the dryer before either "tying up" the conveyor or forcing the dryer to stop because the wet grain conveyor is not available. The surge bin may be instrumented with pressure switches to operate automatically. A surge bin used with a batch dryer also allows a smaller dry grain conveyor to be used.

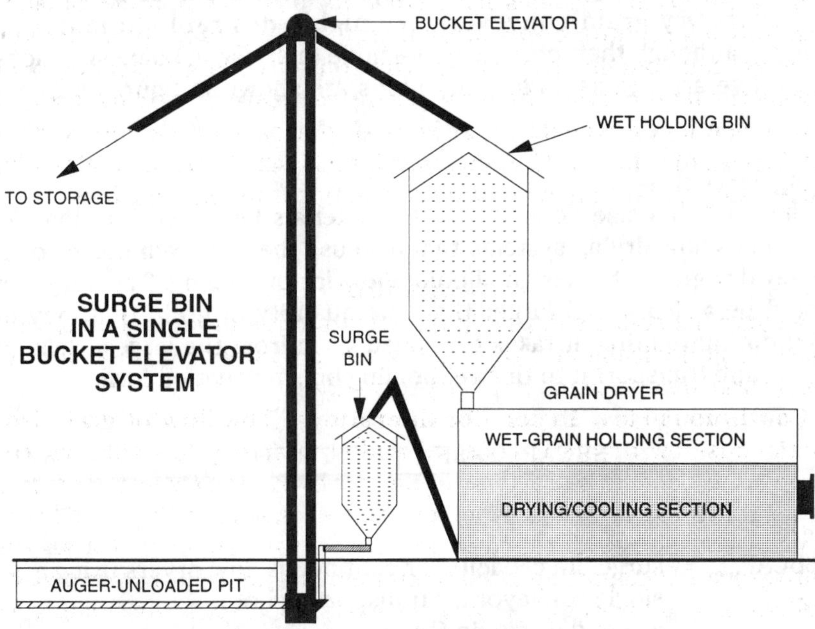

Figure 6.25–Surge bin placement with wet leg.

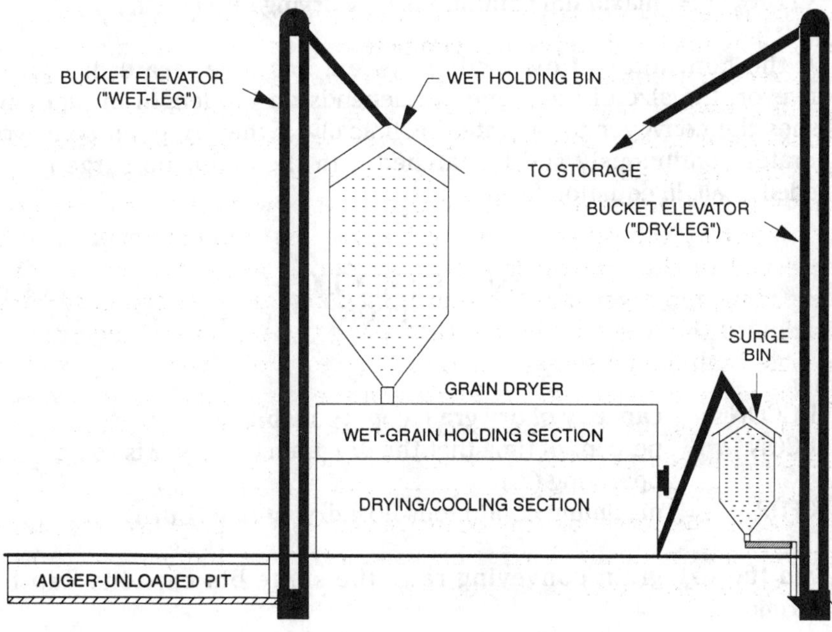

Figure 6.26–Surge bin placement with dry leg.

The use of a surge bin does not prevent the use of a dry grain conveyor. Dry grain conveyors are highly desirable in materials handling although their costs may preclude their use in some situations. Alternative arrangements for surge bins are shown in figures 6.25 and 6.26.

Sizing Surge Bins

The surge bin size depends on the materials flow configuration. For continuous-flow drying systems that also use the wet grain conveyor to convey dry grain, the design should allow for oversizing. The surge bin should be at least 10% larger than the quantity of grain that may be dried during the time it takes to unload grain from the largest delivery vehicle and transport it to the wet holding bin. In equation form:

$$SBCAP_{wet} = SF * T * CFDR / 60 \tag{6.41}$$

where

$SBCAP_{wet}$ = surge bin capacity for continuous-flow dryers using a single conveyor for transporting both wet and dry grain (bu)

SF = safety factor for design (use 1.1 as a minimum)

T = time to unload all the grain from the largest delivery vehicle and place into the wet holding bin (min)

CFDR = maximum continuous flow drying rate (bu/h)

If the continuous-flow drying system has a separate dry grain conveyor, the size of the surge bin depends on the length of time one wishes the conveyor to operate. Technically, if the dry grain conveyor operates continuously and is matched with the dryer, no surge bin is needed at all. In equation form:

$$DGCONR = \frac{CFDR * 100}{PCON} \tag{6.42}$$

where

DGCONR = capacity of dry grain conveyor (bu/h)

PCON = percent of time that the dry grain conveyor is to be operating (%)

CFDR = maximum continuous-flow drying rate (bu/h)

Given the dry grain conveying rate, the surge bin capacity can be determined:

$$SBCAP_{dry} = \frac{(DGCONR - CFDR) * PCON}{100} \tag{6.43}$$

where

$SBCAP_{dry}$ = the surge bin capacity (bu) when a separate dry grain conveyor is used

The previous equation is dimensionally correct because the percentage PCON is actually a fraction of 1 h.

If batch drying is used without a dry grain conveyor, the minimum size of the surge bin should be twice that of the batch and in even increments of the batch size. For example, if the batch size is 100 bu, the surge bin should be sized in 100-bu increments.

If a dry grain conveyor is used, its minimum capacity should be:

$$DGCONR = \frac{SBCAP_{dry}}{(BDDFT/60)} \tag{6.44}$$

where

DGCONR = minimum outflow rate of the dry grain conveyor (bu/h)
BDDFT = minimum time required to fill and dry one batch of grain (min)
$SBCAP_{dry}$ = the surge bin capacity (bu) when a separate dry grain conveyor is used

It should be noted that the above equations represent minimum design specifications. The throughput capacity of a grain dryer increases as the moisture content of the incoming grain decreases. It is important that sizing be made under the conditions of maximum flow to keep from restricting the drying process.

There are other considerations besides moisture content that may influence the surge bin sizing. It may be desirable, for example, to size the bin to be the same as the maximum daily harvest. The wet grain conveyor could then be used exclusively to transfer dry grain to storage after harvesting has stopped for the day (assuming there is not continuous harvesting). With this design, there is a reduced danger of inadvertently transferring grain to the wrong bin, and the surge bin can be used for dryeration. Dryeration can, however, be better conducted with two surge bins, with an "every-other harvest day" filling schedule.

Surge bins receive grain through conveyors, typically an auger connected directly to the dryer. Small bucket elevators also serve this purpose well and are more flexible in that they can be used to transport grain both to and from surge or small dryeration bins. Low capacity legs are disproportionately less expensive than larger units, partially because fewer auxiliary components are required.

Summary

Surge bins are important as an economical means of maintaining a uniform drying rate. They reduce the capacity demands of conveyors and allow for multiple usage of a single receiving conveyor. Surge bins may also serve as dryeration bins.

Pneumatic Conveyers

In recent years, pneumatic grain conveyors have become increasingly popular. Pneumatic systems have been most often used in commercial operations, especially those involving flat storage or barges. However, these types of systems also have a place in on-farm grain handling.

Advantages and Disadvantages

A major advantage of pneumatic conveyors is their ability to reach "out of the way" locations that would otherwise require several augers connected in series. Many times, storage facilities that would otherwise not satisfy the "closed loop" principle can be incorporated into the total materials handling flow using pneumatic conveyors. Pneumatic conveyors are especially suited for flat storage systems in that this type of storage is much more difficult to unload using augers than are conventional grain bins. With pneumatic conveyors, unloading flat storage may be compared to using a portable vacuum cleaner to reach every part of the floor area. Dust and shoveling are minimized and a single individual can operate the unit during unloading. Mechanical parts are at ground level. The units are self-cleaning and relatively safe because moving parts are not exposed (Hellevang, 1985). In short, pneumatic conveyors can be used effectively to enhance what would otherwise be poorly designed grain handling and storage facilities.

A major disadvantage of pneumatic conveyors is that considerably more energy is required per unit of grain transported. Thus, for similar capacities, much larger power requirements are needed, and often the units are powered through tractor driven ptos. Excessive noise also can be a problem.

Grain damage and dust associated with pneumatic conveying are similar to that of bucket elevators and drag conveyors where velocities are 4,000 ft/min or less. However, damage increases exponentially for higher velocities. Feed rate has little effect (Baker et al., 1985; Magee et al., 1983).

Design Considerations

Pneumatic conveyors may utilize positive pressure (push units), negative pressure (vacuum), or combinations of the two systems within the same unit.

Positive Pressure Units. A schematic of a positive pressure (push) unit is shown in figure 6.27. Air is forced through a pipe using either a rotary positive displacement blower or a centrifugal blower. The rotary positive displacement blower delivers relatively high pressure with a nearly constant volume of air. Centrifugal blowers deliver relatively large volumes of air at low pressures. With either blower, grain is passed from a holding tank through an air lock into the moving air stream and on to its final destination.

Negative Pressure Units. A schematic of a negative pressure (vacuum) unit is shown in figure 6.28. In negative pressure units, the blower pulls both grain and air through the receiving tube. Before the grain reaches the blower, a cyclone separator is used to separate the grain and the air.

Combination Units. A combination of these units is shown in figure 6.29. Note that the same blower and air are used to both "pull" and "push" the grain.

Power Requirements. The theoretical pressure drop (equation 6.45; Segler, 1951) to lift grain was reported by Baker et al. (1985) to be about

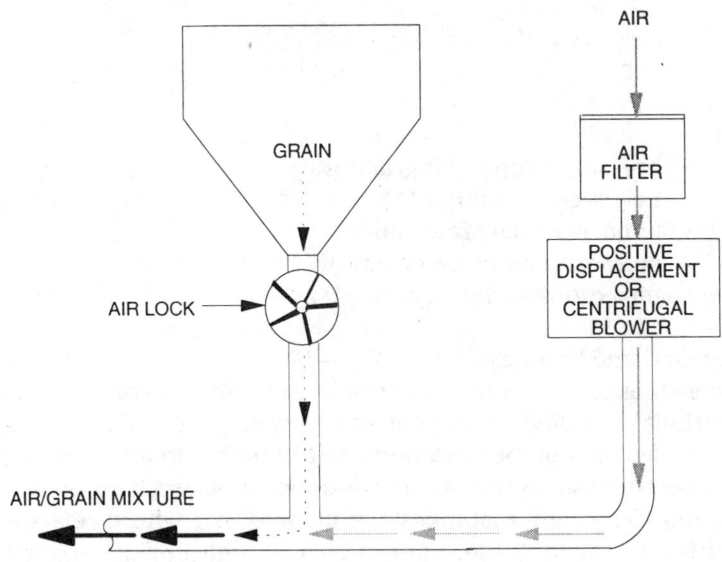

Figure 6.27–Push-type pneumatic conveyor.

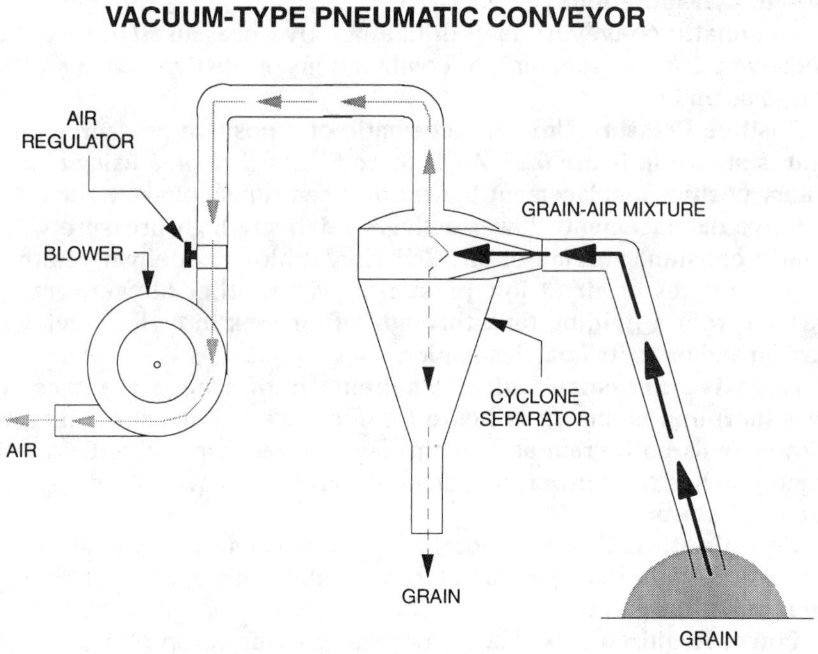

Figure 6.28–Vacuum-type pneumatic conveyor.

one tenth of that actually needed, the difference being associated with frictional energy loss.

$$dP_v = (dh * m_g) / (154 * v_a * A_p) \qquad (6.45)$$

where

- dP_v = pressure drop to lift grain (psi)
- dh = height grain is lifted (ft)
- m_g = grain mass flow rate (bu/h)
- v_a = conveying air velocity (fpm)
- A_p = pipe cross-sectional area (ft^2)

Economic Considerations

The economics of using a pneumatic conveyor, as compared to more conventional handling equipment, is site specific. Pneumatic conveyors are more likely to compete economically in those situations where there is no direct access to the points where grain is received or delivered, where relatively long distances are involved, or where relatively low capacities are acceptable. Energy cost is high because of frictional losses in the delivery tube, and grain damage may be higher than other

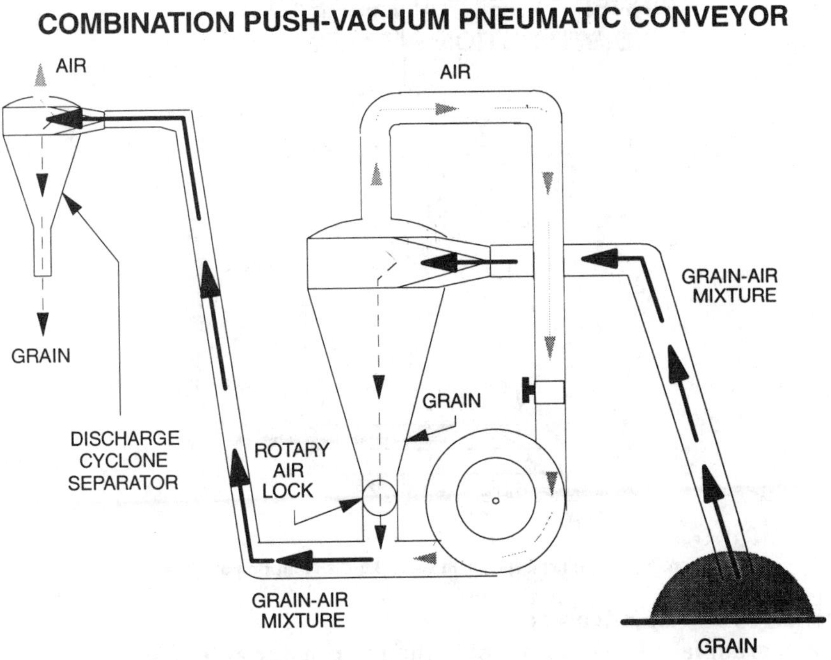

Figure 6.29–Combination push and vacuum pneumatic conveyor.

conveying systems if relatively high air velocity is used. Purchase cost can be reduced significantly if a tractor can be used as a power source rather than purchasing an electric motor.

Summary

Pneumatic conveyors are attractive in certain types of grain handling situations, especially with flat storage systems. Power requirements per unit of grain conveyed are much higher than with augers or bucket elevators.

Grain Spreaders

Grain spreaders commonly are used in round metal grain storage bins. The most common type is often called a spinner-type spreader. Most of these are powered by electric motors, especially in larger bins, although some use the falling grain to drive the unit and distribute the material (fig. 6.30). Another grain spreader design uses an auger to distribute the grain (fig. 6.31).

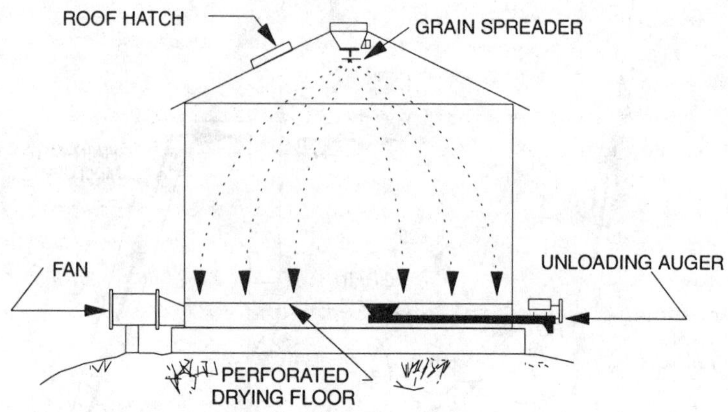

Figure 6.30–Spinner-type grain spreader distribution pattern.

Advantages and Disadvantages

When grain falls freely into a pile, the finer material tends to settle in the center of the grain mass, increasing the bulk density in that portion of the pile. The air tends to bypass the zone where the fine material is concentrated when the air is passed through the grain mass, either when drying or aerating the material. Grain spreaders perform two major functions that make them standard equipment in most grain bins.

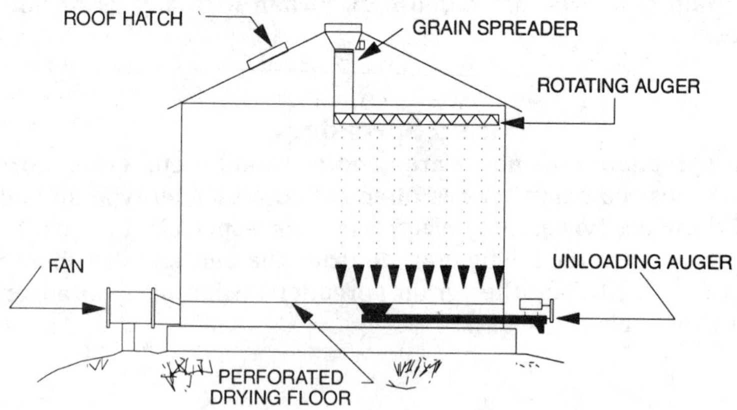

Figure 6.31–Auger-type spreader filling pattern.

First, the device spreads the grain somewhat uniformly over the bin eliminating hand labor for shoveling. Second, spreading the grain more uniformly distributes the fine material leading to greater uniformity in airflow. Spreading the grain, however, also increases its overall bulk density leading to an increased resistance to airflow. Stephens and Foster (1976, 1978) reported increases in bulk density and airflow resistances of up to 20% and 300%, respectively, in shelled corn when using spreaders. However, they observed also that spreaders decreased the uniformity of fine material in sorghum while there was little difference in wheat. Differences between spout and spreader filled bulk densities in sorghum and wheat were 13% and 7%, respectively, and airflow resistances increased 110% and 101%. Thus, spreader performance is somewhat dependent on the type of grain. Also, increases in bulk density lead to greater weight capacity in grain bins.

Summary

Grain spreaders offer more advantages than disadvantages. However, the degree to which they function is somewhat dependent on the type of grain.

Chain Conveyors

Chain conveyors may be used to move grain and similar materials and are further categorized as being either flight or drag conveyors. In both types, a continuous chain moves flighting using pulleys located at each end of the conveyor (fig. 6.32).

If grain is carried along the top of a center petition that divides the chain conveyor, the conveyor is called a flight conveyor. If the grain is dragged along the bottom of the chain conveyor, it is referred to as a drag conveyor. Drag conveyors also are called paddle wheel conveyors, and sometimes the term “flight conveyor” is used interchangeably with “chain conveyor”, “apron conveyor”, or “scraper conveyor”.

Chain conveyors typically operate at speeds of less than 200 ft/min. The power requirements are usually much less than augers, especially when elevating grain. They are also operationally quieter than augers.

Power Requirements

The theoretical power requirements for a chain conveyor are (Henderson and Perry, 1966):

$$hp = \frac{2*V*LC*WC*FC + Q*(LL*FM + HTDC)}{33{,}000} \qquad (6.46)$$

where

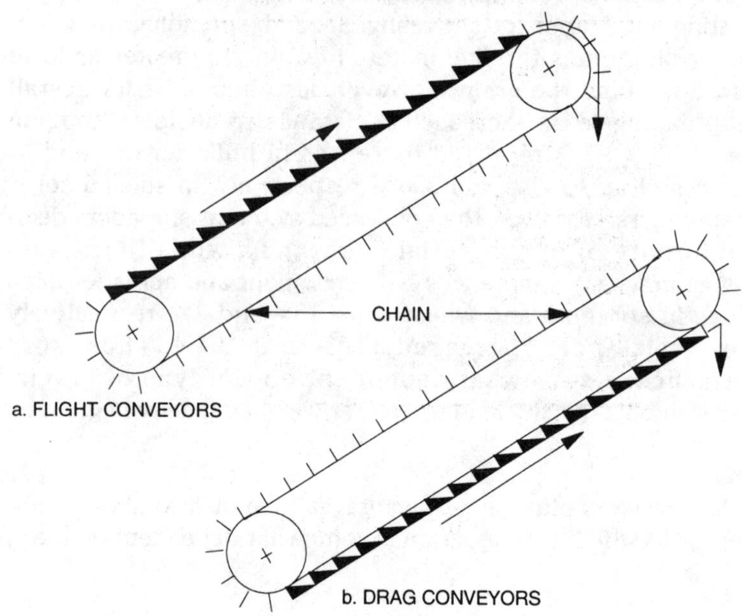

Figure 6.32–Types of chain conveyors.

V = velocity of the chain (ft/min)
LC = horizontal projected length of the conveyor (ft)
WC = weight of flights and chain (lb/ft)
FC = coefficient of friction for chains and flight (see table 6.10)
Q = quantity of material to be handled (lb/min)
LL = horizontal projected length of the part of the conveyor that is loaded (ft)
FM = coefficient of friction for material (table 6.10)
HTDC = discharge height (ft)

The following intermediate equations may be used:

Projected conveyor length:

$$LC = COS(c) * TCL \tag{6.47}$$

where
c = angle of elevation of the conveyor (°)
TCL = total conveyor length (ft)

Material to be handled:

Table 6.10. Sliding friction coefficients for chain conveyors (Henderson and Perry, 1966)

Material	Coefficient of Friction
Metal on oak	0.50 - 0.60
Oak on oak, parallel fibers	0.48
Oak on oak, cross fibers	0.32
Cast iron on mild steel	0.23
Mild steel on mild steel	0.57
Grain on rough board	0.30 - 0.45
Grain on smooth board	0.30 - 0.35
Grain on iron	0.35 - 0.40
Coal on metal	0.60
Dry sand on metal	0.60
Malleable roller chain on steel	0.35
Roller-bushed chains on steel	0.20

$$Q = (BPH * BD) / 48 \tag{6.48}$$

where

BPH = bu/h capacity
BD = bulk density of grain (lb/ft^3) (see table 6.1)

Projected horizontal conveyor length:

$$LL = COS\ (c) * TLL \tag{6.49}$$

Discharge height:

$$HTDC = SIN\ (c) * TLL \tag{6.50}$$

where

TLL = total loaded length of the conveyor (ft)

For horizontal conveying, equation 6.46 can be simplified to equation 6.51 using the following assumptions:

1. The bulk density of grain is 48 lb/ft (60 lb/bu) (i.e., soybeans).
2. Conveyor power losses can be expressed as a single efficiency term.
3. Grain will be conveyed the entire length of the conveyor.
4. Coefficient of friction for grain on metal is 0.6.

$$\text{horsepower} = \frac{\text{BPH} * \text{LCON}}{550 * \text{E}} \tag{6.51}$$

where

LCON = length of conveyor (ft)
E = efficiency of operating the conveyor (85% is a typical value) (%)
BPH = bu/h capacity

For example, what are the horsepower requirements for a 25-ft flight conveyor used to transport soybeans (60 lb/bu) horizontally at the rate of 3,000 bu/h? The following assumptions are made:

1. FC and FM values of 0.35 and 0.4, respectively.
2. The total length of the elevator is to be filled.
3. WC is 10 lb/ft.
4. Velocity is 100 ft/min.

From equation 6.46:

$$\text{horsepower} = \frac{2 * 100 * 25 * 10 * 0.35 + 3{,}000 * (25 * 0.4 + 0.0)}{33{,}000}$$

$$= 1.44$$

Using equation 6.51 with an efficiency of 85%:

$$\text{horsepower} = \frac{3{,}000 * 25}{550 * 85}$$

$$= 1.60$$

In either instance, a 3-hp motor would provide an adequate factor of safety in the design.

If the grain in the above example were to be elevated at a 30° angle, horizontal and vertical displacements must be considered. From equations 6.47 and 6.50:

$$\text{LC} = \text{COS}(30) * 25 \text{ ft}$$
$$= 21.65 \text{ ft}$$
$$\text{HTDC} = \text{SIN}(30) * 25 \text{ ft}$$
$$= 12.5 \text{ ft}$$
$$\text{horsepower} = \frac{2*100*21.65*10*0.35 + 3000*(21.65*0.4 + 12.5)}{33{,}000}$$
$$= 2.38$$

Other Design Considerations

The capacity of a flight conveyor is determined by multiplying its chain speed and holding capacity. In equation form:

$$\text{CAPFLT} = \text{V} * \text{BUPFT} * 60 \quad (6.52)$$

where
CAPFLT = capacity of the flight conveyor (bu/h)
V = velocity of the chain (ft/min)
BUPFT = bu/ft of conveyor length

The bushels per foot of length may be estimated by:

$$\text{BUPFT} = \text{CSA} * \text{BFD} * \text{FFAC} * 0.8 \quad (6.53)$$

where
CSA = cross-sectional area occupied by the grain (ft^2)
BFD = between flight distance (ft)
FFAC = filling factor (see below)

The filling factor is associated with the angle that the grain is being conveyed. For horizontal transport, the filling factor has a value of 1. If the grain is conveyed upward at a greater angle than that of the grain's angle of repose, the grain will spill over the flights (fig. 6.33). The maximum filling factor may be estimated by the following equation:

$$\text{FFAC} = 1.0 - (\text{TAN}(b - a) * \text{BFD}) / (2*\text{HF}) \quad (6.54)$$

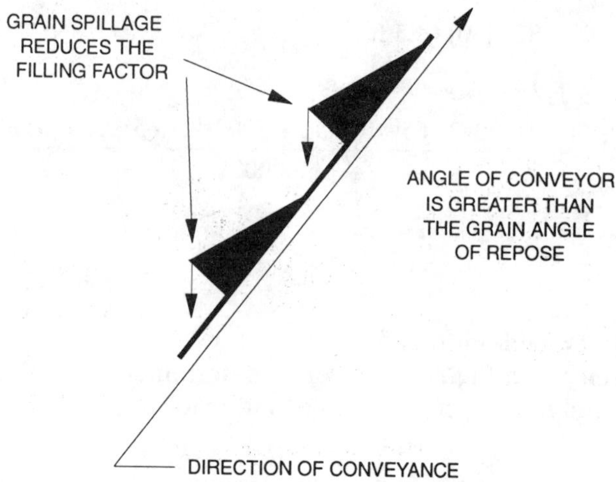

Figure 6.33–Filling factor and grain spillage.

where

FFAC = filling factor applicable for systems where BFD ≤ [HF/TAN(b - a)] and b>=a
BFD = between flight distance
HF = height of flight measured in the same units as BFD
b = angle of chain conveyor
a = effective angle of repose for the material being conveyed

For systems where BFD > [HF/TAN (b - a)] (fig. 6.33):

$$FFAC = \frac{(2*HF - 1.0)}{2*BFD*TAN(b - a)} \quad (6.55)$$

For equations 6.54 and 6.55 to apply, "b" must be greater than "a"; that is, the angle of repose of grain "a" has to be exceeded by the angle of the conveyor "b" before the filling factor would have a value less than 1.0. However, if the conveyor is vibrating, the real angle of repose for the material being conveyed may be less than the static value for a stationary grain mass. The extent to which this angle is reduced depends not only on vibration but also length of travel, chain speed, and the initial velocity and direction of the grain entering the conveyor.

If a drag conveyor is used with a center plate that separates the upper and lower flighting, the filling factor will be 1.0 because the grain

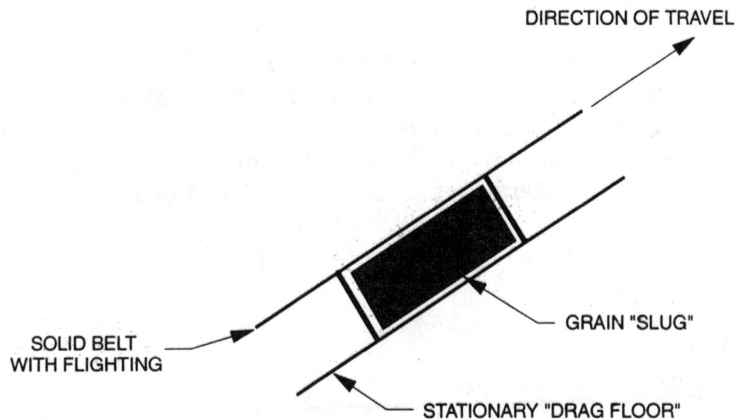

Figure 6.34–Drag conveyor.

will be contained in all directions (fig. 6.34). The grain is "trapped" between the flights so that the angle of lift has little influence on filling efficiency. This principle is utilized in conveying systems where flighting is passed through round tubing like that of a screw conveyor (fig. 6.35). These systems are very flexible in that the tubing may "turn corners" and change elevations within the length of travel. The most common use of these types of systems is to distribute feed in animal housing systems. However, these larger capacity "grain pumps" are becoming popular as an economical method of transporting grain.

Summary

A characteristic of chain conveyors is that the slow movement of the grain relative to the conveyor results in relatively low damage and a quiet operation. Thus, the drag conveyor is especially popular in the

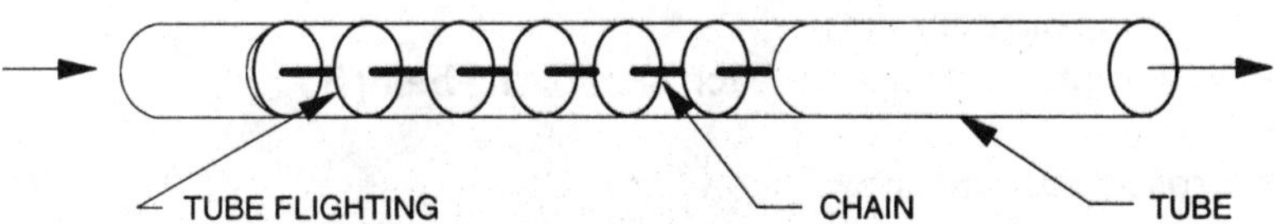

Figure 6.35–Tube-type chain conveyor.

seed industry. Chain conveyors also require relatively less horsepower than augers in transporting the same quantity of material.

Belt Conveyors

Belt conveyors are a very efficient but relatively expensive means of transporting grain. Grain damage is relatively low so this type of conveyor is often used in seed processing and conditioning systems. Belt conveyors are often used in large commercial operations because conveying capacities can be high. Their most common use in on-farm systems is for pit unloading. They are limited in their angle of elevation for grain to 15 to 17°.

Capacity

The capacity of a belt conveyor is:

$$BLTCAP = 48 * CSA * BLTS \quad (6.56)$$

where

BLTCAP = belt conveyor capacity (bu/h)
CSA = cross-sectional area (ft^2)
BLTS = belt speed (ft/min)

The cross-sectional area profile is determined by the type of rollers. Typically, three rollers are used, the center one is horizontal with the other two cupped inward at angles ranging from 20 to 45° (fig. 6.36). Flat belts may be used but at much reduced capacities. The following equations may be used to estimate belt cross-sectional area (fig. 6.37):

Total belt width:

$$THW = CW + COS(rola) * (TBW - CW - 2*M) \quad (6.57)$$

where

THW = total horizontal width of the belt (ft)
TBW = total belt width (ft)
CW = center width (ft)
rola = angle of rollers (°)
M = margin clearance on each side of the belt (ft)

Bottom cross-sectional area:

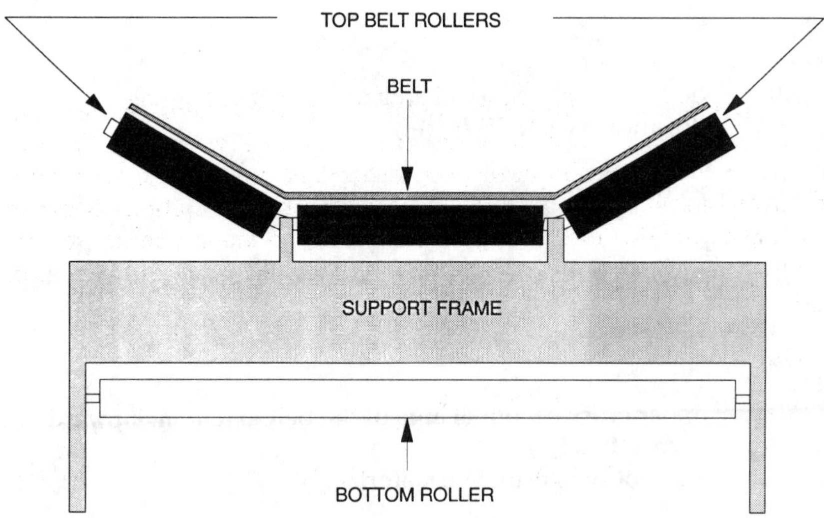

Figure 6.36–Belt conveyor.

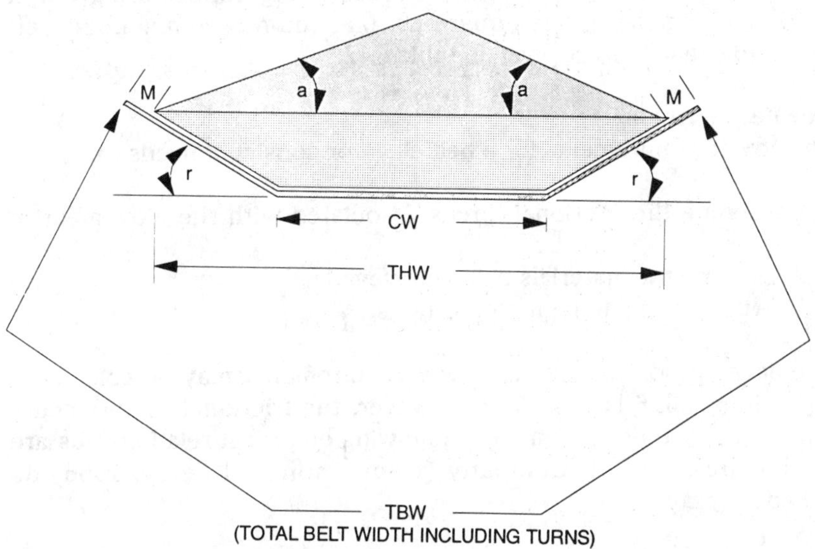

Figure 6.37–Grain geometry on belt conveyor.

$$CSAB = ([THW + CW] * SIN[rola] * [TBW - CW - 2*M]) / 4 \quad (6.58)$$

where

CSAB = bottom cross-sectional area of the belt extending downward from THW (ft^2)

Upper cross-sectional area:

$$CSAU = (THW / [2 * COS\{90 - a\}])^2 * ([a * pi/180] - SIN[2*a] / 2) \quad (6.59)$$

where

CSAU = upper cross-sectional area of the belt extending upward from THW (ft^3)

a = angle of repose for the material (°)

pi = 3.1416

$$CSA = CSAB + CSAU \quad (6.60)$$

where

CSA = total cross-sectional area of the belt (ft^2)

Cross-sectional areas for a variety of roller configurations are given in table 6.11 using the above equations. Maximum recommended belt speeds and capacities are given in table 6.12.

Power Requirements

The power requirements for a belt elevator consist of needs to:

1. Overcome the frictional forces associated with the movement of the belt.
2. Accelerate the materials being conveyed.
3. Lift the material to higher elevations.

The acceleration and lifting power requirements may be calculated by equations 6.6, 6.11, and 6.12. However, the frictional requirements are highly system dependent. The following empirical relationships are given by the Link-Belt Company (Henderson and Perry, 1966) as modified for grain.

$$HPDC_{drive} = BLTS*(C1 + C2*CLG) / 100 \quad (6.61)$$

where

$HPDC_{drive}$ = horsepower to drive the conveyor
BLTS = belt speed (ft/min)
C1, C2 = constants that depend on belt width and are given in table 6.13
CLG = conveyor length (ft)

$$HPCG_{hor} = BPH*BDG*(0.48 + 0.00302*CLGH) / 200{,}000 \quad (6.62)$$

Table 6.11. Cross-sectional areas for various belt sizes, roller angles, and surcharge angles

Belt Width (in.)	Belt Margin (in.)	Roller Angle (°)	Surcharge Angle (°)	Center Width (in.)	Horizontal Width (in.)	Cross-Sectional Area (ft²)
14	1.7	20	10	4.67	10.24	0.0738
		20	20	4.67	10.24	0.0956
		20	30	4.67	10.24	0.1185
		35	10	4.67	9.53	0.1023
		35	20	4.67	9.53	0.1211
		35	30	4.67	9.53	0.1410
		45	10	4.67	8.86	0.1145
		45	20	4.67	8.86	0.1308
		45	30	4.67	8.86	0.1479
16	1.8	20	10	5.33	11.97	0.1017
		20	20	5.33	11.97	0.1315
		20	30	5.33	11.97	0.1628
		35	10	5.33	11.12	0.1409
		35	20	5.33	11.12	0.1666
		35	30	5.33	11.12	0.1936
		45	10	5.33	10.33	0.1575
		45	20	5.33	10.33	0.1797
		45	30	5.33	10.33	0.2030
18	1.9	20	10	6.00	13.71	0.1340
		20	20	6.00	13.71	0.1731
		20	30	6.00	13.71	0.2141
		35	10	6.00	12.72	0.1856
		35	20	6.00	12.72	0.2193
		35	30	6.00	12.72	0.2546
		45	10	6.00	11.80	0.2074
		45	20	6.00	11.80	0.2363
		45	30	6.00	11.80	0.2667

Table 6.11 continues on next page.

where

$HPCG_{hor}$ = horsepower to convey grain horizontally
BPH = bu/h to be conveyed
BDG = bulk density of grain (lb/bu)
CLGH = conveyer length in the horizontal direction (ft)

$$HPLG_{lift} = LIFT*1.015*\left(BPH*BDG / 2{,}000{,}000\right) \tag{6.63}$$

Table 6.11. (Continued)

Belt Width (in.)	Belt Margin (in.)	Roller Angle (°)	Surcharge Angle (°)	Center Width (in.)	Horizontal Width (in.)	Cross-Sectional Area (ft²)
20	2	20	10	6.67	15.44	0.1708
		20	20	6.67	15.44	0.2204
		20	30	6.67	15.44	0.2724
		35	10	6.67	14.31	0.2365
		35	20	6.67	14.31	0.2791
		35	30	6.67	14.31	0.3238
		45	10	6.67	13.27	0.2641
		45	20	6.67	13.27	0.3007
		45	30	6.67	13.27	0.3391
24	2.2	20	10	8.00	18.90	0.2577
		20	20	8.00	18.90	0.3320
		20	30	8.00	18.90	0.4100
		35	10	8.00	17.50	0.3567
		35	20	8.00	17.50	0.4204
		35	30	8.00	17.50	0.4873
		45	10	8.00	16.20	0.3979
		45	20	8.00	16.20	0.4525
		45	30	8.00	16.20	0.5098
30	2.5	20	10	10.00	24.10	0.4214
		20	20	10.00	24.10	0.5421
		20	30	10.00	24.10	0.6689
		35	10	10.00	22.29	0.5830
		35	20	10.00	22.29	0.6863
		35	30	10.00	22.29	0.7947
		45	10	10.00	20.61	0.6497
		45	20	10.00	20.61	0.7380
		45	30	10.00	20.61	0.8307
36	2.8	20	10	12.00	29.29	0.6251
		20	20	12.00	29.29	0.8035
		20	30	12.00	29.29	0.9908
		35	10	12.00	27.07	0.8646
		35	20	12.00	27.07	1.0169
		35	30	12.00	27.07	1.1770
		45	10	12.00	25.01	0.9629
		45	20	12.00	25.01	1.0929
		45	30	12.00	25.01	1.2295

Table 6.12. Maximum belt speeds for certain widths and belt edge clearances (Henderson and Perry, 1966)

Belt Width (in.)	Clearance Margin (in.)	Maximum Belt Speed (ft/min) Non-Abrasive Fine Material	Grain
14	1.7	300	400
16	1.8	300	450
18	1.9	400	450
20	2.0	400	500
24	2.2	500	600
30	2.5	500	700
36	2.8	550	800
42	3.1	600	800
48	3.4	600	800
54	3.7	600	800
60	4.0	600	800

where

$HPLG_{lift}$ = horsepower to lift grain

LIFT = vertical distance to convey grain (ft)

$$HT_{total} = HPDC_{drive} + HPCG_{hor} + HPLG_{lift} \tag{6.64}$$

where

HT_{total} = total horsepower requirements for the conveyor

Table 6.13. Drive horsepower constants for belt conveyors (Henderson and Perry, 1966)

Conveyor Belt Width (in.)	C1	C2
14	0.20	0.00140
16	0.25	0.00140
18	0.30	0.00162
20	0.30	0.00187
24	0.36	0.00224
30	0.48	0.00298
36	0.64	0.00396
42	0.72	0.00458
48	0.88	0.00538
54	1.00	0.00620
60	1.05	0.00765

For example, what are the power requirements for moving 3,000 bu/h of corn (56 lb/bu) along an 80-ft felt conveyor to an elevation of 10 ft? Assume an 18-in. wide belt with a 20° roller angle and 20° surcharge angle is used. Its operating speed may be calculated by rearranging equation 6.56 and utilizing the cross-sectional area given in table 6.11.

$$\begin{aligned}\text{Belt speed (ft / min)} &= \frac{3{,}000 \text{ bu/h}}{48*0.1731 \text{ ft}^2} \\ &= 361 \text{ ft/min}\end{aligned}$$

Given the belt speed and length, the power requirements may now be calculated.

For conveyor operation (equation 6.61):

$$\begin{aligned}HPDC_{drive} &= 361*(0.30 + 0.00162*80) / 100 \\ &= 1.55 \text{ hp}\end{aligned}$$

For horizontal grain movement (equation 6.62):

$$\begin{aligned}HPCG_{hor} &= 3{,}000*56*(0.48 + 0.00302*80) / 200{,}000 \\ &= 0.61\end{aligned}$$

END UNLOADING OF A BELT CONVEYOR

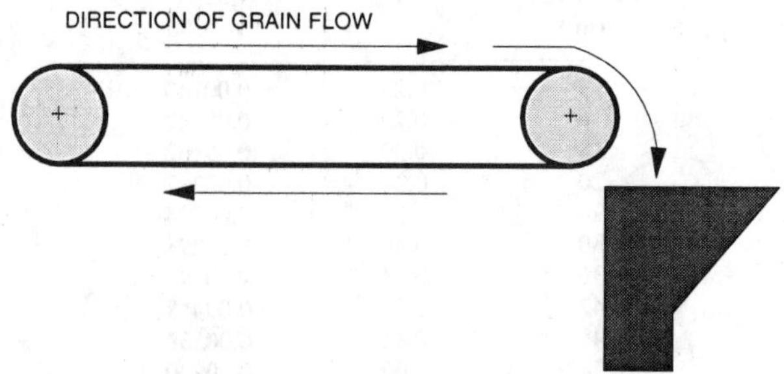

Figure 6.38–End of belt unloading.

For vertical lift (equation 6.63):

$$HPLG_{lift} = 10*1.015*3{,}000*56/2{,}000{,}000$$
$$= 0.85$$

Total power requirements (equation 6.64):

$$HT_{total} = 1.55 + 0.61 + 0.85$$
$$= 3.01$$

A 5-hp motor would be utilized to allow for some margin of safety.

Unloading Methods

Belt conveyors may be unloaded over the end of the belt, by a diagonal scraper or by a tripping mechanism (figs. 6.38 to 6.41). End unloading is the easiest method. This method is best suited to situations where the conveyor has a single unloading point such as when conveying grain from several bins to a pit (fig. 6.38). However, considerable flexibility can be added by placing the conveyor on rollers and using a reversing mechanism so that several end points may be reached (fig. 6.39).

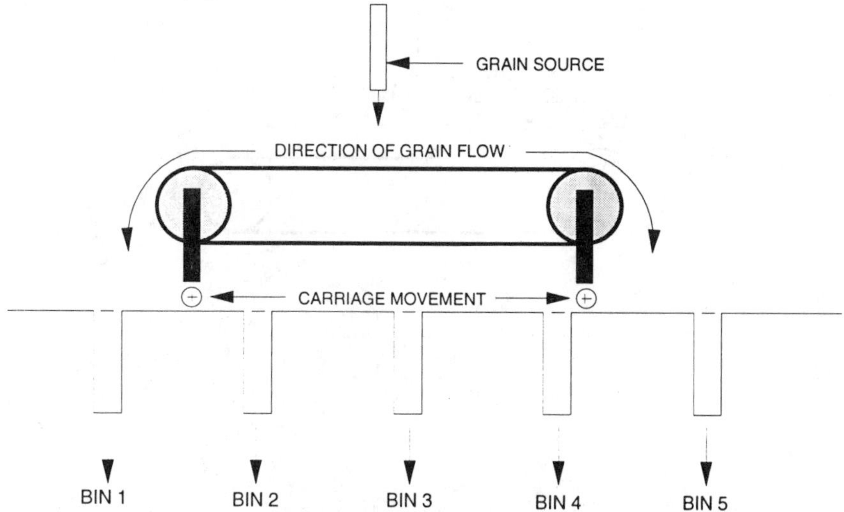

Figure 6.39–Belt conveyor on rollers.

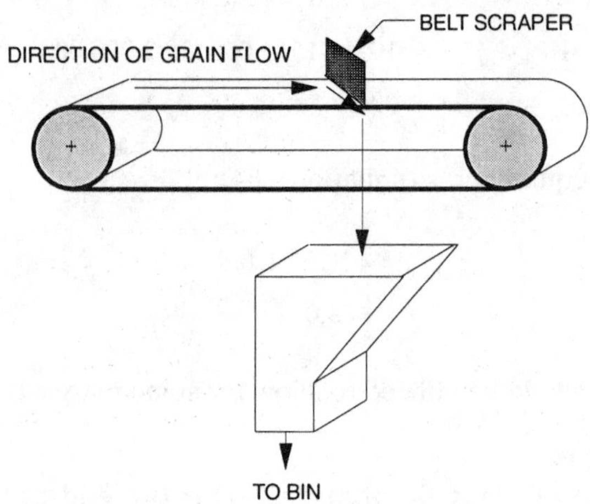

Figure 6.40–Belt conveyor scraper.

A diagonal scraper may be used to remove the material from the belt at any point along its travel. The scraper needs to closely match the contour of the belt and may be incorporated with a flat section of upper rollers (fig. 6.40).

A tripping mechanism may be used (fig. 6.41) to convey the material to the side of the belt. These mechanisms may be made to move along the length of the belt so that material may be removed at any point

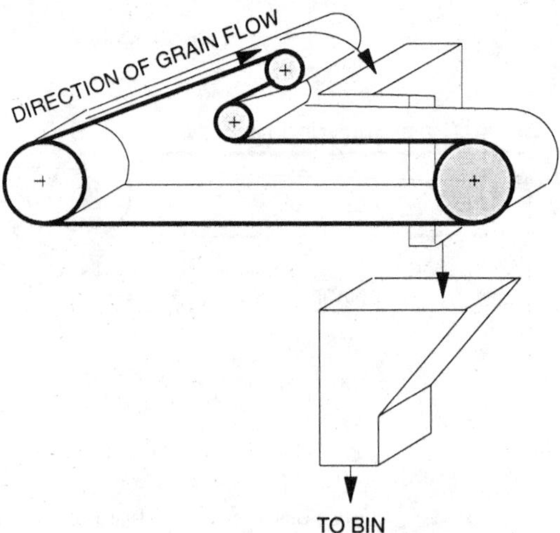

Figure 6.41–Belt conveyor tripping mechanism.

along the belt's length. The tripper requires additional horsepower to overcome added friction and to elevate the material above the belt.

Summary

Belt conveyors are efficient mechanisms for transporting grain. They are quiet, self-cleaning, and readily accessible for purposes of maintenance and repair. Generally, the investment cost is relatively high for on-farm systems.

Problems

6.1 A force of 50 lb moves an object 15 ft in the direction of the force. How much work is done?

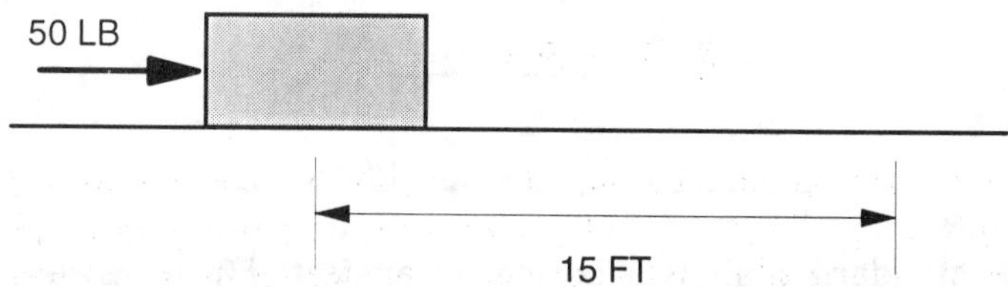

6.2 If the force is applied at an angle of 30° as shown below, how much work is done if the object moves 12 ft horizontally?

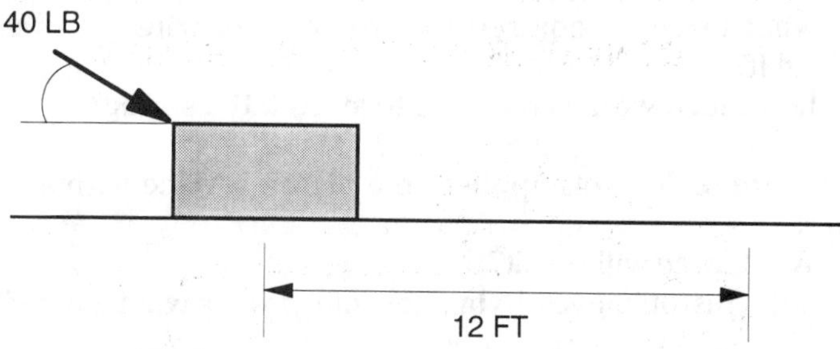

6.3 A force of 100 lb is applied to an object weighing 500 lb. At what rate will the object accelerate?

6.4 The dynamic coefficient of friction between a wooden block and a steel surface is 0.45.

(a) If a force of 50 lb is applied to the block as shown below, what is the magnitude of the force resisting movement of the block in the horizontal direction?

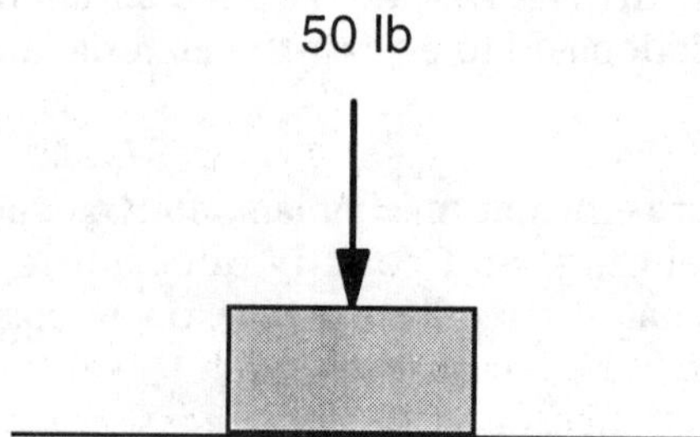

(b) What will be the magnitude of the resisting force if the force is applied at an angle of 45° as shown below?

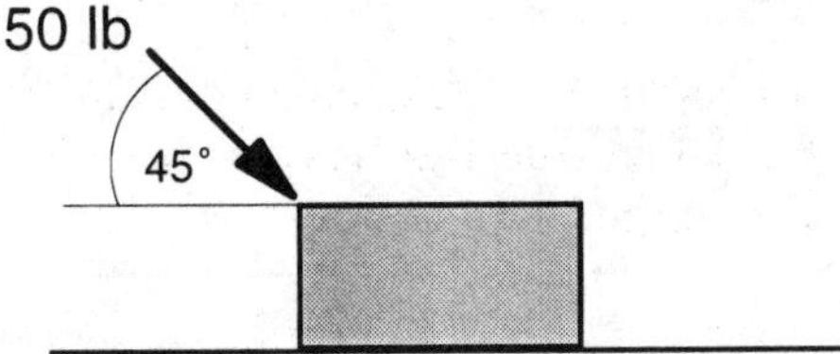

6.5 If 500 standard bushels of corn are transferred from an elevation of 10 ft to an elevation of 20 ft, how much work is done?

6.6 A spring has a spring constant of 25 lb/in. If the spring has an initial length of 1 ft:

(a) what force is required to stretch the spring to a length of 1.4 ft?

(b) how much work is required to stretch the spring?

6.7 A pressure of 30 psi is applied to a piston surface with an area of 27 in.^{2}.

(a) What force will result?

(b) If the piston moves 16 in., how much work will be done?

6.8 One bushel of corn moves along a conveyor belt at an initial velocity of 1 ft/s.

(a) How much work will be required to increase the velocity of the bushel of corn to 2 ft/s?

(b) How many horsepower would be required to change the velocity of one bushel of corn from 1 to 12 ft/s over a period of 30 s.

6.9 A grain elevator lifts corn at a rate of 200 bu/min. If grain enters the elevator at the bottom of an 8-ft pit and leaves at an elevation of 40 ft, what is the horsepower required to overcome the effects of the gravitational pull on the grain?

6.10 What is the theoretical capacity of an 8-in.-diameter auger with a 1-in. diameter shaft and a pitch of 8 in. if the shaft is turning at a speed of 600 rpm?

6.11 An auger with a capacity of 500 ft^3/min moves corn over a horizontal distance of 30 ft. What horsepower motor would be required to power the conveyor?

6.12 A portable auger is to be used to load grain into an 18-ft-diameter bin with an eave height of 15 ft. The bin roof slope is 30° and the angle of the auger is 40°. What length of auger is required?

6.13 Dry wheat is to be discharged from a bucket elevator at a height of 40 ft into a bin 30 ft from the base of the elevator. What length downspout is required?

6.14 How much power is required to operate a bucket elevator 60 ft tall that carries corn at a rate of 750 bu/h? Assume the efficiency of the bucket elevator is 80%.

6.15 If the effective radius of the head pulley of a bucket elevator is 12 in., what should the speed in revolutions per minute of the head pulley be?

6.16 If the diameter of the head pulley of a bucket elevator is 15 in. and the perpendicular distance of the cup to the belt is 5 in., what is the effective radius of the pulley?

6.17 If a 14-in. diameter head pulley turns at 18 rpm, what is the belt speed?

6.18 If the distance between the head and boot pulley is 30 ft and the diameters of the head and boot pulleys are 12 in., what belt length is needed?

6.19 How many times will a belt revolve completely in 5 min if the belt moves at 200 ft/min and the belt length is 60 ft?

6.20 How many cups are required for a belt carrying 10 bu of corn per revolution if the corn weighs 52 lb/bu and each cup has a volume of 400 $in.^3$?

6.21 What is the required spacing between cups if the belt length is 220 ft and the belt contains 220 cups?

6.22 A truck carrying a load of wheat dumps grain into a pit at a rate of 100 bu/min. If the truck is to be unloaded in 5 min and an elevator removes grain from the pit at a rate of 50 bu/min, what size pit is necessary?

6.23 A truck with a bed width of 12 ft unloads grain from a height of 4 ft. If the angle of repose of the wheat is 16°, what are the minimum dimensions of the pit to avoid grain spilling over the pit edges?

6.24 Wheat is unloaded into a conveyor unloading pit with a length of 6 ft.

(a) If the angle of repose of the wheat is 16°, what is the required depth of the lower portion of the pit?

(b) If the conveyor extends along the entire width of the 6-ft-wide pit, what is the capacity of the lower section of the pit?

6.25 What depth upper portion of a pit is needed for a pit with a lower section capacity of 300 bu and a total pit capacity of 800 bu if the pit is 12 ft wide and 6 ft long?

6.26 A gravity unloaded pit is 12 ft wide and 18 ft long. If the pit contains wheat with an angle of repose of 16°, what is the required lower depth of the pit?

6.27 A gravity pit is 12 ft wide and 30 ft long. If the pit contains wheat with an angle of repose of 16°, what is the required lower depth of the pit?

6.28 Grain is harvested from a field over a 19-h period. The drying system is equipped with a 850 bu pit. If grain is harvested at a rate of 1,100 bu/h and the dryer can handle grain at a rate of 950 bu/h, what size wet holding bin is needed?

6.29 A continuous-flow dryer processes grain at a rate of 1,000 bu/h. If 15 min are needed to empty the largest grain truck, what size surge bin is needed if a single conveyor is used to transport both wet and dry grain?

6.30 What capacity dry grain conveyor is needed if the dry grain conveyor operates 70% of the time in a system that has a maximum continuous drying rate of 800 bu/h?

6.31 If the dry grain conveying rate is 1,200 bu/h and the maximum continuous-flow drying rate is 800 bu/h and the dry grain conveyor operates 70% of the time, what size surge bin is needed for dry grain?

6.32 If a 2,000-bu surge bin is used to hold dry grain, what capacity dry grain conveyor is needed if a minimum of 60 min is required to fill and dry one batch of grain?

6.33 A pneumatic conveyor with a diameter of 10 in. is used to lift corn from an elevation of 2 ft to an elevation of 85 ft. If the corn is moved at a rate of 9,000 bu/h using an air velocity of 500 ft/min, what pressure drop would be expected in the system?

6.34 What is the theoretical amount of power required for a chain conveyor to move corn at a rate of 1,000 bu/h if 80 ft of the 100 ft conveyor are loaded with grain? The conveyor is inclined at an angle of 5° from horizontal. The flights and chain weigh 10 lb/ft, the coefficient of friction for the chains and flight is 0.57, and the coefficient of friction between grain and conveyor is 0.3. The chain moves at a velocity of 200 ft/min.

6.35 If 1,000 bu/h are carried over a distance of 100 ft by a horizontal conveyor with an empty efficiency of 85%, what is the expected horsepower needed to drive the conveyor?

6.36 What is the capacity of a horizontal flight conveyor that has a chain velocity of 120 ft/min and carries grain with a cross-sectional area of 1 ft^2 if the distance between flights is 8 in.?

6.37 What is the capacity of a belt conveyor carrying grain if the belt moves at a velocity of 300 ft/min? The belt has a total horizontal width of 24 in. The belt width is 30 in. The center width is 18 in. The roller angle is 15° and the margin on each side of the belt is 0.1 ft. The angle of repose of the grain is 17°.

6.38 How much power is required to drive a belt conveyor moving at a velocity of 350 ft/min if the conveyor is 2 ft wide and 80 ft long. Corn is to be conveyed at a rate of 1200 bu/h. The conveyor is tilted at an angle of 10° from horizontal.

References for all chapters begin on page 541.

7

Selecting Optimum Equipment Sets

Grain processing equipment sets are dynamic in their performance in that the load demands placed on them often vary over time. Thus, static design solutions to dynamic processes should be approached with caution. In this chapter, the principles and methods of selecting optimum equipment sets will be addressed.

Principles of Selecting Optimum Equipment Sets

If profits are to be maximized, resources must be allocated in an optimum manner. Resources include labor, materials and equipment, with the term "equipment set" referring to all of these factors. An "optimum allocation" refers to the production of a given level of output for the least possible costs. Costs include all factors associated with an equipment set, some of which may have a "cost equivalent" basis such as convenience, service, reputation, etc. There are certain principles that should be employed when selecting equipment sets. The objective of this chapter is to describe broadly these principles in reference to a typical farm materials handling situation.

Principle No. 1: There will always be a "bottleneck" in any materials handling system.

A materials handling system, such as the harvesting, delivery, drying and storage of grain, will always be limited in capacity by one of its components. This principle may be compared to the proverbial chain that is only as strong as its weakest link. If at least one of the materials handling components were not limiting, the capacity of the system would be infinite! For the grain harvesting, delivery, drying and storage system, one of the equipment sets must be limiting or all the grain would be harvested instantaneously. The fact that one component is limiting is not important so long as the total system can deliver material at the specified design rate.

Principle No. 2: If the design capacity is not being obtained, always increase resource expenditures (usually money in some form) for increasing the capacity of the "bottleneck" until the cost of adding one additional unit of capacity exceeds the benefits.

Generally, costs for increasing the capacity of the "bottleneck" will exceed benefits when:

1. Some other component becomes the bottleneck
2. The design capacity is obtained.

If combines are the limiting factor, this bottleneck may be alleviated, for example, by adding other machines, working longer hours, or replacing existing machines with larger capacity units. In all cases, however, there is no benefit to be obtained by adding capacity beyond that of the desired design rate. Similarly, no benefit would be realized if combine capacity were increased past the point where delivery vehicles could haul the grain to the dryer.

Principle No. 3: If intelligent decisions are to be made concerning the performance of a materials handling system, the decision maker must be aware of what is happening and its associated time of occurrence within the system.

The decision maker must be thoroughly aware of the total materials handling system if it is to be optimized. Awareness includes not only what is happening but "when". "When" refers to the initiation, cessation, and duration of each activity. Observation of each part of the system will enable the decision maker to determine what is happening. The tools for determining "when" may include a stop watch, calculator, and/or computer.

The stop watch and calculator may indicate quickly the source of the bottleneck. A computer simulation of the particular materials handling system may indicate bottlenecks associated with the more complex interactions of workers and machines. There are two methods for evaluating performance efficiency.

Event Method of Determining Performance Efficiency

The event method for evaluating performance efficiency involves the following steps:

1. Identifying activities (or events) that have a definite starting and ending time and occur sequentially.
2. Determining the starting and ending times for each event in order to obtain the duration of the activity.

3. Calculating the theoretical time required to complete the operation based on summing the times of the associated events assuming no loss of time because of inefficiency.
4. Observing the actual system and determining the actual time required to complete that same operation.
5. Computing the efficiency of the operation as follows:

$$\text{efficiency (\%)} = \frac{\text{theoretical time} * 100}{\text{actual time}} \qquad (7.1)$$

For example, let's determine the efficiency of combine harvesting. The following events are identified as being associated with harvesting and stationary unloading of grain.

1. Combine starts harvesting.
2. Combine stops harvesting when grain tank is filled.
3. Delivery vehicle moves into position after combine stops harvesting.
4. Combine begins to unload grain after vehicle is in position.
5. Combine completes unloading of grain tank.
6. Combine resumes harvesting after the delivery vehicle clears the area.

The times required to complete each of these sequential events can be determined by using a stop watch or making computations using flow rates of grain. The total theoretical time to complete a cycle is the sum of the individual activity times. Assume that the theoretical cycle time is 30 min. Suppose a stop watch is used to compute the average actual cycle time based on field observation, and that this value is 45 min. From equation 7.1:

$$\text{efficiency (\%)} = \frac{30 \text{ min} * 100}{45 \text{ min}} = 66.7\%$$

Another method of determining efficiency is to compare the actual materials processed to the theoretical processing rate. Using the above example, suppose that the combine hopper size is 100 bu and the daily harvesting time is 9 h. The theoretical cycle time of 30 min/load results in 18 loads or 1,800 bu/day being processed. If the actual time required per load is 45 min, only 12 loads (1,200 bu) would be processed. Using equation 7.1, the efficiency is:

$$\text{efficiency (\%)} = \frac{\text{actual bu harvested} * 100}{\text{theoretical bu harvested}}$$

$$= \frac{1{,}200 \text{ bu} * 100}{1{,}800 \text{ bu}}$$

$$= 66.7\%$$

This method of calculating efficiency has several advantages over the first method. Actual bushels delivered to the grain facility over the harvesting day is more easily determined than the measurement of average time per unloading. That is, one can determine more easily the grain delivered to the facility and note the initial starting and final stopping times of the combine rather than take time measurements of each individual unloading.

Cycle-Time Method of Selecting Equipment Sets

The cycle-time method of selecting equipment sets involves the following steps:

1. Determine the design capacity per unit of time for the total materials handling system.
2. Specify each separate step or activity in the materials handling process.
3. By activity, identify each item of equipment and associated labor that could be used in processing the material, and compute the ownership and purchase cost of each item per unit of design capacity.
4. Determining the number of pieces of equipment of each type required for each activity using the following equation:

$$\begin{array}{c}\text{no. of items}\\ \text{required for}\\ \text{a given stage}\end{array} = \frac{\text{design capacity per unit of time}}{\begin{array}{c}\text{(capacity per unit of time}\\ \text{for a particular item)}\end{array}} \quad (7.2)$$

5. Compute the total purchase and ownership cost for each activity by multiplying the respective per unit costs by the number of units required (Step 4).
6. Select the least cost combination of equipment for each activity based on the criteria of ownership or purchase cost.

Let's compute the design efficiency of a hypothetical harvesting, delivery, and drying system. From Step 1, assume the design capacity to

Table 7.1. Dryers used in example problems

Dryer Type	Capacity (bu/h)	Purchase Cost ($)
1	50	4,000
2	100	7,000
3	150	11,000
4	200	13,000

be 2,800 bu/day. The separate steps (or activities) are harvesting, delivery, and drying. For purposes of example, assume that the dryers listed in table 7.1 may be used. Likewise, assume that purchase price will be the criteria by which the decision to acquire a particular dryer will be made. If the dryer(s) are to be operated 12 h/day, the design capacity is equal to 233 bu/h. From Step 4, the combinations of dryers given in table 7.2 may be selected to meet the design criteria. From these choices, the least cost combination is Choice 5.

The design efficiency may be computed using the following equation:

$$\text{design efficiency (\%)} = \frac{\text{design capacity} * 100}{\text{actual capacity}} \tag{7.3}$$

Using this relationship, the efficiencies given in table 7.3 are determined for the alternative equipment sets given in table 7.2. Note that the design efficiencies are the same for several of the choices, yet there is a difference in economic efficiency as defined by purchase cost. This same procedure is followed in determining the optimum equipment set for harvesting and delivery.

Discussion

The cycle-time method of selecting optimum equipment sets will produce an optimum combination of equipment and labor based on a

Table 7.2. Choice combinations to meet design capacity of 233 bu/h given in the example problem

Choice No.	Dryer No. and ID	Capacity (bu/h)	Total Purchase Cost ($)
1	5 type 1 dryers	250	20,000
2	3 type 2 dryers	300	21,000
3	2 type 3 dryers	300	22,000
4	2 type 4 dryers	400	26,000
5	1 type 4 and 1 type 1	250	17,000
6	1 type 3 and 1 type 2	250	18,000
7	1 type 3 and 2 type 1	250	19,000

Table 7.3. Design efficiencies for dryer combinations given in table 7.2

Choice No.	Design Efficiency (%)
1	93.2
2	77.7
3	77.7
4	58.3
5	93.2
6	93.2
7	93.2

design capacity. However, the decision maker should temper the results from this method with judgment associated with determining the design capacity. For the previous example, the feasible set of equipment had to meet or exceed the design capacity. A single No. 4 dryer satisfied 85.8% (200 bu/h/233 bu/h) of the design capacity for $4,000 less than the optimum choice. If the hours allowed for drying had been increased by 2 from 12 to 14 hours, one No. 4 dryer would have been $1,000 cheaper than the next best alternative (2 No. 2 dryers).

The decision maker must determine if excess capacity has any economic value. For the above example, it may be more economical to select Choice 4 in order to reduce future remodeling costs if it is known that drying capacity will have to be increased to 400 bu/h in the near future. Likewise, extending drying time may be the best most economical means of reducing cost. In Choice 5, capacity can be increased from 3,000 to 6,000 bu/day without additional investment cost by lengthening the allowable drying time from 12 to 24 h.

Note that "best fit" tends to dictate economics. The closer an equipment set matches the design capacity, the more likely that set is to be near optimum in terms of cost.

Purchase cost was the criteria used for the dryer selection example. However, the better criteria is the total ownership cost of the equipment set. For the dryer example, total ownership cost would include drying fuel, electricity, interest, depreciation, and other factors discussed in more detail in Chapter 9.

Summary

In conclusion, a design capacity must be stated if an optimum equipment set is to be determined. However, judgment and economic considerations should be used to obtain the most economical design capacity.

Selection Techniques to Minimize Harvesting Bottlenecks

There will be a bottleneck in every harvesting / delivery / handling / drying / storage system; that is, some component of the system will be the limiting factor. In a well-designed system, that component will be the combine in that it is usually the most expensive link in this materials handling system. In the following discussion, each type of operation in a typical harvesting system will be examined and a determination made of how to select equipment to maximize total profit.

Combine Selection

One of the most important economic factors to consider is harvest loss. Harvest losses typically range from 5% to 15% of gross yield (Johnson and Lamp, 1966). They increase as harvest time is extended and are highly dependent on weather conditions. A "rule of thumb" for avoiding "excessive" harvest losses is that there should be sufficient combine capacity to harvest the entire crop in 18 to 22 working days. The first step in combine selection is to calculate the desired daily harvest rate:

$$\begin{array}{c}\text{daily harvest rate}\\ \text{(bu/day)}\end{array} = \frac{\text{(total acres)} * \text{(yield in bu / acre)}}{\text{(actual harvesting work days)}} \qquad (7.4)$$

For example, suppose a farmer has 300 acres of corn that is expected to yield 100 bu/acre. The individual wishes to complete harvesting in 20 working days. Using the equation above:

$$\begin{array}{c}\text{Daily harvest rate}\\ \text{(bu / day)}\end{array} = \frac{\text{300 acres} * \text{100 bu/acre}}{\text{20 harvesting days}}$$

$$= \text{1,500 bu / day}$$

The daily harvesting capacity of a combine may be calculated using the following equation:

$$\begin{aligned}\text{combine capacity (bu / day)} = \Big[&\big(\text{h / day}\big) * \big(\text{yield, bu / acre}\big)\\ &* \big(\text{header width, ft}\big) * \big(\text{miles/h}\big)\\ &* \big(\text{efficiency, \%}\big)\Big] / 825 \qquad (7.5)\end{aligned}$$

Suppose the farmer in our previous example wishes to purchase a 4-row combine that has a width of 12 ft. The performance estimates are that the combine will travel 2.5 mph and operate an average of 6 h/day at 70% efficiency. Using equation 7.5, the minimum daily harvesting capacity of the combine is:

$$\text{combine capacity} = \left[(6\text{ h day}) * (100\text{ bu / acre}) * (12\text{ ft}) * (2.5\text{ mph}) * (70\%)\right] / 825$$

$$= 1{,}527\text{ bu / day}$$

It is expected that the smallest machine capable of harvesting 1,527 bu/6-h workday would be the proper economic choice in that it would most probably cost less than a larger capacity machine and yet meet the design criteria set by the farmer.

The procedure for combine selection usually is to select the fewest number of combines that will meet the design requirements. It is possible, however, that several smaller machines could be less expensive than a single larger machine, all based on the same total harvesting capacity. Harvesting efficiency is also important. For example, if a 3-row machine (9 ft width) were 100% efficient, it could harvest 1,636 bu/day, approximately 7% more than the 4-row machine. The question then is how to measure efficiency. By definition:

$$\text{combine efficiency (\%)} = \frac{(\text{bushels harvested per time period})}{(\text{bushels that could have been harvested if the combine never stopped over the same time period})} \quad (7.6)$$

Thus, for the 4-row combine 1,527 bu may be harvested over the 6-h day given 70% efficiency, whereas 2,181 bu may be harvested if the combine were not idle 30% of the time. This "wasted" time will be spent in unloading, turning at the end of the rows, machine breakdown, or rest breaks. In a poorly designed system, it is also spent waiting in the field to unload. Of course, another option is to increase the working time per day. For example, harvesting 8 h/day compared to 6 h would yield a 33% increase in daily harvest, all other factors remaining the same. Thus, the 3-row combine satisfies the design harvest rate (1,500 bu/day) with its daily harvesting rate now being 1,527 bushels at the 70% harvesting efficiency (the same daily harvesting rate as the 4-row combine if the 4-row machine were to operate 6 h/day at 70% efficiency).

Suppose the 3-row combine mentioned previously has a 100-bu grain tank and unloads at the effective rate of 50 bu/min. Assume that the total turning time and rest breaks are 7.4 min for each hopper load of grain. If there are no machine breakdowns, the effective time between completing hopper loads is:

$$\text{fill time (min)} = \frac{495 * (\text{grain tank capacity, bu})}{(\text{header width, ft}) * (\text{speed, mph}) * (\text{yield, bu / acre})} \quad (7.7)$$

$$= \frac{495 * 100 \text{ bu}}{(9 \text{ ft}) * (2.5 \text{ mph}) * (100 \text{ bu / acre})}$$

$$= 22 \text{ min}$$

$$\text{unloading time (min)} = \frac{(\text{grain tank capacity, bu})}{(\text{effective unloading rate, bu / min})} \quad (7.8)$$

$$= 100 \text{ bu} / (50 \text{ bu / min})$$

$$= 2 \text{ min}$$

$$\text{turning time \& rest breaks (given)} = 7.4 \text{ min}$$

$$\text{total cycle time per hopper (min)} = 22 + 2 + 7.4 = 31.4 \text{ min}$$

The number of hopper loads that may be gathered in 8 h is:

$$\text{no. of filled hoppers (actual)} = \frac{8 \text{ h} * 60 \text{ min / h}}{31.4 \text{ min / hopper}}$$

$$= 15.3$$

If the combine did not have to stop, the theoretical number of hopper loads would be:

$$\text{no. of filled hoppers (theoretical)} = \frac{8 \text{ h} * 60 \text{ min / h}}{22 \text{ min / hopper}}$$

$$= 21.8$$

The combine efficiency is then:

$$\underset{(\%)}{\text{combine efficiency}} = \frac{\text{actual no. of hopper loads} * 100}{\text{theoretical no. hopper loads}} \tag{7.9}$$

$$= \frac{15.3 * 100}{21.8}$$

$$= 70.0\%$$

By another method:

$$\underset{(\%)}{\text{combine efficiency}} = \frac{\text{minutes to fill tank} * 100}{\text{total cycle time (min)}} \tag{7.10}$$

$$= \frac{22.0 * 100}{31.4}$$

$$= 70.0\%$$

Delivery Vehicle Selection

There are many different methods that may be used to deliver grain from the combine to the grain facility. Techniques used include:

1. Single-stage delivery vehicles that deliver grain directly from the combine to the grain facility. This method is used most often for corn, soybean, and wheat harvesting systems.
2. Two-stage delivery systems where grain is transferred from a combine to a wagon which, in turn, transfers the grain to a truck at the edge of the field for final transport to the grain facility. This system is typical of rice harvesting systems where soil conditions do not permit the truck to travel directly to the combine.
3. Multi-wagon transport systems where the wagons are left in the field until loaded and then one or more are hitched to a tractor and transported to the grain facility. In this system, there can be fewer tractors than wagons, and it is used where labor may be the limiting factor.

Of course, there may be combinations of these systems in the field at the same time.

Given these conditions, the question becomes "How many vehicles do I need and how big should they be?". Unfortunately, it is almost impossible to determine the answer from a single formula unless many simplifying assumptions are made. However, there are procedures that result in close approximations of both delivery vehicle size and number.

First, the times to perform the activities of the delivery vehicle must be determined. These are:

1. Positioning time under the combine before receiving grain.
2. Time required for the combine to complete unloading.
3. Time to relocate the vehicle so that the combine can resume harvesting.
4. Travel time to the grain facility.
5. Time to sample and weigh the grain on the vehicle (if applicable).
6. Time spent waiting in line to unload (if applicable).
7. Time to unload.
8. Travel time back to the combine.
9. Miscellaneous time, i.e., rest breaks, minor maintenance or repair.

The only one of these factors that will be changed significantly by vehicle size is the time to unload at the facility (assuming a single-stage delivery system). If travel and unloading times are sufficiently short, relatively few delivery vehicles are needed regardless of vehicle size. This can best be shown through an example.

Suppose the 3-row combine mentioned previously was the only harvester being used. From previous calculations, this machine needs 29.4 min to fill a hopper including 22 min to fill and 7.4 min for rest breaks and turning time. Unloading time from the 100-bu grain tank was computed to be 2 min. The primary consideration is that the delivery vehicle be able to make the round-trip to and from the grain facility before the combine refills, i.e., 29.4 min. Suppose the delivery vehicle is unloading directly into a portable auger at the rate of 10 bu/min. Likewise, assume that total travel, positioning, and miscellaneous time for a delivery vehicle of any size is 19.4 min. Then, the delivery vehicle must unload within 10 min if the combine is not to be left waiting in the field (10 min unloading time + 19.4 min travel time = 29.4 min, the fill time for the combine). If the farmer chooses a delivery vehicle with a 100-bu capacity, the unloading time is 10 min (100 bu/10 bu/min, auger unloading rate). However, if a 200-bu delivery vehicle is selected, the unloading time increases to 20 min and the combine will have to wait an additional 10 min before it can unload. This will occur every other combine unloading, so the effective wait time becomes 36.4 min/hopper (31.4 min originally + 5 min additional wait time) and reduces the combine efficiency from 70% to 60.4% (22 min theoretical/36.4 min actual). The efficiency decreases even further for this example if vehicle size continues to increase.

From the above example, several general observations may be made concerning the selection of delivery vehicles:

1. Again, the combine usually is the most expensive single item of equipment in the system, and, therefore, it should not be made to wait on other components.
2. When bottlenecks occur, it may be better to select a smaller delivery vehicle or to reduce the size of each load, especially when it takes the vehicle a relatively long time to unload.
3. For optimum efficiency, the vehicle should receive an even number of unloadings from the combine. This maximizes the time that the vehicle has to deliver the grain and return before the combine has completed filling.
4. The minimum delivery vehicle capacity should be the largest hopper size of any of the combines in the field so that any delivery vehicle can unload completely any combine.
5. The most expensive type of wet-holding for grain is in the form of delivery vehicles.

The number of vehicles can be estimated from the following equation:

$$\text{no. of vehicles} = \text{no. of combines} * \frac{\left(\begin{array}{c}\text{total time for vehicle traveling,}\\ \text{unloading, and other time spent}\\ \text{at the grain facility}\end{array}\right)}{\left(\begin{array}{c}\text{time to fill the combine hopper}\\ \text{including turning time and rest breaks}\end{array}\right)} \tag{7.11}$$

where fractional numbers of vehicles are rounded to the next highest whole number. This estimate is based in part on the assumption that all combines are identical in performance and capacity. Obviously, adjustments may have to be made in a dynamic field setting.

Pit or Receiving Hopper and Bucket Elevator or Portable Auger

Pits (or receiving hoppers) and bucket elevators (or portable augers) must be considered in combination because the size of one affects the optimum size of the other. For purposes of example, the pit-bucket elevator combination will be addressed but the same logic applies to the receiving hopper-portable auger system.

During harvest, the objective of the pit-bucket elevator combination is to transfer the grain from the delivery vehicle to the wet-holding bin, drying bin or storage bin at a sufficiently fast rate so that the delivery vehicle can unload and return to the field before a combine is forced to wait for unloading. This may be accomplished by either dumping the grain into a relatively large pit or by removing it very quickly using a high capacity bucket elevator. The larger the pit, the smaller that the bucket elevator may be and vice versa so long as the grain may be removed from the field without causing delays in harvesting.

To illustrate this point, assume that a vehicle containing 300 bu must unload in 10 min if it is to arrive back at the combine before its hopper is filled. If the pit is very small (essentially zero capacity), the bucket elevator must be of sufficient size to empty the vehicle in 10 minutes, i.e., a rate of 30 bu/min or 1,800 bu/h (300 bu in 10 min). At the other extreme, assume that the maximum rate that the vehicle could unload under any condition is 100 bu/min, and that the bucket elevator can only remove 10 bu/min. The pit size must then be 270 bu. This is obtained by subtracting the grain that will be removed by the bucket elevator during the 3-min unloading time from the vehicle capacity (300 bu/100 bu/min = 3 min to unload; 3 min to unload * 10 bu/min bucket elevator removal rate = 30 bu; 300 bu truck capacity – 30-bu removed during the time that the vehicle is unloading = 270 bu pit size). In equation form:

$$\begin{array}{c}\text{pit size for a}\\ \text{given bucket elevator}\\ \text{or transport auger (bu)}\end{array} = \begin{array}{c}\text{vehicle}\\ \text{capacity}\\ \text{(bu)}\end{array} - \left[\frac{(\text{vehicle capacity, bu})}{\left(\begin{array}{c}\text{vehicle unloading}\\ \text{rate, bu/min}\end{array}\right)} * \left(\begin{array}{c}\text{bucket elevator}\\ \text{or transport auger}\\ \text{capacity, bu/min}\end{array}\right)\right] \tag{7.12}$$

Note, however, that the optimum combination of pit and bucket elevator is the one that has the least total annual cost when considering the expenses for both items.

There are some additional factors to consider:

1. The cost of a pit varies greatly depending on the contractor.
2. The cost for a bucket elevator increases at a relatively slow rate for capacities above approximately 1,000 bu/h.
3. When selecting the size of a bucket elevator, consider that it may be dumping into a cross-auger. If the bucket elevator has a relatively large capacity, say 5,000 bu/h, the receiving auger must be of sufficient size to accept the grain, thus making the receiving auger more expensive than would otherwise be necessary.
4. Generally, excessively large pits cannot be economically justified.
5. Bucket elevator systems usually cost $0.02 to 0.05 more per bushel per year of annual throughput than portable auger systems.

However, their dependability and convenience are usually worth it in the long run. Notice, for example, that bucket elevators are rarely replaced by portable augers while the reverse is certainly not true.

Wet-Holding Bin

The wet-holding bin is relatively easy to size using the following relationships.

For continuous-flow dryers:

$$\text{wet-holding bin size (bu)} = \left(\text{bushels harvested / day, desired}\right) - \left(\text{bushels dried over the same time period as when combines are operating}\right) - \text{pit size (bu)} \quad (7.13)$$

For automatic-batch dryers:

$$\text{wet-holding bin size (bu)} = \left(\text{bushels harvested / day, desired}\right) - \left(\text{bushels dried over the same time period as when combines are operating}\right) + \left(\text{bushels contained in 1 dump of the dryer}\right) - \text{pit size (bu)} \quad (7.14)$$

The "bushels harvested per day" must reflect the desired harvest rate and not that of an inefficient system. For example, if the combines are forced to wait to unload, the actual bushels harvested per day may be considerably less than the desired level.

Dryer Selection

The drying method to be selected depends on cost, convenience, and the risk that the farmer is willing to take. Regardless, the drying rate should be set equal to the daily harvest rate except in colder growing areas where natural air or low-temperature drying may prove satisfactory.

With layer drying, the grain should be alternated among all the drying bins in order to minimize overall grain depth. This allows the fans to deliver more drying air and increases the drying rate. As the bins are filled, the incoming grain generally will be drier, thus partially off-setting the reduced drying rate associated with a reduction in air flow. It is essential that grain dry as quickly as possible to avoid the possibility of aflatoxin contamination.

In batch-in-bin drying, the grain must be dried, cooled, and unloaded from the drying bin to the storage bin (or for transport to other facilities) before the first load of the day is received. The typical design criteria is to select a bin size so that the daily harvest rate will not exceed a depth of 4 ft. To allow for future growth, a 3 ft-design depth is suggested.

When using continuous-flow or automatic-batch dryers, the dryer size must allow the daily harvest rate to be dried in one day. The cost of the dryer will be less per bushel as operating time increases. However, the longer the operating time, the less room for expansion and the more the owner will have to depend on automatic controls or extra labor.

Generally, batch-in-bin drying and storage systems are the least expensive except for single bin layer drying systems with less than 8,000 bu capacity that utilize layer drying. As size increases, portable dryers become more competitive in price and batch-in-bin systems become more inconvenient to operate. Layer-drying systems over this same range will be considerably higher in price for the same drying capacity. These comparisons are more fully addressed in Chapter 9.

Storage

The size of the storage facility is solely dependent on the maximum quantity and categories (varieties, quality differences, etc.) of grain to be stored at any one time. Any bin arrangement should allow for additional expansion without having to relocate major components. Likewise, it should always be possible to deliver and receive grain from the same location point, commonly called the closed-loop principle.

In terms of cost, it is usually less expensive to build as few bins as possible for a given total storage capacity so long as the eave height does not exceed approximately 24 ft (9 rings). At this height, most companies add reinforcement to the bin structure, thus increasing the per bushel cost. Of course, there may be other considerations, such as storing more than one type of grain or utilization of portable handling equipment, that may require several different bins in the system.

What About Poor Design?

A study was made of the effects of reducing the optimum size of the pit, receiving conveyor, wet-holding bin, and dryer on combine efficiency (Loewer et al., 1980). Some of the results are shown in the table 7.4.

From this study, on the average, drying capacity was the most important factor that affected combine performance while pit size was

Table 7.4. Sensitivity of combine efficiency to sizes of facility components

For a 1% reduction in optimum	Average reduction in combine efficiency (%)
1. Pit capacity	0.042%
2. Receiving conveyor rate	0.300%
3. Wet-holding bin capacity	0.253%
4. Drying capacity	0.496%

of little importance. However, each farm design is unique to the needs of the individual.

An Expert System for Determining Bottlenecks in On-Farm Grain Processing Systems

The computer simulation model SQUASH (acronym for Simulation of Queues involving Unloadings and Arrivals for Systems of Harvesting) was developed to analyze harvesting, delivery, handling, drying, and storage systems with regard to system bottlenecks (Benock et al., 1981). The model was used primarily as an extension tool, especially in a workshop setting, to help farmers evaluate existing or proposed grain systems (Loewer et al., 1977, 1978). It used graphics, histograms, and performance statistics to measure and describe system performance and identify the bottleneck in a given system. In workshop situations, however, even the developers of the model could not always readily identify the system bottleneck, and they were not successful in developing an internal algorithm for SQUASH that would identify the bottleneck in a sufficiently high percentage of the situations. Consequently, they examined output from the model, suggested possible alterations to the system, and then reran SQUASH to see the consequences of their recommendation. In essence, the model developers served as "experts" in evaluating systems bottlenecks using a computer simulation model as a basis for system performance. Expert systems techniques offered a means of identifying system bottlenecks in a systematic and timely manner (Loewer et al., 1990).

The objectives of this section are to:

1. Present an expert system model (EXSQUASH) for identifying bottlenecks in a grain harvesting, delivery, handling, drying, and storage system.
2. Demonstrate the validity of the expert system using the SQUASH simulation model.

Grain System Components

Grain systems are composed of many different components and component combinations. Table 7.5 shows the possibilities considered in the SQUASH model and in the development of the expert system model called EXSQUASH. SQUASH can be used to simulate combinations of harvesting, delivery, materials handling, drying, and storage operations. Combinations include harvesters and delivery vehicles (each may differ as to number and operational characteristics), receiving capacity (pit or hopper on portable conveyor), drying method (layer, low temperature, batch-in-bin, in-bin continuous-flow, portable continuous-flow, and portable batch) and final storage. Both wet and dry grain delivery elevators may be used with any drying method except layer and low temperature where only wet grain delivery capacity is

Table 7.5. Equipment considered in the development of EXSQUASH

Item	Description
Harvesters	Any type of grain harvester.
Delivery vehicles	Types include those that operate between the field and grain facility and receive grain directly from combines.
Receiving Capacity	Types include any size of: a. Pit (gravity flow or conveyor) b. Hopper on portable conveyor
Wet Grain Delivery	Types include: a. Bucket elevators b. Portable conveyors (typically augers)
Drying Method	Drying methods include: a. Layer b. Low temperature c. Batch-in-bin d. In-bin continuous flow e. Portable continuous-flow f. Portable batch
Wet Holding Bin	May be used only with in-bin continuous flow, portable continuous-flow or portable batch dryers.
Dryeration Bin	May be used only with in-bin continuous flow, portable continuous-flow or portable batch dryers.
Surge Bin	May be used only with portable continuous-flow or portable batch dryers.
Dry Grain Delivery	Bucket elevator or portable conveyor that is optional and may be used only with batch-in-bin, in-bin continuous flow, portable continuous-flow or portable batch dryers.
Storage	Any type.

specified. Wet holding bin characteristics must be given for in-bin continuous-flow, portable continuous-flow, and portable batch drying methods. Dryeration bin and/or surge bin characteristics may be specified only for portable continuous flow and portable batch dryer systems.

EXSQUASH Model Overview

The EXSYS expert system package (EXSYS,1985) was used to develop EXSQUASH. EXSQUASH was designed as a completely deterministic expert system; that is, each bottleneck was identified with a probability of 1. Bottlenecks may be identified by answering appropriate questions using complete "decision tree" logic as is shown in figure 7.1. In essence, figure 7.1 is the EXSQUASH expert system expressed in decision tree form rather than as a listing of rules. A detailed listing of EXSQUASH Version 1.0 rules is given by Loewer et al. (1988).

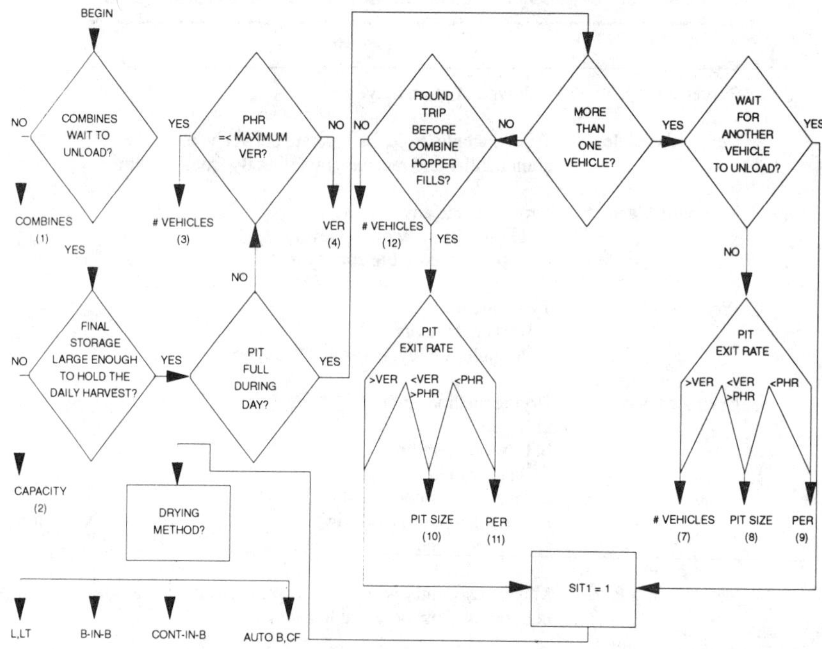

Figure 7.1–Flowchart of decision tree used in EXSQUASH (rule numbers, as given by Loewer et al., 1988), denote possible pathways and follow the stated bottleneck; VER - minimum effective vehicle grain exit rate considering pit capacity; PHR - potential harvesting rate for all combines; PER - pit exit rate; ER - exit rate; WH - wet holding; DRYTION - dryeration; L, LT - layer or low temperature drying; B-IN-B - batch in bin dryer; CONT-IN-B - continuous-in-bin dryer; AUTO B, CF - automatic batch or continuous flow dryer; SIT1, SIT2A, and SIT3 - logic variables that represent a combination of previous logic choices in order to simplify rule development (Loewer et al., 1990).

(Figure 7.1 continued on next page)

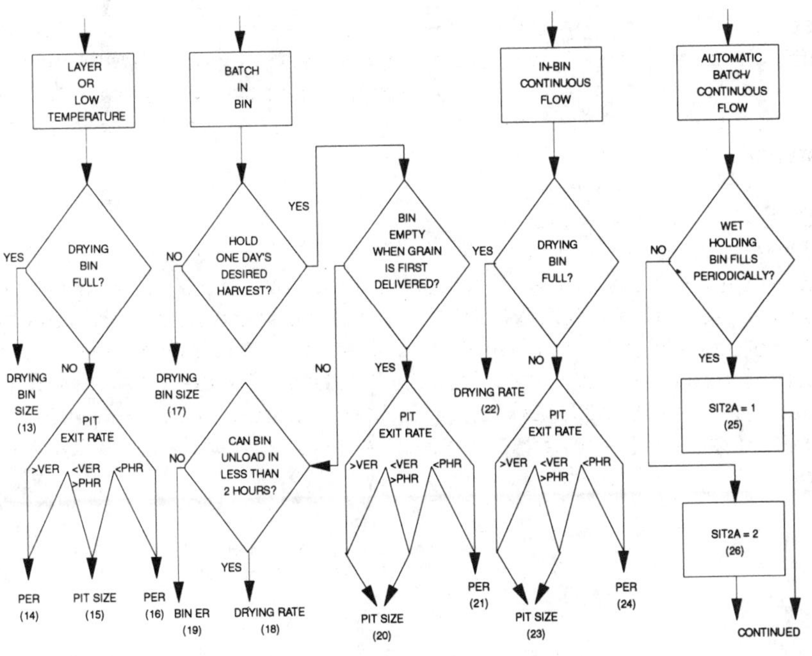

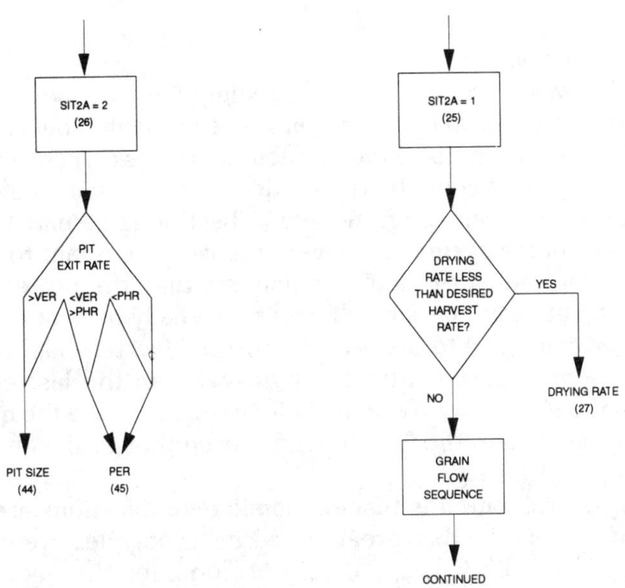

Figure 7.1–Continued.

(Figure 7.1 continued on next page)

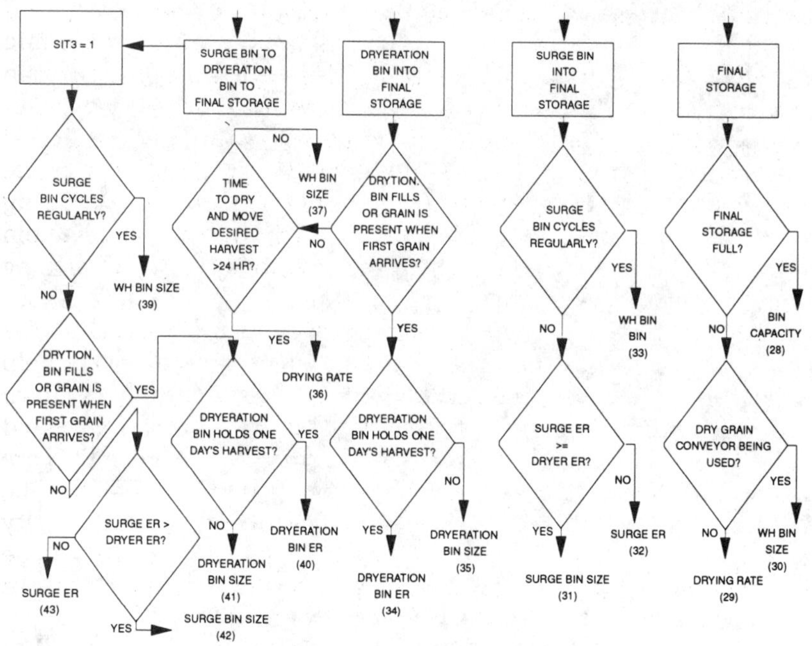

Figure 7.1–Continued

System Bottlenecks

EXSQUASH was developed with the assumption that one and only one item will be the physical bottleneck in a given system. Thus, EXSQUASH will return its first solution as the system bottleneck. Technically, more than one component could be the bottleneck at the same time, but the chances of this happening are remote at best, in large part because of the dynamics of the system. However, it is very important to recognize that increasing the capacity of components that are located in series before the actual bottleneck may increase the daily harvesting capacity. This may be compared to the periodic surges of water that flow though a series of containers controlled by a valve in the last container. Increasing the size of any container will directly increase the quantity of water that passes into the first container even though the final valve is the limiting bottleneck.

Another consideration is that economic determinations are site and equipment specific so that broad-based decision rules are difficult to formulate. Thus, EXSQUASH does not consider the economics of increasing the capacity of one of the "series" components although this may, in fact, be a better economic decision than increasing the capacity of the physical bottleneck.

To illustrate the concepts of series components and economic determination, consider a continuous-flow dryer with a pit, wet grain bucket elevator, wet holding bin, dryer, surge bin, dry grain bucket elevator, and dryeration bin, all in series. Assume that EXSQUASH identifies correctly the drying rate of the continuous-flow dryer as the bottleneck keeping the desired daily harvesting rate from being achieved. Still, increases in vehicle size, pit capacity or wet holding bin capacity will each allow more grain to be harvested per day [assuming less than a 24-h/day operation by the combine(s)] because of increases in wet grain holding capacity. Components such as these are all in the "up side" series before the identified bottleneck and thus may help alleviate the system bottleneck. Components in series on the "down side" of the bottleneck (surge bin and dryeration bin) may be increased in capacity without altering the daily harvesting rate. It may be more economical to increase the size of one of the "up series" components, such as the wet holding bin, rather than to increase the drying capacity of the dryer. In these situations, dynamic system performance should be evaluated extensively, perhaps using a model such as SQUASH, to make sure of the benefits of increasing "secondary" bottlenecks.

Another consideration in identifying bottlenecks is the extent to which increased capacity influences system performance. Bottleneck identification reflects a point solution in time. EXSQUASH does not compute the capacity increase necessary for the bottleneck to be shifted to another component in the system. Iterative runs of SQUASH are one means of determining when other factors begin to become limiting.

Model Evaluation

The expert system was tested with hypothetical grain harvesting, drying, and handling systems and the SQUASH model. The approach is somewhat similar to that used by Khuri et al. (1988) in examining citrus harvesting systems. In SQUASH, each hypothetical system utilized one combine harvesting continuously (unloading on-the-go) at 540 bu/h, from 10:00 A.M. until 6:00 P.M. There are 45 pathways through figure 7.1 that result in identification of the system bottleneck. Combinations of equipment types and/or characteristics were made so that each pathway in figure 7.1 was tested. Characteristics of system components are given in table 7.6.

Each independent system (a combination of equipment and associated characteristics) was adjusted to achieve an "optimum" efficiency, defined as maximizing the utilization of each operational component with the harvesters not having to wait to unload. The optimum systems processed the potential daily harvest in less than 24 h. The test systems were designed from these optimum systems by shrinking one element of the optimum system at a time to generate a "known" bottleneck. Test systems were simulated with the SQUASH program to yield behavioral data that were used as input to the EXSQUASH expert system. The bottleneck obtained from the expert system evaluation was compared to the "known" bottleneck to determine the accuracy of the expert system.

EXSQUASH determined all bottlenecks to be the same as the designed bottlenecks, thus confirming the logic in each of the 45 pathways shown in figure 7.1. However, special attention should be given to component combinations involving grain holding capacity and the associated grain exit rate, i.e., pit and pit exit rate, wet holding bin and drying rate, and surge bin and surge bin exit rate. In these situations, either the grain

Table 7.6. Sizes of elements in the optimum systems for EXSQUASH validation

No.	System Element	Optimum System Components
1	Harvester - volume/h each*	540
1	Delivery vehicles:	
or	- capacity, volume each	180
2	- unloading rate (volume/min each)	80
1	Layer drying bin: volume	3,925
1	Batch-in-bin dryer	
	- drying rate by 7:20 A.M.	3,925
	- unloading rate (volume/h)	1,500
1	Continuous-in-bin	
	- capacity (volume)	1,110
	- maximum drying rate (volume/h)	500
	- minimum drying rate (volume/h)	200
1	Continuous-flow Dryer:	
	- volume	400
	- drying rate (volume/h)	400
1	Wet holding bin: volume	1,095
1	Dryeration Bin:	
	- capacity (volume)	3,925
	- exit rate (volume/h)	1,500
1	Storage bin: volume	3,925
1	Pit or receiving hopper: volume	126
1	Wet grain conveying rate: volume/h	1,300
1	Dry grain conveying rate - volume/h	400

* Volume may be in any unit but "bushel" is most often used.

capacity or the grain exit rate could be considered the bottleneck because one directly influences the other. For example, increasing the wet holding bin size reduces the required hourly capacity of the dryer (assuming all grain can be processed before the next day's grain arrives) and vice versa. Thus, EXSQUASH could select either wet holding capacity or drying rate and be correct in that both are operating at full capacity so that increases in either one would relieve the bottleneck. Again, economics of selection within a limiting series of components is not considered.

As with dryers and wet holding bins, it is often difficult to distinguish between the wet holding bin size and pit size bottlenecks because both elements serve the same function of storing wet grain prior to drying. Conversely, the need for larger wet holding or pit size can often be relieved by increasing the drying rate so that drying rate can be interpreted as being the bottleneck.

One simulation was performed on a system designed with two simultaneous bottlenecks to evaluate the capability of the expert system to find two bottlenecks. The expert system correctly identified the first bottleneck in its line of questioning. The program then stopped running, so the second bottleneck was not identified. The EXSYS program can be set to stop after either single or multiple solutions are reached. In EXSQUASH, it was determined that a single solution was preferable.

The best way of evaluating a proposed or existing system is to run both the expert system EXSQUASH and the SQUASH simulation program. The expert system can be used to determine the bottleneck. Data describing the system, with increases in the size of the bottleneck (as determined by EXSQUASH), can be used with SQUASH to simulate system behavior to see if the bottleneck has been relieved as evidenced by increases in harvesting capacity. This iterative process can be used to remove all but the harvesting bottleneck from the system (when all other bottlenecks have been removed from the system, the combines (harvesting capacity) will be identified as the bottleneck).

Summary

The expert system EXSQUASH has been presented as a means of identifying bottlenecks in a grain harvesting, delivery, handling, drying, and storage system. The validity of the "expert rules" relating to selected types of systems has been examined and confirmed through the use of the computer simulation model SQUASH.

Procedures for Selecting and Evaluating Seed Processing Equipment Including Computer Modeling

A seed processing and bagging plant represents a large investment; therefore, it is important that an optimum design be selected. A typical seed processing and bagging system is composed of cleaner(s), spiral separators, bagging machine(s), forklift(s), seed conveyors, and labor. Labor is composed of bagging persons and pallet persons. Bagging persons obtain an empty bag from a local bag storage and position it on the bagging machine. Pallet persons perform three duties: (1) receiving a full bag from the end of the bagging belt and placing it on a pallet, (2) operating a forklift, and (3) retrieving an empty pallet from the local pallet storage and positioning it in the pallet site area.

Typically, seed flows in a continuous path through a seed processing system as shown in figure 7.2. The unit with the least capacity controls the flow through the processing plant. Thus, the cycle time method discussed earlier in this chapter may be used to select optimum equipment and labor sets for seed processing facilities.

Cycle-Time Method for Optimization

The cycle-time method of selecting equipment or evaluating efficiency is best used in the original design of the seed plant. The steps in this process are:

1. Determine the cycle-time for various operations such as:
   a. Time required to retrieve and place an empty bag on the bagging machine.

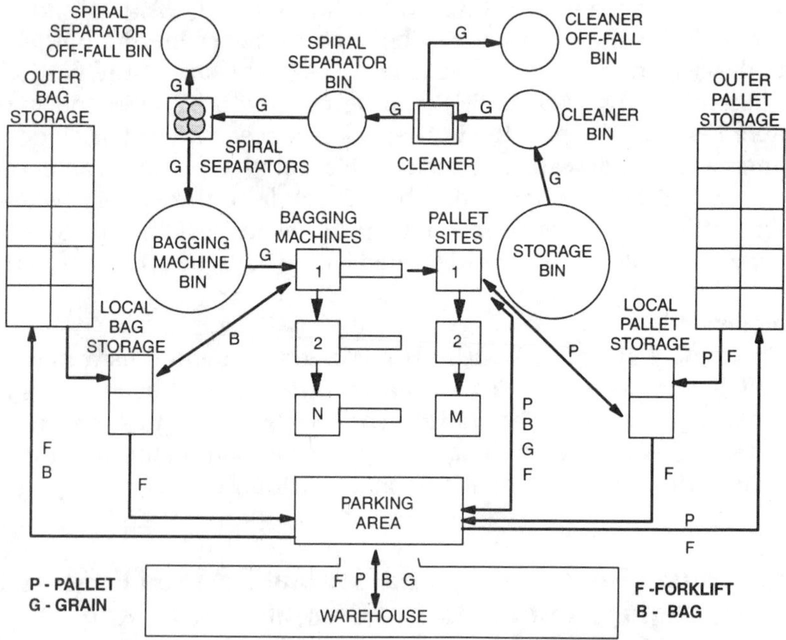

Figure 7.2–Seed flow in a seed processing plant (Bucklin et al., 1989).

   b. Time required for the bagging machine to fill and place a bag on the bagging belt.
   c. Time required for a person to receive a full bag, place it on a pallet and return to the bagging belt.
   d. Time required for the forklift to transport a loaded pallet from the pallet site area to the warehouse and return.
2. Convert all times measured previously to bags/min/person or machine unit.
3. For an existing system:
   a. Multiply the bags/min/machine or man unit by the number of machines or people actually used in that operation.
   b. The type of operation that has the lowest capacity is the one that is limiting and is defined as being 100% efficient.
   c. Efficiency can be determined using the following equation:

$$\underset{(\%)}{\text{efficiency}} = \frac{(\text{capacity of the limiting operation}) * 100}{(\text{capacity of an operation})} \qquad (7.15)$$

   d. Overall efficiency may be improved by continuing to eliminate one unit of a non-limiting component until it would become limiting when an additional unit is removed.
4. For optimization of a system:
   a. Determine the desired daily processing rate and convert to a bags/min rate.
   b. Compute the cycle times of each operation as discussed previously.
   c. Continue to add one unit of a component until it meets or exceeds the desired design capacity.
   d. Evaluate the cost consequences and benefits of adding or removing one unit of each component.

A Computer Model Utilizing the Cycle-Time Method

Overview of JAWS. The cycle-time method has been employed in a computer program called JAWS (Joint Analysis of Warehousing and Sacking) (Bucklin et al., 1982). The function of JAWS is to provide a list of equipment and costs for a new seed conditioning facility based on the specifications of the contractor, equipment dealer, or engineer. JAWS designs a seed plant based on the seed flow shown in figure 7.2. The seed plant designer may provide part or all of the information shown in table 7.7. Equipment performance and cost data are stored within JAWS (table 7.8) but may be changed easily to reflect more current information.

The discrete nature of the equipment characteristics also affects the economic output of JAWS. For instance, the cost per bagged unit declines as people and equipment operate at larger fractions of their capacity, and then levels off or increases with the addition of an individual or piece of equipment. Also, JAWS does not give direct information about the dynamic performance of the selected system. However, another computer program (to be discussed later) can be used to calculate the dynamic efficiencies of the various components of the system.

JAWS Program Information. The program JAWS was developed to assist with the design of new facilities and considers both functional specifications and economic factors. Input and output data are arranged so that interested parties can readily examine and compare each component and economic factor related to the system. The user specifies the design parameters needed to describe the overall operation as illustrated in table 7.7. The user-supplied information is

Table 7.7. Specifications used in JAWS for seed plant design

I. Scheduling Information

Amount of Material to be Bagged per day (Bu)	Length of Work Day (h)	Number of Work Days per Year
2,000.0	8.0	200.0

II. Sizes, Capacities, and Initial Values

Bag Size	No. of Full Sacks Per Pallet	Average Height of the Stack (ft)	Maximum No. of Pallets Allowed in a Stack	Original No. of Bags in Local Storage	Original No. of Pallets in Local Storage	No. of Empty Bags Per Pallet
1.0	64	4.0	3	120	5	1,500

III. Distances

Bagging Pallet Site to Warehouse (ft)	Bagging Pallet Site to Outer Bag Storage (ft)	Bagging Pallet Site to Outer Pallet Storage (ft)	Storage Bin to Cleaner (ft)	Cleaner to Spirals (ft)	Spirals to Bagging Machines (ft)
300.0	300.0	300.0	10.0	10.0	10.0

IV. Economic Information

Interest Rate (%)	Annual Charge for Taxes and insurance (%)
15.0	1.5

Table 7.7. Continued

V. Worker Characteristics

A. Bagging People

Wage ($/h)	Time from Arrival at Local Bag Storage to Arrival at Bagger and Vice Versa (min)
4.0	0.04

B. People Who Carry Sacks

Wage ($/h)	Travel Time from Belt to Pallet when Carrying a Sack (min)	Travel Time from Pallet to Belt (min)	Travel Time from Local Pallet Storage Back to the Pallet Area (min)	Travel Time from End of the Belt to Local Pallet Storage (min)
4.0	0.2	0.15	0.2	0.2

VI. Material Removed During Cleaning

Material Removed in Spirals (%)	Material Removed in Cleaners (%)
5.0	10.0

used to determine the distances between physical components of the facility and specifies the minimum overall flow rate required in the system. The operational size and cost specifications for typical machinery are stored in the program. If the user desires, the specifications for other machinery types can be substituted.

Performance and economic characteristics of available equipment types are stored in the program. The basic concept of JAWS is to calculate the rate of material flow required to achieve the specified output of bagged material per work day. Each component is then sized so that it operates at no less than this rate. If more than one item is needed to obtain the material flow rate, the minimum number of similar sized machines necessary is selected. For example, if it is determined that a material flow rate of 400 units/h was necessary for the design while the selected line of equipment contains cleaners of 350 and 450 units/h of actual capacity, then the 450 unit/h cleaner would be selected. If the required material flow rate were 600 units/h, two 350 units/h cleaners would be selected. The number of personnel is calculated in a similar manner so that the required rate of flow of material is maintained.

The purchase cost of each piece of equipment is calculated in JAWS along with the annual cost based on the units of production depreciation method, an estimated life and rate of repair and constant interest, tax and insurance rates. Bagging, pallet, and labor costs are added to the annual costs. Property, construction, and energy costs are not considered.

Event Method of Determining Efficiency

The event method of evaluating performance efficiency is best suited to dynamic systems such as found in the seed plant bagging operations. The steps of this process are:

1. Identify each major event that occurs within the operation.
2. Trace each event based on the previous event.

Table 7.8. Equipment performance and cost data used in JAWS

I. Cleaners

Cleaner Type	Rate that Cleaner Processes Material (Bu/min)	Height of Unit (ft)	Cost per Unit ($)	Functional Life (h)	Repairs as a Percentage of List Price
1	1.25	6.0	15,400	20,000	60.0
2	2.50	7.0	16,565	20,000	60.0
3	3.75	8.0	17,890	20,000	60.0
4	6.67	8.0	20,740	20,000	60.0
5	9.17	9.6	22,630	20,000	60.0

II. Spiral Separators

Spirals Type	Rate at which Spirals Process Material (Bu/min)	Height of Unit (ft)	Cost per Unit ($)	Functional Life (h)	Repairs as a Percentage of List Price
1	0.50	6.0	570	18,000	60.0
2	0.83	6.0	1,141	18,000	60.0
3	1.67	7.0	2,151	18,000	60.0
4	3.33	7.0	4,055	18,000	60.0

III Bagging Machines

Bagging Machine Type	Height of Machine (ft)	Time to Receive, Fill and Deliver a Bag (min)	Cost Per Unit ($)	Functional Life (h)	Repairs as a Percentage of List Price
1	4.0	0.15	6,125	15,000	60.0
2	4.0	0.083	6,600	15,000	60.0

(Table 7.8 continued on next page)

3. Identify the areas where one event cannot take place because of having to wait for another event to occur. For example, a person must wait for a filled bag to arrive (an event) before this individual can deliver a bag to a pallet (another event).
4. Calculate the waiting or idle time for each person or machine.
5. Determine the system component efficiency using the following equation:

$$\text{efficiency } (\%) = \frac{(\text{total time of operation} - \text{waiting time}) * 100}{(\text{total time of operation})} \tag{7.16}$$

The event method of evaluating performance is best handled by a computer program model. A model developed for this process is called PACASACS (Process Analysis Combining A Sacking And Cleaning System) (Loewer et al., 1979; Bucklin et al., 1989).

Table 7.8. Continued

IV. Elevators

Elevator Type	Capacity of Elevator (Bu/min)	Size of Elevator Cups (in.^{3})
1	1.00	3.0
2	1.42	3.0
3	2.83	3.0
4	2.92	4.0
5	3.25	4.0
6	4.50	4.0
7	4.58	6.0
8	5.83	6.0
9	7.00	6.0

V. Forklifts

Fork-Lift Type	Grd. Speed (ft/min)	Spd. That Will Lift Base (ft/min)	Cost Per Unit ($)	Funct-ional Life (h)	Repair as a Percent-age of List Price	No. of Empty Plets. That May be Hauled in One Load	No. of Plets. Filled with Bags That May Hauled in One Load	Load Carrying Capacity (lb)
1	730.4	96.0	14,700	12,000	60.0	5.0	1.0	2,000
2	730.4	95.0	14,800	12,000	60.0	5.0	1.0	2,500
3	721.6	94.0	15,100	12,000	60.0	5.0	1.0	3,000
4	765.6	88.0	18,800	12,000	60.0	5.0	1.0	3,500
5	765.6	88.0	19,000	12,000	60.0	5.0	1.0	4,000
6	765.6	88.0	19,200	12,000	60.0	5.0	1.0	4,500
7	756.8	88.0	19,500	12,000	60.0	5.0	1.0	5,000

A Computer Model Utilizing the Event Method

PACASACS Program Information. The PACASACS model is composed of the following geographic locations, equipment types and job functions (fig. 7.2):

1. Storage Bin - contains the stored seed that may be processed. No additional seed may be added to this bin over the simulation period.
2. Conveyor SC - conveys seed from the storage bin to the cleaner bin.
3. Cleaner Bin - serves as a surge bin between the storage bin and the cleaner.
4. Cleaner - divides the seed into two categories, cleaned seed and cleaner off-falls.
5. Conveyor CO - conveys off-falls from the cleaner to the cleaner off-falls bin.
6. Cleaner Off-falls Bin - contains the off-falls from the cleaner.
7. Conveyor CS - conveys clean seed from the cleaner to the spiral separator bin.
8. Spiral Separator Bin - serves as a surge bin between the cleaner and the spiral separators.
9. Spiral Separators - divides the seed into two categories, seed suitable for bagging and spiral separator off-falls.
10. Conveyor SO - conveys off-falls from the spiral separators to the spiral separator off-falls bin.
11. Spiral Separator - contains the off-falls from the off-falls bin spiral separators.
12. Conveyor SB - conveys seed suitable for bagging from the spiral separators to the bagging machine bin.
13. Bagging Machine Bin - serves as a surge bin between the spiral separators and the bagging machine(s).
14. Bagging Machine - receives an empty bag from a bagging person and terminates its activity after it directs a full bag onto the bagging belt.
15. Bagging Person - obtains an empty bag from the local bag storage and positions it on the bagging machine.
16. Local Bag Storage - an area relatively close to the bagging machine(s) where empty bags are stored before being positioned on the bagging machine.
17. Bagging Belt - conveys bags from a single bagging machine to the pallet site area.
18. Pallet Site Area - where the pallet person receives a full bag from the end of the bagging belt and places it on an unfilled pallet if available.

19. Pallet - a structure on which bagged material can be placed and the whole structural unit can then be transported by forklift.
20. Pallet Person - performs three duties: receives a full bag from the end of the bagging belt and places it on the pallet; drives a forklift; and retrieves a pallet from the local pallet storage and positions it in the pallet site area.
21. Local Pallet Storage - an area relatively close to the pallet site area where empty pallets are stored prior to being used to transport full bags.
22. Forklift - may be used for three activities: conveying full pallets from the pallet site area to the warehouse; transporting empty pallets from the local pallet storage areas; and delivering a pallet filled with empty bags from the outer bag storage area to the local bag storage area. An activity is completed when the forklift returns to the pallet site parking area. Each activity must be completed before a new one is undertaken.
23. Warehouse - contains all the full pallets of seed.
24. Outer Pallet Storage - where all the unused pallets are stored except for those in the local pallet storage area.
25. Outer Bag Storage - where all the empty bags are stored except for those in the local bag storage area.

Conceptually in the program, seed is conveyed during conditioning from the storage bin through the cleaner and then through the spiral separators into the bagging machine bin (fig. 7.2). After a worker places a bag on the bagging machine, conditioned seed is passed from the bagging machine bin into the bag. The bag is sealed automatically and ejected onto the bagging belt. The full bag is conveyed to the pallet site area where a pallet person removes the bag from the belt and places it on an unfilled pallet. When the pallet contains the proper number of bags, the forklift, driven by a pallet person, transports the loaded pallet to the warehouse and returns to the pallet site area. Upon arrival, if there are insufficient bags in the local storage area, the forklift transports a pallet filled with empty bags from the outer bag storage area.

PACASACS was developed so that the user of the model could define the existing or proposed system, and the model would simulate the plant activities and present a summary set of statistics indicating the performance and efficiency of the system. PACASACS is a "consequences of actions" type model; that is, the model simulates a given system but makes no attempt to optimize individual components.

Input information to PACASACS is shown in table 7.9. The user may specify either bushels, pounds, or kilograms as the basic units to be used. A 24-h clock is used for starting and stopping the simulation; however, the plant activities may not be extended past midnight unless the quitting time is given as an accumulated time rather than restarting the 24-h clock.

Table 7.9. User input information for PACASACS

I. Output Control Information

Define Units (0 = bu, 1 = lb, 2 = kg)	Time Between Status Reports on Material Quantity and Location (min)
0	10

II. Scheduling Information

Time of Day that Bagging Operation Begins (24-h Clock)		Time of Day that Bagging Machine Quits (24-h Clock)	
h	min	h	min
8.0	0.0	8.0	30.0

III. Description of Storage Bin

Initial Amount of Material in Storage Bin (units)	Maximum Storage Bin Capacity (units)	Capacity of Bucket Elevator Taking Grain Out of Storage (units)
5,000.0	6,000.0	1,200.0

IV. Description of Cleaner

Initial Amount of Material in Bin above Cleaner (units)	Maximum Cleaning Bin Capacity (units)	Rate that Cleaner Processes Grain (units/h)	Lower Limit of the Cleaning Bin When the Elevator taking Grain Out of Storage can Turn on (units)	Percentage Loss in the Cleaner
80.0	80.0	180.0	20.0	10.0

V. Description of Spiral Separator

Material in the Bin Above Spiral Separator (units)	Rate that the Spiral Spearator Processes Grain (units/h)	Maximum Spiral Separator Bin Capacity (units)	Lower Limit of Spiral Bin When Cleaner Turns On (units)	Percentage Loss in the Spiral Separation
100.0	150.0	150.0	20.0	5.0

Table 7.9 continued on next page

Only one bin of each type may be specified; that is, only one bin for storage, cleaning, cleaner off-falls, spiral separation, spiral separator off-falls, and bagging may be used. However, if more than one bin of a given type is used in the actual system, their capacities may be added together to approximate a single bin system.

Table 7.9. Continued

VI. Description of Bagging Machine Bin

Material in the Bin Above Bagging Machines (units)	Lower Limit of Bagging Bin When Spirals Turn On (units)	Maximum Bagging Bin Capacity (units)	Level to Which Bin Must Refill After Emptying for Bagging Machines to Restart (units)
100.0	20.0	200.0	0.0

VII. Description of Bagging Machine

Bagging Machine ID	Time to Receive Fill and Deliver a Bag to the Delivery Belt (s)	Time to Travel from Bag Loading Point to the End of the Belt (s)
1.0	5.0	13.80

VIII. Description of Bags

Bag Size (units)	Original No., of Bags in Local Storage	Minimum No. of Bags that area Desired in Local Storage	Original Number of Pallets Loaded with Bags in Outer Storage	No. of Empty Bags/ Pallet
1.00	120	25	48	1500

IX. Description of Pallets

Original No. of Pallets in Local Storage	Minimum No. of Pallets in Local Storage	Original No. of Pallets in Outer Storage	Original No. of Pallet Sites in System	No. of Full Sacks Per Pallet	Maximum No. of Pallets Allowed in the Stack	Height of Each Pallet (ft)
5	0	2,000	1	64	3	2.67

X. Randomness Associated with Bagging

Standard Deviation of Times for People who do the Bagging
0.0

Table 7.9 continued on next page

In the model, when the seed capacity in a cleaner, spiral separator, or bagging machine bin reaches its maximum, the conveyor delivering the incoming grain is turned off. It is not restarted until the seed level reaches a lower restart point specified by the model user. In an actual system, this would correspond to placing a "turn-on" pressure switch near the bottom of the bin. These arrangements ensure that seed overflows do not result from the capacity of one system component being greater than another.

The model user does not specify the capacity of the cleaner off-falls bin and the spiral separator off-falls bin. In effect, the model computes the quantity of off-falls present in each bin but assumes that there will always be sufficient storage space to maintain cleaner and spiral separator operation.

Table 7.9. Continued

XI. Description of People who do the Bagging

ID of Person Who Does the Bagging	Time from Arrival at Local Bag Storage to Arrival at Bagger and Vice Versa (s)	Placement Time if Bag is Being Filled When Person Arrives (s)
21.0	3.00	0.85

XII. Randomness Associated with Carrying Sacks

Standard Deviation of Times for People who Carry the Sacks
0.0

XIII. Description of People who Carry the Sacks

ID of Person Who Carries the Sacks	Travel Time from the Belt to the Pallet when Carrying a Sack (s)	Travel Time From the Pallet to the the Belt (s)	Travel Time from the Pallet Area to the Local Pallet Area (s)	Travel Time From the Local Pallet Storage Back to the Pallet Area (s)	Travel Time From the End of the Belt to the Local Pallet Storage (s)
41.0	12.0	9.0	12.0	12.0	12.0
42.0	12.0	9.0	12.0	12.0	12.0

XIV. Randomness Associated with Forklift Travel Time

Standard Deviation of Times for Forklift Operation
0.0

Table 7.9 continued on next page

Table 7.9. Continued

XV. Description of Forklifts

ID of Forklift	Travel Time to Warehouse From Bagging Pallet Site When Carrying A Full Pallet (s)	Rate that Full Pallet Moves Upward (ft/s)	Travel Time From Warehouse to Bagging Site When Traveling Empty (s)	Travel Time from Bagging Belt Area to Outer Pallet Storage (s)	Travel Time From Outer Pallet Storage to Local Pallet Storage Including Loading (s)
61.0	23.5	1.47	23.5	23.5	23.5

Travel Time From Local Pallet Storage to Bagging Pallet Area Unloading Time (s)	No. of Empty Pallets that May Be Hauled in One Load	Travel Time From Bagging Pallet Area to Outer Bag Storage (s)	Travel Time From Outer Bag Storage to Local Bag Storage Including Loading (s)	Travel Time From Local Bag Storage to Bagging Pallet Area Including Unloading Time (s)	No. of Pallets Filled that May be Carried by the Forklift
2.40	5.0	23.5	23.5	6.0	1.0

The model user must state the percentage of seed entering the cleaner or spiral separators that will be placed in the off-falls bin. The model considers this to be constant over the simulation period.

Any number of bagging machines may be used in the model, each with its own individual performance characteristics. PACASACS, however, considers only one bagging machine bin which serves all the bagging machines collectively but supplies the demands of each individual machine. Each bagging machine has its own individual belt that is used to transport a filled bag to the pallet site area. If, in an actual system, the pallet person picked up the filled sack directly from the bagging machine rather than off the end of the belt, the user would simply specify the time difference between when the bag is filled and when it may be transported by the pallet person rather than the travel time on the bagging belt. Provision is also made for automatic bagging machines.

The user has complete flexibility in defining the bag size, pallet capacity, the number of pallets positioned for filling, local storage parameters and the warehouse loading plan in terms of pallet height. However, PACASACS does not maintain an accounting of the geometry of the warehouse. Also, only one pallet at a time may receive full bags.

The model user may specify the individual performance characteristics of any number bagging persons, pallet persons and forklifts.

PACASACS does not use speed and distance to compute operation times but uses the stated work time directly. Likewise, the model user may elect to use either the average time for each operation or a randomly generated time based on a normal distribution with user specified means and standard deviations.

PACASACS Events. PACASACS is a 4,000 statement combined continuous-discrete FORTRAN simulation that utilizes the GASP IV simulation language (Pritsker and Alan, 1974). The following discrete events have been defined:

1. Bag filling begins.
2. Bag filling ends.
3. Bagging person arrives at the local bag storage site.
4. Full bag arrives at the end of the bagging belt.
5. Forklift arrives at the pallet site area.
6. Forklift departs from pallet site area.
7. Forklift arrives at the warehouse unloading site.
8. Forklift departs from the warehouse after unloading.
9. Forklift arrives at the outer pallet storage area.
10. Forklift arrives at the local pallet storage area.
11. Forklift arrives at the outer bag storage area.
12. Forklift arrives at the local bag storage area.
13. Pallet person arrives at the local pallet storage area.
14. Pallet person arrives at the pallet site area and places a full bag on the pallet.
15. Pallet person arrives at the bagging belt.
16. Pallet person arrives at the pallet site area and positions a pallet.

Several events are related to the quantities of seed in the storage, cleaning, spiral separator, and bagging machine bins. These are called state events and are triggered by a bin reaching either its maximum fill point, lower desired limit, or zero capacity at which time the appropriate conveyors are started or stopped.

When a discrete event occurs, a situation is usually created whereby another event is scheduled or the entity involved is placed in a waiting line. For example, when a pallet person arrives at the end of the bagging belt (a discrete event), the individual will pick up a full bag and be scheduled to arrive at the pallet site (another discrete event). However, if there is no full bag available, the person will have to wait, i.e., be placed into a waiting line (queue).

Waiting Lines. The following waiting lines are used in PACASACS.

1. Bagging persons waiting to place a bag on a bagging machine.

2. Bagging persons waiting to get a bag from the local bag storage.
3. Pallet persons waiting to receive a bag from the end of the bagging belt.
4. Forklift waiting for its pallet to become loaded.
5. Forklift waiting for a pallet site position to become open.
6. Forklift(s) waiting to unload at the warehouse.
7. Bagging machine(s) waiting to receive an empty bag from a bagging person.
8. Full bags that have reached the end of the bagging belt but have not been picked up by a pallet person.

Generally, an entity (bag, bagging person, pallet person, bagging machine, or forklift) will be placed into a waiting line if (1) nothing is available for additional transport, e.g., no full bags at the end of the bagging belt, or (2) another entity has arrived earlier and has not cleared the area, such as a forklift arriving at the warehouse but having to wait for an earlier forklift that has not yet finished unloading its pallet.

Output and Efficiency. The output from PACASACS begins by listing the input information. The user has the option of obtaining a total or partial event tracing of the activities that occurred over the simulation period including a status report of seed quantities and locations. If any entity had to wait, a set of summary statistics and a histogram is printed.

The computation of system component efficiencies are of primary importance (table 7.10). The theoretical capacities of the cleaner, spiral separator(s), bagging machine(s), bagging person(s), pallet person(s), and forklift(s) are computed for the time that the bagging machine operates. This capacity is divided into the quantity of seed that was actually processed or transported by each entity to obtain the component efficiency.

Table 7.10. Efficiencies determined by PACASACS based on sample inputs

System Component	Actual Amount of Material Processed (units)	Amount of Material that would have been Processed if the Component had Operated Continuously (units)	Component Efficiency (%)
Cleaner	479.7	480.0	99.9
Spiral Separator	469.9	540.0	87.0
Bagging Machine	365.1	400.0	91.3
People who do Bagging	365.0	600.0	60.8
People who Carry Sacks	361.0	1,115.4	32.4
Forklift(s)	360.0	5,106.4	7.0

Table 7.11. Example PACASACS efficiencies when adding another bagging person

System Component	Actual Amount of Material Processed (units)	Amount of Material that would have been Processed if the Component had Operated Continuously (units)	Component Efficiency (%)
Cleaner	479.7	480.0	99.9
Spiral Separator	469.9	540.0	87.0
Bagging Machine	365.1	400.0	91.3
People who do Bagging	365.0	1,200.0	30.4
People who Carry Sacks	361.0	1,115.4	32.4
Forklift(s)	360.0	5,106.4	7.0

The final output is a graphical plot of the material flow over the simulation period. Also, included in the plot are the number of entities in some of the waiting lines.

The PACASACS model may be used to evaluate an existing or proposed processing and storage system. For example, the efficiencies shown in table 7.10 were derived from the example inputs given in table 7.9. If an additional bagging person of similar ability is added, the systems efficiencies would be as shown in table 7.11. Results are also shown for adding a similar bagging machine, a similar pallet person, and a similar forklift (tables 7.12 to 7.14). From these examples, the best alternative in terms of the actual amount of material processed was to add a similar bagging machine. This may not necessarily be the best economic choice.

Using Computer Models

Two seed processing computer models have been presented, each based on a different principle of determining seed plant efficiency. Both of these models have been tested and validated under actual seed plant operations (Loewer et al., 1979; Bucklin et al., 1989). How can these

Table 7.12. Example PACASACS efficiencies when adding another bagging person

System Component	Actual Amount of Material Processed (units)	Amount of Material that would have been Processed if the Component had Operated Continuously (units)	Component Efficiency (%)
Cleaner	479.8	480.0	99.9
Spiral Separator	470.0	540.0	87.0
Bagging Machine	465.2	800.0	58.2
People who do Bagging	465.0	600.0	77.5
People who Carry Sacks	461.0	1,096.6	42.0
Forklift(s)	440.0	5,106.4	8.6

Table 7.13. Example PACASACS efficiencies when adding another pallet person

System Component	Actual Amount of Material Processed (units)	Amount of Material that would have been Processed if the Component had Operated Continuously (units)	Component Efficiency (%)
Cleaner	479.7	480.0	99.9
Spiral Separator	469.9	540.0	87.0
Bagging Machine	400.0	400.0	100.0
People who do Bagging	400.0	600.0	66.7
People who Carry Sacks	397.0	2,315.4	17.1
Forklift(s)	360.0	5,106.4	7.0

models be best used? The answer depends on the objectives of the user. However, one should consider using the programs in combination to answer the following types of questions:

1. If the work day length or days spent bagging are altered, how will this affect equipment selection, purchase cost, annual cost, or labor demand?
2. If another bagging machine is added, how will output be affected?
3. What is the bottleneck in the bagging operation?
4. How much unused capacity exists in the different components of the system?
5. How much would bagging capacity change if labor were added, removed or switched?
6. How would an increase in bagging bin capacity change daily productivity?

Summary

Examples of existing computer programs for design and analysis of seed conditioning plants have been presented. When using computer

Table 7.14. Example PACASACS efficiencies when adding another forklift

System Component	Actual Amount of Material Processed (units)	Amount of Material that would have been Processed if the Component had Operated Continuously (units)	Component Efficiency (%)
Cleaner	479.7	480.0	99.9
Spiral Separator	469.9	540.0	87.0
Bagging Machine	365.1	400.0	91.3
People who do Bagging	365.0	600.0	60.8
People who Carry Sacks	361.0	11,155.4	32.4
Forklift(s)	360.0	10,212.8	3.5

programs of any type, the user should become familiar with the basic inputs, assumptions, and decision making criteria contained within the program. Do not expect the output from the program to be any better than its inputs or structure. Computer models cannot consider factors that are not programmed into them. Computer models are simply extensions of how programmers view the "real world" as altered by their specific objectives. For example, PACASACS does not consider machinery breakdowns, not because they do not occur, but because this is not an important consideration in fulfilling the objectives of this particular program.

Computer models provide a relatively inexpensive means of gaining experience quickly. For examples, JAWS or PACASACS can provide insights and information into the design and inter-workings of a seed plant that would otherwise come only with great expenditures of time and money.

In conclusion, computers cannot do more than someone well-trained with a stopwatch and calculator who possesses the skills necessary to make intelligent decisions. Computer models generally can, however, provide complete detailed analysis more quickly, accurately, and economically than this trained individual. A computer model does not make the final decisions concerning the system, but it can provide information that may improve significantly the efficiency of a seed plant.

Performance Cost and Design Considerations in Seed Processing Facilities

One of the most important cost considerations in designing a seed processing and bagging plant is the quantity of grain to be processed each day. Other factors that affect the economics of the system significantly include the number of days the plant is to be used per year and the number of hours per day that the plant will operate. The relative economic importance of these factors can be evaluated by examining a range of values for each factor and the resulting effect on the cost of the system.

Example Design Specifications

Some of the considerations in selecting seed processing equipment sets may be best shown through example. Suppose the designer were given the following assignment:

1. Determine the optimum sizes of equipment components used in seed processing systems ranging in capacity from 500 bu to 10,000 bu/day for 50 to 200 days of operation and with 8, 16, and 24 h/day work shifts.

2. Compute the operating efficiency of each major equipment and labor component in the optimum equipment set.
3. Calculate the total and annual cost for each of the optimum systems.

In addition, soybeans are specified as the seed to be processed. Available warehouse space is assumed adequate for the desired system capacity. Efficiencies are to be calculated using the dynamic simulation PACASACS, and optimum equipment sets and costs are to be determined using the computer program JAWS (these programs are discussed in detail earlier in this chapter).

Processing Plant Concepts

Typically, seed processing and bagging systems consist of cleaner(s), spiral separator(s), bagging machine(s), and their respective surge bins, and elevators, forklift(s), pallets, bags, and the necessary personnel (fig. 7.1). Personnel are divided into two categories, bagging worker(s) who place the empty bags on the bagging machine, and pallet worker(s) who have three responsibilities: (1) placing bags from the end of the bagging belt onto a pallet, (2) driving the forklift, and (3) replacing full pallets with empty ones from the local pallet storage.

JAWS determines the material flow rate through a system network required to provide the desired daily output. It then sizes each component so that it may handle a flow rate equal to or greater than that of the design.

For this example, performance and economic characteristics of the various system components were obtained from lines of equipment commercially available. These values were stored in JAWS and are given in table 7.8. The pieces of equipment necessary to sustain the required design material flow rates were selected from these equipment options. Equipment for the system was chosen so that the lowest capacity of any component was equal to or in excess of the required capacity.

The base set of processing plant characteristics are given in table 7.7. These values are assumed to be constant for all systems except as otherwise noted.

Program Economics

The system components that are considered for cost are the same as those listed in table 7.8. Pallet, bag, and energy costs are not included. Labor, other than that required to bag and deliver the grain, is not considered. The purchase, fixed, and annual costs are calculated for each item using representative manufacturers suggested list prices at the time of the analysis.

Annual costs are calculated using the units of production depreciation method with no salvage value. Repair costs are determined as a percent of list price over the life of the machine. A constant percent of the purchase price is used to determine the charges for interest, taxes, and insurance. Property costs and costs of construction are not considered.

Results

Optimum Equipment and Manpower Sets. For this example, the daily processing capacities range from 500 to 10,000 bu. Daily operating times for the equipment are 8, 16, and 24 h. The number of work days per year do not influence the optimum equipment set because design is based on a daily processing rate. Equipment and manpower needs for least cost systems are given in table 7.15. Again, these sets are designed to meet the minimum specifications given by the daily capacity and usually will

Table 7.15. Optimum equipment set and associated efficiencies for a range of daily processing rates and operating times*

	500 Bu		1,000 Bu		1,500 Bu		2,000 Bu		2,500 Bu	
Item	%	#(t)	%	#(t)	%	#(t)	%	#(t)	%	#(t)
					8-h					
BM	47	1(1)	47	1(1)	73	1(1)	73	1(1)	73	1(1)
F	4	1(5)	4	1(5)	6	1(5)	6	1(5)	6	1(5)
BP	63	1	63	1	97	4	97	1	97	1
PP	96	1	96	1	85	2	85	2	85	2
SS	62	1(3)	63	1(4)	94	1(4)	63	2(4)	78	2(4)
C	83	1(1)	83	1(2)	83	1(3)	62	1(4)	78	1(4)
					16-h					
BM	47	1(1)	47	1(1)	47	1(1)	47	1(1)	47	1(1)
F	4	1(5)	4	1(5)	4	1(5)	4	1(5)	4	1(5)
BP	63	1	63	1	63	1	63	1	63	1
PP	96	1	96	1	96	1	96	1	96	1
SS	63	1(2)	62	1(3)	94	1(3)	63	1(4)	78	1(4)
C	42	1(1)	83	1(1)	63	1(2)	83	1(2)	69	1(3)
					24-h					
BM	47	1(1)	47	1(1)	47	1(1)	47	1(1)	47	1(1)
F	4	1(5)	4	1(5)	4	1(5)	4	1(5)	4	1(5)
BP	63	1	63	1	63	1	63	1	63	1
PP	96	1	96	1	96	1	96	1	96	1
SS	69	1(1)	84	1(2)	62	1(3)	83	1(4)	52	1(4)
C	28	1(1)	56	1(1)	83	1(1)	55	1(2)	69	1(2)

* Operating rates are in bushels per day. Equipment and labor components use the following abbreviations: BM - bagging machine; F - forklift; BP - bagging person; PP - pallet person; SS - separator; C - cleaner. Efficiency is followed by the number (#) and type code (t) of each component as defined by table 7.8.

Table 7.15 continued on next page

provide some excess capacity owing to the discrete sizes of equipment. Likewise, the optimum set is determined from a given set of equipment that does not include all possible sizes available from all manufacturers.

Efficiency of Optimum Set. Each optimum equipment and labor set is evaluated in terms of its efficiency using the PACASACS program to determine performance of the bagging machine(s), forklift, and associated labor. For each of these components, efficiency is defined as the ratio of the quantity of grain that was processed as compared to the quantity that could have been processed over the same time period. In the case of cleaners and spiral separators, efficiency is defined as the ratio of the desired daily design capacity to the rated capacity as specified in table 7.8. This efficiency rating does not consider that the optimum equipment set usually will have some excess capacity, and that the design capacity is for processed material. Hence, the excess material that must be handled by the cleaner and spiral separators in

Table 7.15. Continued*

	3,000 Bu		4,000 Bu		5,000 Bu		7,500 Bu		10,000 Bu	
Item	%	#(t)	%	#(t)	%	#(t)	%	#(t)	%	#(t)
					8-h					
BM	74	1(1)	41	1(2)	41	1(2)	41	2(2)	41	2(2)
F	6	1(5)	6	1(5)	6	1(5)	13	1(5)	13	1(5)
BP	98	1	98	1	98	1	99	2	99	2
PP	57	3	43	4	43	4	58	6	43	8
SS	94	2(3)	83	3(4)	78	4(4)	94	5(4)	89	7(4)
C	68	1(5)	45	2(5)	57	2(5)	85	2(5)	76	3(5)
					16-h					
BM	73	1(1)	73	1(1)	73	1(1)	41	1(2)	41	1(2)
F	6	1(5)	6	1(5)	6	1(5)	6	1(5)	6	1(5)
BP	97	1	97	1	97	1	98	1	98	1
PP	85	2	85	2	85	2	58	3	43	4
SS	94	1(42)	63	2(4)	78	2(4)	78	3(4)	78	4(4)
C	83	1(3)	62	1(4)	78	1(4)	85	1(5)	57	2(5)
					24-h					
BM	47	1(1)	73	1(1)	73	1(1)	73	1(1)	41	1(2)
F	4	1(5)	6	1(5)	6	1(5)	6	1(5)	6	1(5)
BP	63	1	97	1	97	1	97	1	98	1
PP	96	1	85	2	85	2	85	2	58	3
SS	63	1(4)	83	1(4)	52	2(4)	78	2(4)	70	3(4)
C	83	1(2)	74	1(3)	52	1(4)	78	1(4)	76	1(5)

* Operating rates are in bushels per day. Equipment and labor components use the following abbreviations: BM - bagging machine; F - forklift; BP - bagging person; PP - pallet person; SS - separator; C - cleaner. Efficiency is followed by the number (#) and type code (t) of each component as defined by table 7.8.

the form of screenings tends to offset the excess capacity associated with the incremental selection of optimum equipment sets.

Efficiencies are shown in table 7.15. For the system sizes shown, there is a component that is limiting. Bagging or pallet person labor is usually the limiting component depending upon which one first approaches 100% utilization. Generally, if the equipment component number or type changes with an increase in design capacity, the efficiency of its utilization will decrease. Likewise, increased capacity for the same number and type of component will increase efficiency.

Costs of Equipment. An index of annual and purchase costs of equipment is given in table 7.16. The cost index rating is based on a comparison with the least expensive system in terms of annual cost.

As would be expected, both purchase cost and annual cost decrease rapidly with increases in daily processing capacity and with the number of days worked per year. However, purchase cost does not change greatly with increases in length of the work day. The magnitude of these

Table 7.16. Relative annual and purchase cost for optimally designed processing and bagging facilities*

H/Day	Daily Design Capacity (Bu)									
Days/ YR	500	1,000	1,500	2,000	2,500	3,000	4,000	5,000	7,500	10,000
8/50	7.3	3.9	2.5	2.0	1.9	2.0	1.7	1.4	1.4	1.4
	25.3	13.8	9.6	8.3	5.8	5.8	6.4	5.6	4.2	4.0
16/50	11.5	5.8	3.9	3.0	2.4	2.5	2.0	1.6	1.3	1.3
	24.7	12.7	8.7	6.9	5.7	4.8	4.2	3.4	2.6	2.8
24/50	15.6	7.9	5.3	4.0	3.3	2.8	2.6	2.2	1.5	1.4
	24.4	12.4	8.4	6.5	5.5	4.6	3.6	3.3	2.3	1.9
8/100	5.8	3.0	2.5	2.0	1.6	1.6	1.6	1.3	1.2	1.1
	12.7	6.9	4.8	4.2	3.4	2.9	3.2	2.8	2.1	2.0
16/100	10.0	5.0	3.4	2.6	2.1	2.2	1.8	1.4	1.2	1.1
	12.4	6.3	4.3	3.4	2.8	2.4	2.1	1.7	1.3	1.4
24/100	14.2	7.1	4.8	3.6	3.0	2.5	2.4	2.0	1.3	1.3
	12.2	6.2	4.2	3.3	2.8	2.3	1.8	1.6	1.1	1.0
8/150	5.3	2.8	2.3	1.8	1.5	1.5	1.4	1.2	1.1	1.0
	8.4	4.6	3.2	2.8	2.3	1.9	2.2	1.9	1.4	1.3
16/150	9.5	4.8	3.2	2.5	2.0	2.1	1.7	1.3	1.1	1.1
	8.2	4.2	2.9	2.3	1.9	1.6	1.4	1.1	0.9	0.9
24/150	13.7	6.9	4.6	3.5	2.9	2.4	2.3	1.9	1.3	1.2
	8.1	4.1	2.9	2.2	1.8	1.5	1.2	1.1	0.8	0.6
8/200	5.0	2.6	2.2	1.8	1.4	1.4	1.4	1.1	1.1	1.0
	6.3	3.5	2.4	2.1	1.7	1.4	1.6	1.4	1.1	1.0
16/200	9.2	4.7	3.1	2.4	2.0	2.1	1.6	1.3	1.1	1.1
	6.2	3.2	2.2	1.7	1.4	1.2	1.0	0.8	0.6	0.7
24/200	13.4	6.8	4.6	3.4	2.8	2.3	2.3	1.9	1.3	1.2
	6.1	3.1	2.1	1.6	1.4	1.2	0.9	0.8	0.6	0.5

* Ratio of annual (top value) and purchase cost (bottom value) to the annual and purchase cost, respectively, of the system with the least annual cost.

changes are related primarily to charges for interest, taxes, and insurance and in the depreciation method that is used.

The units of production method of computing depreciation, used in JAWS, assigned a depreciation cost for each hour (or unit) of production associated with the machine. With this method, there is no difference in the total depreciation taken for one or multiple machines so long as the sum of the production units is the same. For example, one cleaner operating for 24 h/day, 200 work days would have the same depreciation as three similar cleaners working 8 h/day, 200 days/year. However, the investment in the three cleaners would be three times as large, thus greatly affecting interest, taxes, and insurance costs. The units of production depreciation method is reflective of the actual "wear-out" cost of the machine. However, it may or may not be acceptable as a viable depreciation method by the Internal Revenue Service depending on current tax law.

As design capacity increases, usually it is possible to select machinery that more nearly fits the desired work rate. Thus, smaller systems typically are more expensive per unit of production, partially because they contain inherently excess capacity and partially because of the economics of scale of larger machines.

Another important observation is that a ranking of systems with least annual costs does not correspond directly with a ranking of systems according to purchase costs as shown by the cost index values used in table 7.16. Again, the reasons for this are associated with the choice of depreciation method.

Equipment and Labor Zones. It is important to note that most system components do not change with incremental increases in design capacity. Another consideration is that surge bins may be used to altar effectively the utilization of cleaners and spiral separators. For example, these latter components may operate for 16 h to generate enough capacity so that the bagging machine can function for 8 h. If the cleaner and spiral separators are excluded from the analysis, a more limited number of system component combinations is sufficient to process the grain. Forklifts may also be excluded in that the same number and type were required for the entire range of design capacities. The result is that seven equipment and labor "zones" describe the modified design criteria as shown in table 7.17.

Summary

This example design has presented an optimum equipment set for a broad range of daily and annual capacities. Performance efficiencies have been determined as have absolute and relative costs. Similar procedures could be followed for other equipment sets or design assumptions.

Table 7.17. Optimum equipment set zones when considering only bagging machines, bagging machine persons, and pallet persons

H/ Day	Equipment Type	Daily Design Capacity (bu) 500	1,000	1,500	2,000	2,500	3,000	4,000	5,000	7,500	10,000
	BM		1(1)	1(1)			1(1)	1(2)		2(2)	2(2)
8	BP		1	2		Zone	1*	1	Zone	2**	2***
	PP		1	2		2	3	4	4	6	8
	BM									1(2)	Zone
16	BP									1	4
	PP			Zone	1			Zone	2	3@	
	BM										Zone
24	BP										7
	PP										

BM - number of bagging machines and type.
BP - number of bagging persons.
PP - number of pallet persons.
* Zone 3.
** Zone 5.
*** Zone 6.
@ Zone 7.

In designing a network type system, such as a seed processing and bagging facility, it is important to recognize that one system component will always be the limiting factor when expanding capacity. An optimum equipment and labor set almost always has some excess capacity in all or part of the system. The key to proper economic decision making is to recognize the limiting factor and continue enhancing the performance of this bottleneck until either the desired performance of the total system is attained, until another factor becomes limiting, or until the marginal cost of increasing performance is greater than the associated marginal benefits.

Problems

7.1 A combine harvesting grain has a 120-bu hopper. The theoretical time to fill the hopper is 45 min. If the hopper is filled 10 times during a 10-h work day, what is the efficiency of the harvesting operation?

7.2 A grain harvesting and storage operation is designed to handle 3,500 bushels of soybeans per 12-h day. Using the information given in table 7.1:

(a) Find the combinations of drying equipment that can be used to dry the soybeans.

(b) Calculate the design efficiency and total cost for each combination.

7.3 A farmer wants to harvest a 250-acre field of corn in 15 working days. If the expected yield is 120 bu/acre, what is the required daily harvest rate?

7.4 A combine with a 10-ft header is used to harvest a corn crop with a yield of 110 bu/acre. If the ground speed of the combine is 2.75 mph and the machine is operated for 8 h/day at an efficiency of 67%, what is the combine capacity in bushels per day?

7.5 A 12-ft wide combine with a 110-bu hopper is operated at a groundspeed of 2.67 mph. The hopper unloads at a rate of 40 bu/min. Turning time and rest periods average 8.3 min for each hopper load of grain. If the combine is operated for 9 h each day to harvest grain with a yield of 105 bu/acre, what is the combine efficiency?

7.6 A grain harvesting operation uses 3 combines. The actual time to fill a hopper with grain is 18 min. The total travel time for a vehicle hauling grain to the storage facility is 15 min. If the time needed to unload the vehicle at the storage facility is 14 min, how many vehicles are needed?

7.7 A vehicle containing 250 bu of grain must unload in 8 min. The vehicle can be unloaded at a rate of 90 bu/min. If the bucket elevator has a capacity of 12 bu/min, what size receiving pit is needed?

7.8 A continuous-flow dryer handles grain from an operation harvesting 2,000 bu/day. The dryer is able to dry 500 bu/day during the time the combines are in operation. If the pit size is 100 bu, what size wet holding bin is needed?

7.9 An automatic batch dryer handles grain from an operation harvesting 2,000 bu/day. The dryer is able to dry 500 bu/day during the time the combines are in operation. If the pit size is 100 bu and 100 bu of grain are contained in one dump of the dryer, what size wet holding bin is needed?

7.10 Use the computer simulation model SQUASH (see Chapter 1) to design a complete harvesting, delivery, handling, and continuous-flow drying system that will result in 3,800 to 4,000 bu being placed into storage in not less than 11 h nor more than 12 h from the time harvest begins. Combines are not to have to wait to unload because a delivery vehicle is not available to receive grain.

7.11 Run SQUASH with an increase in combine harvesting capacity until at least one combine has to wait to unload because a delivery vehicle is not available to receive grain. Use EXSQUASH to determine the probable source of the bottleneck. Run SQUASH again with an increase in capacity in the source of the bottleneck and determine the extent to which combine waiting time decreases.

7.12 Use the computer model JAWS (see Chapter 1) to design a complete seed processing facility capable of bagging 2,300 bu in a 10-h day. The facility is to operate 170 days/year.

7.13 Use the computer simulation model PACASACS (see Chapter 1) to simulate the facility designed in the previous problem. Determine the extent to which the reduction of the capacity of selected component within the facility adversely impacts design capacity.

References for all chapters begin on page 541.

8

Layout and Design of Grain Storage Systems

The layout and design of grain storage systems is a long term investment. Mistakes in planning are relatively difficult to overcome, both economically and physically. Temporary solutions often become long-term problems.

Principles of System Layout and Design

The construction of farm storage for grain may be one of the largest investments that a farmer ever makes. Because of this, it becomes very important that the grain handling system be completely and thoroughly planned. Constructing the wrong size bin or placing a bin in an awkward location could necessitate very costly remodeling or modification when the facility is expanded or when additional mechanization is desired.

Principles of Grain Handling Facilities Design

Although there are many guidelines to be followed when designing a grain handling facility, there are two principles that must be applied if the facility is to be utilized efficiently, conveniently, and profitably. These principles are:

Closed-Loop Principle	It must always be possible to easily bring grain back to its point of delivery.
Expansion Principle	The system should have the potential for easy expansion from two to four times the original capacity of the facility while not violating the first principle.

The closed-loop principle allows a truck or wagon to deliver or receive grain from the same location. It permits the moving of grain from one bin to another or back to the same bin if desired. Any bin to which grain cannot be unloaded and transported easily is not usually included in the facility.

The expansion principle allows for additional storage space while keeping to a minimum the equipment required for handling. It should be noted that many times the "best" location for a bin in the initial facility layout may block expansion paths when additional bins are added. The expansion principle allows the grain storage facility to grow with the farming operation it serves.

Site Selection

The most important step in laying out a grain storage facility is in selecting the proper site. A well-drained, accessible area should be chosen with plenty of room to expand. Many people make the mistake of locating a single bin in an isolated spot "between the old barn and the pond and out of the way", overlooking the fact that no expansion can occur. The end result is that their investment is not utilized efficiently. The key is to "keep all the options open" and this begins with selecting a site large enough to accommodate future as well as present storage needs.

Although the area required for storage certainly will vary with the size of the facility, a minimum space of approximately 100 ft by 200 ft should be reserved for the site if at all possible. The ground should be smooth with adequate drainage away from the facility. Other factors to consider are the:

1. Availability of fuel, electrical power, and all-weather roads.
2. Ability to maneuver and park trucks and wagons in and around the receiving and shipping point.
3. Relative location of residential housing (because of noise, dust, and security).
4. Physical security of the stored grain.

Site Coordinate System

The site should be divided into quadrants for purposes of locating and planning system components. Basically, the site is made into a grid with the center point (or origin) serving as the reference for all components of the system. The center point typically would be located near the point where grain is to be received. Two perpendicular lines should be marked to pass through the center point (fig 8.1). The horizontal line, depending on the frame of reference, is referred to as the "X-axis". The vertical line is called the "Y-axis". From figure 8.1, points (or locations) that lie to the right of the Y-axis have positive "X-values" while points to the left of the Y-axis have negative "X-values". Similarly, points above the X-axis have positive "Y-values" while points below have negative "Y-values". If a bin location center were given as 35 ft in the X-direction and 25 ft in the Y-direction, it lies in Quadrant 1. Similarly, if the center were specified as (–35 ft, –25 ft) the point is located in Quadrant 4. Note that the X values are always given first; that is (X value, Y value).

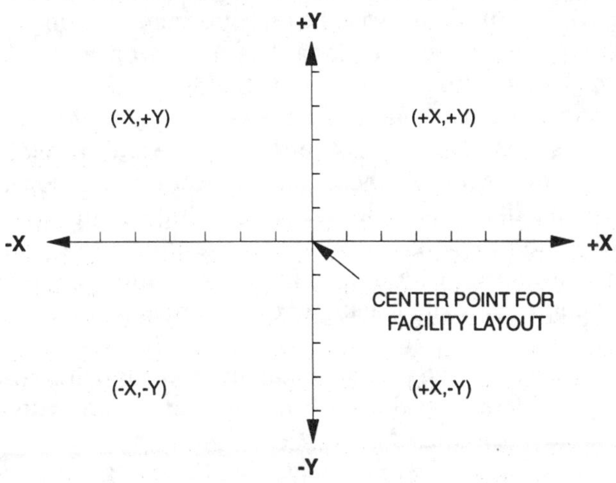

Figure 8.1–Quadrant system used for bin layouts.

A third dimension, referred to as the "Z-axis" is perpendicular to both the X and Y axis. A positive Z value extends above and perpendicular to the ground at the center point. A negative Z value results when a point is lower in elevation than the center point. If "Z" is equal to zero, it lies on the same plane as the center point.

The Z dimension is useful in referencing points that are above ground such as the filling point of a grain bin. A reference of (35 ft, 25 ft, 24 ft) might refer to the filling point being 24 ft above the center of a bin located in Quadrant 1. The Z coordinate follows the X and Y values in the coordinate specifications.

The above coordinate system is referred to as Cartesian coordinates. If only X and Y values are used, the system is often called rectangular coordinates. Other reference systems may be used. However, the Cartesian reference system will be used in this text for location specifications.

Developing a Plan

A scaled drawing of the site should be made showing the X-Y coordinates. Prepare paper "cut-outs" of all structures and system components to the same scale as that of the site. The "cut-outs" can be arranged easily in all possible configurations within the boundaries of the scaled site drawing. Planning and moving facilities that exist only on paper are much easier than altering actual facilities! Cut-outs should include bins of different diameters, center buildings, pit, bucket elevator(s), dryer(s), augers, wet holding bin(s), surge bin(s), dryeration bin(s), single axle trucks, tandem trucks, trailer trucks, and tractors with wagons. Site restrictions may dictate changes in sizes and capacities from what would otherwise be an "optimum" design.

After preparing the scaled drawings, the next step is to plan the handling system. The location of the center building, the dump pit, and the bucket elevator should be made. This is not to say that these items are constructed first (or at all) but rather that space will be allotted for them when expansion occurs. If the dump pit and bucket elevator are not constructed the first year, alternate handling methods must be considered. After completing the plans for handling the grain, the location of each bin should be determined. The following questions should be asked concerning the feasibility of each plan:

1. Can all types of delivery vehicles maneuver easily with regard to loading and unloading grain, both inside and around the center building, and into and out of the site area?
2. Does the closed-loop principle apply? That is, can grain be loaded easily back onto a delivery vehicle without moving it? This does not mean that all the handling equipment to accomplish this will be purchased immediately. Rather, it helps locate "dead-end" storage bins that will be used because of their isolation only as a last resort rather than as part of a functional system.
3. Can electrical power be provided while not creating safety hazards associated with portable augers, dump beds, etc., coming into contact with overhead power lines?
4. Is there space for an LP gas tank? Similarly, biomass burners may come into more use in the future and will need to be located close to drying units and in areas sufficiently large so that adequate biomass may be provided for burning.
5. If solar drying is to be considered, does the orientation of the bins allow for adequate collection and utilization of solar energy?
6. In the event of major rains or flooding, will the grain in storage and the grain handling and drying equipment remain above water? Likewise, can delivery vehicles move into and out of the site area without being stuck in mud or creating impassible situations for other vehicles that may arrive later?
7. Will the direction of the drying fans create unacceptable noise levels for family, employees, or neighbors?
8. Is there sufficient room for feed processing equipment should it be added at a later time?
9. Will the facility process adequately the design daily harvest rate?
10. Are there adequate safeguards with regard to mismatching of materials handling capacities? For example, a bucket elevator may be designed to handle significantly more grain than an in-line auger to a storage bin. This design may be desirable if the bucket elevator unloads directly into either wet holding or other storage

bins, or if two or more bins are to be unloaded into truck on wagon at once. However, precautions must be taken when the auger-fed storage bin is filled.

11. Is the arrangement safe? Does it create "blind-spots", fire hazards, open pit areas, etc., that might result in injury or loss of life?
12. Lastly, can the facility be expanded easily to accommodate increased materials handling capacity, larger storage volumes, and greater daily drying rates with a minimum of cost?

After several alternative plans have been developed, the "best" alternative must be selected. By using wooden pegs and string, the future facility can be marked completely at the site of construction, thus eliminating much of the guesswork with regard to space and arrangement. After the facility has been marked or "staked out", delivery vehicles of different types and capacities should be driven through the yet to be constructed facility to ensure against planning error.

Summary

1. Plan the system carefully. It will be utilized in some form for the next 20 to 30 yr.
2. Allow for expansion of the grain facility of from 2 to 4 times the present need. Most farmers say they'll never get any bigger but most of them do.
3. Remember, there's always going to be one bottleneck in the system, and preferably, it should be the combine.
4. Don't be hesitant to move an existing metal storage bin if it's in the wrong place. After all, what is a few days of extra work when compared to a 20-year mistake.
5. Do not locate the grain facility "between the old barn and the pond and out of the way". Consider that it may be one of the farmer's most important investments, and the facility may need to be expanded.
6. Before purchasing equipment, first identify the bottlenecks in the operation. Then purchase those items that will increase system efficiency the greatest for the least amount of money. Sometimes, $2,000 spent on a larger receiving auger can increase harvesting efficiency many times more than $20,000 spent on a larger combine.
7. And most importantly, the design should be as "safe" as possible in terms of both layout and management. Flowing grain and grain harvesting machinery both present potentially hazardous situations to workers. A person can be submerged in flowing grain in a bin and suffocate in less than 6 s, truly a tragic mistake.

Centralized Systems

There are two basic schemes commonly used in designing a grain storage facility (centralized and circular systems) each with its own advantages and disadvantages. Centralized layouts are very popular because of flexibility, ease of expansion, and relatively low-labor requirements (fig. 8.2).

A major advantage of the centralized layout is that it meets the needs of both the large and small operator. It is adapted easily to include any number of bins and any type of drying method. Bins of different sizes may be placed easily in the system. An orderly expansion can occur with minimum duplication of existing grain handling equipment (this to be discussed in detail later). Labor is kept to a minimum by mechanization, and the system offers several conveniences in that grain may be transported easily to and from a single sending and receiving point.

Disadvantages of this system are that more skill is required in construction and a larger initial investment may be required for handling equipment.

Layout and Expansion of Centralized Facilities

To illustrate the flexibility of the centralized arrangement, assume that the facility will begin with one bin, use layer drying, and expand to six bins using a portable dryer.

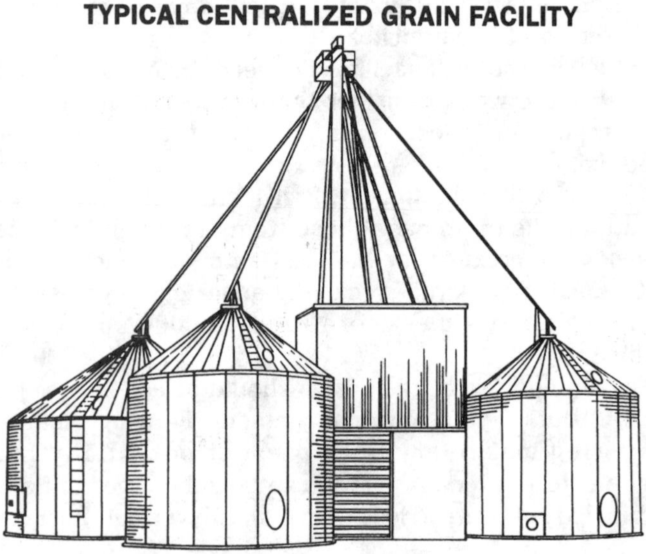

TYPICAL CENTRALIZED GRAIN FACILITY

Figure 8.2–Centralized grain facility.

Where will the first bin be located? If two perpendicular lines were to be drawn through the center of the future building site such that the facility area would be divided into four quadrants, the first bin to be constructed would be in the same quadrant as the proposed bucket elevator (Quadrant 1), as shown in figures 8.3 and 8.4. The grain in this bin would be dried using the layer drying technique; that is, drying a 10 to 12-in. layer of grain per day using a drying air temperature 15 to 20° F higher than the outside temperature. (See Chapter 3 concerning drying methods and schedules.)

Assuming that more storage and additional drying capacity are needed, the next two bins to be built would be located in quadrants 2 and 3 (fig. 8.5). Batch-in-bin drying techniques then could be used in Bin 1 instead of layer drying. The batch-in-bin method allows up to 4 ft of grain to be dried overnight in Bin 1 with a heated air temperature of approximately 140° F. (See Chapter 3 for detailed information.) An unloading auger could be extended from Bin 1 to a location where it could feed the proposed bucket elevator. At this location, a single portable auger can convey the dried grain to either Bin 2 or Bin 3. After Bins 2 and 3 are filled and the bottom 4 ft of Bin 1 is dried, the remaining grain to be placed in Bin 1 is conditioned using the layer drying method.

The decision to expand to four bins generally requires the addition of new handling equipment. However, if planned correctly from the beginning, very little alteration of the existing facility will be needed. The new equipment required will include a bucket elevator with downspouting to each bin, a dump pit, and extension of the unloading augers from Bins 2 and 3 to the pit so that the "closed-loop" principle

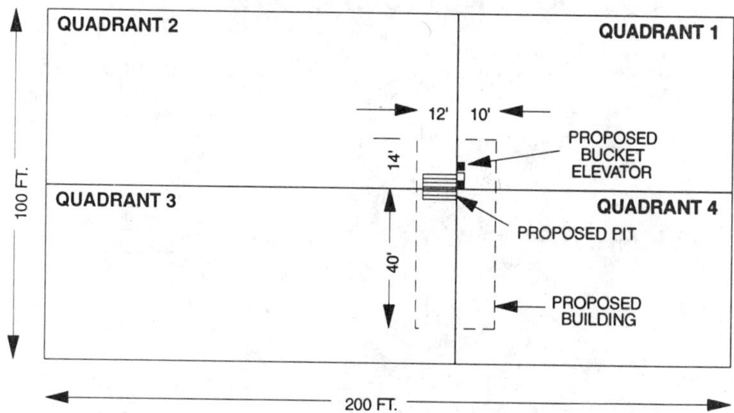

Figure 8.3–Bucket elevator location.

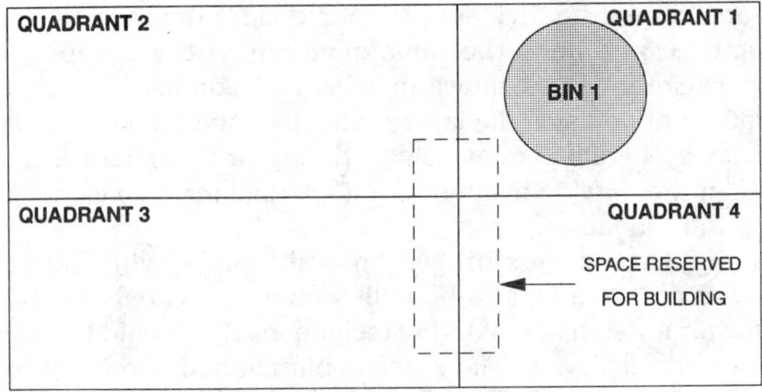

Figure 8.4–Location of initial bin.

can be applied. Bin 4, located in Quadrant 4, also becomes part of the closed loop (fig. 8.6).

Suppose that the capacity of the batch-in-bin dryer is too small. Because of prior planning, a portable batch or continuous flow dryer may be placed between Bins 1 and 4. A minimum space of 14 ft between these two bins is recommended. This dryer may also be placed in front of the building near Bin 1 (fig. 8.7). Selection of a portable batch or continuous-flow dryer usually dictates an overhead wet holding bin positioned over the dryer for gravity feeding. In addition, a continuous-flow dryer usually would include a surge bin for the dried grain.

What if additional storage space is required? If the additional storage is needed, it may be placed in line with Bins 2 and 3 with the "closed-loop" accomplished through overhead and unloading augers feeding

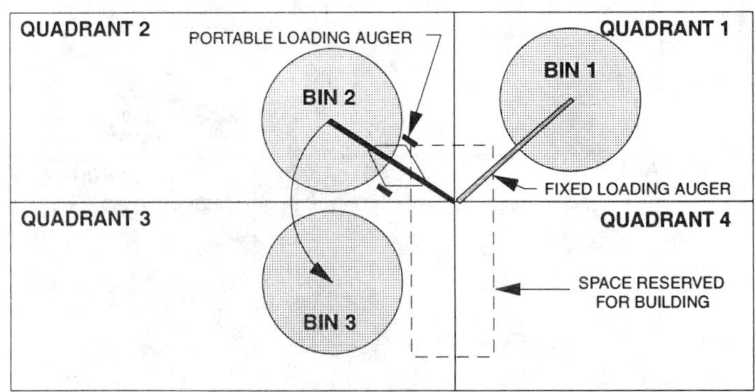

Figure 8.5–Location of second and third bins.

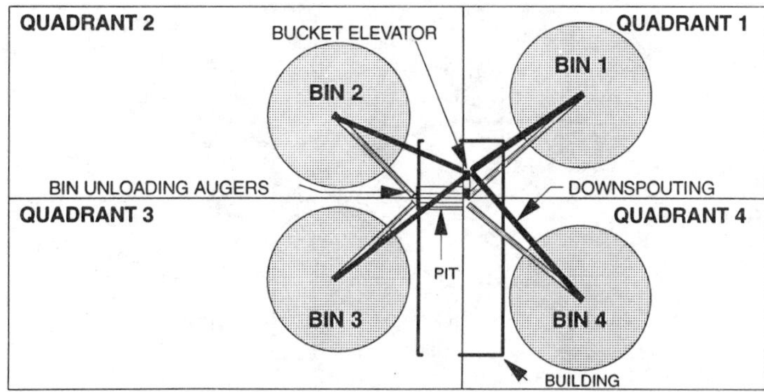

Figure 8.6–Location of fourth bin.

grain to and from the expansion bins (fig. 8.8). Expansion may continue in several alternate directions, depending primarily on the available space. At this point in facility expansion, the cost per bushel of storage decreases because the major materials handling items (bucket elevator, pit, dryer, etc.) have already been purchased.

If sufficient space is allowed for the building, the option of installing feed processing equipment can be taken. Alternate locations include the area to the side of the driveway, overhead above the driveway, or perhaps in one of the quadrants previously reserved for a storage bin.

Summary

The centralized scheme for bin layout offers a number of advantages. But regardless of the layout, the "closed-loop" and "expansion"

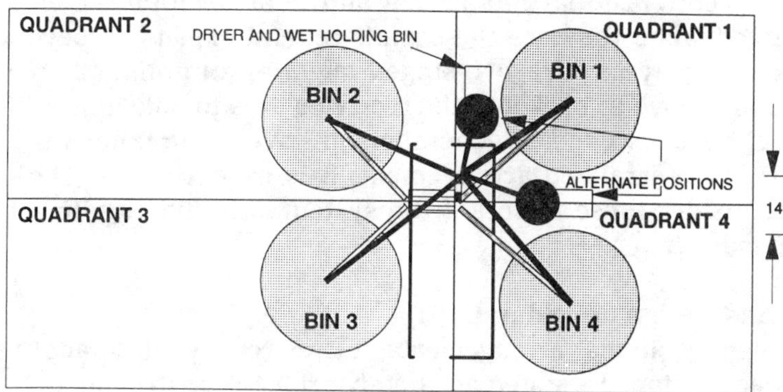

Figure 8.7–Alternative locations of the portable dryer.

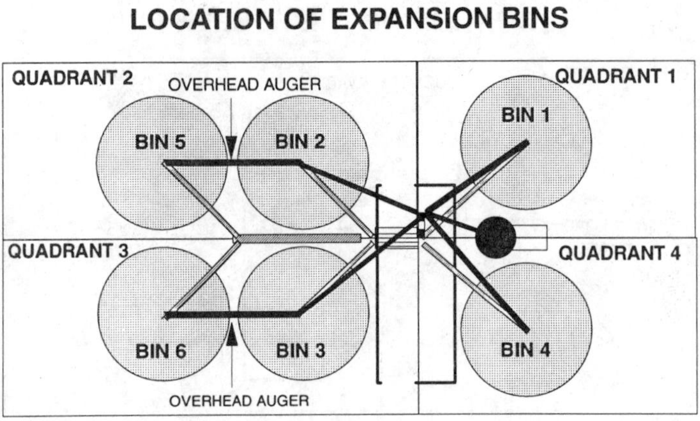

LOCATION OF EXPANSION BINS

QUADRANT 2
OVERHEAD AUGER
QUADRANT 1
BIN 1
BIN 5
BIN 2
QUADRANT 3
QUADRANT 4
BIN 6
BIN 3
BIN 4
OVERHEAD AUGER

Figure 8.8–Location of expansion bins.

principles must not be violated if the grain facility is to produce maximum profit and convenience. The surest way to fail in designing a grain handling facility is to select a poor site without sufficient space for expansion and centralization of storage and equipment. Add to this an inadequate grain handling method (such as moving large quantities of grain by scoop shovel) and the result will be an expensive but poorly functioning system. If one phrase could express the key to designing grain handling facilities, it would be to "keep as many options open as possible".

Circular Grain Storage Systems

A popular grain drying and storage facility configuration is the circular arrangement (fig. 8.9). The primary advantage of this system is the efficient handling of grain with portable conveying equipment. Disadvantages include: (1) expansion of the facility is more difficult than with centralized layouts in that all the bins within the circle are best served when they are the same size, and expansion beyond the initial circle may require two-staged augering of grain; (2) delivery vehicles may have to back into the receiving area to unload; and (3) if a bucket elevator is used, it generally will have to be taller than for a comparable centralized arrangement. It is essential that the initial design consider these factors if the system is to function in a highly efficient manner.

Layout and Design Considerations

In designing circular arrangements, it is extremely important that the basic specifications be stated accurately prior to construction to obtain optimum efficiency (Loewer et al., 1986). Optimum efficiency is based

TYPICAL CIRCULAR CONFIGURATION

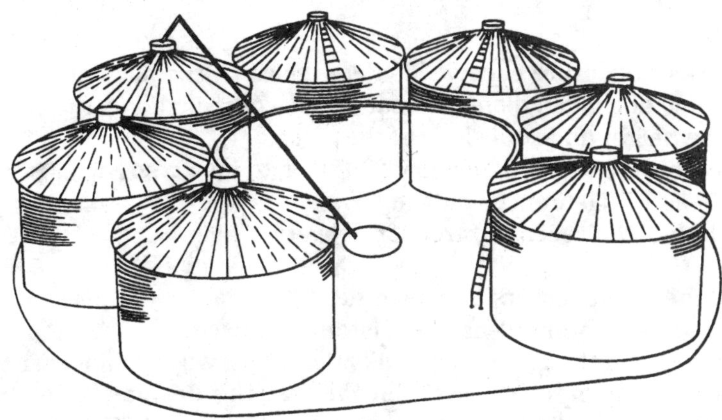

Figure 8.9–Typical circular configuration for grain storage facilities.

on the priorities assigned to factors that dictate the design; that is, the designer must decide if system capacity (both present and future), land area or delivery conveyor considerations (auger or bucket elevator) will govern the design. Many of these factors interact. For example, auger power requirements increase disproportionately with increases in elevating height, other factors remaining constant. In addition, the length of auger needed to reach the top center of a bin may be influenced by the eave height as well as the total height of the bin. Thus, it is not sufficient to base the design on only the minimum area needed to place the bins.

Factors such as bin size and the dynamics of materials handling in the design of grain storage systems are given by Loewer et al. (1976a, b; 1980), Bridges et al. (1979a, b) and Benock et al. (1981). The circular system is designed most efficiently when (1) all bins are the same size and (2) all bins can be filled by a portable conveyor that extends directly from a central unloading pit or receiving hopper. Similarly, the unloading auger from each bin should be directed to a point at which it can feed directly a second portable conveyor that conveys grain back to the central receiving area. Using this configuration, grain that is being unloaded into the receiving pit may be placed directly into delivery vehicles or into other storage bins. Bin unloading augers usually are designed to reach the central receiving point directly (fig. 8.10), or they can be placed so that two bin augers may feed the portable receiving auger at the same time. Simultaneous unloading doubles the unloading rate and/or reduces the number of times that the receiving conveyor must be moved.

Given the influence of the above considerations, the following items must be specified before designing the circular layout for space requirements:

1. Present or future capacity as determined by the maximum number of bins of a given diameter that can be constructed on the circle.
2. Eave height and roof slope of the bins.
3. Delivery conveyor geometry (if this is to be a governing factor) as to lengths, clearances and angles.
4. Width of the receiving area for delivery vehicles.

All of the above factors influence the center radius of the facility. The center radius is defined as the distance between the centers of the receiving pit and the grain bins and is the primary design dimension in a circular arrangement. It is essential that the designer specify the conditions under which auger geometry will dictate the center radius and vice versa. The following discussion indicates the design criteria under which the center radius is computed.

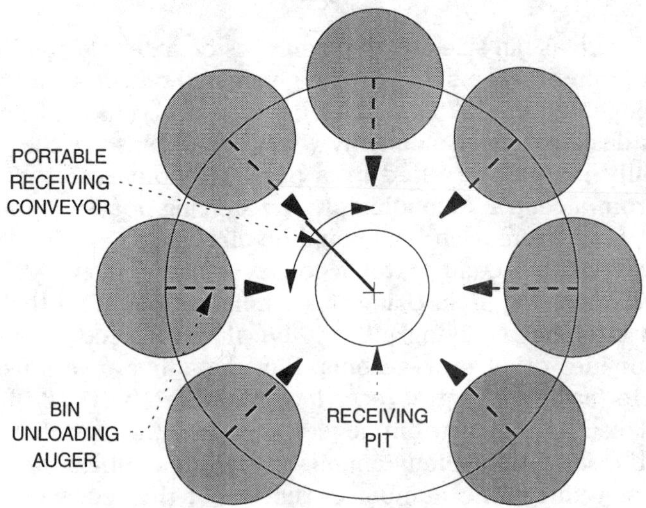

Figure 8.10–Unloading configuration for a circular arrangement.

Design Procedure

The design procedure for locating storage bins in a circular arrangement begins by determining the center radius. The designer must specify:

1. Minimum and maximum storage capacities on the center radius.
2. The closest distance between the walls of any two bins except those on either side of the receiving area.
3. The minimum distance allowed between two bins for entry by delivery vehicles to the receiving pit.

The layout of the facility, including possible expansion bins, would appear as shown in figure 8.11 for the situation in which all bins are the same size.

There are several geometrical considerations that must be incorporated into a circular system design. An iteration procedure in design is required because a series of design parameters must be satisfied before the system will function within the specified criteria. This requires the changing of each design component until the minimum sizing of the most limiting component is obtained. The basic criteria for

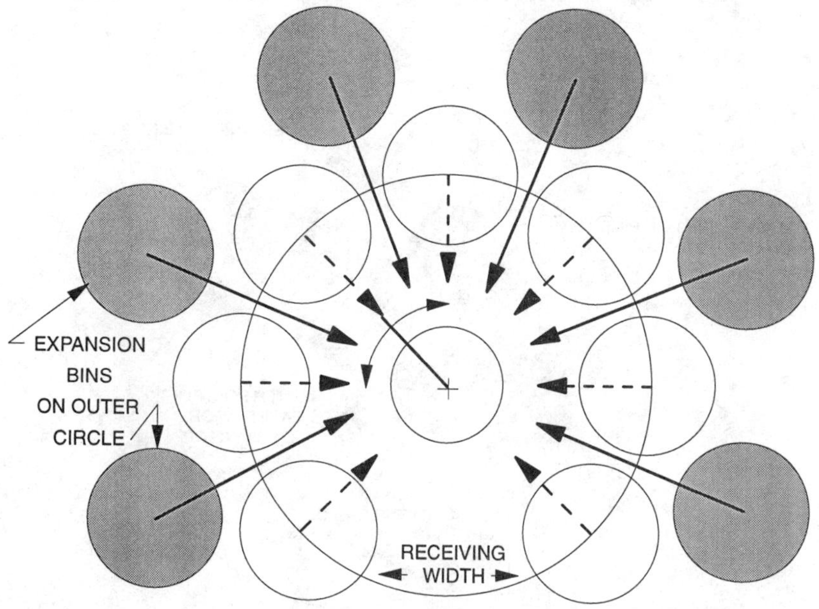

Figure 8.11–Circular system expansion.

circular systems is that grain must be delivered and received in a timely manner and that there must be sufficient storage capacity.

The first geometric consideration is that any two bins may be used to mark a line on which the center point of the circle will lie. The center point falls on a line that bisects and is perpendicular to another line that joins the bin centers (fig. 8.12). The addition of a third bin will define the exact center location of the receiving pit regardless of the diameters of any of the bins. If the three bins do not have the same dimensions, the conveying auger may need to be altered to reach the top center of each bin. Once the conveying pit center has been established, all future bins should have their center unloading wells located on the center radius .

The design of circular layouts is simplified when all bins to be constructed on the circle have the same dimensions. Regardless, the center radius must be of sufficient length so that all the bins can be placed on the circle while allowing sufficient room for unloading and the movement of conveying equipment through the receiving area. The center radius must also allow for the operation of conveying equipment within the design criteria related to power, capacity, and grain damage. Either conveying criteria or storage capacity will be the limiting factor in defining the center radius.

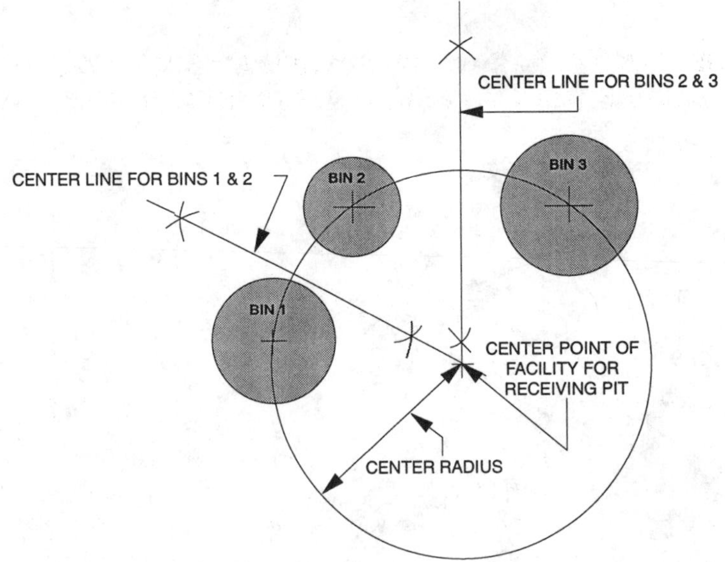

Figure 8.12–The centers of any three bins define the center radius in a circular configuration.

The central conveyor may be either a bucket elevator or a portable auger, the latter being most common for circular layouts. If a bucket elevator is to be used, the center radius is limited by bin diameter and space between bins while allowing for sufficient width to receive and deliver grain. The height of the bins generally is not a factor in that bucket elevators may be built high enough for filling. Portable augers present a different type of problem in terms of power requirements, capacity, and grain damage. Because of these factors, a general guideline for circular systems is that the angle of delivery for the portable auger be no greater than the roof slope of the grain bin (usually about 30°). However, there is a savings in auger length as the unloading angle of the auger approaches 45°. Another factor influencing auger length is that the auger must extend into the receiving pit, usually several feet below the ground surface.

Given the appropriate design criteria, the center radius can be computed as follows:

for a ≤ b (fig. 8.13)

$$H = G + \left(B/2\right) * TAN(b) + T \tag{8.1}$$

$$R = \left(H + D\right) / TAN(a) \tag{8.2}$$

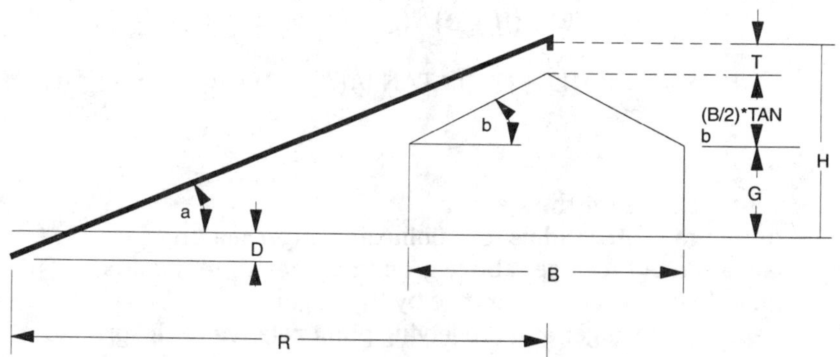

Figure 8.13–Center radius "R" for systems where the slope of the auger (a) is less than or equal to the bin roof slope (b) (Loewer et al., 1986).

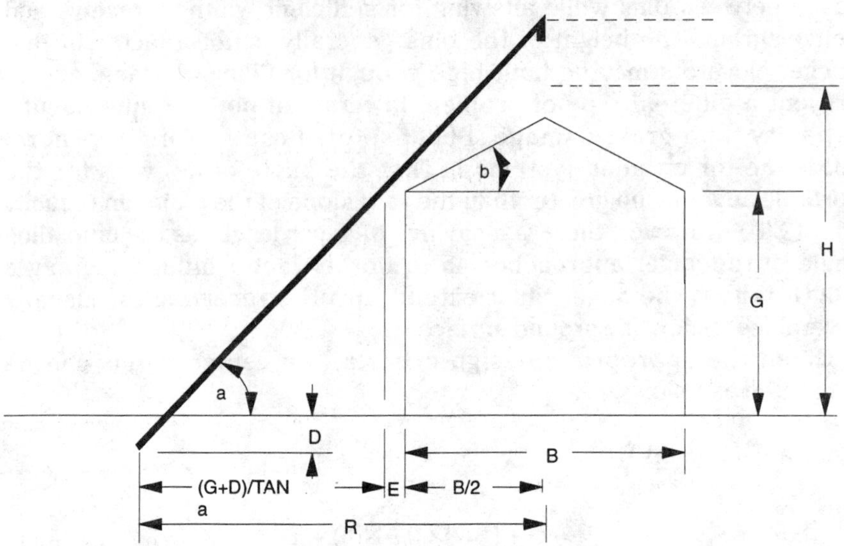

Figure 8.14–Center radius "R" for systems where the specified slope of the auger (a) is greater than or equal to the bin roof slope (b) and overrides the specified values of "H" (minimum fill height) as a design consideration (Loewer et al., 1986).

for a > b where "a" overrides "H" as a constraint (fig. 8.14)

$$R = E + (B/2) + (G + D) / TAN(a) \tag{8.3}$$

for a > b where "H" overrides "a" as a constraint (fig. 8.15)

$$R' = (H + D) / TAN(a') \tag{8.4}$$

$$a' = arc\ tan\left[(T + (B/2) * TAN(b)) / ((B/2) + E)\right] \tag{8.5}$$

where

R = center radius for the system

R′ = optimum center radius for minimizing auger length

H = elevation of the auger above ground at the point of unloading to include vertical bin clearance by the auger

D = depth of the auger at its receiving point relative to the ground level of the bin

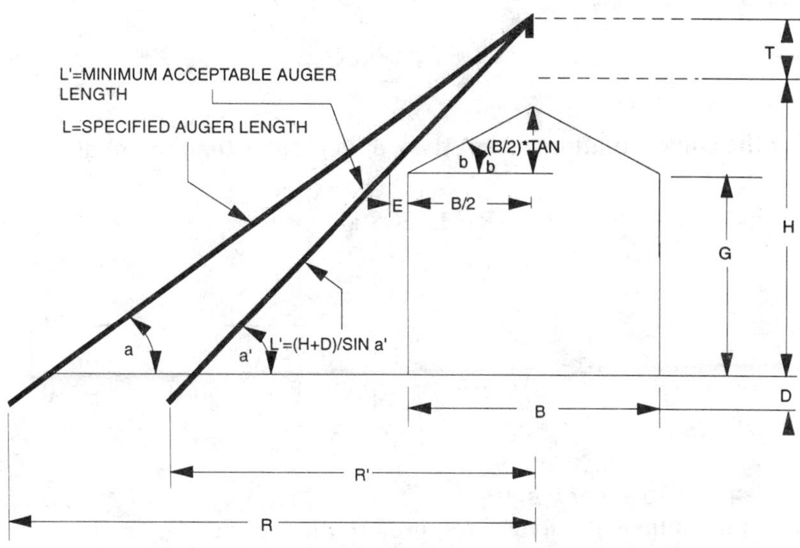

Figure 8.15–Relationships among the minimum auger length (L′) and its associated center radius (R′), and a specified auger length (L) and its associated center radius (R) (Loewer et al., 1986).

a = desired angle of elevation of the auger (°)
a′ = optimum angle of elevation for the auger for minimizing auger length (°)
b = slope of the bin roof (°)
E = horizontal clearance to the closest point of the bin
B = bin diameter
G = distance from the ground to the bin eave
T = vertical clearance between the unloading auger and the bin roof

If the length of the conveying auger is to govern the design, if its specified length "L" is greater than or equal to the minimum feasible design length "L′", and if the desired auger elevation angle "a" is less than the optimum angle of elevation "a′" for "L′", then "R" may be computed as follows (fig. 8.15):

$$H = L * SIN(a') - D \qquad (8.6)$$

$$R = \left(L^2 - (H + D)^2\right)^{0.5} \tag{8.7}$$

where

$$L' = (H + D) / SIN(a') \tag{8.8}$$

For the same situation except that "a" is greater than or equal "a′":

$$R = L*COS(a) \tag{8.9}$$

where

$$L' = (G + D) / SIN(a) + (((B/2) + E) / COS(a)) \tag{8.10}$$

and
L = specified auger length
L′ = minimum auger length required to fill the bin

Once the center radius has been determined, the next step is to determine the number of bins that will fit in the circle and whether this number will allow sufficient room for unloading vehicles. This process begins by determining chord length. The chord length joining two adjacent bins of the same diameter is computed as follows (fig. 8.16):

$$P = B + S \tag{8.11}$$

where
P = chord length
S = distance between adjacent bin walls

Equation 8.11 allows the computation of the maximum number of bins that may be placed in the circle by computing the angle between adjacent bin centers followed by determining the arc length of a segment between two adjacent bins (fig. 8.16). The circumference divided by the arc length gives the maximum number of bins that may be placed in the inner circle. In equation form:

GEOMETRIC CONSIDERATIONS IN CIRCULAR SYSTEMS

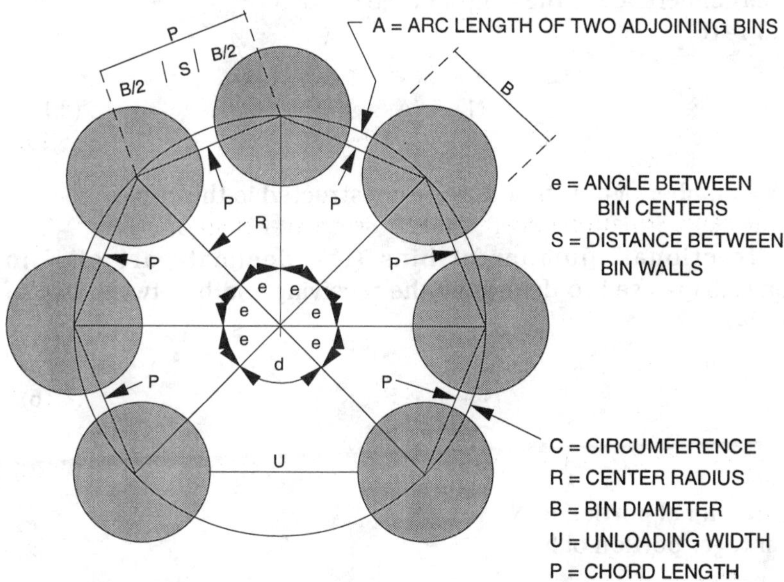

Figure 8.16–Relationship among center radius, chord length, arc length, number of bins, and unloading space (Loewer et al., 1986).

$$e = \text{ARC COS}\left[\left(2*R^2 - P^2\right) \;/\; \left(2*R^2\right)\right] \quad (8.12)$$

or

$$e = 2*\text{ARC SIN}\left(P/\left(2*R\right)\right)$$

where

e = angle (in rad) between two adjacent bin centers (Note: There are 2 pi rad in a circle. A circle has 360°. Therefore, a rad = 57.3°.)

$$A = R * e \quad (8.13)$$

where

A = arc length of two adjoining bin centers

$$C = 2*pi*R \quad (8.14)$$

where

C = circumference of the center circle

pi = 3.1416

$$N = C/A \tag{8.15}$$

where N = number of bins that may be constructed in the circle.

The "fractional" number of bins (the decimal part of N in equation 8.15) is used to determine the receiving width between two of the bins.

$$N' = N - F \tag{8.16}$$

where

F = fractional portion of N

N′ = integer portion of N

$$d = \left(C - (N' - 1) * A\right) / R \tag{8.17}$$

where d = angle (in rad) between adjacent bin centers that divide the receiving area.

$$U = 2*R*SIN(d / 2) - B \tag{8.18}$$

where U = unloading width.

If "U" is less than the minimum width allowed for unloading, the number of bins (N′) is reduced by one (equation 8.17) and the new spacing is reevaluated (equation 8.18) to see if it now meets the design criteria. This process continues until the space criteria for unloading are satisfied.

If a bucket elevator is to be located in the center of the facility, its discharge height may be calculated by the following expression:

$$Q = TAN(c)*R + G + TAN(b)*(B/2) \tag{8.19}$$

where

Q = discharge height for the bucket elevator

c = angle for downspouting (see Chapter 6 for recommended angles)

Center Radius Versus System Capacity

In the previous discussion, the emphasis was on determining the geometry of a circular bin arrangement given that the storage capacity requirements were satisfied. Capacity and center radius can be related directly by using geometrical ratios of selected design criteria. For example, given a specified depth of grain, the number and diameter of bins determine the total system capacity. In turn, the number and diameter of bins plus the receiving width and the distance between adjacent bins can be used to determine the minimum center radius. The relations among these variables can be simplified by representing each bin as a circle with a diameter of "B + S" (bin diameter plus distance between adjacent bins). Then, the adjacent circles are contiguous, and the unloading distance between the end bins is "U - S". For further simplification, the adjusted unloading distance and the minimum center radius may both be expressed in ratio to "B + S". The resulting equations, given below, summarize equations 8.11-8.19 for various configurations. Because the equations cannot be solved directly for "R′", a numerical method of solving for "R′" is used for specified values of "N."

If bins occupy more than "pi + e" radians of the circle (fig. 8.16), the following equation applies:

$$N = \frac{\left[\text{pi-ARC TAN}\left((1 + U')\right) / \left(4R'**2\text{-}(1 + U')**2\right)**0.5\right]}{\text{ARC TAN}\left(1/(4R'**2-1)**0.5\right)} + 1 \quad (8.20)$$

where

N = number of bins

U′ = (U-S)/(B+S)

R′ = R/(B+S)

and R′ > (U′+1)/2 with pi being measured in radians.

For situations in which bins occupy more than "pi" but less than "pi + e" of the circle (Situation A in fig 8.17), the following equation applies:

$$N = \frac{\left\{\dfrac{\text{pi-ARC TAN}\left(\left[\left(1+U'\text{-}2R'\right) * \left(1 + 2R'\text{-}U'\right)\right]**0.5\right)}{\left(1+2R' + U'\right) * \left(2R' + U'\text{-}1\right)}\right\}}{\text{ARC TAN}\left(1/\left(4R'**2\text{-}1\right)**0.5\right)} + 1 \quad (8.21)$$

where $(U'^2+1)^{0.5}/2 < R' < (U'+1)/2$.

SPECIAL CIRCULAR CONFIGURATIONS

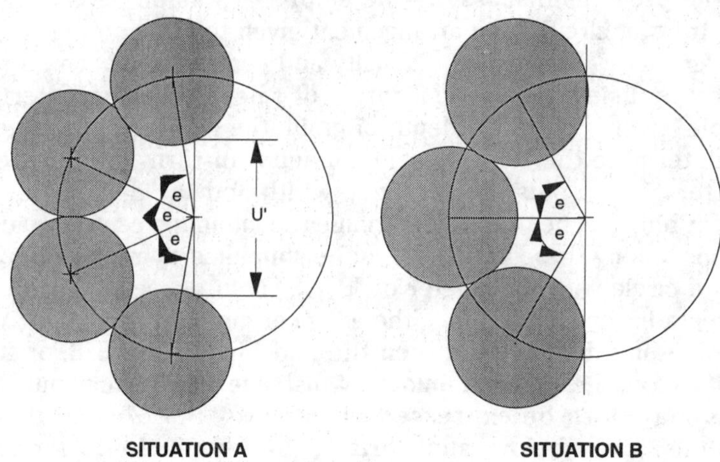

Figure 8.17–Bin configurations that may be used in estimating dimensions and system capacity (Loewer et al., 1986).

If bins occupy exactly "pi" of the circle, the unloading space is not limited by bin location, and the center radius is determined entirely by the number, diameter, and distance between adjacent bins (Situation B in fig. 8.17). Then:

$$N = \frac{pi}{2\ ARC\ TAN\left(1/\left(4R'**2\text{-}1\right)**0.5\right)} \qquad (8.22)$$

where $R' < (U'^2+1)^{0.5}/2$.

The equation to use for calculating N is dictated by the value of R′ in the range specified for each equation.

Table 8.1 gives the minimum center radius ratios (R′) for several values of N and U′ so that the minimum R can be estimated from known values of N, U, and S. The ratios were calculated through iteration of equations 8.20 through 8.22 with varying values of R′ until integer values of N were obtained. To use table 8.1, assume a bin diameter, space between adjacent bins, unloading space, and number of bins, for example B=28, S=6, U=20 and N=7 (any consistent units will suffice). Then, U′ = (20 – 6)/(28 + 6) = 0.4117, and an estimate of R′= 1.223 can be obtained by interpolation in table 8.1. Thus, R = 1.223 * (28 + 6) = 41.6.

Example Designs

An example design configuration is shown in table 8.2 based on equations 8.1 through 8.19. These equations were incorporated into a spreadsheet program to demonstrate some of the dynamics of the design calibrations. The scenario was computed with the constraint that the minimum required auger length would be rounded to the next largest integer value. A first and second design approximation is made based on the iteration process required to ensure that there is sufficient width to unload a vehicle directly into the receiving pit.

In table 8.2, the computed auger length is based on a minimum auger slope of 20°. The design auger length is given as 10 ft to ensure that the program will calculate a minimum auger length greater than the initial

Table 8.1. Minimum system center radius as influenced by the unloading space ratio and the maximum number of bins to be located on the circle

Unloading Space Ratio (U′)	Minimum System Center Radius Ratio (R′)					
	Maximum Number of Bins (N)					
U′	3	4	5	6	7	8
0.20	0.625	0.745	0.886	1.034	1.186	1.340
0.25	0.641	0.756	0.896	1.043	1.195	1.348
0.30	0.658	0.767	0.905	1.052	1.203	1.357
0.40	0.700	0.791	0.925	1.071	1.221	1.374
0.50	0.748	0.816	0.946	1.090	1.239	1.392
0.60	0.790	0.845	0.969	1.110	1.258	1.410
0.80	0.861	0.913	1.017	1.152	1.297	1.447
1.00	0.915	1.000	1.072	1.198	1.338	1.485
1.25	0.963	1.106	1.153	1.261	1.393	1.537
1.50	0.991	1.184	1.253	1.332	1.454	1.591
2.00	1.000	1.281	1.457	1.510	1.593	1.712
2.50	1.000	1.307	1.571	1.727	1.769	1.856
U′	10	12	14	16	20	24
020	1.651	1.964	2.279	2.595	3.228	3.863
0.25	1.659	1.972	2.287	2.603	3.236	3.871
0.30	1.667	1.981	2.296	2.611	3.244	3.879
0.40	1.684	1.997	2.312	2.628	3.260	3.895
0.50	1.701	2.014	2.328	2.644	3.277	3.911
0.60	1.718	2.031	2.345	2.660	3.293	3.927
0.80	1.753	2.065	2.378	2.693	3.326	3.959
1.00	1.789	2.099	2.412	2.727	3.358	3.992
1.25	1.836	2.144	2.456	2.769	3.400	4.033
1.50	1.885	2.190	2.500	2.812	3.442	4.074
2.00	1.989	2.286	2.591	2.901	3.527	4.158
2.50	2.104	2.389	2.688	2.994	3.615	4.243

* U′ = (U-S) / (B + S) where U = unloading width, S = distance between adjacent bin walls, B = bin diameter.

specification but based on top and side clearances rather than on slope of delivery (fig. 8.15). The values in table 8.2 reflect the situation in which auger length, center radius, and total capacity are minimized given the design specifications. The table values give design information for a bin arrangement in which auger length is not an overriding consideration or in which a bucket elevator is to be added to the system after portable augers have been used. What are the consequences of altering the table 8.2 design specifications in the spreadsheet?

Setting the minimum slope of the portable auger to 45° gives the shortest possible auger under the condition in which vertical and horizontal clearance do not dictate a longer auger length. Likewise, the minimum auger lengths, center radius and capacity are reduced when

Table 8.2. Example analysis of circular grain system design

Design Specifications for Circular Bin Arrangement

1.000	-	Minimum horizontal distance from auger to bin wall (ft) (top clearance has priority)
45.000	-	slope of downspouting if a bucket elevator is placed in the center of the bin.
1.500	-	height of foundation ring (ft)
6.000	-	number of rintgs per bin
2.667	-	height of each storage ring (ft)
17.500	-	bin eave height (ft) (computed from above information)
30.000	-	bin roof slope (°)
10.000	-	auger length (ft) (used only if it determines center radius)
2.000	-	depth of the receiving pit (ft)
2.000	-	space between walls of adjacent bins (ft)
12.000	-	minimum acceptable space between two of the bins for vehicle
1.000	-	clearance between top of bin and the unloading auger discharge point
20.000	-	minimum slope of the unloading auger (°)

Bin Diameter (ft)	Minimum Fill Height (ft)	Minimum Auger Length	Required Auger Length	Center Radius (ft)	Auger Slope (°)	Actual Fill Height (ft)
12	21.96	44.57	45.00	37.94	32.53	22.20
15	22.83	46.74	47.00	39.82	32.09	22.97
18	23.70	48.79	49.00	41.65	31.78	23.81
21	24.56	50.76	51.00	43.46	31.55	24.69
24	25.43	52.68	53.00	45.25	31.38	25.60
27	26.29	54.56	55.00	47.03	31.24	26.52
30	27.16	56.42	57.00	48.80	31.12	27.46
33	28.03	58.25	59.00	50.56	31.03	28.41
36	28.89	60.07	61.00	52.32	30.95	29.37
39	29.76	61.88	62.00	53.21	30.88	29.82
42	30.62	63.68	64.00	54.96	30.82	30.79
45	31.49	65.47	66.00	56.71	30.77	31.76
48	32.36	67.25	68.00	58.46	30.72	32.74

Table 8.2 continued on next page

the 45° slope specification is used. This says nothing about the power requirements and capacity of the auger at 45° as compared to lesser slopes. However, this design is optimum for systems in which bucket elevators are to be installed initially in that a 45° downspouting slope is specified. The constant auger slope of 45° for all diameter bins makes this specification the dominant factor in determining the center radius rather than bin top clearance.

Suppose a design constraint was the minimum allowable length of the filling auger. In a situation where no less than a 60-ft portable auger is to be used and the minimum slope is not to be less than 20°, the 60-ft auger governs the design until a 33-ft-diameter bin is selected. The design is essentially like that in table 8.2 for bins equal to or greater than 30 ft in

Table 8.2. (Continued)

Bin Diameter (ft)	Chord Length (ft)	Angle Between Bins (°)	Arc Lg. Between Bins (ft)	Center Radius Circum-ference	Fraction of Bins Allowed	Bucket Elevator Discharge (ft)
12	14.00	21.26	14.08	238.39	16.93	58.91
15	17.00	24.65	17.13	250.19	14.60	61.65
18	20.00	27.78	20.20	261.71	12.96	64.35
21	23.00	30.69	23.28	273.06	11.73	67.02
24	26.00	33.39	26.37	284.31	10.78	69.68
27	29.00	36.92	29.48	295.48	10.02	72.32
30	32.00	38.28	32.60	306.59	9.40	74.96
33	35.00	40.50	35.74	317.67	8.89	77.58
36	38.00	42.59	38.89	328.71	8.45	80.21
39	41.00	45.32	42.09	334.34	7.94	81.97
42	44.00	47.19	45.27	345.34	7.63	84.59
45	47.00	48.96	48.46	356.33	7.35	87.20
48	50.00	50.64	51.67	367.30	7.11	89.81

	First Approximation			Final Decision			
Bin Diameter (ft)	Max. No. of Bin in Circle	Receiving Angle (°)	Unloading Space (ft)	Max. No. of Bins in Circle	Unloading Space (ft)	Per Bin Capacity (bu)	Total Capacity (bu)
12	16	41.05	14.60	16	14.60	1448	23162
15	14	39.53	11.93	13	27.31	2262	29405
18	12	54.39	20.07	12	20.07	3257	39086
21	11	53.12	17.86	11	17.86	4433	48768
24	10	59.47	20.88	10	20.88	5791	57906
27	10	36.74	2.64	9	28.72	7329	65958
30	9	53.74	14.11	9	14.11	9048	81430
33	8	76.49	29.59	8	29.59	10948	87583
36	8	61.87	17.78	8	17.78	13029	104231
39	7	88.09	34.99	7	34.99	15291	107035
42	7	76.86	26.32	7	26.32	17734	124136
45	7	66.23	16.97	7	16.97	20358	142503
48	7	56.17	7.04	6	45.86	23162	138974

diameter. If the 60-ft auger is extended at a minimum angle of 45°, the specified auger length controls the design for all bins less than 45 ft in diameter. For diameters of 45 ft or more, the top and horizontal clearances require a longer auger than that specified.

Expansion

One of the primary criteria that must be met in specifying a design is that of potential capacity. Because capacity is affected greatly by the center radius, design should be based on capacity rather than on the availability of a portable auger that may restrict future expansion. Circular designs may, however, allow for expansion beyond the initial circle. Expansion of circular systems beyond the inner circle involves two-stage grain handling when portable augers are used. The usual procedure is to add an outer ring of bins that may be unloaded directly into the center pit. Roof augers are used to receive grain from the inner bin circle and convey the material to the outer bin circle (fig. 8.11). Roof augers also can be used with inner circle bins if required to meet power and height constraints.

The replacement of the portable auger by a bucket elevator is a design alternative. Likewise, a portable dryer may be used with either a separate wet holding bin or with one of the storage bins used temporarily for wet grain storage. The receiving pit also can be used as a surge bin.

Summary

Circular arrangements of grain drying and storage bins offer an efficient means of handling and storing grain. This arrangement is best suited to systems in which portable handling equipment is used and in which expansion beyond the inner circle of bins is not needed. Total storage capacity is very much related to the center radius. A minimum auger slope of 45° will give the shortest portable auger length but may result in reduced conveying capacity or greater energy input. Using a specified auger length or delivery angle may dictate the entire design. The combinations of design considerations result in a situation more complex than would be expected initially.

Flat Storage Systems

The term "flat storage" generally refers to rectangular-shaped structures with relatively low height-to-width ratios (Loewer et al., 1988, 1989). Flat storage systems are often multipurpose structures that utilize permanent floors capable of supporting the weight of a truck or tractor. Generally, portable conveying equipment is used for filling and emptying these types of facilities, and duct-type aeration systems are employed, most often above the floor. Usually, grain is not placed directly against the wall of the building but against a portable structure indented from the edge of the facility. At the farm level, flat storage systems often are viewed as being "temporary", "stop gap", or "last choice" storage measures rather than as permanent "first choice" types of facilities. However, flat storage is often used commercially by utilizing a relatively large floor area with the grain covered by plastic rather than a conventional roof. Although the thrust of the following discussion will be directed toward farm-size facilities, the principles apply equally to the larger scaled units (Siebenmorgen et al., 1986). Thus, the objectives of this chapter are to:

1. Discuss the advantages and disadvantages of farm-sized flat storage systems.
2. Present geometric design considerations for flat storage facilities.
3. Describe the influence of geometric configurations on cost of construction.

Advantages and Disadvantages

Flat storage systems generally are viewed at the farm level as being multipurpose facilities. For example, the same structure might be used at different times as a shop, as warehouse storage for sacked material (fertilizer, feed, grain, etc.) or as bulk grain storage. Scheduling problems arise when there are conflicting demands on the facility. For example, the farmer may have need for a shop facility at the same time that grain should be stored. Thus, it is highly desirable that flat storage design be able to satisfy both of these demands. For example, one end of the building may be used as a shop, while the other is used to store grain, with the understanding that one use will have priority over the other.

If grain is to be stored directly against the walls of the flat storage facility, added wall strength must be incorporated into the design. Especially important are the shear forces near the floor. The forces against the building walls or intermediate partitions are much greater than would appear to the casual observer. Adding to the structural strength problem is accessibility in that flat storage systems generally have relatively large doorways. These doorways are either allowed to open freely or portable partitions are placed across them. In either case, there must be sufficient strength across the doorway to keep the building from pulling apart.

Aeration of flat storage systems is more difficult than with conventional storage bins because duct systems must be used in combination with varying grain depth. Portable aeration systems generally are preferred because they can be moved readily when the building is used for other purposes. However, permanent ducting flush with the floor may be incorporated into the design, so long as provisions are made to support vehicle weight during filling or unloading.

A major consideration in using flat storage is filling and unloading the structure. In farm systems, portable augers or flight conveyors are used most commonly. However, pneumatic conveyers are a viable alternative when the additional cost can be justified. Care must be taken to fill the vertical building space completely before moving the conveyor because it may be impossible to reach certain portions of the structure after grain occupies all of the floor area.

Insect and rodent control is more difficult than when using round metal bins. These pests have better access to the grain, and treatments such as fumigation are comparatively more difficult.

A very important consideration is that flat storage should be used only with dry grain. Flat storage systems have little or no capacity to dry grain so that introduction of any excessively wet grain can present major problems in locating, isolating, and removing the affected material.

Despite its disadvantages, flat storage is considered a viable alternative because:

1. It may offer cost advantages in some instances, especially if the structure can be used for many purposes.
2. If it's built exclusively for grain, flat storage can be designed to avoid many of the aeration and materials handling problems associated with multipurpose structures.

The key to economic feasibility of flat storage lies in being able to store a large quantity of grain per unit of floor area.

Design Considerations

The capacity of a flat storage system (fig. 8.18) may be computed by adding together the several geometric shapes that compose a pile of grain while recognizing that the grain slopes may not be uniform. Note that capacity is measured volumetrically rather than as a function of grain weight. Thus, this design procedure does not account for "packing" of the grain over time. (See Thompson et al., 1990, 1992 for a discussion of the effects of packing upon flat storage capacity and the software package WPACKING.) Either length or width of the pile will be the limiting factor governing grain height. The following equation may be used to determine grain volume (table 8.3, fig. 8.19 - 8.21):

$$\begin{aligned} BVC = \ & C * \left[(L*W*V) + \left(W^3 * TAN(t)\right)/6 \right. \\ & \left. + \left((L-W) * W^2 * TAN(t)\right)/4 \right] \end{aligned} \tag{8.23}$$

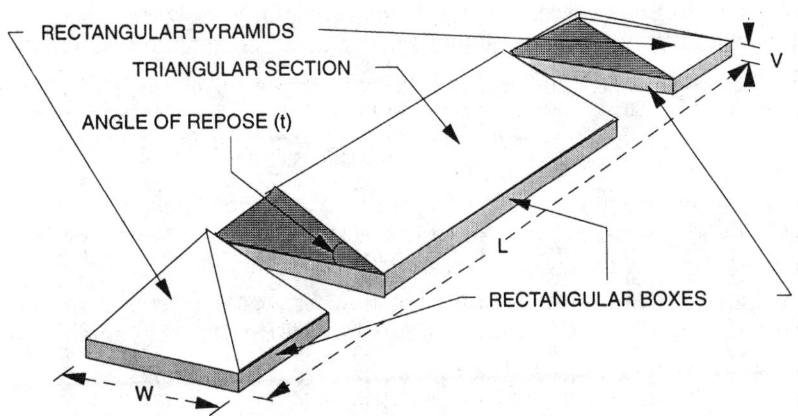

Figure 8.18–Flat storage facility.

where

BVC = volumetric capacity for rectangular flat storage
C = conversion factor (1 for SI units, 0.8 for bu if L, W and V in ft)
L = length of structure, where L ≥ W
W = width of structure
t = angle of repose for the type of grain being stored (°)
V = useable wall height

The "shape" of the grain pile used in equation 8.23 is composed of a rectangular solid, a triangular solid, and two rectangular pyramids (fig. 8.18). The base formed when the two rectangular pyramids are placed together forms a square. The slope of the grain is referred to as the angle of repose. If the structure is filled uniformly, the longitudinal measurement of the base of each rectangular pyramid is one half of the width of the structure (assuming the width is less than or equal to the length). This is because the width governs the height that the grain may be placed in the structure. Therefore, the length of the triangular solid is the length of the facility minus its width. The dimensions and volumes of the different sections may be computed as follows:

$$H = TAN(t) * (W/2) \tag{8.24}$$

$$T = V + H \tag{8.25}$$

Table 8.3. Capacities of various sizes of flat storage systems

Repose Angle	Dimensions (ft)			Total Capacity		Ratios of Volume (ft^3) to	
(°)	Length	Width	Wall	(ft^3)	(bu)	Width	Area
25	100	50	0	24,287	19,429	486	4.86
25	100	50	1	29,287	23,429	586	5.86
25	100	50	2	34,287	27,429	686	6.86
25	100	50	3	39,287	31,429	786	7.86
25	100	50	4	44,287	35,429	886	8.86
25	100	75	0	49,181	39,345	656	6.56
25	100	75	1	56,681	45,345	756	7.56
25	100	75	2	64,181	51,345	856	8.56
25	100	75	3	71,681	57,345	956	9.45
25	100	75	4	79,181	63,345	1056	10.56
25	100	100	0	77,718	62,174	777	7.77
25	100	100	1	87,718	70,174	877	8.77
25	100	100	2	97,718	78,174	977	9.77
25	100	100	3	107,718	86,174	1077	10.77
25	100	100	4	117,718	94,174	1177	11.77
30	100	50	0	30,070	24,056	601	6.01
30	100	50	1	35,070	28,056	701	7.01
30	100	50	2	40,070	32,056	801	8.01
30	100	50	3	45,070	36,056	901	9.01
30	100	50	4	50,070	40,056	1001	10.01
30	100	75	0	60,892	48,714	812	8.12
30	100	75	1	68,392	54,714	912	9.12
30	100	75	2	75,892	60,714	1012	10.12
30	100	75	3	83,392	66,714	1112	11.12
30	100	75	4	90,892	72,714	1212	12.12
30	100	100	0	96,225	76,980	962	9.62
30	100	100	1	106,225	84,980	1062	10.62
30	100	100	2	116,225	92,980	1162	11.62
30	100	100	3	126,225	100,980	1262	12.62
30	100	100	4	136,225	108,980	1362	13.62
30	100	50	0	30,070	24,056	601	6.01
35	100	50	1	41,469	33,175	829	8.29
35	100	50	2	46,469	37,175	929	9.29
35	100	50	3	51,469	41,175	1029	10.29
35	100	50	4	56,469	45,175	1129	11.29
35	100	75	0	73,850	59,080	985	9.85
35	100	75	1	81,350	65,080	1085	10.85
35	100	75	2	88,850	71,080	1185	11.85
35	100	75	3	96,350	77,080	1285	12.85
35	100	75	4	103,850	83,080	1385	13.85
35	100	100	0	116,701	93,361	1167	11.67
35	100	100	1	126,701	101,361	1267	12.67
35	100	100	2	136,701	109,361	1367	13.67
35	100	100	3	146,701	117,361	1467	14.67
35	100	100	4	156,701	125,361	1567	15.67
25	200	50	0	53,431	42,745	1069	5.34
25	200	50	1	63,431	50,745	1269	6.34
25	200	50	2	73,431	58,745	1469	7.34
25	200	50	3	83,431	66,745	1669	8.34
25	200	50	4	93,431	74,745	1869	9.34
25	200	75	0	114,755	91,804	1530	7.65
25	200	75	1	129,755	103,804	1730	8.65
25	200	75	2	144,755	115,804	1930	9.65

Table 8.3 continued on next page.

Table 8.3. (Continued)

Repose Angle	Dimensions (ft)			Total Capacity		Ratios of Volume (ft³) to	
(°)	Length	Width	Wall	(ft³)	(bu)	Width	Area
25	200	75	3	159,755	127,804	2130	10.65
25	200	75	4	174,755	139,804	2330	11.65
25	200	100	0	194,295	155,436	1943	9.71
25	200	100	1	214,295	171,436	2143	10.71
25	200	100	2	234,295	187,436	2343	11.71
25	200	100	3	254,295	203,436	2543	12.71
25	200	100	4	274,295	219,436	2743	13.71
30	200	50	0	66,155	52,924	1323	6.62
30	200	50	1	76,155	60,924	1523	7.62
30	200	50	2	86,155	68,924	1723	8.62
30	200	50	3	96,155	76,924	1923	9.62
30	200	50	4	106,155	84,924	2123	10.62
30	200	75	0	142,082	113,666	1894	9.47
30	200	75	1	157,082	125,666	2094	10.47
30	200	75	2	172,082	137,666	2294	11.47
30	200	75	3	187,082	149,666	2494	12.47
30	200	75	4	202,082	161,666	2694	13.47
30	200	100	0	240,563	192,450	2406	12.03
30	200	100	1	260,563	208,450	2606	13.03
30	200	100	2	280,563	224,450	2806	14.03
30	200	100	3	300,563	240,450	3006	15.03
30	200	100	4	320,563	256,450	3206	16.03
30	200	50	0	66,155	52,924	1323	6.62
35	200	50	1	90,232	72,186	1805	9.02
35	200	50	2	100,232	80,186	2005	10.02
35	200	50	3	110,232	88,186	2205	11.02
35	200	50	4	120,232	96,186	2405	12.02
35	200	75	0	172,317	137,853	2298	11.49
35	200	75	1	187,317	149,853	2498	12.49
35	200	75	2	202,317	161,853	2698	13.49
35	200	75	3	217,317	173,853	2898	14.49
35	200	75	4	232,317	185,853	3098	15.49
35	200	100	0	291,753	233,403	2918	14.59
35	200	100	1	311,753	249,403	3118	15.59
35	200	100	2	331,753	265,403	3318	16.59
35	200	100	3	351,753	281,403	3518	17.59
35	200	100	4	371,753	297,403	3718	18.59
25	300	50	0	82,575	66,060	1652	5.51
25	300	50	1	97,575	78,060	1952	6.51
25	300	50	2	112,575	90,060	2252	7.51
25	300	50	3	127,575	102,060	2552	8.51
25	300	50	4	142,575	114,060	2852	9.51
25	300	75	0	180,330	144,264	2404	8.01
25	300	75	1	202,830	162,264	2704	9.01
25	300	75	2	225,330	180,264	3004	10.01
25	300	75	3	247,830	198,264	3304	11.01
25	300	75	4	270,330	216,264	3604	12.01
25	300	100	0	310,872	248,697	3109	10.36
25	300	100	1	340,872	272,697	3409	11.36
25	300	100	2	370,872	296,697	3709	12.36
25	300	100	3	400,872	320,697	4009	13.36
25	300	100	4	430,872	344,697	4309	14.36
30	300	50	0	102,239	81,791	2045	6.82

Table 8.3 continued on next page.

where

H = height of grain within the triangular section

T = total height of the grain pile relative to the floor

Rectangular Box:

$$VR = C * L * W * V \tag{8.26}$$

where VR = volume of the rectangular section.

Triangular Section:

$$VT = C * (L - W) * W * H / 2 \tag{8.27}$$

where VT = volume of triangular section.

Table 8.3. (Continued)

Repose Angle	Dimensions (ft)			Total Capacity		Ratios of Volume (ft³) to	
(°)	Length	Width	Wall	(ft³)	(bu)	Width	Area
30	300	50	1	117,239	93,791	2345	7.82
30	300	50	2	132,239	105,791	2645	8.82
30	300	50	3	147,239	117,791	2945	9.82
30	300	50	4	162,239	129,791	3245	10.82
30	300	75	0	223,272	178,618	2977	9.92
30	300	75	1	245,772	196,618	3277	10.92
30	300	75	2	268,272	214,618	3577	11.92
30	300	75	3	290,772	232,618	3877	12.92
30	300	75	4	313,272	250,618	4177	13.92
30	300	100	0	384,900	307,920	3849	12.83
30	300	100	1	414,900	331,920	4149	13.83
30	300	100	2	444,900	355,920	4449	14.83
30	300	100	3	474,900	379,920	4749	15.83
30	300	100	4	504,900	403,920	5049	16.83
30	300	50	0	102,239	81,791	2045	6.82
35	300	50	1	138,995	111,196	2780	9.24
35	300	50	2	153,995	123,196	3080	10.27
35	300	50	3	168,995	135,196	3380	11.27
35	300	50	4	183,995	147,196	3680	12.27
35	300	75	0	270,783	216,627	3610	12.03
35	300	75	1	293,283	234,627	3910	13.03
35	300	75	2	315,783	252,627	4210	14.03
35	300	75	3	338,283	270,627	4510	15.03
35	300	75	4	360,783	288,627	4810	16.03
35	300	100	0	466,805	373,444	4668	15.56
35	300	100	1	496,805	397,444	4968	16.56
35	300	100	2	526,805	421,444	5268	17.56
35	300	100	3	556,805	445,444	5568	18.56
35	300	100	4	586,805	469,444	5868	19.56

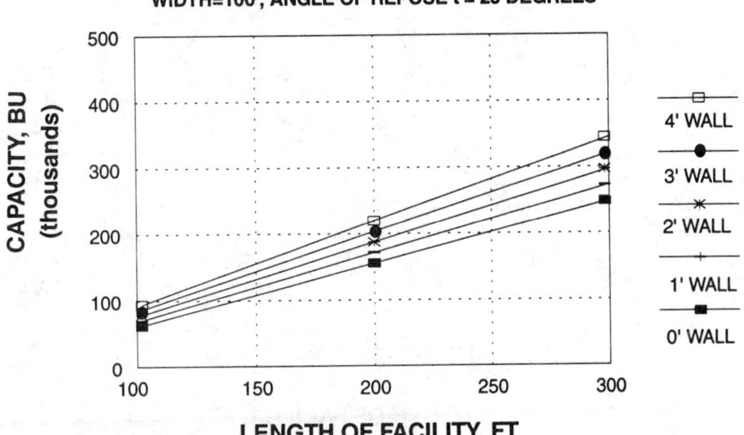

Figure 8.19–Bushel capacity of a flat storage facility with various building dimensions and a grain angle of repose of 25° (Loewer et al., 1989).

Rectangular Pyramid Section:

$$VP = C * W^2 * H / 6 \tag{8.28}$$

where VP = volume of each of the pyramid sections.

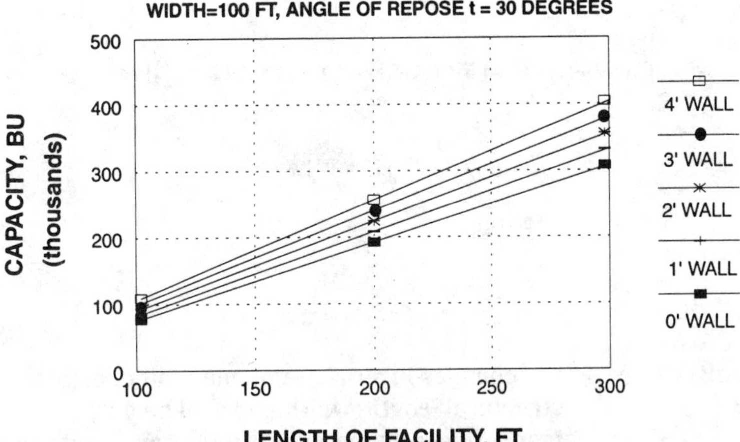

Figure 8.20–Bushel capacity of a flat storage facility with various building dimensions and a grain angle of repose of 30° (Loewer et al., 1989).

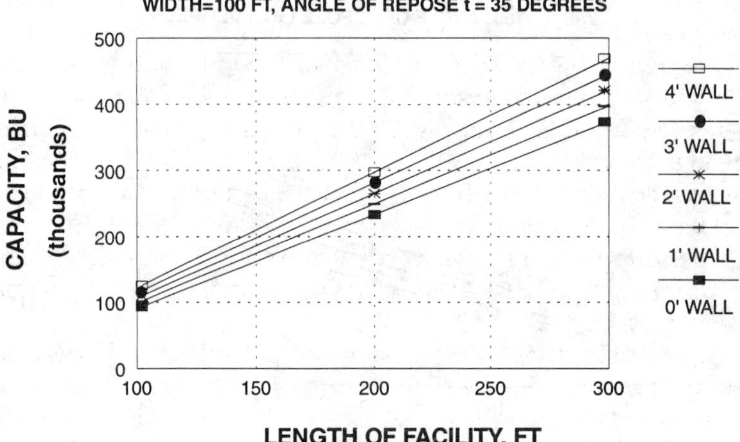

Figure 8.21–Bushel capacity of a flat storage facility with various building dimensions and a grain angle of repose of 35° (Loewer et al., 1989).

Equation 8.23 may be derived by substituting the value for "H" in equations 8.27 and 8.28, and summing equations 8.26 to 8.28 allowing for two rectangular pyramids (fig. 8.18). The incremental changes in the storage volume (BVC), as influenced by changes in the system length (L), width (W), and wall height (V), may be found by differentiating equation 8.23 with respect to L, W, and V.

$$\mathrm{dBVC/dL} = \left[(W * V) + \left(W^2 * \mathrm{TAN}(t)\right)/4\right] * C \tag{8.29}$$

$$\mathrm{dBVC/dW} = \left[(L * V) + \left(\mathrm{TAN}(t) * \left(2 * L * W - W^2\right)\right)/4\right] * C \tag{8.30}$$

$$\mathrm{dBVC/dV} = [L * W] * C \tag{8.31}$$

where

$W \leq L$

dBVC/dL,
dBVC/dW
and dBVC/dV = the changes in storage volume /unit change in structural length, width and wall height, respectively, as measured in the same units as L, W, and V with C = 1 (C=0.8 for obtaining changes in bushel capacity when lengths are measured in ft)

Equations 8.24 - 8.26 allow for the examination of the effects of changing "L", "W", and "V" on total grain volume. The unit change in volume per unit change in length (dBVC/dL) for a given flat storage system depends on its initial width, wall height, and grain angle of repose with changes becoming more pronounced as system width increases (equation 8.29 and fig. 8.22). The unit change in volume per unit change in width (dBVC/dW) is influenced also by the initial length of the system with the effects of width becoming relatively less pronounced as it increases (equation 8.30 and fig. 8.23). Unit volume change per unit increase in wall height (dBVC/dV) increases linearly in proportion to the initial system values of length and width (equation 8.31 and fig. 8.24).

An important consideration in selection of facility dimensions is that the optimum storage volume to floor surface area ratio is obtained when a square floor area is used. This may be verified mathematically by ignoring "C" and substituting "L=A/W" in equation 8.23 ("A" is floor area), differentiating with respect to "W," and solving for the optimum condition (i.e., where dBVC/dW=0). However, in most flat storage situations, width is less than length. The extent to which the volume to surface area ratio is influenced may be determined by rearranging equation 8.23 and dividing both sides by the floor area "L∗W" giving the following expression:

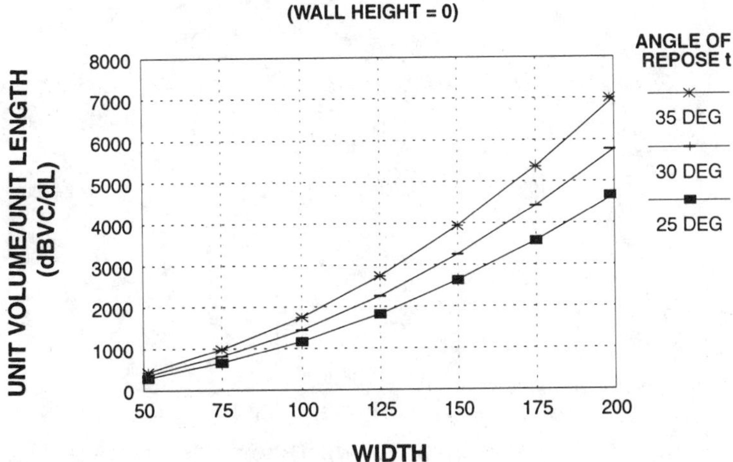

Figure 8.22–Change in volume per unit length for a flat storage facility and various angles of repose for grain (Loewer et al., 1989).

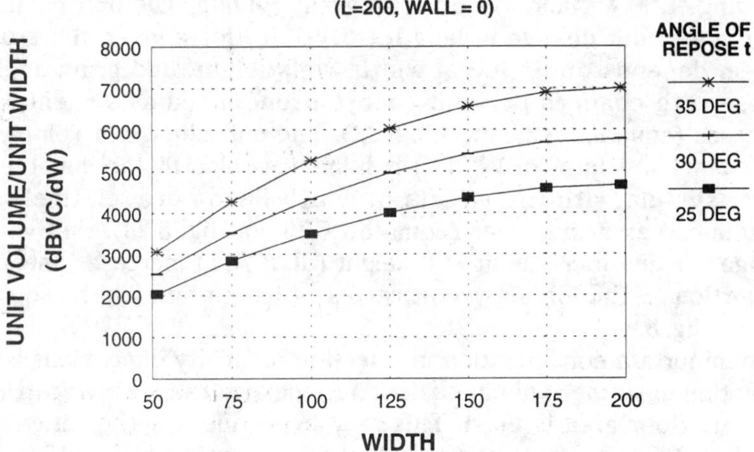

Figure 8.23–Change in volume per unit width for a flat storage facility and various angles of repose for grain (Loewer et al., 1989).

$$F = V + \left(TAN(t)\right) * W^2 / \left(6L\right) + \left(TAN(t)\right) * W / 4 - \left(TAN(t)\right) * W^2 / \left(4L\right) \quad (8.32)$$

where

F = volume to surface area ratio [BVC/(L*W)]

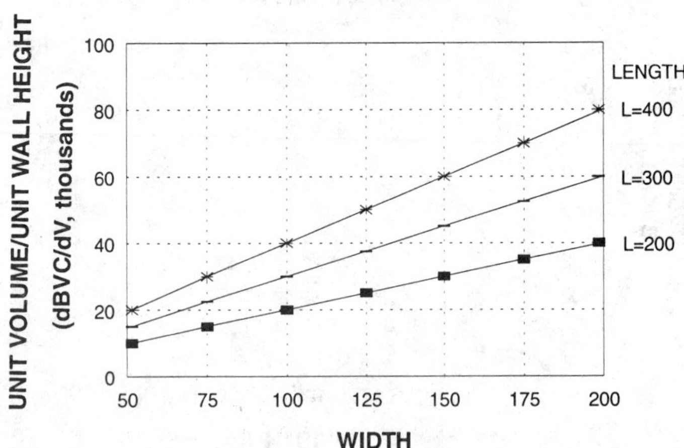

Figure 8.24–Change in volume per unit wall height for a flat storage facility with various lengths of building (Loewer et al., 1989).

Taking partial derivatives of equation 8.32 gives the following:

$$dF/dL = -(TAN(t)) * W^2 / (6L^2) + (TAN(t)) * W^2 / (4L^2)$$
$$= (TAN(t)) * W^2 / (12L^2) \quad (8.33)$$
$$dF/dW = (TAN(t)) * W / (3L) + (TAN(t)) / 4 - (TAN(t)) * W / (2L)$$
$$= (TAN(t)) * ((1/4) - (W/6L)) \quad (8.34)$$
$$dF/dV = 1 \quad (8.35)$$
$$dF/dt = (SEC^2(t)) * (3W * L - W^2) / (12L) \quad (8.36)$$

where dF/dL, dF/dW, dF/dV, dF/dt are the changes in the volume to floor surface area ratio as influenced by a unit change in length L, width W, wall height V, and angle of repose t.

Equation 8.33 (fig. 8.25) shows that the unit change in volume to floor surface area ratio per unit of length (dF/dL) decreases with facility length and increases with width and the grain angle of repose. Equation 8.34 indicates that the unit change in volume to floor surface area ratio per unit increase in width (dF/dW) increases with the grain angle of repose and length and decreases with width and reaches a maximum when "W" is equal to "L" (by definition "W" must be no greater than "L"). (figure 8.26 illustrates this relationship for one angle of repose).

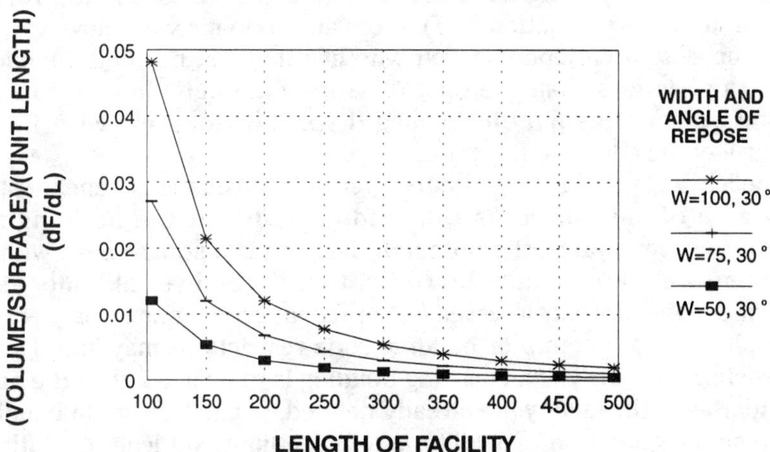

Figure 8.25–Volume to surface ratio as affected by length of building (Loewer et al., 1989).

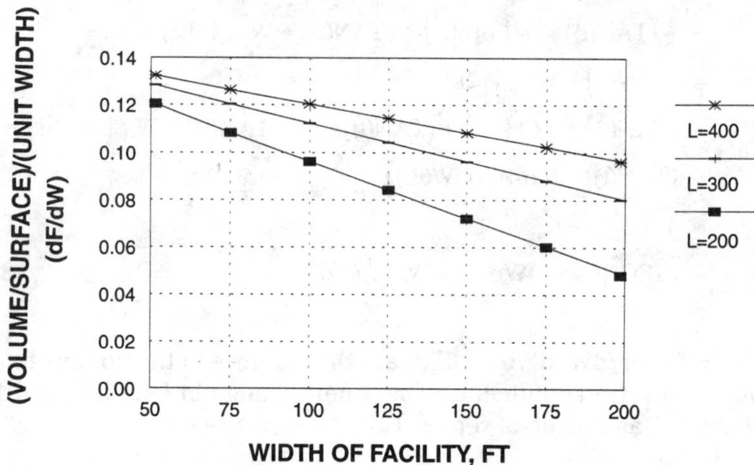

Figure 8.26–Volume to surface ratio as affected by width of building (Loewer et al., 1989).

Equation 8.35 indicates that a one-unit increase in wall height will also increase the volume to floor area ratio by one, which is by far the largest of the responses. The unit change in volume to floor surface area ratio per unit increase in grain angle of repose (dF/dt) increases with grain angle of repose and initial system length (equation 8.36). It increases and then decreases as initial facility width increases (the conditions in figure 8.27 show only the increase).

Retaining Walls and Economic Considerations

Increases in wall height result in the greatest increase in the volume to floor area ratio (equation 8.35). From an economic view, however, the optimum design is dependent on whether the cost per unit volume of "going up" is less than the cost of "spreading out". In addition, the "going up" cost usually increases disproportionately with height because of the added wall loads.

The least-cost flat storage system per unit of volume is dependent on the value of base area. As the land available at the building site approaches zero value, the optimum design has a square base with no retaining wall. Frequently, however, there are physical limits as to spatial expansion that, in essence, represent an economic barrier. For example, land ownership or building code restrictions may limit facility dimensions. Similarly, if an existing building is to be used, then the outer boundaries of the facility are already defined. In addition, grain handling equipment or aeration capability may place limits on length, width, or height of the grain mass. When facility width is the limiting factor,

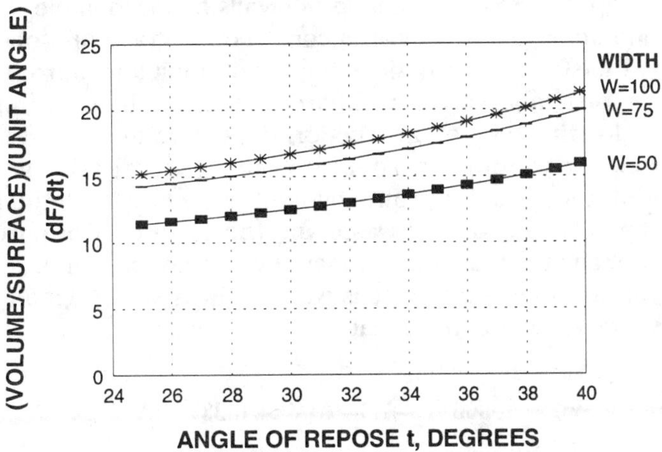

Figure 8.27–Volume to surface ratio as affected by grain angle of repose (Loewer et al., 1989).

additional capacity may be obtained by increasing the height of grain on the retaining wall (the retaining wall may be the outside wall of the facility or an internal wall). The force of grain against a wall increases disproportionately with wall height. Thus, the cost of the wall will also increase disproportionately. The economically optimum wall height is reached when one of the following conditions is first met:

1. Design capacity is satisfied.
2. Maximum allowable filling height is reached.
3. The marginal cost of increasing retaining wall height exceeds the cost of other storage alternatives.

Force Considerations. Determining lateral static wall pressure for grain in shallow bins involves a detailed engineering analysis and is beyond the scope of this book. At the time of this writing, the American Society of Agricultural Engineers (ASAE) is preparing a new engineering standard dealing with this issue. The concept that needs to be understood is that force along the walls increases disproportionately with height of grain. That is, doubling the height of grain against the grain wall significantly will more than double the force against the wall. In the idealized case, the cost of a cantilever wall for resisting lateral grain forces is proportional to the (5/2) power of maximum grain depth.

Structural Cost Considerations. For an estimate of cost versus wall height, the actual construction costs of 100-ft lengths of movable grain storage walls of four heights were estimated from plans prepared by the Midwest Plan Service (1983, 1986) that used the ASAE standard that was current at that time (Loewer et al., 1989). Overturning moments due to lateral grain forces are resisted by transmitting vertical forces of the grain on floor panels to the adjoining sidewalls by diagonal tie rods. The sidewalls and floor panels consist of construction-grade plywood nailed to studs and joists, with provisions for distributing forces from the tie-rods. Prices for lumber, steel, and hardware were obtained from local vendors, and estimates of construction time were multiplied by local hourly rates for welders, carpenters, and general laborers. Cost estimates did not allow for profits by contractors, but the totals were increased by 10% to allow for waste and the extra cost of purchasing materials in standard lots. The estimated costs are shown in table 8.4. The cost per unit wall length versus wall height is shown graphically in figure 8.28 and represents the equation:

$$CWL = 7.6866*V - 1.3436*V^2 + 0.28741*V^{(5/2)} \qquad (8.37)$$

where
CWL = cost (\$/ft of wall length)
V = wall height up to 12 ft

This least-squares regression model crosses the origin, implying that the cost of "no wall" is zero. For short walls, costs should be nearly proportional to height because standard lumber dimensions and spacings are more than adequate for the grain loads. As wall height increases in the range of 6 to 10 ft, more of the wall components are loaded to near their design capacity, providing more economy relative to height. With greater heights, the load bearing components contribute a greater proportion of the total cost and can be designed nearly to match their required capacity. The third term in equation 8.37 has V to the 5/2 power to estimate the disproportionate increase of materials required for higher walls.

Table 8.4. Cost for vertical wall, ($) based on a 100-ft length*

	Wall Height (ft)			
Source of Expense	6	8	10	12
Steel and hardware	220.15	248.26	371.75	505.51
Lumber and plywood	934.33	1191.83	1653.00	2216.67
Tie rods @ $28 each	700.00	700.00	700.00	700.00
Carpentry and assembly	270.00	306.00	360.00	396.00
TOTAL	2124.48	2446.09	3084.75	3818.17
TOTAL + 10%	2336.93	2690.70	3393.22	4199.99
Per unit length per ft	23.37	26.91	33.93	42.00

* Loewer et al., 1989.

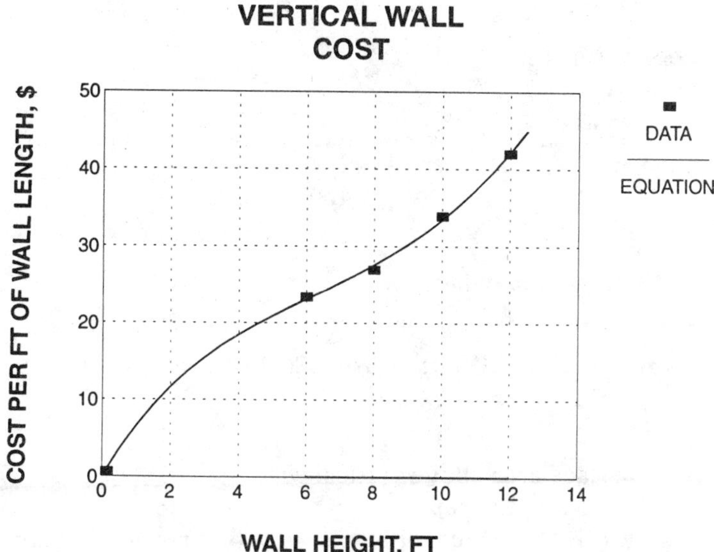

Figure 8.28–Approximate vertical wall cost (Loewer et al., 1989).

Economic Comparisons. Economic caparisons are dependent, in part, on establishing a "base line" among various alternative systems. The base line conditions may not be applicable to all situations, and, as with any economic comparison, changes in relative and absolute prices tend to make absolute price determinations obsolete rather quickly. Given these conditions, for this analysis it will be assumed that:

1. Lengths and widths of the floor are predetermined.
2. Factors to be considered in computing the per-volume cost of the storage facility are the vertical wall, concrete floor, and plastic cover for the grain.
3. Only purchase prices will be compared in that other economic factors (life of the facility, interest rates, alternative uses of the structure, etc.) will be considered the same for all systems.
4. Additional costs that might be associated with corners of the floor will not be considered.

Given the volume (equation 8.23), perimeter (equation 8.38), floor area (equation 8.39), and top surface area (equation 8.40) of a facility with a specified length, width, wall height, and grain angle of repose, the cost per unit volume may be determined.

$$P = 2*L*W \tag{8.38}$$

where

P = floor perimeter
L = floor length ($L \geq W$)
W = floor width

$$A_{floor} = L*W \tag{8.39}$$

where A_{floor} = floor area of facility.

$$A_{surface} = \left(W^2/COS\left(t\right)\right) + \left(W/COS(t)\right) * (L - W) + E*P \tag{8.40}$$

where

$A_{surface}$ = surface area above vertical wall
t = grain angle of repose
E = width of grain cover in excess of that required to cover the surface area above the vertical wall

The costs of the floor and grain cover may be estimated by multiplying the areas of each by the respective cost per unit area of material. The total purchase cost of the facility may be obtained by adding the costs of the floor, grain cover, and vertical wall. Division of total cost by equation 8.23 gives total purchase cost per unit volume of grain storage capacity. The following examples illustrate the influence of design on the economic desirability of flat storage systems given the costs and design criteria at the time this study was conducted.

Flat Storage in Existing Structures. In situations where flat storage structures are used also for other purposes, such as a farm shop, the cost of adding a vertical wall to contain the grain is the major additional expense associated with grain storage. That is, the basic structure already exists, most often with walls that are not designed to carry the loads associated with grain. When considering only the additional cost of the vertical wall (fig. 8.29), there are three economic "zones" to consider. Based on the economic conditions used in the study by Loewer et al. (1989), Zone 1 is for the vertical wall height ranging from 0 to 4 ft, Zone 2 is from 4 to 8 ft, and Zone 3 is above 8 ft. The cost per bushel increases when a vertical wall is added if there are no other cost considerations such as floor and grain cover. This cost continues to increase throughout Zone 1, stays the nearly same or decreases in Zone 2, and increases exponentially in Zone 3. The following general conclusions can be drawn with regard to flat storage in existing multi-use structures:

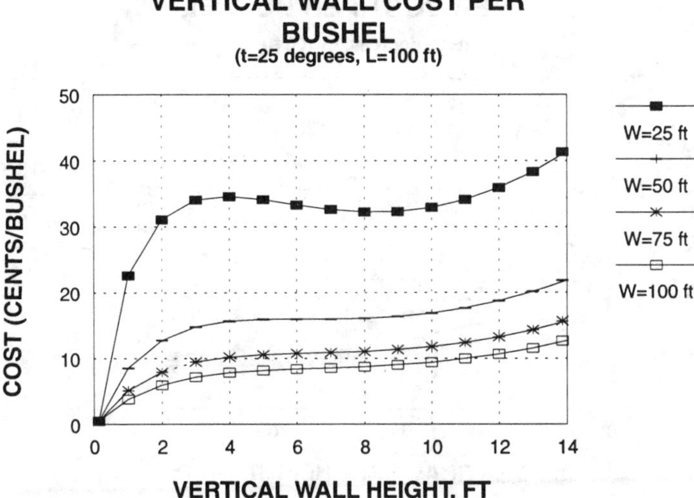

Figure 8.29–Vertical wall cost per bushel (Loewer et al., 1989).

1. The cost of adding a vertical wall cannot be justified economically if there is sufficient volume within the facility to hold the grain and the amount of space saved does not have higher alternative value.
2. On a per-bushel basis, vertical wall cost decreases or stays essentially the same for wall heights in the range of wall height from approximately 4 to 8 ft (Zone 2).

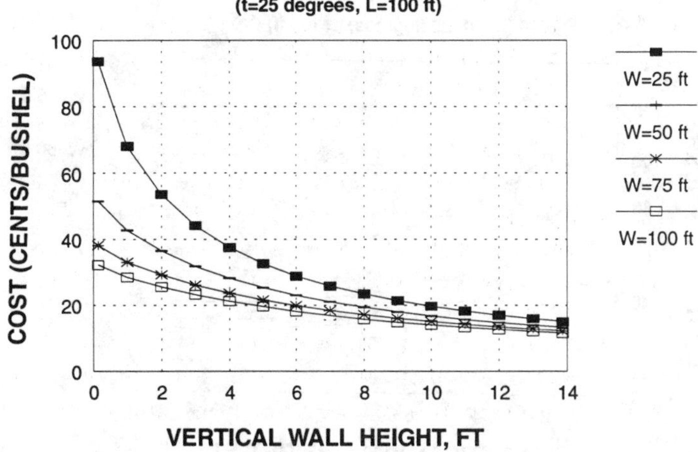

Figure 8.30–Concrete floor cost per bushel (Loewer et al., 1989).

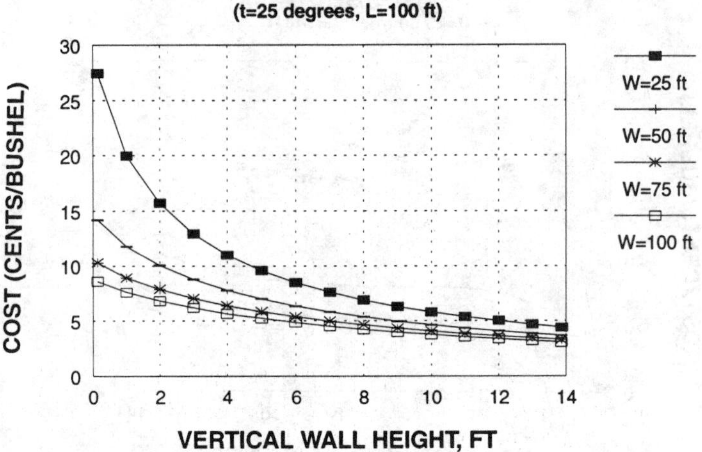

Figure 8.31–Cover cost/bushel (Loewer et al., 1989).

Temporary Flat Storage Structures. Temporary flat storage structures are defined for purposes of this discussion as being used only for grain storage. The structures consist of a floor, plastic grain cover, and optional vertical wall. In figure 8.30, the cost of a concrete floor is shown using a cost of \$2.00 /ft^2. Similarly, figure 8.31 shows the cost of a high quality PVC covering priced at \$0.45 /ft^2 and allowing 2 ft of material to extend past the perimeter of the facility. Both these costs decrease exponentially per bushel of grain as vertical wall height increases. Total wall, floor, and cover costs (fig. 8.32) continue to

FLOOR, COVER & WALL COST PER BUSHEL

(t=25 degrees, L=100 ft)

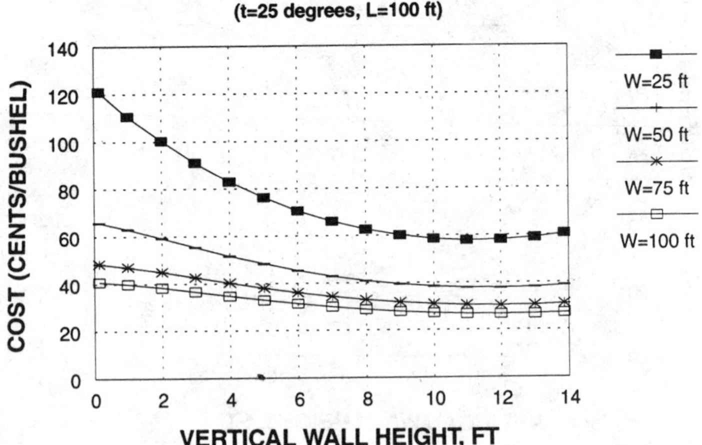

Figure 8.32–Floor, cover, and wall cost/bushel (Loewer et al., 1989).

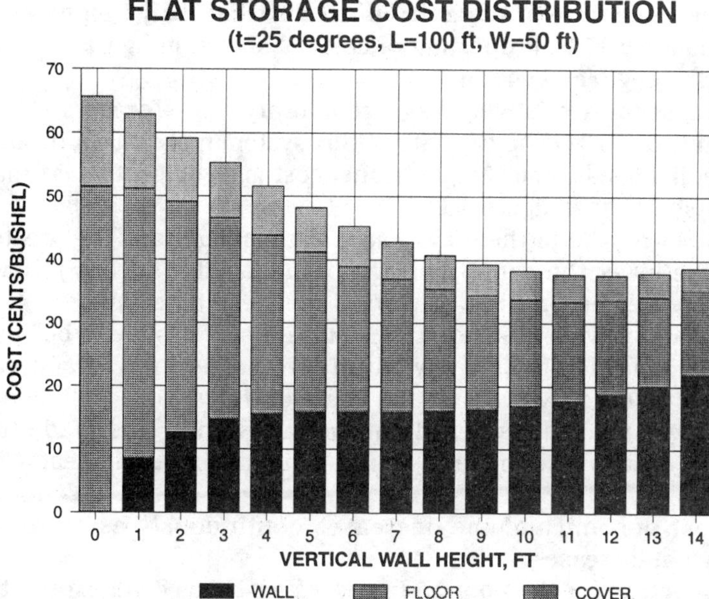

Figure 8.33–Flat storage cost distribution for a relatively small storage structure (Loewer et al., 1989).

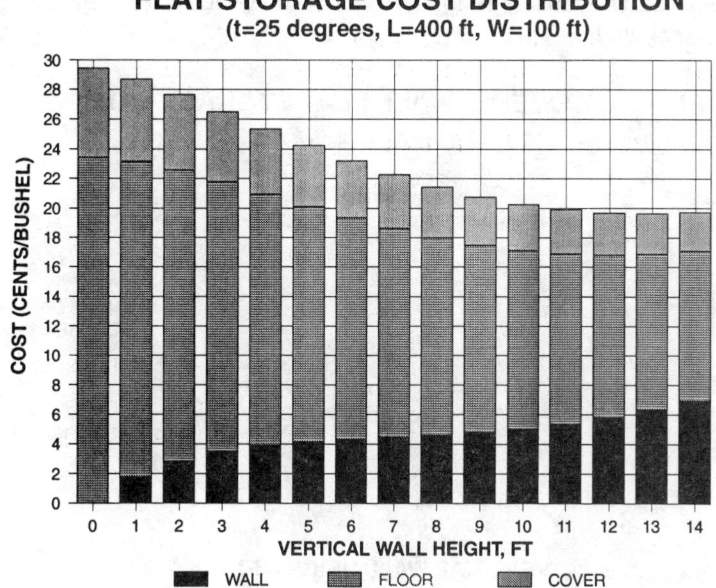

Figure 8.34–Flat storage cost distribution for a relatively large storage structure (Loewer et al., 1989).

decrease per unit of capacity until vertical wall height exceeds approximately 11 ft. Proportion of costs is shown in figure 8.33 for one particular length and width.

The cost for a relatively large, temporary, flat storage structure is given in figure 8.34. The capacity of this system makes it more suited to commercial needs, and the per-bushel cost is significantly less than the smaller structure in figure 8.33.

Angle of repose of the grain also is a consideration. The greater the angle, the lower the per-bushel cost of the facility (fig. 8.35). However, the differences become less as vertical wall height increases.

In situations where temporary flat storage systems are to be built, the following considerations apply, given the range of prices and vertical wall heights used. Again, only the costs of the vertical wall, floor, and grain cover were considered, and no allowance was made for the influence of structure shape on filling and unloading equipment.

1. Cost per unit volume decreases continuously as vertical wall height increases.
2. The costs for the floor and grain cover usually will exceed that of the vertical walls.
3. Increasing wall height decreases the floor area and maximizes the marginal increases in capacity (equation 8.35). Even in Zone 3 where the cost of the wall increases exponentially, the cost per unit volume continues to drop over the range tested, in part, because the cost savings from the floor may be applied to the vertical wall.

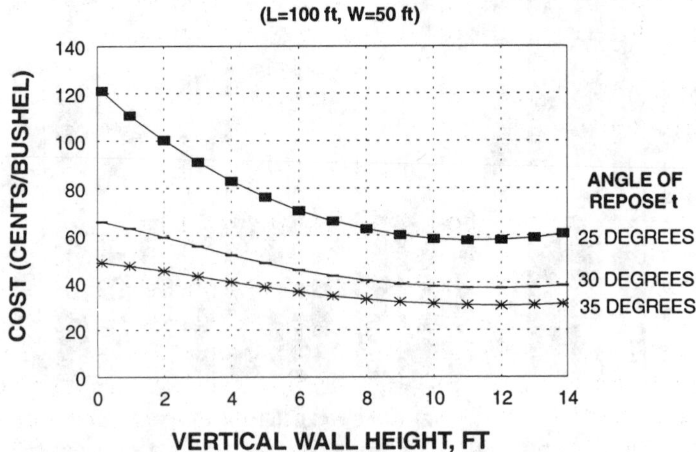

Figure 8.35–Floor, cover, and wall flat storage cost for a relatively small storage structure as affected by vertical wall height and grain angle of repose (Loewer et al., 1989).

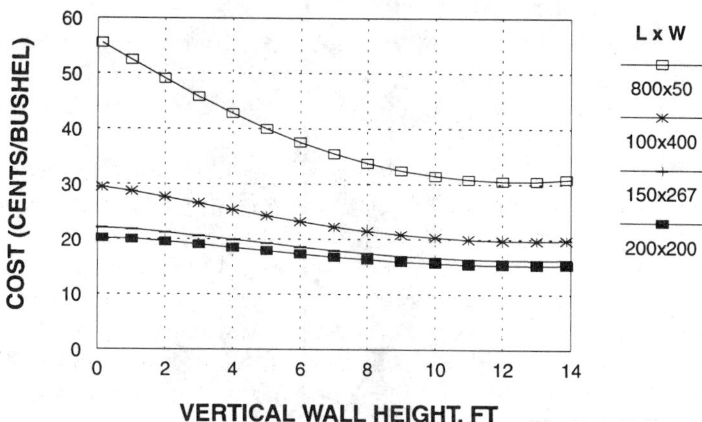

Figure 8.36–Floor, cover, and wall flat storage cost as influenced by vertical wall height and combinations of length and width that provide the same floor surface area (Loewer et al., 1989).

4. Cost per unit volume decreases generally with increases in floor area; the bigger the facility, the lower the per unit volume cost.
5. For the given wall height and length, cost per bushel decreases as width approaches length. For a given width and wall height, cost per bushel decreases as length increases. However, the lowest cost per bushel is for a square floor area (fig. 8.36).

Summary

Flat storage systems, as compared with conventional metal grain bins, are difficult to fill and unload. Grain is difficult to aerate, and there is very little drying capacity. Pest control is also more difficult. However, this analysis has shown that flat storage may have some cost per unit volume advantages, especially if the facility is to be a multipurpose structure.

Expansion of Existing Systems

Earlier in this chapter, the cardinal rules of grain handling were given concerning the "closed-loop" and "expansion" principles. Once a storage facility is constructed, expansion and closed-loop options become more limited and site-specific. When planning expansion, however, one should always consider the possibility of "starting over" in terms of location and arrangement in that this may be the best long term economic decision. Once the decision is made to expand an existing bin type facility, there are several options available.

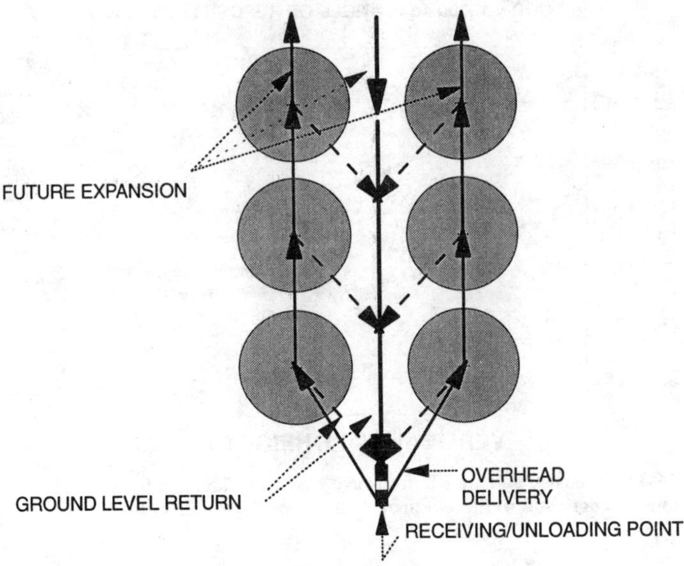

Figure 8.37–Adding expansion bins in pairs.

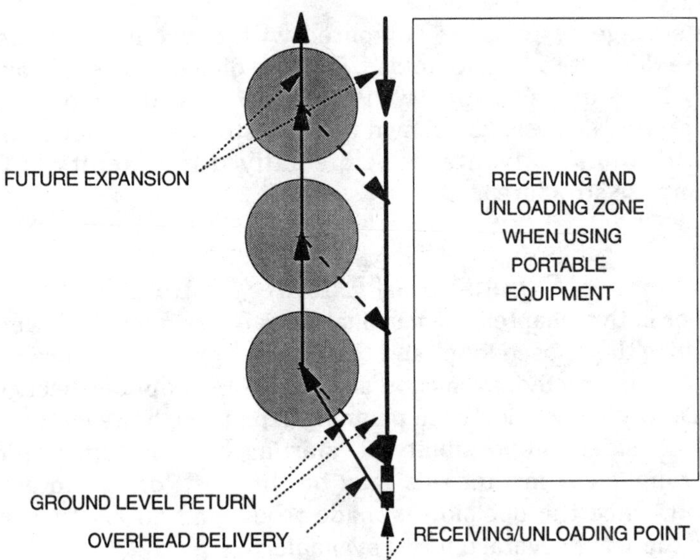

Figure 8.38–Adding expansion bins in line.

In-pairs Expansion

The most desirable type of expansion is to allow for adding bins in pairs (fig. 8.37). The primary advantage is that a common return conveyor may be used that minimizes conveying distance back to a single receiving - unloading point.

In-line Expansion

In-line expansion works best when bins are to be filled individually and unloaded with portable handling equipment (fig. 8.38). Otherwise, this method of expansion maximizes conveying distance and requires more area devoted to receiving - unloading than other approaches.

Terminal Expansion

Terminal expansion is when an expansion bin is placed in the path of the return conveyor (fig. 8.39). This approach should be avoided unless there is no possibility of adding additional bins at a later time. It has the short-term advantage of minimizing conveying length and avoiding two-stage conveying. However, it limits future options severely.

Temporary Storage

Occasionally, temporary storage is needed that may be obtained by modifying existing facilities or using low-cost short-life systems. Options that may be considered include the following:

Liners for corn cribs
Structural modifications to barns or similar facilities
Plastic covering for outdoor storage of grain

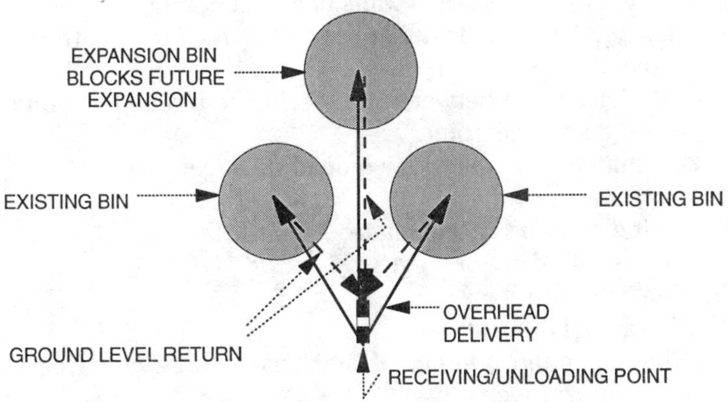

Circular metal sidewalls inside existing buildings
Silos

Most of these types of facilities are harder to fill and unload than conventional storage systems. More importantly is that the stored grain is comparatively difficult to aerate, leading to a greater risk of spoilage. These types of facilities require adequately dried grain free of insects to reduce spoilage risk and, hence, reduce actual storage cost.

Summary

The ability to expand storage capacity is very important in design. Short-term economic gain may lead to long-term economic loss when routes of expansion are blocked. Temporary storage facilities often turn out to be primary storage facilities, so great care should be taken with grain placed in temporary storage to reduce the risk of spoilage to acceptable levels.

Problems

8.1 Given the following design specifications for a circular grain system design:

(a) 1-ft minimum horizontal distance from auger to bin wall.
(b) 45° downspouting slope if a bucket elevator is placed in the center of the bin.
(c) 1.5-ft height of foundation ring.
(d) 6 rings/bin.
(e) 2.667-ft ring height.
(f) 30° roof slope.
(g) 10-ft auger length.
(h) 2.0-ft deep receiving pit.
(i) 2.0-ft distance between walls of adjacent bins.
(j) 12-ft minimum acceptable space between two of the bins for vehicle.
(k) 1.0-ft clearance between the top of the bin and the unloading auger discharge point.
(l) 20° minimum slope of the unloading auger.

For a 21-ft diameter bin, calculate:

(a) The required auger length.
(b) The center radius.
(c) The actual fill height.
(d) The maximum number of bins that can be constructed in a circular arrangement

(e) How do these answers compare with those shown in table 8.2?

8.2 How many bins can be constructed in a system with a center radius of 20 ft, an unloading width of 12 ft, 3-ft space between adjacent bin walls, and 15-ft-diameter bins?

8.3 What is the capacity in bushels of a flat storage structure 200 ft long and 75 ft wide containing corn if the structure has a 4-ft effective side wall height? The corn has an angle of repose of 28°.

8.4 Estimate the cost per unit length of a flat storage with an effective wall height of 4 ft?

8.5 A flat storage structure containing wheat is 100 ft long, 50 ft wide, and with no walls. If the building costs $\$4/ft^2$ of floor area and grain cover costs $\$0.50/ft^2$, what is the cost of the structure per bushel of storage capacity? The wheat has an angle of repose of 28° and the cover extends 2.5 ft past the perimeter.

References for all chapters begin on page 541.

9

Economics of On-Farm Drying, Storage, and Feed Processing

Consideration of economics forms the basis for most, if not all, decisions regarding selection of grain drying and storage systems. The following discussion begins with a presentation of the basic principles of production economics. This is followed by an examination of specific considerations relevant to most on-farm situations.

Principles of Production Economics

What is production economics, why is it important, and how does it relate to grain storage and handling systems? Production economics relates the benefits of selling a product to the physical and economical costs of production. All decisions are based on benefits and costs recognizing that individuals may differ in opinion concerning these factors.

Disclaimer for Price and Tax Information

This chapter presents an overview of the concepts of production economics and decision making, and how they are related. Only the competitive market situation will be presented. The primary assumption is that prices of all factors of production remain constant during the time that decisions are made.

Grain, energy, and equipment prices are used throughout this book, especially in this chapter. Relative and absolute prices are always subject to change, most notably during periods of inflation. Similarly, tax laws change which can alter a given economic analysis. Specifically, regulations relating to depreciation, investment credit, and tax brackets may differ year-to-year. As a result, the use of price or tax information often dates a publication. The purpose of this chapter is to present concepts of economic analysis in decision-making as related to grain drying and storage systems. Therefore, the reader is encouraged strongly to consider the principles used in the comparative economic analysis rather than the actual dollar values that may be given.

Optimizing Economic Return

Why is optimization important? Optimization is the process of allocating resources in order to obtain the greatest possible difference between benefit and cost. The optimum system represents the most efficient utilization of resources to accomplish an objective. The goal of most decision makers is to obtain the greatest possible total net return for personal investment.

Certain preconditions must be met in order to optimize the allocation of resources. First, there must be a value-based common denominator that has interpersonal validity. Second, there must be a set of decision rules governing the procedures for optimization. And third, mathematical second-order conditions must exist.

Common Denominator. A value-based common denominator relates benefit to cost (both are expressed in the same units for purpose of value comparison). For example, if one wishes to determine the quantity of fertilizer needed to obtain the optimum yield of corn, all other factors held constant, a common base of comparison must be established. The common measures of benefit and cost, e.g., bushels of corn and pounds of fertilizer, may not relate the factors in a meaningful way for purposes of optimization. If this is the case, "bushels" of corn and "pounds" of fertilizer must be converted to terms that are directly comparable. For example, both corn and fertilizer could be expressed in terms of energy content (thermal energy for corn, embedded energy in fertilizer), if the objective of the optimization is to maximize energy efficiency. The problem of optimization would involve maximizing the difference between the benefit (or output), e.g., thermal energy in the corn, and the cost (or input), e.g., embedded energy in the fertilizer, other factors remaining constant.

The most frequently used normative common denominator is money, and it will be the common denominator for our discussion. The reason for this choice is that the marketplace functions by assigning a relative value to all goods and services based on monetary units. For example, both benefit and cost could be expressed in dollars of corn produced and dollars of fertilizer used.

Interpersonal Validity. The value-based common denominator also must be interpersonally valid; that is, it must have the same value for everyone. In the case of money, the inference is that monetary gain or loss would affect each person in the same way regardless of individual station in life or cultural background. In other words, a dollar would represent the same value to everyone. The difficulty in using money as a value-based common denominator lies in converting many personal values to equivalent monetary values. For example, how much is a human life worth when determining an optimum engineering design for a grain bin safety ladder?

Decision Rules. Decision rules are essential for optimization. These rules relate the objectives of the optimization (generally to make the greatest possible profit) to the mathematics of the procedure. It is essential that all benefits and costs be included in quantitative terms because the objective of optimization is to maximize the difference between the two. Once the benefit and cost factors have been determined, the particular set of decision rules (i.e., optimization procedure) is used, and mathematical second-order conditions are inferred.

Second-order Conditions. Second-order conditions refer to an analysis of the rate at which a rate of change is changing. Confusing? Before examining the mathematics of this concept, it is essential that we investigate the key to profit maximization and second-order conditions — The Law of Diminishing Returns.

Law of Diminishing Returns

The *Law of Diminishing Returns* states that the addition of a variable input to fixed inputs results first in total output that increases at an increasing rate; second, in total output that increases at a decreasing rate; and third, in total output that decreases with further increases in variable input (Bradford and Johnson, 1953).

Graphically, the *Law of Diminishing Returns* appears as in figure 9.1 where "Y" equals output and "X" equals a single input, all other inputs remaining fixed. For simplicity, let's assume that input and output are

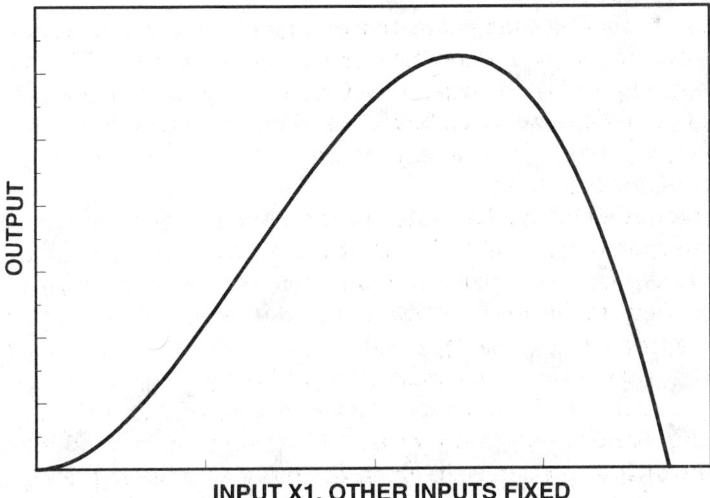

Figure 9.1–Graphical presentation of the Law of Diminishing Returns as influenced by a single input X1, all other inputs remain constant.

measured in the same units. Using our previous example, suppose that benefit (or output) is measured in dollars of corn produced while cost (or input) is measured in dollars of fertilizer applied in producing the corn. All other inputs, such as land area, weather, soil properties, crop management, etc., remain fixed. For our example, the hypothetical output-input data is given in table 9.1 and figure 9.2.

How much fertilizer should be applied to optimize the value received from corn? Figure 9.2 shows that $100 of fertilizer input results in $1,000 of corn production output. The maximum difference between benefit (corn value) and cost (fertilizer value) is $900. Let's examine the data

Table 9.1. Dollars of corn produced as a function of fertilizer dollars applied (all other inputs remain fixed)

Fertilizer Applied ($)	Corn Produced ($)	Fertilizer Increment ($)	Corn Increment ($)	Marginal Value of Fertilizer in Terms of Extra Corn Produced ($)	Average Value of Fertilizer in Terms of Extra Corn Produced ($)	Rate at which the Marginal Value of Fertilizer is Changing ($)
0.00	0.00				0.00	
		10.00	28.00	2.80		
10.00	28.00				2.80	0.48
		10.00	76.00	7.60		
20.00	104.00				5.20	0.36
		10.00	112.00	11.20		
30.00	216.00				7.20	0.24
		10.00	136.00	13.60		
40.00	352.00				8.80	0.12
		10.00	148.00	14.80		
50.00	500.00				10.00	0.00
		10.00	148.00	14.80		
60.00	648.00				10.80	−0.12
		10.00	136.00	13.60		
70.00	784.00				11.20	−0.24
		10.00	112.00	11.20		
80.00	896.00				11.20	−0.36
		10.00	76.00	7.60		
90.00	972.00				10.80	−0.48
		10.00	28.00	2.80		
100.00	1000.00				10.00	−0.60
		10.00	−32.00	−3.20		
110.00	968.00				8.80	−0.72
		10.00	−104.00	−10.40		
120.00	864.00				7.20	−0.84
		10.00	−188.00	−18.80		
130.00	676.00				5.20	−0.96
		10.00	−284.00	−28.40		
140.00	392.00				2.80	−1.08
		10.00	−392.00	−39.20		
150.00	0.00				0.00	

presented in table 9.1 to obtain a clearer understanding of how we could determine the correct answer mathematically rather than graphically.

The incremental levels of fertilizer input and corn output are given in the third and fourth columns of table 9.1, and the marginal value of fertilizer is given in column 5. Marginal value refers to the change in output value associated with the value of the last unit of input. Note that the corn value per dollar of fertilizer (column 5) increases at an increasing rate until an amount of fertilizer equal to $50 has been applied. This represents the first phase of the *Law of Diminishing Returns*. The average value of the corn per dollar of fertilizer input (column 6) also increases during this phase.

As the value of fertilizer input increases from $50 to $100, the marginal (or added) value of the fertilizer remains positive. That is, the additional money spent on the last dollar of fertilizer still yields a value in corn that is greater than the cost of the additional fertilizer. However, the magnitude of the rate is decreasing. This corresponds to the second phase of the *Law of Diminishing Returns*. Note that the average value of corn per dollar of fertilizer begins to decrease after the peak value is obtained at either $70 or $80 of input. Thus, the maximum average ratio of output and input does not yield the level of input required for optimum distribution of resources.

Increases in fertilizer input greater than $100 result in a decreased total value of corn. Physically, this might correspond to the beginning of toxicity associated with the fertilizer. The decrease in corn value

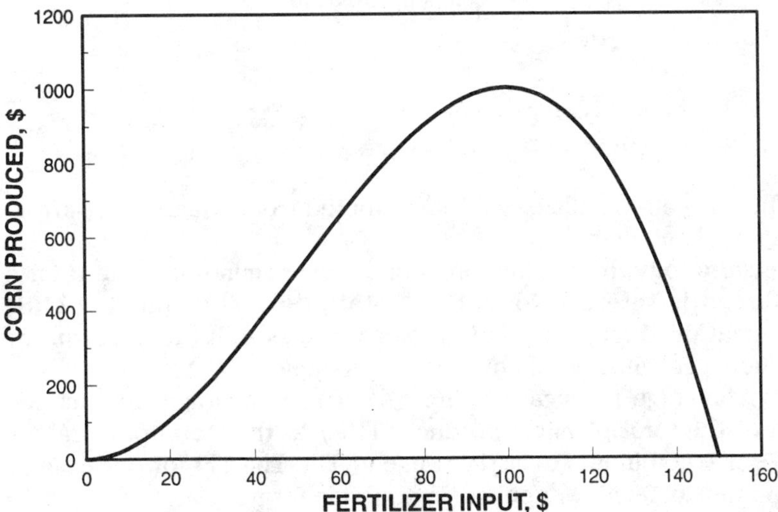

Figure 9.2–Example relationship between the value of corn produced and value of fertilizer applied, all other inputs are assumed to remain constant.

associated with an increase in fertilizer input corresponds to the third phase of the *Law of Diminishing Returns*. Notice, however, that the total value of the corn remains positive until an amount of fertilizer equal to $150 has been applied.

Given this example, what decision rule should be followed? The basic rule of optimization is that we should continue to add a unit of input (or cost) so long as we continue to receive even greater output (or benefit). The slope at any point of the curve shown in figure 9.2 corresponds to the rate at which benefit, e.g., value of corn, is changing with added cost, e.g., value of fertilizer. For the above example, additional fertilizer should not be added beyond the point where the slope of the curve becomes zero; that is, where additional benefit is not received. We will now examine what economists refer to as the stages of production and how they influence decision-making.

Production Functions

A production function relates the quantity of a product produced to the level of inputs associated with variable factors of production. Production functions are based on the *Law of Diminishing Returns*, and decisions are based on this law.

The units of a production function represent physical measures of output and input. Using our earlier example, the production function could relate bushels of corn produced per pound of fertilizer applied (all other factors of production remaining constant). Suppose the following production function described this relationship:

$$Y = 0.00375\ p^2 - 0.000005\ p^3 \qquad (9.1)$$

where

Y = bushels of corn produced
p = pounds of fertilizer applied

The corn and fertilizer values computed from equation 9.1 are given in table 9.2 and figure 9.3.

Terminology. In production economics terminology, the total corn produced is referred to as the "total physical product" (TPP) or "physical product". The TPP is measured in units of the item being produced, i.e., bushels of corn for our example.

The "average physical product" (APP) or "average product" is the ratio of the total physical product (TPP) to the "total physical input" (TPI) or "total input" over the range of TPI. The TPI for our example is the pounds of fertilizer applied. In equation form:

$$APP = TPP / TPI \qquad (9.2)$$

The "marginal physical product" (MPP) or "marginal product" is the ratio of the incremental change in total physical product and the incremental change in total physical input. In equation form:

$$MPP = dTPP / dTPI \qquad (9.3)$$

Mathematically, the marginal physical product is the first derivative of the production function. Hence, MPP is the slope of the production function curve.

Table 9.2. Bushels of corn produced/pound of fertilizer applied (all other inputs remaining constant)

Pounds of Fertilizer Applied	Bushels of Corn Produced	Fertilizer Increment (lb)	Corn Increment (bu)	Marginal Product of Fertilizer (bu/lb)	Average Product of Fertilizer (bu/lb)
0	0.00				
		50.0	8.75	0.175	
50	8.75				0.175
		50.0	23.75	0.475	
100	32.50				0.325
		50.0	35.00	0.700	
150	67.50				0.450
		50.0	42.50	0.850	
200	110.00				0.550
		50.0	46.25	0.925	
250	156.25				0.625
		50.0	46.25	0.925	
300	202.50				0.675
		50.0	42.50	0.850	
350	245.00				0.700
		50.0	35.00	0.700	
400	280.00				0.700
		50.0	23.75	0.475	
450	303.75				0.675
		50.0	8.75	0.175	
500	312.50				0.625
		50.0	−10.00	−0.200	
550	302.50				0.550
		50.0	−32.50	−0.650	
600	270.00				0.450
		50.0	−88.75	−1.175	
650	211.25				0.325
		50.0	−83.75	−1.775	
700	122.50				0.175
		50.0	−122.50	−2.450	
750	0.00				0.000

Stages of Production

The production function may be divided into three stages in terms of allocating the optimum input resources. The production stages are commonly referred to as Stage 1, Stage 2, and Stage 3.

Stage 1 Production. Stage 1 production is that portion of the production function where the marginal physical product is greater than the average physical product (fig. 9.3). The APP always lags MPP in response to increased input. Conceptually, if MPP is greater than the APP, APP will increase with additional input. Likewise, if MPP is less than APP, APP will decrease with additional input.

In economics, the quantity of product being produced should always exceed the quantity associated with the maximum Stage 1 production level assuming constant input and output prices. The price assumptions are inherent to microeconomics and state that supply-demand relationships for the inputs-outputs do not change based on the production-utilization decisions of one individual decision maker. That is, no single farmer is going to affect significantly the overall price of corn or fertilizer. Another interpretation of Stage 1 is that the average physical product is increasin continuouslyg in this production zone. Production should always be increased to at least the point where the average physical product begins to decrease. Using our example, we

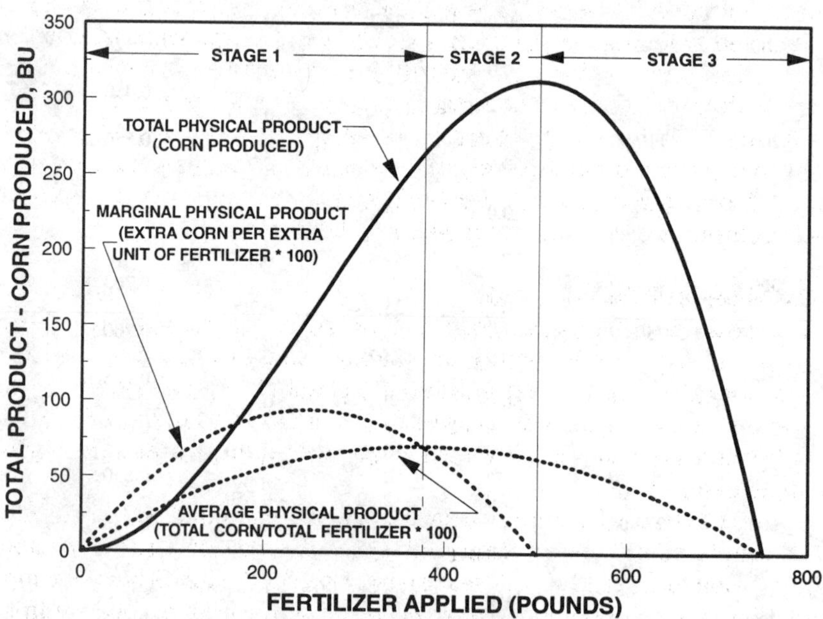

Figure 9.3–Relationship between total, marginal, and average physical product and Stage 1, 2, and 3 production zones for the example given.

should continue to apply fertilizer to obtain more corn so long as the return from the sale of corn is increasing even faster than the cost of the fertilizer that is applied.

Stage 3 Production. Like Stage 1, Stage 3 represents an economically infeasible zone of production but for entirely different reasons. In Stage 3, the marginal physical product becomes negative. This results in a decrease in total physical product with further increases in total physical input. For our example, Stage 3 begins where additional fertilizer inputs result in a lower corn yield. This might be explained as the point where additional fertilizer starts to become toxic to the corn, thus reducing yield. Obviously, an optimum level of fertilizer input would preclude a production zone where increased input cost would reduce gross return.

Stage 2 Production. Stages 1 and 3 have been eliminated as economically feasible zones of production. By process of elimination, the optimum input allocation must lie in Stage 2. The value of both inputs and outputs must be known in order to determine the optimum production levels. In essence, a method is needed for converting the physical input-output relationships given in table 9.2 and figure 9.3 to the value relationships given in table 9.1 and figure 9.2. The question is that of determining when an additional unit of input will cost more than the benefit received from the associated marginal increase in output. Using our example, we must ask at what point the cost of applying an additional pound of fertilizer is greater than the value of the extra yield associated with that application. Certainly we would want to continue adding fertilizer so long as the value of the extra corn produced is greater than the cost of the extra fertilizer. That is, one should always continue to increase production in Stage 2 until the value of the marginal physical product equals the cost of the marginal physical input. If prices are constant for both inputs and outputs, the optimum level of input is at the boundary between Stages 2 and 3.

Substitutes and Complements

An assumption in production economics is that the variable inputs may be substituted for each other. For our example, one could increase the amount of corn being produced by simply increasing the corn acreage rather than the amount of fertilizer. This leads to the production economics concepts of perfect substitutes and perfect complements.

Perfect Substitutes. Input variables may be classified as "substitutes" or "complements". Most inputs fit both of these categories to some extent. However, in the strictest sense, "perfect" substitutes are input variables that may be substituted for each other with no change in the physical productivity of a given product. An example would be the use of either natural or LP gas in grain drying. Each of these energy sources may be substituted readily for the other. The production function is linear (in a straight line) when relating two inputs such as these, as is shown below:

$$G = c*L + d*N \tag{9.4}$$

where

G = energy required for grain drying
c = conversion coefficient for LP gas
d = conversion coefficient for natural gas
L = amount of LP gas used
N = amount of natural gas used

The linear production function means that the optimum production level lies everywhere along the function. However, the optimum profit is associated with using only one of the two energy sources. This is because the two inputs are perfect substitutes so each can be substituted for the other with no penalty in production, i.e., for this example the quantity of grain being dried. Therefore, the least expensive input always should be selected in order to maximize profits. That is, if two inputs may be substituted readily for each other, always purchase the least expensive one.

Perfect Complements. "Perfect" complements are variable inputs that must be used in fixed ratios if either of the inputs are to be effective in producing a product. For example, if water is to be produced, two parts of hydrogen are required for each unit of oxygen. The quantity of water resulting from this mix of ingredients may be limited by either one or both of the inputs. Price plays no part in selection of the optimum ratio of inputs that are perfect complements; that is, the ratio of hydrogen and oxygen remains fixed regardless of the price of each.

Using a machinery example, a combine contains a "complement" of components, each of which contributes to the continued performance of the machine. The combine parts are complementary in that one part of the combine may not readily be substituted for another. Yet, one model or make of combine may be substituted for another so that they become substitutes.

If one combine is a "perfect" substitute for another, why then does one see different makes of machines on what appears to be the same type of farm? The reason is that the basic assumptions of production economics (interpersonally valid, value-based, common denominator) are rarely true. Individuals are different, and their production systems are all somewhat unique. Yet, for purposes of further discussion, we will assume that individuals wish to maximize the difference between benefit and cost even if they differ as to the respective values of certain inputs and outputs. One of the important aspects of decision making and maximizing profit is in determining the value of money over periods of time.

Cost Considerations in Decision Making

A decision maker always selects the alternative that provides the greatest difference between benefit and cost. Sometimes benefits and costs are not directly measured in dollars. However, for purposes of further discussion, it is assumed that decisions are based on monetary differences unless stated otherwise. Even then, the method of selecting an alternative may vary. In the following sections, we will examine firstly the costs to be considered when making a purchase. This will be followed by a discussion of various methods of selecting the proper course of action.

Costs

Each of the following items is a cost that should be considered when evaluating a possible purchase. Not all of these factors are relevant to every purchase.

Interest. Interest represents the cost of borrowing money for the purchase of an item. Money is always borrowed, even if it is obtained from one's own savings. Interest charges may be expressed in several forms. However, the most common are simple interest and compound interest. In both of these forms, the rate is expressed as a percentage for some specified time period.

Simple Interest. Simple interest refers to a method of calculating by which the charge rate is based on the original quantity of money borrowed (called the principal). For example, suppose the principal is $100 and money is to be paid back in one sum in 5 years. If the simple interest is 10% per year, an interest charge of $10 (10% * $100) would be assessed at the end of each year. At the end of five years, the total amount due would be the original principal ($100) plus $50 in interest ($10 for each of five years) for a total of $150. The simple interest method is rarely used.

Compound Interest. Compound interest refers to a method of calculating interest in which the charge rate is based on the remaining principal plus any interest not paid to that point in time. In effect, interest is paid on the interest due as well as the principal. Referring to the previous example, the compound interest rate yields a final payment due of $161.05 rather than the $150 associated with the simple interest method. The final amount due would be computed according to the schedule given in table 9.3. Obviously, compound interest results in a larger final amount due. It will be this method of determining interest that will be used in further discussion.

Table 9.3. Example of compound interest calculations

End of Year No.	Beginning of Year Principal and Interest ($)	Interest ($)	Amount Due ($)
1	100.00	10.00	110.00
2	110.00	11.00	121.00
3	121.00	12.10	133.10
4	133.10	13.31	146.41
5	146.41	14.64	161.05

Interest rates usually are expressed as a percentage per year. However, the period over which interest is calculated may be less than one year. When the interest rate for periods less than one year is converted to a yearly or annual rate while not considering the effects of compounding, the calculated rate is said to be the nominal annual rate. For example, if the interest rate were 1%/month, the nominal annual rate would be 12% (1%/month * 12 months/year). When the effects of compounding are considered, the annual rate is referred to as the effective annual rate. The following conversion formula may be used:

$$\text{effective interest rate per year (\%)} = \left[\left\{\left(1 + (1/100)\right) **M\right\} - 1\right] *100 \qquad (9.5)$$

where

I = interest rate per interest period (%)
M = number of interest periods per year

Using the above example for a 1%/month interest charge compounded monthly:

$$\text{effective interest rate per year (\%)} = \left[\left\{\left(1 + (1/100)\right) **12\right\} - 1\right] *100$$

$$= 12.68\%$$

The effective annual interest rate is the value that should be used for comparative purposes. It is often referred to as the APR (Annual Percentage Rate).

Other Methods for Calculating Interest. There are several other methods by which interest is calculated. Many of these infer an interest rate that is, in fact, less than the effective annual rate. For example, suppose that $100 were to be borrowed at a supposed annual rate of 10%. One method of lending is to calculate the charge for interest first, for our example $10 ($100 * 10%). The lender would immediately deduct the interest from the loan leaving a balance to the borrower of $90 to be used for the year. Effectively, $10 has been charged for the $90 loan. In this situation, the effective annual rate is 11.11% ($10/$90 * 100) rather than the supposed 10%.

Another misleading method is to quote the "average" interest rate. When using compound interest, the interest rate is charged on the unpaid balance which includes both unpaid principal and unpaid interest. If a uniform series of repayments are made, the first payment is largely made up of interest while the last payment is mostly principal. Suppose that $1,000 is borrowed at an effective compound interest rate of 10%. The repayments are to be made uniformly once each year for 10 years. Each payment would be $162.75 (we'll examine the method for determining this value later). This means that the average interest charge would be $62.75 based on an average principal payback of $100 ($1,000/10 years). The original principal was $1,000. Therefore, the "average" interest rate was 6.275% ($62.75/$1,000 * 100). Obviously, the "average" is a false representation, the true effective interest rate is 10%.

Depreciation. Depreciation refers to the cost of an item that results from its loss of value over time. Loss of value is related to a reduction in the item's ability to produce in an economical manner. Effectively, the item becomes worthless when it has no remaining value that may be depreciated even though it still may be functional. For example, horse-drawn wagons were replaced by modern trucks even though the wagons may have been entirely functional. Most of the time, however, an item is said to be totally depreciated when it is decided that the benefits associated with replacing it exceed the costs for further maintenance.

The time period over which the item retains its value is referred to as the life of the item. In the United States, the tax laws governing depreciation change from time-to-time. The "tax life" of an item refers to the time period over which the government allows depreciation to be considered when computing tax liability. The particular guidelines regarding governmental regulations concerning depreciation are beyond the scope of this book. It will be assumed, however, that the true economic life and tax life, as defined previously, are in fact the same for purposes of discussion. The main point to remember is that depreciation is a very real expense even though there may be some tax allowance for it. This point will be discussed in greater detail later in the chapter.

Investment Credit. At various times the U.S. tax law has allowed for a tax credit when certain items are purchased as part of a business operation. This represents a direct reduction in one's income taxes during the year that the investment credit was obtained. When, in effect, investment credit serves as a reduction in purchase price.

Repair and Maintenance. All items have some repair and maintenance costs associated with them. Typically, these costs increase as the item gets older. However, these costs are relatively unpredictable insofar as time of occurrence is concerned. For planning purposes, a typical method of estimating repair cost is to assign a percentage of the purchase cost to repair and maintenance over the useful life of the item. A uniform annual cost is obtained by dividing the total repair and maintenance cost by the useful life to establish a uniform annual cost. One may wish to do this directly by assigning an annual percentage of the purchase price to repair and maintenance.

Housing. The cost for housing may not apply to all items, or it may be considered as part of maintenance. The typical procedure is to assign an annual cost based on a percentage of the purchase price.

Insurance and Risk. The cost for insurance is referred to as the insurance premium. Premiums are based on a percentage of the value of the item. An item is subject to destruction or theft according to some probability based on the type of equipment, geographic location, etc., irrespective of whether it is actually insured. Thus, the cost of insurance is the probability of loss times the replacement cost. The source of the insurance may be an insurance company that receives actual premiums. However, an individual may choose to bear the risk in which case this person is said to be self-insured. Regardless, the cost of insurance should be considered irrespective of whether a payment is actually made.

Miscellaneous Taxes. Taxes may take several forms including sales tax, property tax, and income tax (to be discussed later). The magnitude of taxes is dependent on a particular geographic location. The sales tax is a one-time expense that occurs at the time of purchase. Property taxes generally are based on an annual percentage of the current value of the item.

Income Tax. Income tax considerations are always important to the decision maker, the goal is usually to pay as little as is legally required. Another way of stating this concept is that an individual will usually try to maximize personal accumulation of wealth. In terms of accounting, this requires a maximization of net income. However, for any level of gross income, one generally will try to minimize the income tax liability.

Income taxes are assessed on the difference between one's income and expenses. Expenses may include costs for depreciation, interest, repair, maintenance, insurance, and other taxes. The tax rate is graduated so that higher incomes pay disproportionately higher tax rates. Historically, if investment credit is allowed, it is subtracted directly from the income tax obligation.

Expenses represent actual monetary transactions (with depreciation somewhat the exception) for purposes of computing net income. For example, insurance premiums would be considered an expense while one who is self-insured would not be able to treat this risk as an expense. Depreciation does not usually coincide with cash payments used for purchasing the item. This discrepancy may result in a significant difference between income as reported correctly to the U.S. government, and actual cash on hand. Thus, cash flow may be critical to the economic success of an investment even though the investment would otherwise be profitable.

Net Worth. Net worth refers to the difference between one's assets and liabilities. Assets are the value of all the items owned by the business as determined by standard accounting procedures. Likewise, liabilities are the debts owed currently. Net worth represents an accumulation of wealth over the life of a business. It is usually a better index of overall profitability than is income for any single year.

Intangibles. Two individuals, both faced with what appears to be identical situations, will choose different courses of action. Yet, both have made the correct decision. How can this be true? The answer to this apparent paradox lies in the relative value that each places on what might be termed "intangible" considerations. In truth, the individuals are not in the same situation as evidenced by their difference in choice. In fact, so called intangibles may be the single most important factor in decision making and certainly the most difficult to define quantitatively (see the discussion of the value-based common denominator). Many times the term "intangible" could be translated to mean "convenience". For example, two farmers with the same resources wish to purchase a replacement for their current methods of handling grain. A detailed economic analysis might indicate that a bucket elevator system will cost $0.05/bu/year more than a comparable portable auger system. Yet, one farmer may purchase the bucket elevator system. When asked why, the person might mention factors such as convenience of handling grain, especially if future labor supply is viewed as being limited. In essence, this person is willing to insure against the risk of inadequate labor in future years, the cost of insurance being $0.05/bu/year. The other farmer presently envisions the four sons in the family working together with the parent(s) on the farm, so future labor needs for grain handling are not viewed as an important consideration. Further probing has led us to discover that the values of these farmers in this one area are sufficiently different so that they reached different economic decisions. It does not say that the farmers, individually or collectively, were correct in their value system but rather that concepts of future situations are important to decision making.

Fixed and Variable Costs

Fixed costs refer to costs that occur each year regardless of whether any production occurs. For example, once grain storage facilities have been purchased, a financial obligation is incurred that is independent of whether any grain is actually stored in the facility, hence the term "fixed" cost.

Variable costs are those costs that are associated directly with production or sale of a product. For example, fossil fuel for drying and handling grain is a variable cost.

Some items may be placed in both categories. Insurance for the storage structure is a fixed expense while insurance for the grain itself is a variable cost.

Given sufficient time, fixed costs may be converted to variable costs. This conversion occurs when the life of an item is given in terms of the units of production it can generate. For example, assume that the production unit associated with a vehicle is measured in "miles" and that a particular vehicle costing $20,000 has a "production life" of 100,000 miles. The fixed cost, considering only purchase price, may be converted to a variable cost by dividing the purchase cost by the production units. For this situation, $20,000 divided by 100,000 miles yields a "variable" cost of $0.20 per mile for the purchase.

The following set of decisions as to when to operate a business is based on the concept of fixed and variable costs.

1. If the income from a business, generated by the sale of products or services, is greater than the variable cost, consider the continuation of production in that at least some of the fixed cost is being paid. Perhaps even all the fixed cost is being satisfied so that the firm is profitable.
2. Cease production when the income does not exceed the variable cost in that "zero production" represents the optimum, although negative, economic position.

Value of Money Over Time

Money normally increases in value over time in that it may be invested to generate even more money. The return per unit of time is normally expressed as a percentage and called the effective compound interest rate. The following terminology is used to describe the relationships among investment and return:

- i = interest rate or expected rate of return (expressed as a decimal for the formulas below) to be compounded each time period.
- n = the number of time periods for which the investment is to be considered, usually expressed in years.
- P = present sum of money to be invested.
- S = the sum of money that will exist at the end of "n" periods from the present date and equivalent to investing "P" with an expected rate of return of "i".

R = a uniform series of payments occurring at the end of each "n" period and equivalent to "P" invested at "i".

The factors may be related using the following expressions:

Given "P" to find "S",

(Single Payment Compound Amount Factor, SPCAF)

$$S = P * (1 + i) ** n \tag{9.6}$$

Given "S" to find "P",

(Single Payment Present Worth Factor, SPPWF)

$$P = S* \frac{[1]}{(1 + i) ** n} \tag{9.7}$$

Given "S" to find "R",

(Uniform Series Sinking Fund Factor, USSFF)

$$R = S * \frac{[i]}{[(1 + i) ** n] - 1} \tag{9.8}$$

Given "P" to find "R",

(Uniform Series Capital Recovery Factor, USCRF)

$$R = P * \frac{i*[(1 + i) **n]}{[(1 + i) **n] - 1} \tag{9.9}$$

Given "R" to find "S",

(Uniform Series Compound Amount Factor, USCAF)

$$S = R* \frac{[(1 + i) **n] - 1}{i} \tag{9.10}$$

Given "R" to find "P",

(Uniform Series Present Value Factor, USPVF)

$$P = R* \frac{[(1 + i) **n - 1]}{i(1 + i)^n} \tag{9.11}$$

These formulas often are incorporated into tables for a range of interest rates and payback periods. However, for the following examples, the formulas will be used directly.

Example 1. Given P to find S: $10,000 is to be invested in a saving account drawing 8% interest compounded yearly for a period of 10 years. How much money will be in the account at that time? The single payment compound amount factor, SPCAF, may be calculated from equation 9.6 to equal 2.159. Thus,

$$S = P * (SPCAF) = \$10{,}000 * 2.159 = \$21{,}590$$

Example 2. Given S to find P: A farmer anticipates the need to purchase a replacement for current on-farm grain handling system in five years. The expected price will be $20,000 at that time. How much money would have to be set aside today drawing 10% interest compounded yearly to have the $20,000 available in five years? The single payment present value factor is determined from equation 9.7 to equal 0.6209.

$$P = S * SPPWF = \$20{,}000 * 0.6209 = \$12{,}418$$

Example 3. Given S to find R: Using Example 2, suppose the farmer wished to set some money aside for each of the five years to make the purchase. How much would have to be deposited each of the 5 years? Using equation 9.8,

$$R = S * USSFF = \$20{,}000 * 0.16380 = \$3{,}276$$

Example 4. Given P to find R: A grain dryer is purchased for $15,000. Money for the purchase is obtained from a local bank at an interest rate of 15% compounded every six months. Payments are to be scheduled uniformly every six months for the next eight years. How much are the payments? (Note that in this situation the number of payments "n" equals 16.) From equation 9.9,

$$R = P * USCRF = \$15{,}000 * 0.16795 = \$2{,}519.25$$

Example 5. Given R to find S: Money is to be set aside in an account that pays 10% annually for the purchase of an office building. If $3,000 is to be placed in the account each year, how much money would be available for the office at the end of seven years? Using equation 9.10,

$$S = R * USCAF = \$3{,}000 * 9.487 = \$28{,}461$$

Example 6. Given R to find P: It is anticipated that a grain storage structure will return an average net profit of $0.05/bu of capacity for each of its 25 years of projected life. Current construction loans have an interest charge of 12%/year. If the payback period may be as long as the life of the structure, i.e., 25 years, what is the maximum amount that may be invested per bushel in the structure? Using equation 9.11,

$$P = R * USPVF = \$0.05 * 7.843 = \$0.39$$

Present-Value Method of Evaluating Alternatives

The present-value (also referred to as present-worth) method of evaluating investments provides a framework for selecting the most economical alternative. The basics for this system are that every cost or return associated with the alternative may be converted to a present-day dollar equivalent. The alternative with the greatest present-day dollar equivalent value (present value) would be the correct choice. This concept assumes that there must be at least two alternative courses of action, one being to retain the status quo. Otherwise, there is no choice to be made. The steps for this procedure are:

1. Identify all possible alternatives. Some alternatives may include a combination of other alternatives.
2. Identify all the expenses associated with the alternative as to amount and time of occurrence.
3. Convert these to an equivalent present-day value using equations 9.7 and 9.11 as required.
4. Repeat Steps 2 and 3 for expected return.
5. Subtract the present value of the expenses from that of the return.
6. Repeat Steps 2 to 5 for each alternative.
7. The proper alternative is the one that has the greatest present value, i.e., the greatest present-day equivalent value.

Alternatives. The following example will illustrate these concepts. Suppose a farmer was limited to selecting one of the following alternatives:

Alternative 1 - Retain the status quo.
Alternative 2 - Select Drying Alternative 2.
Alternative 3 - Select Drying Alternative 3.

Alternative 1 is a natural air drying system. Electrical costs are estimated to be \$2,500 per year for the 15-year life of the drying fans. Purchase price of the unit is \$1,500. Grain losses and other remaining costs associated with natural air drying are estimated to be \$2,000/year. Net return to this on-farm drying system, as compared to not having any on-farm drying, is \$6,000/year.

Alternative 2 is a batch-in-bin drying system costing $3,200. Fossil fuel costs will be $3,000/year for the 12-year life of the fan. Grain losses and other remaining costs are estimated to be $1,000/year. Net return to this on-farm drying system, as compared to not having any on-farm drying, is $7,000/year.

Alternative 3 is a continuous-flow dryer requiring an investment of $10,000. The life of the dryer is 10 years. Fossil fuel costs will be $8,000/year with other costs being $1,500/year. Expected return, as compared to not having any on-farm drying, is $14,000/year in that the farmer may now do some custom work.

Money may be borrowed for 10%/year for any of the alternatives.

Alternative 1:

Expenses

Annual Costs = electrical	$2,500
+ misc. other	2,000
	$4500
P = R * USPVF = $4,500 * 7.606 =	$34,227
+ purchase cost	1,500
	$35,727

Gross Returns

P = R * USPVF = $6,000 * 7.606 =	$45,636

Net Present Worth

Gross Return	$45,636
– Expenses	– 35,727
Net Present Value =	$ 9,909

Alternative 2:

Expenses

Annual Costs = fossil fuel	$3,000
(Each of 12 years) + misc. other	1,000
	$4,000
P = R * USPVF = $4,000 * 6.814 =	$27,256
+ Purchase Cost	3,200
	$ 30,456

Gross Returns

P = R * USPVF = $7,000 * 6.814 =	$47,698

Net Present Worth

Gross Return	$47,698
– Expenses	–30,456
Net Present Value =	$17,242

Alternative 3:

Expenses

Annual Costs = fossil fuel	$8,000
(each of 10 years) + misc. other	1,500
	$9,500
P = R * USPVF = $9,500 * 6.144 =	$58,368
+ Purchase cost	10,000
	$68,368

Gross Returns

P = R * USPVF = $14,000 * 6.144 =	$86,016

Net Present Worth

Gross Return	$86,016
– Expenses	–68,368
Net Present Value =	$17,648

For this example, the most profitable alternative was No. 3 with a net present value of $17,648 followed by No. 2 ($17,242) and the existing system, No. 1 ($9,909). Several points should be made. First, the problem would have been considerably more complex had each expense item mentioned previously in the chapter been treated separately. Second, some judgement is always necessary when making a decision regardless of the "answer". For the example, Alternatives 2 and 3 are sufficiently close so that one must reconsider closely the cost assumptions before making a final decision. Third, Alternatives 1 and 3 have very nearly the same benefit to cost ratio (1.26) based on dividing the present value of gross returns by the present value of expenses. Alternative 2 has a considerably higher value, 1.56, thus the alternative with the highest rate of return may not be the one yielding the greatest absolute return. Fourth, net present value analysis does not address the problems associated with cash flow.

Cash Flow

An alternative that may be the most economically feasible choice, as defined by present value analysis, may cause the firm to fail because of cash flow problems. For the previous example, a payback period of two years for the dryer in Alternative 3 would result in annual payments of $5,762 (R = P * USCRF = $10,000 * 0.57619 = $5,762). The total cash outflow during each of the first two years is $15,262 ($9,500 + $5,762) while gross income during that same period was $14,000 per year. This means that the decision maker must subsidize Alternative 3 by $1,262 for the first 2 years even though in the long run it is clearly the best alternative.

Failure of the firm will result if cash flow for the total firm cannot be satisfied. Managers become confused because the "books" indicate that the firm is doing well. In fact, even the opposite may occur in that standard accounting principles indicate that the firm is losing money while cash itself continues to accumulate. The manager has a tendency in the latter case to ignore totally what the accounting records are saying, a very costly mistake. One of the primary causes of these apparent discrepancies is when the depreciation schedule for an item differs significantly from the purchase payment schedule, thus altering income tax payments. If the repayment schedule is larger initially, the firm will tend to be short of cash while showing relatively higher profits, all other factors the same. Conversely, when depreciation occurs at a faster rate than purchase payments, the firm will tend to accumulate cash because of relatively less profitability early in the process. As depreciation decreases, profits go up (all other factors remaining the same) requiring a proportionately larger outlay of cash for income tax, thus depleting the cash reserves.

In summary, cash flow is an important consideration. One should construct, insofar as possible, a chart depicting cash flows in, cash flows out, and balance on hand over time in order to identify possible problem areas.

Summary

The process of decision making is associated with the *Law of Diminishing Returns* and how it relates to a value-based common denominator. This common denominator usually is expressed as money but may include other factors. The key to decision making is to identify correctly and quantify all expenses and returns for a particular alternative. Present value analysis may be used to select the most profitable alternative. However, cash flow should be considered also.

Economics of Harvest Losses

On-the-farm grain drying and storage facilities may result in a reduction of harvest losses under some situations, thus enhancing the economic feasibility of the system. In this section, the impact of harvest losses (mostly in reference to corn) on the economics of on-farm drying and storage facilities will be addressed.

Types of Harvest Losses

Many farmers are not aware of the magnitude of their harvest losses. Often times, 5 to 15% of the potential corn harvest may be left in the field.

Field losses may be categorized as either being "visible" or "invisible". Invisible losses may be sub-divided further into three categories: maturity losses, imperfect shelling losses, and scavenger losses. A maturity loss results when grain is harvested before the maximum amount of dry matter has accumulated. Imperfect shelling losses relate primarily to corn and involve kernel tips that remain on the cob and chips and are lost in separation or handling. These losses in corn are greatest at moisture contents that exceed 30%. Scavenger losses are associated with birds and other wildlife. These losses are small during the early harvest season but may average 1% for a delayed harvest. Overall, invisible losses may range from approximately 2 to 5%. Generally, dry matter accumulation is at its highest point when corn is at approximately 26% moisture content (fig. 9.4), the near optimum range is 24 to 28%.

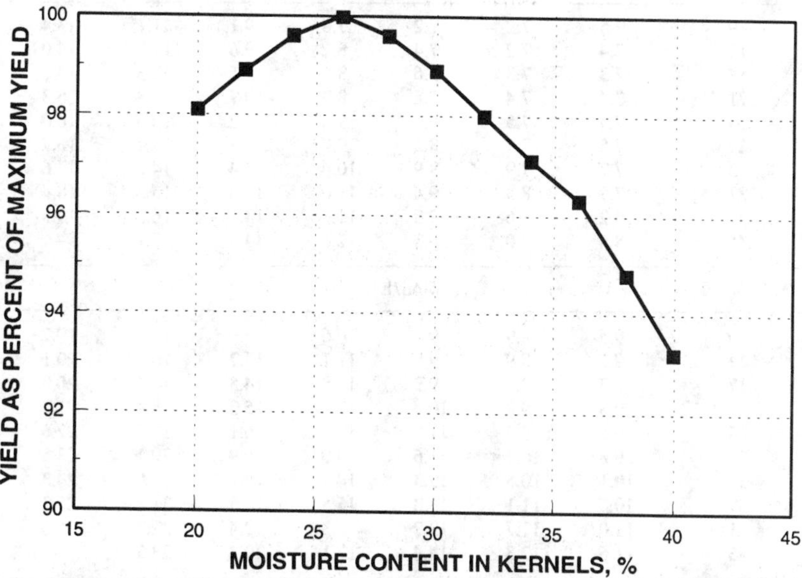

Figure 9.4–Relationship between moisture content and yield (Johnson and Lamp, 1966).

Visible losses are those losses that can be observed and measured under typical farming conditions. These include preharvest ear loss, ear loss during gathering, shelled corn gathering loss, cylinder loss, separation loss, and cleaning loss.

Magnitude of Losses

There is limited information concerning the magnitude of harvest losses. Generally, harvest losses increase as harvest time increases after grain has reached some established level of maturity. Likewise, moisture content is an index of maturity level, and usually decreases over the harvest season. The rate of decrease is somewhat dependent upon the weather. Thus, harvest losses are often referenced to both moisture content and length of harvesting season and weather. Harvest losses also increase with an increase in harvester speed (past some minimum speed necessary for gathering). These factors for corn are incorporated into the harvest loss summaries shown in tables 9.4 to 9.6. A conversion

Table 9.4. Average total field losses of corn as influenced by length of the harvest and moisture content at the beginning of the harvest (Johnson and Lamp, 1966)*

Length of Harvest (calendar days)	Moisture Content at the Beginning of Harvest % Kernel Moisture (w.b.)						
	35	33	30.2	27.5	25.1	22.9	20.8
	(Average Total Field Loss, % of Total Yield)						
	2 mi/h						
9	7.5	7.2	7.2	7.8	9.1	11.1	13.4
13	7.4	7.2	7.4	8.2	9.6	11.7	14.0
17	7.3	7.3	7.6	8.5	10.1	12.3	14.6
21	7.4	7.4	7.8	8.9	10.7	12.8	15.3
25	7.4	7.5	8.1	9.4	11.2	13.5	16.0
29	7.5	7.7	8.5	9.9	11.8	14.1	16.7
33	7.7	7.9	8.9	10.4	12.4	14.8	17.6
37	7.9	8.3	9.4	11.0	13.0	15.5	18.4
41	8.2	8.6	9.8	11.5	13.7	16.3	19.3
45	8.6	9.0	10.3	12.1	14.4	17.1	–
	5 mi/h						
9	9.5	8.9	9.1	10.6	12.9	16.0	19.2
13	9.3	8.9	9.5	11.1	13.7	16.8	20.1
17	9.3	9.1	9.9	11.8	14.5	17.6	20.9
21	9.3	9.3	10.4	12.5	15.3	18.4	21.7
25	9.5	9.6	11.0	13.2	16.1	19.3	22.6
29	9.8	10.1	11.6	14.0	16.9	20.1	23.5
33	10.1	10.5	12.3	14.7	17.7	21.0	24.3
37	10.5	11.1	13.0	15.5	18.5	21.8	25.2
41	11.0	11.7	13.7	16.3	19.4	22.7	26.0
45	11.6	12.3	14.4	17.1	20.2	23.5	–

*Cool, humid season, picker-sheller with spiraled, lugged snapping rolls.

of this data to a daily rate of loss as reflected by moisture content is given in table 9.7. The tabular harvest loss data is in the range of losses found by agricultural engineers in Ohio (Byg et al., 1966) as shown in table 9.8. Of particular interest are the effects of a single rain and snow storm shown in table 9.9 illustrating that harvest losses tend to occur quickly in relatively large quantities rather than as a gradual process.

Management Implications

There always will be harvest losses, regardless of whether a farmer has on-the-farm grain drying and storage facilities. However, the economics of owning these facilities change if the farmer alters harvesting strategies to affect harvest losses. If the farmer does not (or would not) change the harvesting schedule because of farm drying and storage, harvest losses play no part in evaluating the economics of the facilities. This is not to say that harvest losses are not important economically to total farm systems.

Table 9.5. Average total field losses of corn as influenced by length of the harvest and moisture content at the beginning of the harvest (Johnson and Lamp, 1966)*

Length of Harvest (calendar days)	Moisture Content at the Beginning of Harvest % Kernel Moisture (w.b.)								
	35	33	31	29	27	25	23	21.3	19.7
	(Average Total Field Loss, % of Total Yield)								
	2 mi/h								
9	9.5	8.5	7.8	7.5	7.6	8.3	9.2	10.3	11.3
13	9.1	8.2	7.7	7.7	8.0	8.8	9.8	10.7	11.7
17	8.7	8.1	7.8	8.0	8.5	9.3	10.2	11.2	12.0
21	8.5	8.1	8.1	8.4	9.0	9.8	10.7	11.7	12.3
25	8.4	8.2	8.4	8.8	9.4	10.2	11.1	12.0	12.6
29	8.5	8.5	8.7	9.2	9.9	10.6	11.4	12.2	12.9
33	8.7	8.8	9.1	9.6	10.2	11.0	11.7	12.5	13.2
37	9.0	9.2	9.5	10.0	10.6	11.3	12.1	12.8	13.4
41	9.3	9.5	9.9	10.3	10.9	11.6	12.3	13.0	13.7
45	9.6	9.8	10.2	10.7	11.3	11.9	12.6	13.3	13.9
	5 mi/h								
9	12.0	10.7	10.4	10.7	11.5	12.7	14.1	15.7	17.4
13	11.6	10.7	10.7	11.2	12.1	13.4	14.9	16.5	18.2
17	11.3	10.8	11.0	11.7	12.8	14.2	15.7	17.3	18.8
21	11.4	11.1	11.5	12.4	13.6	15.0	16.5	18.0	19.3
25	11.5	11.5	12.1	13.1	14.4	15.8	17.2	18.5	19.9
29	11.8	12.1	12.8	13.9	15.2	16.5	17.8	19.1	20.3
33	12.3	-12.7	13.5	14.6	15.8	17.1	18.3	19.6	20.8
37	12.8	13.3	14.2	15.3	16.4	17.7	18.9	20.1	21.2
41	13.4	14.0	14.8	15.8	17.0	18.2	19.4	20.6	21.7
45	14.0	14.6	15.4	16.4	17.6	18.7	19.9	21.1	22.2

*Warm, dry season, picker-sheller with spiraled, lugged snapping rolls.

Table 9.6. Average total field losses of corn as influenced by length of harvest and moisture content at the beginning of the harvest (Johnson and Lamp, 1966)*

Length of Harvest (calendar days)	Moisture Content at the Beginning of Harvest % Kernel Moisture (w.b.)							
	35	33	30.2	27.1	24.7	22.2	20.1	18.1
	(Average Total Field Loss, % of Total Yield)							
	2 mi/h							
9	7.0	6.4	6.5	7.4	8.6	9.9	11.0	12.0
13	6.8	6.7	7.0	8.0	9.3	10.4	11.5	12.3
17	6.9	7.1	7.6	8.6	9.8	10.9	11.8	12.6
21	7.2	7.5	8.2	9.2	10.3	11.3	12.1	12.8
25	7.6	8.1	8.6	9.7	10.7	11.6	12.4	13.1
29	8.0	8.6	9.2	10.1	11.1	11.9	12.7	13.3
33	8.5	9.0	9.7	10.5	11.4	12.2	12.9	13.6
37	8.9	9.4	10.1	10.9	11.8	12.5	13.2	13.8
41	9.3	9.8	10.4	11.2	12.0	12.8	13.4	14.0
45	9.6	10.2	10.8	11.5	12.3	13.0	13.7	14.3
	5 mi/h							
9	10.1	9.7	10.1	11.5	13.3	15.5	17.4	19.0
13	10.1	10.1	10.9	12.5	14.4	16.4	18.1	19.5
17	10.3	10.7	11.8	13.5	15.3	17.2	18.7	19.9
21	10.6	11.5	12.8	14.4	16.2	17.8	19.2	20.3
25	11.5	12.4	13.7	15.3	16.9	18.4	19.6	20.7
29	12.2	13.2	14.5	16.0	17.5	18.9	20.1	21.1
33	13.0	14.0	15.2	16.6	18.0	19.3	20.5	21.5
37	13.7	14.7	15.9	17.2	18.5	19.8	20.9	21.8
41	14.4	15.4	16.5	17.7	19.0	20.2	21.2	22.2
45	15.0	15.9	17.0	18.2	19.4	20.6	21.6	22.6

*Cool, dry season, picker-sheller with spiraled, lugged snapping rolls.

Table 9.7. Daily machine and field losses as a function of moisture content as estimated from data by Johnson and Lamp (1966)*

Field Moisture Content (%)	Grain Field Losses (%/day)	Machine Losses (%/day)
18	0.3163	0.850
22	0.1687	0.728
26	0.0600	0.668
30	0.0500	0.678

* Linear interpolation was used for intermediate moisture contents when various systems were analyzed.

Table 9.8. Total machine losses and range of losses (bu/acre)*

	Average Loss		Range of Losses	
Year	Picker	Combine	Picker	Combine
1964	4.8	6.4	0.3 to 14	2.3 to 29.4
1965	4.3	6.5	0.9 to 12.4	2.2 to 19.3
1966	5.3	9.3	0.7 to 13.0	3.4 to 22.3

* Byg et al., 1996.

Why would on-farm drying and storage alter harvesting strategy? The reason lies in the rate that grain may be harvested. Producers without on-farm storage and drying may be limited in the quantity of grain that they harvest each day. In some locations, waiting lines at the receiving elevator may restrict severely the farmers harvesting rate. This "slow-down" increases the total time required for harvesting which results in greater harvest losses. Some farmers attempt to avoid this bottleneck by using a sufficient number of grain delivery vehicles so that the combines can continue harvesting. In this situation, harvest losses are not affected. However, there are the additional expenses associated with extra labor and vehicles.

There are other considerations. Farmers with on-farm drying and storage are free to choose the moisture content at which they will harvest grain whereas a commercial elevator may limit the moisture content of the incoming grain it will receive. Generally, changing the delivery location for grain results in a different part of the total harvesting system becoming the bottleneck. When using commercial facilities, grain delivery may regulate harvest. With on-farm facilities, the problem usually is drying capacity. Remember, during the peak of harvest a poorly designed farm facility may result in a greater bottleneck than direct delivery to a commercial facility!

Table 9.9. Harvest losses before and after storm (bu/acre)*

	Average Loss		Range of Losses	
Time	Picker	Combine	Picker	Combine
Before November 2	4.0	6.8	1.0- 9.0	3.4-16.4
After November 2	8.1	11.9	0.7-13.0	4.1-22.3

* Byg et al., 1996.

Economic Savings

The economic savings associated with harvesting losses can be found by:

1. Determining the net amount of grain saved by altering management strategy as reflected by the addition of on-farm drying and storage.
2. Computing the value of these "saved" bushels.

The first step requires that the farmer establish two positions with regard to the individual's production system. First, how would the harvesting operation be managed if there were no on-farm storage and drying. Second, what changes would occur if there is on-farm storage and drying. In both situations, three questions must be answered. At what moisture will harvesting begin? How many calendar days will harvesting occupy? How fast will the harvester be driven?

For example, suppose a farmer would have to deliver grain to a commercial elevator some 10 miles away if on-farm storage and drying were not available. In this situation, corn harvesting would be expected to begin when the corn reached 24% moisture. Harvesting would continue for 33 calendar days and combine speed would be 2 mph.

If on-farm drying and storage were available, the farmer would anticipate a quicker harvest lasting only 25 calendar days. Harvesting

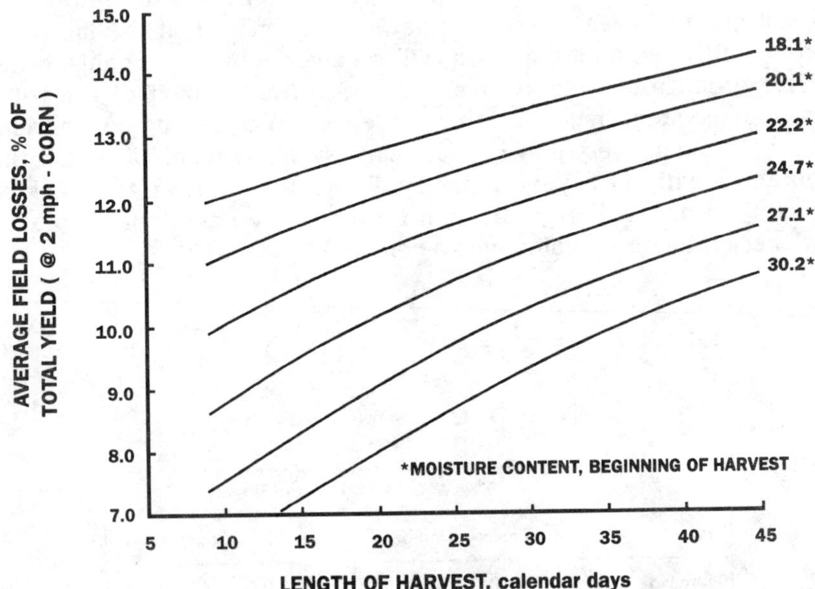

Figure 9.5–Average field losses, as a percentage of potential yield, influenced by the moisture content at the beginning of harvest and the calendar days required for harvesting at a combine operating speed of 2 mph (Johnson and Lamp, 1966).

would begin earlier when the moisture content reached 30%. Combine speed would still be 2 mph. In both instances, a cool dry season is anticipated.

Table 9.4 has been shown graphically in figure 9.5 to visualize better the two situations. Notice that the vertical axis is expressed as a percentage loss of total (or potential) yield. This refers to the yield as if no harvest losses had occurred. In the "no-facility" situation, 11.6% of the total yield would be lost on the average (fig. 9.6). If on-farm facilities were added, 8.7% would be lost. Thus, the addition of on-farm facilities results in a saving, on the average, of 2.9% of the potentially harvestable grain.

This "saved" grain is not all profit. It will require storage space, drying, and handling. Likewise, it will change in value as does the other stored grain. Still, it can represent a significant economic saving. For our example, suppose (1) the potential yield of corn is 120 U.S. No. 2 bu/acre, (2) the farmer has 400 acres of corn, (3) the net value of each bushel is $3.00, and (4) these conditions remain constant over the period in question. The following annual savings are:

Total savings from harvest losses ($/yr) = [2.9% * 120 bu/acre * 400 acres * $3.00/bu]/100

= $4,176

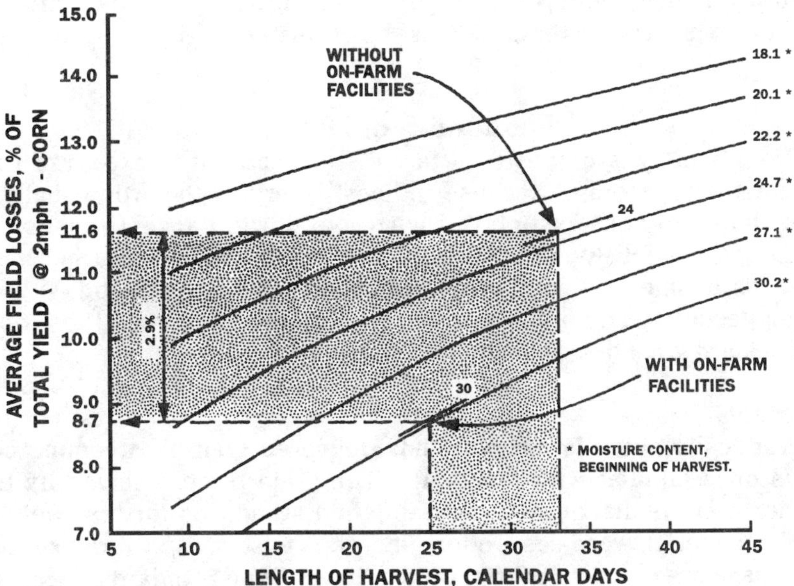

Figure 9.6–Example problem using data reflecting different harvesting strategies depending on the farmer's on-farm grain drying and storage facilities.

If the annual interest rate is 8% and the facility life is 20 years, the annual savings of $4,176 yield a present value of $41,000. The savings in harvest losses therefore contribute $41,000 towards the amount that may be invested in grain storage and drying facilities for this particular example.

Discussion of Assumptions

In the above analysis, average weather, harvesting, and economic conditions for corn were assumed. Certainly, each year and harvest situation are unique. One must make some set of assumptions when planning for the future. However, it is important that the producer recognize the limits of these assumptions. It is advisable to evaluate the "best" and "worst" situations and weigh these carefully against potential cash flow problems.

Harvest losses will not enter into some farm situations with regard to facility economics, if these farmers manage their harvesting operations in the same way regardless of whether on-farm drying and storage are available. Again, each farm system is unique. Note also in the examples that linear interpolation between starting moisture content data points was used. Interpolation could be used also in the harvest loss tables to estimate the effects of intermediate combine ground speeds.

Summary

Harvest losses may contribute to the economics of on-farm storage facilities depending on changes in harvesting strategy. However, it is only one of several factors that must be evaluated.

Economics of Drying

Grain usually is harvested at a moisture content in excess of that acceptable for storage, because the cost of drying the crop usually is less than the cost associated with avoidable harvest losses. As discussed previously in the section on harvest losses, the farmer must view his business in each of two positions; that is, with and without drying facilities. The first situation to be examined is when the farmer has on-farm grain dried commercially.

Discounts

Practically every farmer who has delivered grain to a commercial elevator is familiar with discounts. The grain price received by the farmer may be discounted if established standards for test weight, foreign material, weed seed, odors, insects, moisture, and various types of damage are not met (Official United States Grain Standards, 1988). Of these items, discounts for excessive moisture are often the dominant factor in determining grain prices and will be the only discount considered in the following discussion. However, a more detailed discussion of discounts will be given in a later section of this chapter.

Why do elevators discount grain for excessive moisture? Mainly, because the excess moisture means more water is being purchased relative to dry matter. Rather than adjust for dry matter content, the elevator adjusts the price per bushel (or pound) while maintaining a constant weight per bushel. The price reduction allows the commercial elevator to dry the grain to a safe storage content, recover their costs, and make a certain amount of profit. But, if commercial elevators make a profit on drying grain, could the farmer dry the grain, thus avoiding the moisture discount and make a profit on drying? Usually the answer is "yes", but to do so the farmer must be aware of what to do and the risks that are involved.

Moisture Content Considerations

The moisture content of grain may be defined as the percentage of a given sample of grain that is water. After maturity, the dry matter associated with a given sample of grain remains essentially the same regardless of the change in moisture content. The water that is lost when grain is dried to a lower moisture content is called "shrink". In the case of drying, the shrink represents a loss in weight that has no economic value in itself. Because this loss represents a reduction in saleable material, the price of undried grain is reduced accordingly through discounts to reflect shrinkage loss, drying cost, handling expense, and drying profit. The market is concerned primarily with the available dry matter. Since it is neither practical nor desirable to reduce the moisture content of grain to zero, certain marketing standards have evolved which specify the moisture content at which grain is considered to have no excessive moisture. This standard is termed the "base moisture content". Until 1987, the base moisture content for U.S. No. 2 corn was 15.5% with other grains having base moisture contents ranging from 13.0 to 15.5%. The U.S. Grain Standards changed in 1987 so that moisture content no longer determines grade. Presently, the policy of the local elevator determines the base moisture content for purposes of dockage for moisture. In effect, the base moisture content is the percentage moisture level in grain when a "wet" bushel is considered a "dry" bushel.

Discount and Base Moisture Content Relationship

Discounts may be defined as that charge which buyers or commercial elevators assess the farmer for the moisture in the grain above some established base moisture content. The elevator is telling the farmer that as a condition for buying grain, the excess water will have to be removed by the firm. To cover the elevator's expenses both for drying and for losses due to shrinkage and handling losses, the farmer will have to accept a lower price for the product. The price the farmer receives for grain after discounts are taken into account is normally referred to as the "value sold wet". For example, suppose corn is selling for \$3.00/bu with a 15.5% base moisture content. If the discount for moisture for a particular sample were equal to \$0.40, the "value sold wet" would equal \$2.60/wet bushel.

How are discounts calculated? This depends on the type of grain, the base moisture content, the points (or percentage) of moisture to be removed, the charge per point, and the policy of the local elevator. Typically discounts for moisture follow two forms (even though it may not be obvious when examining the discount schedule). One method involves a fixed charge per point of moisture removed, for example, a \$0.05 discount per point of moisture over a base moisture content level of 15.5%. A second method assesses a charge based on a stated percentage of the selling price for each point of moisture above the base level. For example, 2% of the selling price per point of moisture above the base moisture content level is charged as dockage.

To illustrate these methods, assume wet corn is brought in at 25.5% moisture, the base moisture content being 15.5%. If \$0.06/point of moisture is charged, the discount charge would be:

$$(\$0.06/\text{point}) * (25.5\% - 15.5\%) = \$0.60/\text{wet bu}$$

If the second method is used, the price of corn must be known. Suppose the selling price for corn is \$3.00/bu and the discount rate is 2% per selling point above a base moisture content of 15.5%. The discount is:

$$(2\%/100) * (25.5\% - 15.5\%) * \$3.00/\text{bu} = \$0.60/\text{wet bu}$$

In both instances, the dockage is \$0.60/wet bu because of the values selected. The "value sold wet" is equal to \$2.40/bu (\$3.00 – \$0.60). Note again that we are referring to a "wet" bushel. A "wet" bushel of corn represents the pounds of grain delivered by the vehicle to the elevator and divided by 56 lb before any drying occurs. If the elevator weighs the grain after drying to determine the initial number of bushels, it is in effect, discounting grain for moisture twice!

There is an economic comparison to be made between the "with on-farm drying" and "without on-farm drying" that is not apparent. This concerns the difference in average moisture content that may exist between the options. The "with on-farm drying" option should include savings associated with avoiding the dockage schedule for moisture based on the average moisture content of grain for the "without on-farm drying" option. The drying expenses should be based on the average moisture content of the "with on-farm drying" option. The reasoning behind this becomes clearer if one recognizes that the savings from discounts increase as moisture content increases. If the "with on-farm drying" average moisture content were used, it would always be desirable to increase the average by harvesting earlier and at a faster rate.

Gross Returns to Drying

What if the farmer chooses to dry the grain? In this case, the producer will suffer the loss due to shrinkage and the expense of drying but will not receive moisture discounts if the grain is dried to the base moisture content. If the farmer dries the grain and sells it to the elevator, the price received is referred to as the "value sold dry". Using the previous example, corn at 25.5% will shrink to 87.7% of its original weight after to 15.5%, including a 0.05% invisible loss (table 9.10). This means that the farmer has fewer bushels to sell after drying than before but will also get a higher price for the product. For this example, the "value sold dry" equals 0.877 bu times $3.00 or approximately $2.63 per original undried bushel. The difference between the "value sold dry", $2.63, and the "value sold wet", $2.40, is referred to as the "added gross returns for drying", which for this example equals $0.23 per undried bushel. Out of this gross return, the farmer must pay all of the drying expenses.

Drying Expense

The expense for drying includes costs for labor, electricity, fuel, and equipment. Of these items, labor usually is the least costly in that most drying systems require little on-the-scene manual handling or regulation. For example, during harvest, layer drying systems require practically no labor, while batch-in-batch systems may need manual assistance in unloading. Portable dryers, both automatic-batch and continuous flow, may require someone to regulate the flow of wet and drying grain. In addition, a manual batch dryer will involve physical regulation of the drying time for each batch. Another point to consider is that much of the labor that is involved in drying may also be used for other harvesting operations. For example, the driver of the grain truck or delivery wagon may also regulate the grain flow direction and check the dryer while the vehicle is unloading.

Electricity costs for fan and conveyor operation in a drying system may be approximated by estimating the average kilowatt demand and multiplying this by the total drying time in hours to obtain the kilowatt-hours of energy used. One method of estimating the average kilowatt demand for motor loads is to assume it is equal to the sum of the horsepower rating of electric motors that operate more or less continuously during the drying operation. The per bushel cost of electric energy can be calculated by the equation:

Table 9.10. Base moisture content and field mosture content – weight relationships*

Field Moisture Content	Fraction of Wet Weight Remaining After Drying From Various Field Moisture Contents to the Indicated Base Moisture Content†			Pounds (kg) of Wet Grain at Various Field Moistures Required to Obtain One Pound (kg) at the Indicated Base Moisture Content‡		
(%)	13.0%	14.0%	15.5%	13.0%	14.0%	15.5%
13.00	0.995	–	–	1.005	–	–
13.50	0.989	–	–	1.011	–	–
14.00	0.984	0.995	–	1.017	1.005	–
14.50	0.978	0.989	–	1.023	1.011	–
15.00	0.972	0.983	–	1.029	1.017	–
15.50	0.966	0.978	0.995	1.035	1.023	1.005
16.00	0.961	0.972	0.989	1.041	1.029	1.011
16.50	0.955	0.966	0.983	1.047	1.035	1.017
17.00	0.949	0.960	0.977	1.053	1.041	1.023
17.50	0.943	0.954	0.971	1.060	1.048	1.029
18.00	0.938	0.948	0.965	1.066	1.054	1.036
18.50	0.932	0.943	0.959	1.073	1.060	1.042
19.00	0.926	0.937	0.954	1.079	1.067	1.048
19.50	0.920	0.931	0.948	1.086	1.074	1.055
20.00	0.915	0.925	0.942	1.093	1.080	1.062
20.50	0.909	0.919	0.936	1.100	1.087	1.068
21.00	0.903	0.914	0.930	1.107	1.094	1.075
21.50	0.897	0.908	0.924	1.114	1.101	1.082
22.00	0.892	0.902	0.918	1.121	1.108	1.089
22.50	0.886	0.896	0.912	1.128	1.115	1.096
23.00	0.880	0.890	0.906	1.136	1.122	1.130
23.50	0.874	0.885	0.900	1.143	1.130	1.110
24.00	0.869	0.879	0.894	1.150	1.137	1.117
24.50	0.863	0.873	0.888	1.158	1.145	1.125
25.00	0.857	0.867	0.883	1.166	1.152	1.132
25.50	0.851	0.861	0.877	1.174	1.160	1.140
26.00	0.846	0.855	0.871	1.182	1.168	1.148
26.50	0.840	0.850	0.865	1.190	1.176	1.155
27.00	0.834	0.844	0.859	1.198	1.184	1.163
27.50	0.828	0.838	0.853	1.206	1.192	1.171
28.00	0.823	0.832	0.847	1.214	1.200	1.179
28.50	0.817	0.826	0.841	1.223	1.209	1.188
29.00	0.811	0.821	0.835	1.231	1.217	1.196
29.50	0.805	0.815	0.829	1.240	1.226	1.205
30.00	0.800	0.809	0.823	1.249	1.235	1.213
30.50	0.794	0.803	0.817	1.258	1.244	1.222
31.00	0.788	0.797	0.812	1.267	1.253	1.231
31.50	0.782	0.792	0.806	1.276	1.262	1.240
32.00	0.777	0.786	0.800	1.286	1.271	1.249
32.50	0.771	0.780	0.794	1.295	1.280	1.258
33.00	0.765	0.774	0.788	1.305	1.290	1.268
33.50	0.759	0.768	0.782	1.315	1.300	1.277
34.00	0.754	0.762	0.776	1.325	1.310	1.287
34.50	0.748	0.757	0.770	1.335	1.320	1.297
35.00	0.742	0.751	0.764	1.345	1.330	1.307

* Includes 0.5% invisible losses.

† (100 - initial MC) / (100 - final MC) - 0.005.

‡ (100 - final MC) / (100 - initial MC) + 0.005.

$$\text{electric cost / bu} = \frac{(\text{kW-h} * \text{cost / kW-h})}{\text{bushels dried}} \qquad (9.12)$$

For drying operations utilizing the pto from a tractor, the electricity cost for fan operation would be replaced by tractor fuel cost. This item could be estimated by the following relationship:

$$\text{tractor fuel cost} = \text{fuel factor} * \text{pto hp} * \text{drying h} * \text{cost / gallon} \qquad (9.13)$$

where
fuel factor = 0.109 for LP gas
0.069 for diesel
0.076 for gasoline

The major drying expense is for fuel used to heat the air. The amount of fuel burned per bushel of grain depends on many things, including the initial and final moisture contents of the grain, the drying air temperature, the outside air conditions, the drying method, and the type of dryer. Chapter 2 examined the energy usage associated with different drying methods. Tables 9.11 - 9.12 may be used in combination with figures 3.12 and 3.21 to estimate the energy required for typical drying systems. The cost of fuel may then be estimated by the following equation:

$$\text{fuel cost (\$ / bu)} = \frac{\left(\begin{array}{c}\text{Btu / lb water} * \text{lb water removed / bu} \\ * \text{ cost of fuel per unit} * 100\end{array}\right)}{(\text{Btu / unit of fuel} * \text{burning efficiency (\%)})} \qquad (9.14)$$

where LP gas has 92,000 Btu/gal, natural gas has 1,000 Btu/ft^3, electricity has 3,413 Btu/kW-h, and burning efficiency = 80% for LP and natural gas and 100% for electricity.

Again, note that a bushel is defined differently depending on its use. For purposes of this chapter, a bushel of corn is 56 lb of corn regardless of moisture content.

Annual equipment costs include depreciation, taxes, interest, insurance and repairs for all items used in drying including a portable dryer, in-bin drying fans, drying floors, handling equipment, etc. After calculating the annual cost for the drying equipment, the cost per bushel may be found by the following relationship:

$$\text{equipment cost (\$ / bu)} = \frac{\text{annual cost for drying equipment}}{\text{bushels dried / year}} \quad (9.15)$$

Table 9.11. The weight of water to be removed per unit weight of wet grain when drying to several base moisture content levels

Moisture Content of Incoming Grain (%)	lb (or kg) water removed per lb (or kg) of wet material = (initial MC - base MC) / (100 - base MC)							
	12.0	12.5	13.0	13.5	14.0	14.5	15.0	15.5
12.0	0.000	–	–	–	–	–	–	–
12.5	0.006	0.000	–	–	–	–	–	–
13.0	0.011	0.006	0.000	–	–	–	–	–
13.5	0.017	0.011	0.006	0.000	–	–	–	–
14.0	0.023	0.017	0.011	0.006	0.000	–	–	–
14.5	0.028	0.023	0.017	0.012	0.006	0.000	–	–
15.0	0.034	0.029	0.023	0.017	0.012	0.006	0.000	–
15.5	0.040	0.034	0.029	0.023	0.017	0.012	0.006	0.000
16.0	0.045	0.040	0.034	0.029	0.023	0.018	0.012	0.006
16.5	0.051	0.046	0.040	0.035	0.029	0.023	0.018	0.012
17.0	0.057	0.051	0.046	0.040	0.035	0.029	0.024	0.018
17.5	0.063	0.057	0.052	0.046	0.041	0.035	0.029	0.024
18.0	0.068	0.063	0.057	0.052	0.047	0.041	0.035	0.030
18.5	0.074	0.069	0.063	0.058	0.052	0.047	0.041	0.036
19.0	0.080	0.074	0.069	0.064	0.058	0.053	0.047	0.041
19.5	0.085	0.080	0.075	0.069	0.064	0.058	0.053	0.047
20.0	0.091	0.086	0.080	0.075	0.070	0.064	0.059	0.053
20.5	0.097	0.091	0.086	0.081	0.076	0.070	0.065	0.059
21.0	0.102	0.097	0.092	0.087	0.081	0.076	0.071	0.065
21.5	0.108	0.103	0.098	0.092	0.087	0.082	0.076	0.071
22.0	0.114	0.109	0.103	0.098	0.093	0.088	0.082	0.077
22.5	0.119	0.114	0.109	0.104	0.099	0.094	0.088	0.083
23.0	0.125	0.120	0.115	0.110	0.105	0.099	0.094	0.089
23.5	0.131	0.126	0.121	0.116	0.110	0.105	0.100	0.095
24.0	0.136	0.131	0.126	0.121	0.116	0.111	0.106	0.101
24.5	0.142	0.137	0.132	0.127	0.122	0.117	0.112	0.107
25.0	0.148	0.143	0.138	0.133	0.128	0.123	0.118	0.112
25.5	0.153	0.149	0.144	0.139	0.134	0.129	0.124	0.118
26.0	0.159	0.154	0.149	0.145	0.140	0.135	0.129	0.124
26.5	0.165	0.160	0.155	0.150	0.145	0.140	0.135	0.130
27.0	0.170	0.166	0.161	0.156	0.151	0.146	0.141	0.136
27.5	0.176	0.171	0.167	0.162	0.157	0.152	0.147	0.142
28.0	0.182	0.177	0.172	0.168	0.163	0.158	0.153	0.148
28.5	0.188	0.183	0.178	0.173	0.169	0.164	0.159	0.154
29.0	0.193	0.189	0.184	0.179	0.174	0.170	0.165	0.160
29.5	0.199	0.194	0.190	0.185	0.180	0.175	0.171	0.166
30.0	0.205	0.200	0.195	0.191	0.186	0.181	0.176	0.172
30.5	0.210	0.206	0.201	0.197	0.192	0.187	0.182	0.178

The total cost for drying is the sum of the per bushel costs for labor used exclusively in drying, electricity, fuel, and equipment. If this sum is less than the "added gross returns for drying", the drying enterprise is profitable. However, drying usually is only part of the total drying and storage system. Therefore, additional analysis is needed before the total system can be justified regardless of whether drying itself is profitable.

Table 9.12. The weight of water to be removed to obtain one unit weight of dry grain when drying to several base moisture content levels

Moisture Content of Incoming Grain (%)	lb (or kg) water removed per lb (or kg) of wet material = (initial MC - base MC) / (100 - base MC)							
	12.0	12.5	13.0	13.5	14.0	14.5	15.0	15.5
12.0	0.000	–	–	–	–	–	–	–
12.5	0.006	0.000	–	–	–	–	–	–
13.0	0.011	0.006	0.000	–	–	–	–	–
13.5	0.017	0.011	0.006	0.000	–	–	–	–
14.0	0.023	0.017	0.012	0.006	0.000	–	–	–
14.5	0.029	0.023	0.018	0.012	0.006	0.000	–	–
15.0	0.035	0.029	0.024	0.018	0.012	0.006	0.000	–
15.5	0.041	0.036	0.030	0.024	0.018	0.012	0.006	0.000
16.0	0.048	0.042	0.036	0.030	0.024	0.018	0.012	0.006
16.5	0.054	0.048	0.042	0.036	0.030	0.024	0.018	0.012
17.0	0.060	0.054	0.048	0.042	0.036	0.030	0.024	0.018
17.5	0.067	0.061	0.055	0.048	0.042	0.036	0.030	0.024
18.0	0.073	0.067	0.061	0.055	0.049	0.043	0.037	0.030
18.5	0.080	0.074	0.067	0.061	0.055	0.049	0.043	0.037
19.0	0.086	0.080	0.074	0.068	0.062	0.056	0.049	0.043
19.5	0.093	0.087	0.081	0.075	0.068	0.062	0.056	0.050
20.0	0.100	0.094	0.087	0.081	0.075	0.069	0.063	0.056
20.5	0.107	0.101	0.094	0.088	0.082	0.075	0.069	0.063
21.0	0.114	0.108	0.101	0.095	0.089	0.082	0.076	0.070
21.5	0.121	0.115	0.108	0.102	0.096	0.089	0.083	0.076
22.0	0.128	0.122	0.115	0.109	0.103	0.096	0.090	0.083
22.5	0.135	0.129	0.123	0.116	0.110	0.103	0.097	0.090
23.0	0.143	0.136	0.130	0.123	0.117	0.110	0.104	0.097
23.5	0.150	0.144	0.137	0.131	0.124	0.118	0.111	0.105
24.0	0.158	0.151	0.145	0.138	0.132	0.125	0.118	0.112
24.5	0.166	0.159	0.152	0.146	0.139	0.132	0.126	0.119
25.0	0.173	0.167	0.160	0.153	0.147	0.140	0.133	0.127
25.5	0.181	0.174	0.168	0.161	0.154	0.148	0.141	0.134
26.0	0.189	0.182	0.176	0.169	0.162	0.155	0.149	0.142
26.5	0.197	0.190	0.184	0.177	0.170	0.163	0.156	0.150
27.0	0.205	0.199	0.192	0.185	0.178	0.171	0.164	0.158
27.5	0.214	0.207	0.200	0.193	0.186	0.179	0.172	0.166
28.0	0.222	0.215	0.208	0.201	0.194	0.188	0.181	0.174
28.5	0.231	0.224	0.217	0.210	0.203	0.196	0.189	0.182
29.0	0.239	0.232	0.225	0.218	0.211	0.204	0.197	0.190
29.5	0.248	0.241	0.234	0.227	0.220	0.213	0.206	0.199
30.0	0.257	0.250	0.243	0.236	0.229	0.221	0.214	0.207
30.5	0.266	0.259	0.252	0.245	0.237	0.230	0.223	0.216

Determination of the cost of drying equipment illustrates the difficulty in trying to divide equipment costs among functions.

For systems where the farmer dries and sells grain without storing it, the cost of equipment is straightforward. However, when storage is involved, the cost determination for drying equipment becomes more difficult because many items of equipment are used in both storage and drying, and the proportionate costs are not divided easily. When this occurs and the producer is evaluating a total drying, storage, and feed processing system, it would be better probably to assign the complete cost of equipment to storage. Although the relative return for storage and drying will be altered, the net return for the entire system will not be affected.

Overdrying Considerations

Overdrying is the main pitfall for farmers to overcome when drying their own grain and selling it immediately to the elevator. Grain that is dried below the base moisture content rarely receives a premium and, in addition, the fuel expense for drying also increases. However, it should be noted that the moisture requirements for long-term storage may require drying of the grain to a level below the marketing base moisture content. This loss should be considered as a necessary storage expense.

Summary

The net return to drying depends on many factors, including the discount rate, base moisture content, moisture content of the grain, and the costs for labor, fuel, and equipment. In examining the expected return for the entire farming operation, the producer should be aware of how drying fits into the total on-farm economic system.

Economics of Storage

Although drying grain and selling it directly to an elevator is certainly an option, most farmers who dry their grain store it as well. The reasons for farm storage are numerous and include:

1. More flexibility in marketing.
2. Maintenance of quality.
3. Utilization of available labor and equipment.
4. Inadequate commercial facilities.
5. More efficient on-the-farm feeding.
6. Other conveniences.

Whatever the reason for storing grain on the farm, there is always a need for proper planning and a complete analysis of the economic implications of the storage system. The primary factors to be considered are addressed in the following discussion.

Cost of Storage Facilities

What does farm storage cost? Many factors must be considered to answer this question. Farm storage involves more than just storage bins. Grain handling equipment must also be included. In addition, certain items that are necessary in order to use a particular drying method might be considered as part of storage. For example, perforated floors are required for layer drying but not for continuous flow drying, yet the floors are permanently attached to the storage structure and aid in aeration. Likewise, drying fans may serve a dual role as aeration fans. Grain handling presents a similar situation in that the same equipment may be used in both drying and storage. These examples illustrate the difficulty in separating equipment costs for drying and storage. Therefore, it is better usually to assign the complete facility cost for both drying and storage to storage, keeping in mind that the net profit will not be affected when the entire system is analyzed. It should also be noted that the per bushel cost for the overall grain handling, drying, storage, and feed processing system usually will decrease when utilizing dual-role types of equipment so long as the timeliness of the overall operation is not affected.

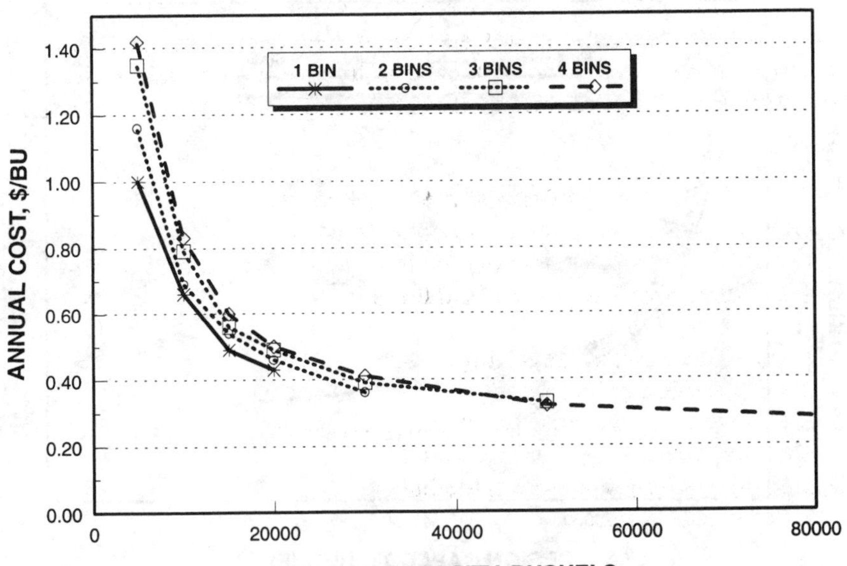

Figure 9.7–Annual cost comparisons for grain drying and storage facilities as a function of the number of bins in the system: Layer drying (Loewer et al., 1976a).

ANNUAL COST OF BATCH-IN-BIN SYSTEMS

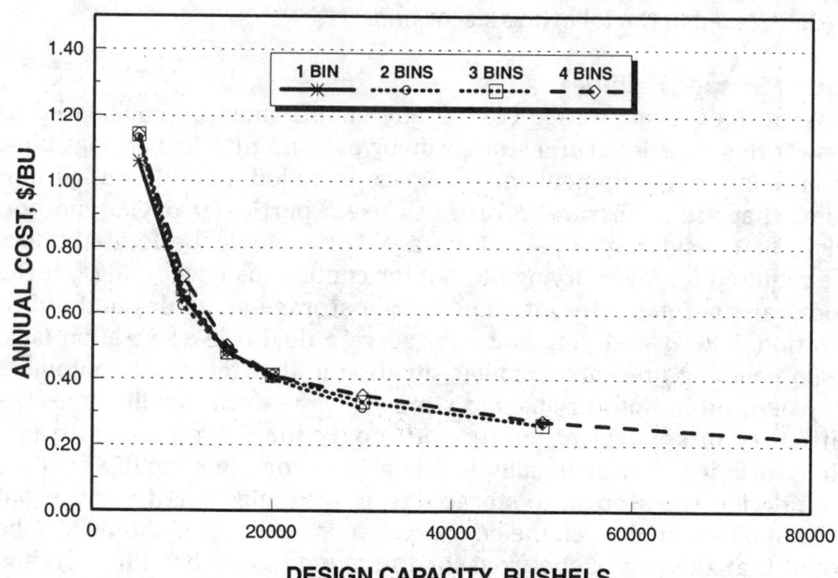

Figure 9.8–Annual cost comparisons for grain drying and storage facilities as a function of the number of bins in the system: Batch-in-bin drying (Loewer et al., 1976a).

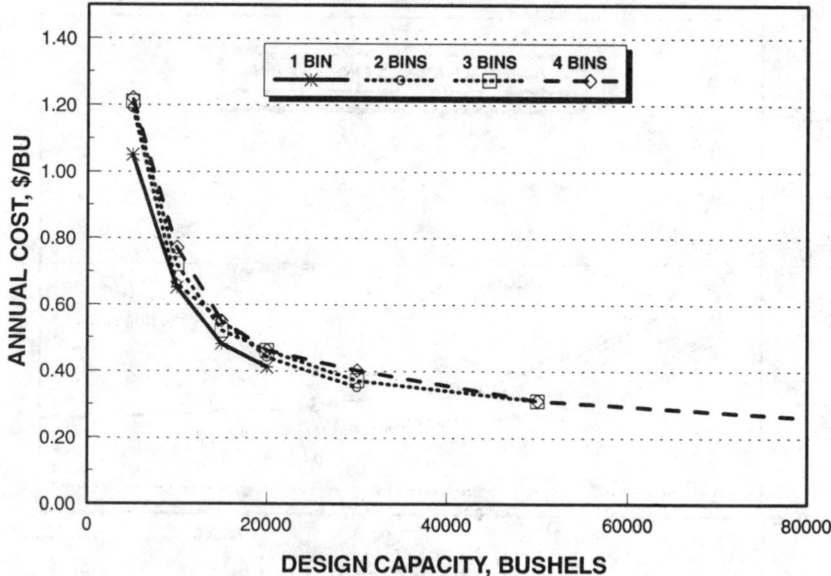

Figure 9.9–Annual cost comparisons for grain drying and storage facilities as a function of the number of bins in the system: Portable drying (Loewer et al., 1976a).

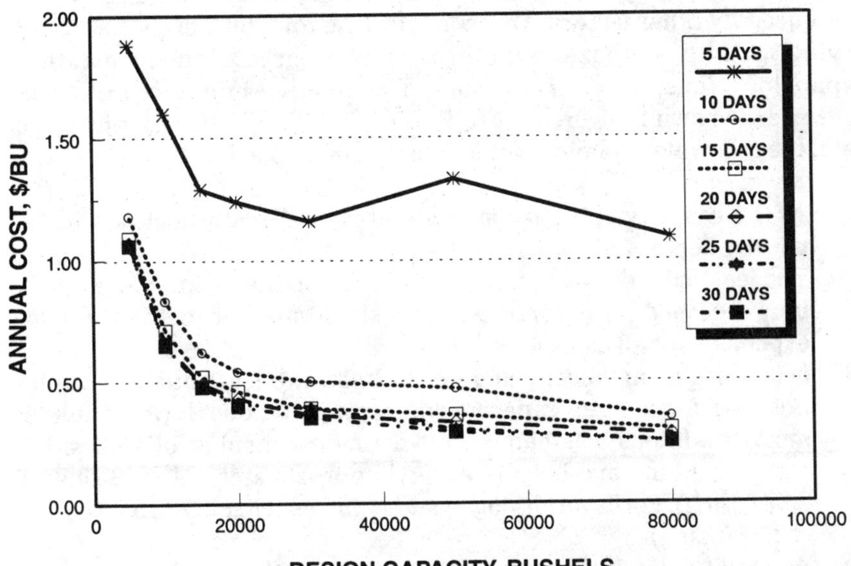

Figure 9.10–Annual cost comparisons for grain drying and storage facilities as a function of drying rate expressed as the number of days that harvesting will occur for a constant volume of grain: Layer drying (Loewer et al., 1976a).

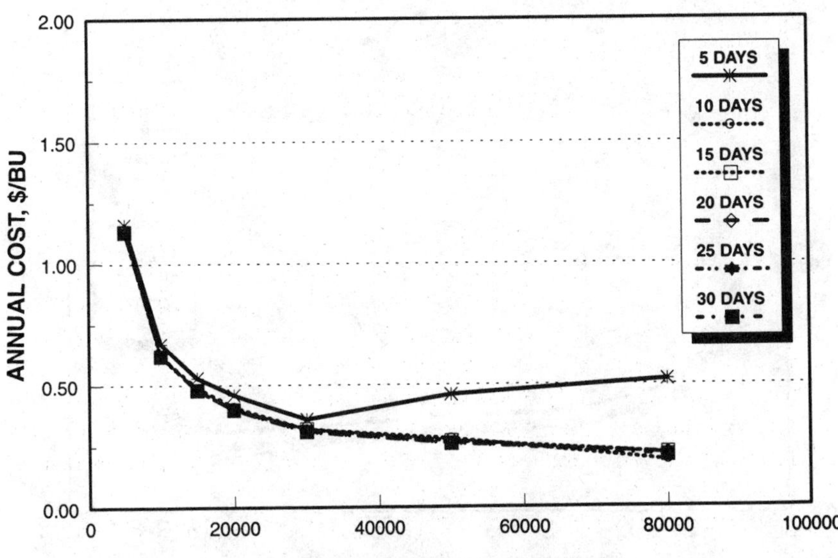

Figure 9.11–Annual cost comparisons for grain drying and storage facilities as a function of drying rate expressed as the number of days that harvesting will occur for a constant volume of grain: Batch-in-bin drying (Loewer et al., 1976a).

The costs for drying, storage, and handling equipment also will be influenced by other factors. These include the total bushels to be stored, drying method, bin size, number of bins, degree of mechanization, expansion plans, and harvest rate. The relative influences of these factors are shown in figures 9.7 to 9.13 (Loewer et al., 1976a). Given this data, the following economic guidelines can be applied:

1. As the capacity of the facility increases, the facility cost per bushel usually decreases.
2. The least cost drying technique depends on the quantity of grain to be processed and stored with no single method being the least expensive for all capacities.
3. It is more economical generally to build the least number of bins possible for a given capacity and farm situation. However, under most conditions it is better to increase the number of bins rather than have bin eave heights exceed approximately 24 ft (9 rings @ 2.67 ft/ring). For layer drying systems the usual limit is 16 ft (6 rings @ 2.67 ft/ring).
4. As facility size increases, a smaller proportion of the cost is for handling equipment.

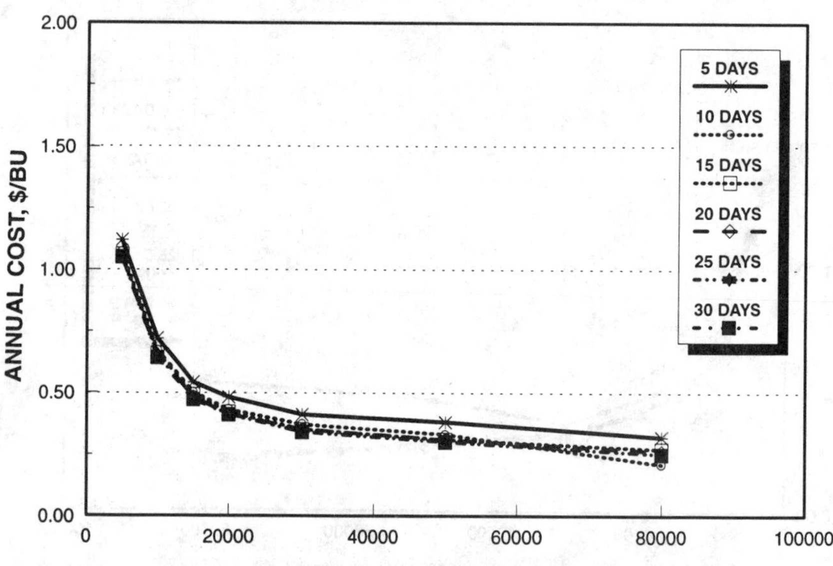

Figure 9.12–Annual cost comparisons for grain drying and storage facilities as a function of drying rate expressed as the number of days that harvesting will occur for a constant volume of grain: Portable drying (Loewer et al., 1976a).

5. If the facility is designed with expansion in mind, the potential increase in capacity may warrant a different drying technique in the original structure.
6. Harvest rate is the single most important factor in selecting a drying technique. Note, however, that daily drying rate is defined as the total quantity of grain to be dried divided by the number of days that harvesting occurs. As a result, the bushels dried per day increase disproportionately as the number of days allocated by harvesting decreases, thus resulting in a disproportionate increases in system cost.

Considering all these factors, the annual cost for grain storage facilities varies over a large range and each farmer must evaluate the system to determine annual cost.

Overdrying Cost

It's best not to dry the grain below the base moisture content (which for corn has traditionally been 15.5%) when drying and selling grain to the local elevator immediately after harvest. Unfortunately, most grain cannot be stored at this base moisture content for long periods of time without assuming some risk of spoilage (see Chapter 4). The suggested

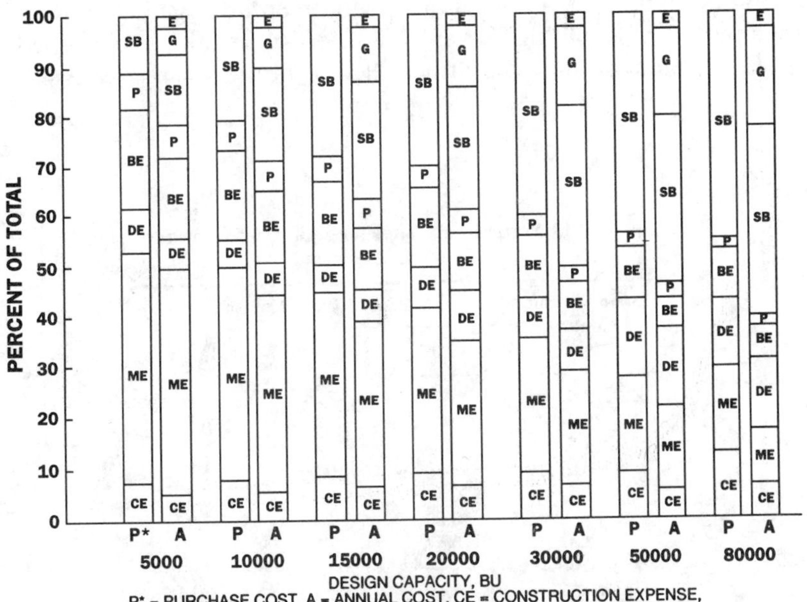

Figure 9.13–Proportionate annual cost comparisons for different components of a layer drying system as influenced by storage capacity (20 harvest days) (Loewer et al., 1976a).

long-term storage moisture content of grain varies with the geographic area of the country. For corn stored in the northern states, a storage moisture content of 14% is suggested while a 12% moisture content is recommended in the southern states. What this means to the farmer is that if the intention is to store grain until the following spring or summer, the corn may have to be dried below the base moisture content and, therefore, lose some crop value. For corn dried from 15.5% to 13.5% where the base moisture content is 15.5%, the loss would be 2.31% (table 9.13) or approximately $0.07/bu for $3.00 corn. For the same corn dried to 10%, the loss would be slightly over $0.18/bu. The point to remember is that drying to moisture levels below the base moisture content results in a reduction of saleable product and this loss must be viewed as a necessary storage expense. However, drying grain below the moisture content required for safe storage is a waste of fuel and will result in unnecessary economic loss.

Interest, Insurance, and Taxes

Interest may be the single most expensive storage cost. If grain had been sold at harvest, interest could have accumulated on the sale, or existing debts could have been paid. If an effective annual interest rate of 12% is assumed, the monthly storage charge is 1%. Grain harvested on the first day of September and stored until July 1 (10 months) would have an interest charge of $0.30/bu based on an average value over the storage period of $3.00/bu. For an 18% interest rate the expense would be $0.45/bu under similar conditions. Note that the interest rate should

Table 9.13. Percent weight loss for grain dried below the base moisture content*

Grain Moisture Content (%)	Base Moisture Content		
	13.0%	14.0%	15.5%
15.5	–	–	0.00
15.0	–	–	0.59
14.5	–	–	1.17
14.0	–	0.00	1.74
13.5	–	0.58	2.31
13.0	0.00	1.15	2.87
12.5	0.57	1.71	3.43
12.0	1.14	2.27	3.98
11.5	1.69	2.82	4.52
11.0	2.25	3.37	5.06
10.5	2.79	3.91	5.59
10.0	3.33	4.44	6.11

* [(base MC - final MC) / (100 - final MC)] * 100.

be assessed against the average value of the grain over the entire storage period rather than on the value at harvest or at the time of sale.

Insurance and taxes on stored grain will vary with local conditions but usually will be no more than 1% of the grain's value. For $3.00 corn, this would mean that somewhat less than $0.03/bu in insurance and taxes must be charged to storage. See figure 9.14 for the interest charges associated with stored grain.

Grain Stored for Feed Purposes

If the grain is for feed, the expenses for storage should be adjusted because losses from overdrying are not considered. This is because there is little difference in dry matter content or nutritional value for overdried grain as compared to grain at the base moisture content. In addition, savings resulting from farm-stored grain, as compared to grain purchased off the farm, should be considered. Also, grain usually may be purchased at a reduced price during the harvest period.

For grain stored and fed on the farm, the gross return for storage is the average value of the stored grain over the storage period minus its value when placed in storage. This assumes that grain will be fed uniformly over the storage period.

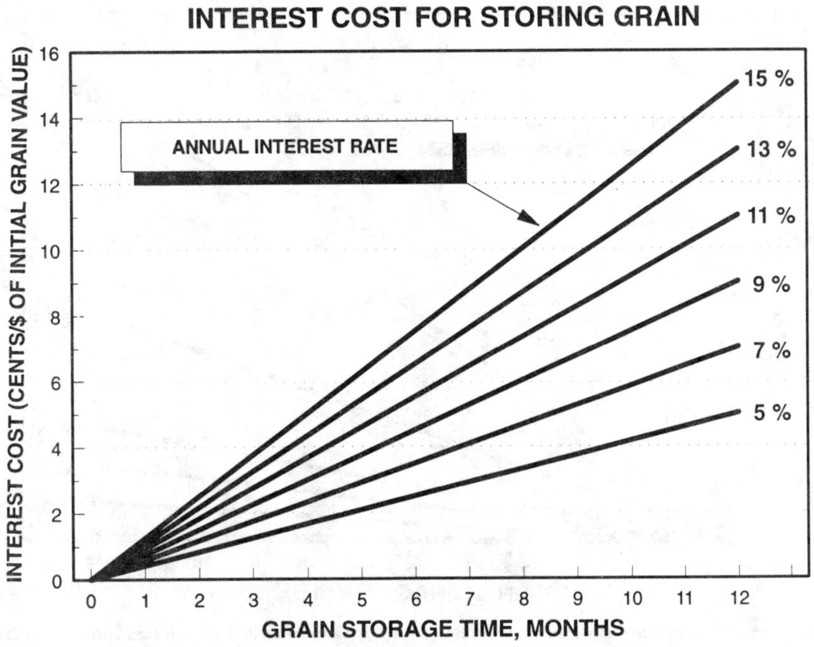

Figure 9.14–Relationship between storage time and the interest cost for various interest rates.

There is also an interest charge for grain that will be fed. The interest charge should be assigned to storage if the savings associated with on-farm feeding are credited to storage.

Increase in Value of Stored Grain

The expenses of farm storage and drying have been discussed. What also needs to be considered is how much the stored grain increases in value over the storage period. This increase varies from year to year and is a function of local, national, and world conditions. In years with a grain surplus, the increase in value may be slight. In years when the demand is great, the increase in value of stored grain is large. If it is anticipated that the value of stored grain will decrease in value over the storage period, it should be sold at harvest to avoid as much of the interest cost (a variable expense) as possible. In the long run, gross returns to storage must exceed the variable and fixed costs associated with storage. Otherwise, no grain would be stored. See figure 9.15 for an estimate of the gross return for several storage expectations.

Summary

When determining the feasibility of grain storage, many factors must be considered. These include facility cost, overdrying cost, interest,

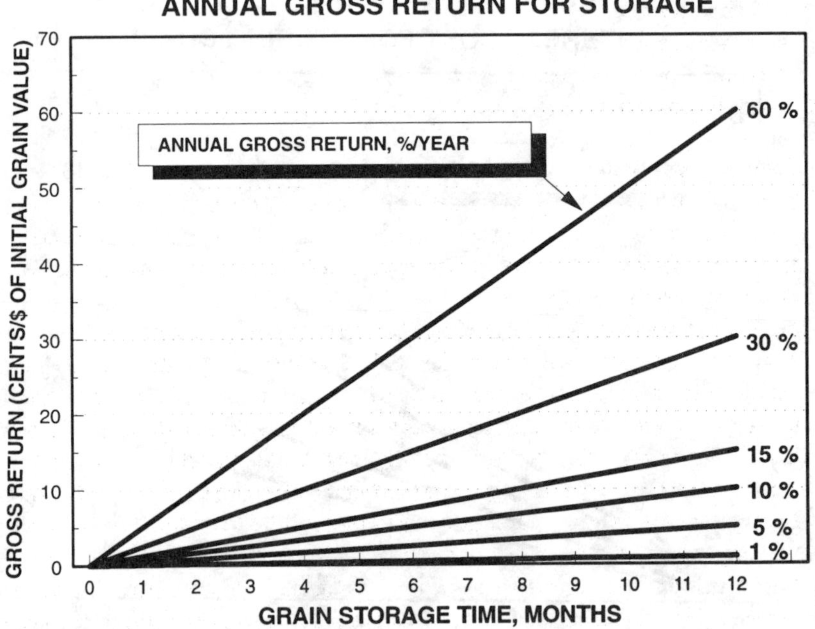

Figure 9.15–Relationship between gross return to storage and storage time as influenced by different rates of return.

taxes, and insurance costs and expected gross return. The expected net return to storage is dependent on whether the grain is sold from storage or fed on the farm and whether all drying, handling, and dual purpose processing equipment is charged to storage. This again emphasizes the importance of considering the total farm system when evaluating the economics of storing grain.

Economics of Feed Processing

In most livestock enterprises, a major portion of the cost of production is associated with the cost of feed. Therefore, any significant reduction in the cost of obtaining feed also represents a significant reduction in the total cost of production and an increased profit margin for the farmer.

Many livestock producers use on-farm feed processing systems to reduce the expense associated with purchasing feed. The move to an on-farm system requires expenditure of capital and selection of appropriate equipment. Factors to be considered include:

1. Quality and quantity of feed required.
2. Cost and advantages of commercial services.
3. Availability of ration ingredients.
4. Availability of labor.
5. Hauling expenses.
6. Equipment costs.
7. Net return on investment.
8. Overall convenience.

Production Systems

There is a wide variety of situations from which a farmer may begin consideration of on-farm feed processing. Three general types of production systems will be examined that describe typical on-farm livestock enterprise situations.

The first system is that of a grain farmer who already has a livestock enterprise or is considering the addition of such an enterprise. Basically, this individual wants to market part or all of the farm grain through livestock. This person grows grain; has grain storage, drying, and handling facilities; and has previously been having livestock feed processed at a local mill or by a custom operator. The return to on-farm feed processing is the difference between the combined fixed and operating cost of the on-farm processing unit and the cost of commercial processing. However, on-farm storage must also be considered. To evaluate return to storage, this producer must consider the difference in price between the average value of grain purchased off the farm and the average value of grain in on-farm storage.

A second system is one in which livestock is the only farm enterprise. There is neither grain production nor grain storage on the farm, and there are no plans to change that situation. Feed ingredients are purchased from a neighbor and/or feed store as needed, and the feed is processed commercially. More typically, the producer may buy a complete ground and mixed ration delivered from a single source. For this system, the economic evaluation would be a comparison between the cost of having a ration prepared commercially and the cost of owning and operating the on-farm system. In both cases, ingredients would be purchased off the farm as required.

The third production system is one in which the farmer does not produce any grain and has no storage or feed processing equipment for the on-farm livestock enterprise. However, this person is willing to consider the addition of these facilities. Presently this farmer is purchasing a commercially prepared ration for livestock but is willing to consider storage as part of on-farm feed processing system, believing that the lowest ration cost can be obtained by purchasing grain during harvest time and processing it on the farm. This farmer is concerned with the difference between the price of a commercially prepared ration and a farm-prepared ration, with on-farm storage considered as a part of the feed processing expense.

All of these producers are concerned primarily with the difference between on-farm costs and commercial costs. However, these costs will not be the same in each case. To determine the net return for each of the above situations, the expenses for processing farm and commercial feed must be determined.

On-Farm Feed Processing and Equipment Expenses

The equipment used for farm feed processing usually consists of a portable or stationary mill of either the roller or hammer type. The mill may be powered by either an electric motor or a tractor. The stationary electric grinder-mixer ranges from 2 to 10 hp. Portable mills usually are powered by the tractor pto. More labor and energy inputs will be required when a portable mill is used. Although higher labor and energy inputs are a disadvantage, this type of mill is portable and very appropriate for many feeding operations, especially those that utilize roughages such as hay or corn cobs.

Other items may also be required in addition to the portable or stationary processing mill. These include regular storage bins, hopper bottom bins, augers, electric motors, wiring and controls. Labor and energy costs must be added to the annual fixed costs to obtain the total annual costs.

The cost for storing and drying grain to be used in farm feed processing includes expenses for depreciation, fuel, repair, insurance, taxes, and interest. Drying returns and reduced harvest losses should be considered in addition to savings associated with feed processing.

Commercial Feed Processing Expenses

Expenses for commercial processing vary with each location and with the services rendered. The cost of a complete commercial ration (ingredients plus processing) is typically 10 to 20% greater than the cost of the ingredients alone when obtained from a farm source. Part of this price difference exists because the commercial facility must store as well as process the ration components. Other costs to consider are the charges for labor and hauling associated with delivery to and from the commercial facility.

Break-even Point Determination

Once the expenses associated with both farm processing and commercial processing have been established, it is possible to calculate the annual feed tonnage required to make on-farm feed processing profitable. This is called the break-even point and can be calculated using the following formula:

Table 9.14. Items to be considered when determining fixed cost

Item	Calculation
Depreciation*	= (purchase price – salvage value) / (years of life)
Interest	= (purchase price + salvage value) * interest rate, %/200
Taxes	= (purchase price + salvage value) * tax rate, %/200
Insurance	= (purchase price + salvage value) * insurance rate, %/200
Repairs and Maintenance†	= (purchase price * % total repairs of purchase) / ((years to be used) * 100)

* Depreciation is a real cost. In terms of physical utility, it should be related to the item's physical life (sometimes referred to as "units of life" method). For example, a $20,000 car that is expected to last 100,000 miles could be said to depreciate at the rate of 5¢/mile. However, depreciation is most normally associated with tax liability and is considered a fixed cost. The method shown is the straight line calculation assuming no salvage value.

† As with depreciation, repair costs are best related to the item's physical life. The formula given assumes that years of life is highly correlated to the physical usage of the machine. It does not address the fact that repair costs tend to increase later in the life of the machine. Fortunately, the combined use of both line depreciation and uniform repair costs tends to produce somewhat offsetting errors giving a truer picture of annual costs than either approach does when used separately.

Table 9.15. Items to be considered when determining operating cost of a stationary mill

Stationary Mill Example Problem	
Item	**Assumptions and Calculations**
Mill capacity	2 tons/h
Purchase cost	Base mill with accessories, $8,000
Estimated life	12 years
Depreciation	On average, 8.33% of purchase cost per year based on an expected life of 12 years
Interest	10% annual rate
Insurance and taxes	2% of the depreciated purchase price per year
Repairs	100% of purchase price spread uniformly over the life of the mill
Labor (maintenance and supervisory time)	$1.50/t based on a labor rate of $6/h, a processing rate of 2 t/h, and assuming that 50% of the labor will be used elsewhere during the time the mill is functioning
Electricity cost	$0.21/ton based on 3.5 kWh/ton at $0.06/kWh
Salvage value	$0.00

Table 9.16. Items to be considered when determining operating cost of a protable grinder-mixer (example problem)

Item	Assumptions and Calculations
Mill capacity	6 tons/h
Purchase cost	Base mill with accessories, $12,000
Estimated life	8 years
Depreciation	On average, 12.5% of purchase cost per year based on an expected life of 8 years
Interest	10% annual rate
Insurance and taxes	2% of the depreciated purchase price per year
Repairs	120% of purchase price spread uniformly over the life of the mill
Labor (maintenance and supervisory time)	$1.00/t based on a labor rate of $6/h, a processing rate of 6 t/h
Tractor cost	$1.67/t based on $10/h value placed on the tractor for all costs and a 6 t/h processing rate
Fuel cost	$1.83/t based on an 80-hp tractor using 11 gal of diesel fuel/h, a fuel price of $1.00/gal, and a processing rate of 6 t/h
Salvage value	$0.00

$$\text{break-even point} = \frac{\text{total annual fixed cost for farm processing}}{\text{(commercial processing cost / ton} - \text{farm operating cost / ton)}} \quad (9.16)$$

The break-even-point will vary with each individual farming operation. To illustrate this point, three example calculations will be made to represent each of the three previously described production systems. The set of assumptions used in all the example systems is given in tables 9.14 to 9.16.

Example 1: The first system is that of a producer who grows grain, has a grain storage facility, and wants to determine the break-even point for installing farm feed processing utilizing a stationary electric mill. This person estimates that an additional 200 bu of supplement and ration storage and 150-ft 4-in. auger (with a 2-hp electric motor) will be required to utilize this system as compared to commercial processing. Referring to tables 9.14 and 9.17, the individual calculates the fixed and variable cost to be $1,813.34 and $1.71/ton, respectively (fig. 9.16).

If feed is processed commercially at the local feed mill, the estimated cost is $8.80/ton ($0.44/cwt) for grinding and mixing and $0.60/ton for hauling and labor resulting in a total cost of $9.40/ton. The break-even point can now be calculated using equation 9.16 as follows:

Table 9.17. Example problem determinations of stationary mill cost

	Tons of Feed Per Year					
Item	50	100	200	400	800	1200
Depreciation	666.67	666.67	666.67	666.67	666.67	666.67
Interest	400.00	400.00	400.00	400.00	400.00	400.00
Insurance and taxes	80.00	80.00	80.00	80.00	80.00	80.00
Repairs	666.67	666.67	666.67	666.67	666.67	666.67
Total Fixed Cost	1,813.34	1,813.34	1,813.34	1,813.34	1,813.34	1,813.34
Labor	75.00	150.00	300.00	600.00	1200.00	1800.00
Electricity	10.50	21.00	42.00	84.00	168.00	252.00
Total Operating Cost (%/t)	1.71	1.71	1.71	1.71	1.71	1.71
Total cost ($)	1898.84	1984.34	2155.34	2497.34	3181.34	3865.34
Total cost ($/t)	37.98	19.84	10.78	6.24	3.98	3.22

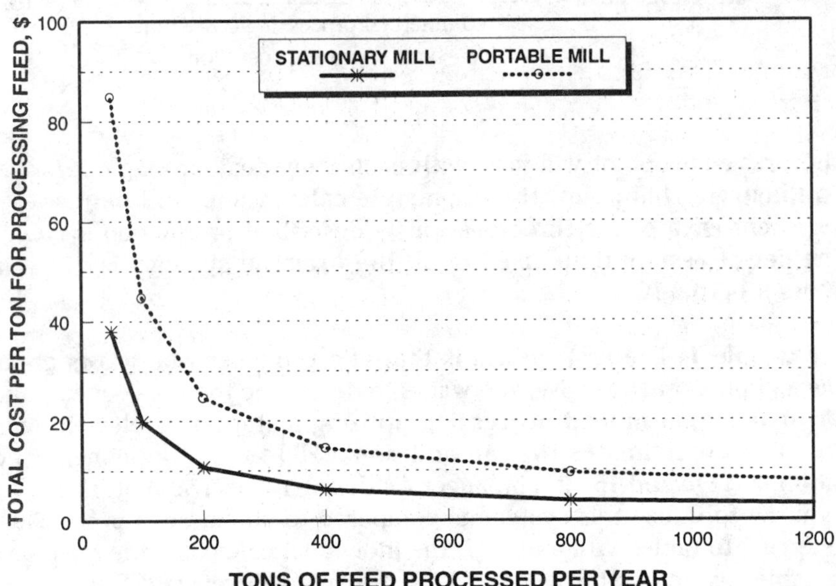

Figure 9.16–Comparative costs of the stationary and portable mills as used in the example problem.

$$\text{break-even point} = \; = \frac{\$1,813.34}{(\$9.40\,/\,\text{ton} - \$1.71\,/\,\text{ton})}$$

$$= 236 \text{ tons / year}$$

Table 9.18. Example problem determinations of portable mill cost

Item	Tons of Feed per Year					
	50	100	200	400	800	1200
Depreciation	1500.00	1500.00	1500.00	1500.00	1500.00	1500.00
Interest	600.00	600.00	600.00	600.00	600.00	600.00
Insurance and taxes	120.00	120.00	120.00	120.00	120.00	120.00
Repairs	1800.00	1800.00	1800.00	1800.00	1800.00	1800.00
Total Fixed Cost	4,020.00	4,020.00	4,020.00	4,020.00	4,020.00	4,020.00
Labor	50.00	100.00	200.00	400.00	800.00	1200.00
Diesel fuel	91.50	183.00	366.00	732.00	1464.00	2196.00
Tractor	83.50	167.00	334.00	668.00	1336.00	2004.00
Total Operating Cost (%/t)	4.50	4.50	4.50	4.50	4.50	4.50
Total cost ($)	4245.00	4470.00	4920.00	5820.00	7620.00	9420.00
Total cost ($/t)	84.90	44.70	24.60	14.55	9.53	7.85

Example 2: In our second system, the farmer produces no grain. This producer has no grain storage and doesn't intend to purchase any. All the grain for livestock feed is purchased from neighbors. Unlike the first producer, this farmer wishes to use a portable grinder-mixer. It is estimated that it will require 30 min to deliver 3 tons of the processed feed after it has been prepared. To support the on-farm electric mill, the plans are to purchase 500 bu of hopper-bottom storage for supplement, ration and ingredients. Cost calculations are shown in table 9.18 and figure 9.16. Fixed cost is $4,020 and variable cost is $4.50/ton.

The farmer presently has grain both ground and mixed at a cost of $10.80/ton. Because the grain has to be hauled regardless of whether commercial or on-farm processing is used, hauling expense would be the same in both cases; thus, this factor is not included in the cost calculations. For this example, then:

$$\text{break-even point} = \frac{\$4{,}020.00}{(\$10.80\ /\ \text{ton} - \$4.50\ /\ \text{ton})}$$

$$= 638 \text{ tons / year}$$

Example 3: In this system, the producer wishes to purchase dry grain during harvest and store and process it on the farm. This farmer will consider storage as part of on-farm feed processing operation. The estimate of the total stationary mill fixed cost is $1,813.34/year plus an operating cost of $1.71/ton. The estimate of annual storage cost is approximately $1,500. Based on local conditions, the farmer estimates that it will cost an additional $16.07/ton ($0.45/bu) to buy a commercially prepared ration instead of purchasing grain during harvest and storing it on the farm. In this case, commercial processing cost includes the increased cost of commercially stored ingredients. Therefore, for this situation:

$$\text{break-even point} = \frac{\$1{,}813.34 + \$1{,}500}{(\$16.07\ /\ \text{ton} - \$1.71\ /\ \text{ton})}$$

$$= 231 \text{ tons / year}$$

If the additional cost of the commercially prepared ration had been $25/ton, the break-even point would have been 142 tons. For this producer, the grain storage return is assumed to be part of the feed processing return.

Summary

The economics of feed processing depend greatly on the particular situation of the producer. Generally, feed processing enhances greatly the economic attractiveness of grain storage and drying facilities, especially for individuals who produce their own grain. Again, this says nothing about the desirability of the feeding operation as a separate economic enterprise.

Table 9.19. Base factors used in analysis

Description	Base 1	Base 2	Base 3
I. Harvesting Strategies			
A. If on-the-farm grain drying and storage is not available			
1. Calendar days required for harvesting	20.0	30.0	45.0
2. Moisture content when harvest begins (%)	25.0	20.0	30.0
3. Speed of corn harvester (mph)	3.5	3.5	2.5
B. If on-the-farm drying and storage is available			
1. Calendar days required for harvesting	20.0	20.0	35.0
2. Moisture content when harvest begins (%)	25.0	25.0	30.0
3. Speed of corn harvester (mph)	3.5	2.5	2.5
II. Facility Management Strategies			
A. Interest, taxes and insurance as a % of average grain value	10.0	8.0	8.0
B. Average number of months that grain is stored	6.0	5.0	4.0
C. Moisture content of stored corn	13.5	13.5	13.5
III. Market Conditions			
A. Price of corn at harvest ($/bu)	2.50	2.50	2.50
B. Expected increase in the value of stored grain (%)	10.0	15.0	15.0
C. Base Moisture content of corn (%)	15.5	15.5	15.5
IV. Energy Considerations			
A. LP gas used per point of moisture (gal)	0.02	0.02	0.02
B. Price of LP gas ($/gal)	0.40	0.35	0.40
C. Average drop per day in field moisture content after harvesting begins (points/day)	0.25	0.25	0.25
V. Facility Design			
A. Potential yield, no harvest losses (bu/acre)	100.0	110.0	120.0
B. Corn to be dried and stored (acres)	100.0	250.0	300.0
C. Drying technique (1=layer; 2-batch-in-bin; 3=portable)	2.0	2.0	3.0
D. Degree of mechanization (%) (0=portable handling system; 50=bucket elevator and pit; 100=bucket elevator and pit, center building, scale)	0.0	0.0	50.0
VI. Net Return per Dry Unit (15.5% w.b.) cents/bushel*	–25.91 (–32.67)	11.89 (6.14)	3.12 (16.13)

* The first value was computed using one set of cost for the facility and used in figures 9.17 to 9.30. The second value was computed with updated values.

Influence of Harvesting Strategies on Economic Feasibility

The economic feasibility of an on-farm grain drying and storage facility reflects the interaction of management decisions and price expectations. The following discussion presents the influence of many of the above mentioned factors on the economic return from on-the-farm grain storage facilities as determined by using the CACHE computer model (Loewer et al., 1976b). The term "storage facility" is used to designate grain drying, handling, and storage equipment and structures associated with a centralized grain facility. Corn is the only grain considered and all of the harvested grain is dried and stored.

Procedure

The factors considered are given in table 9.19 along with various levels for a selected "base" condition. A base condition is defined as a stated set of resource and management condition inputs and may be viewed as an example farm system. A range of input conditions was selected for each of the input factors considered, while holding each of the other base inputs constant. This is commonly referred to as a sensitivity analysis.

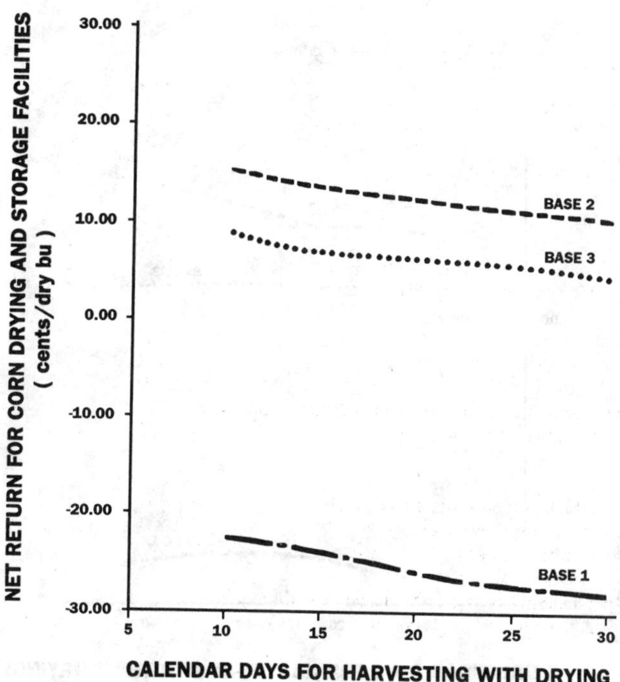

Figure 9.17–Net return to on-farm corn drying and storage facilities as influenced by the calendar days required for harvesting if on-farm drying and storage facilities are available (Loewer et al., 1980).

As defined by expected economic return, Base Condition 1 might be described as the "pessimistic" view, Base Condition 2 as the "optimistic" view with Base Condition 3 the "middle" view. In most cases, however, the effects of changing the values of any factor affect all three conditions in much the same way. In other words, the rate of change is very nearly equal in most instances, and the relative importance of this quantity will be emphasized rather than the absolute values as obtained from the sensitivity analysis.

Harvesting Strategy Factors

Field losses are a function of moisture content, yield, and length of harvest (Johnson and Lamp, 1966). The CACHE model utilizes the data shown in table 9.6 to describe the relationships among these factors. Generally, harvest losses increase with lower beginning harvesting moisture contents and longer harvest times. The economic return to grain storage and drying involves a comparison of a farm with and without a grain facility. In other words, an evaluation will be made of harvesting strategies "with" and "without" the facility. It should be noted that the expected net return per unit volume is based on the quantity of dry grain (15.5% w.b.) actually harvested when grain drying

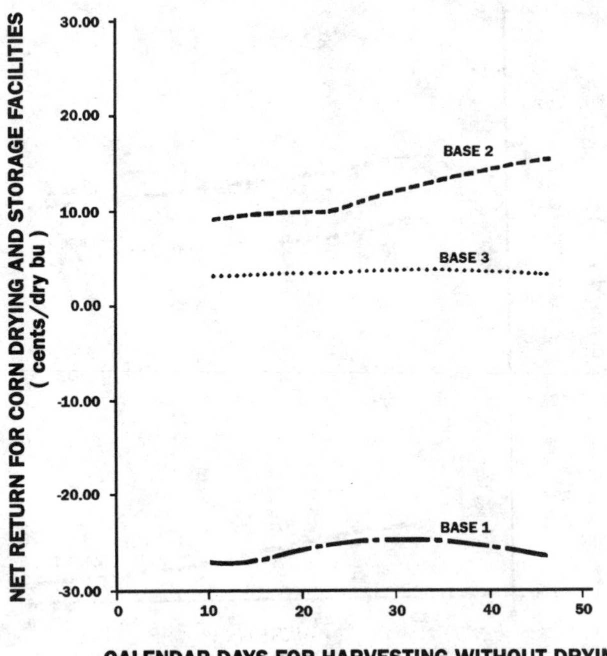

Figure 9.18–Net return to on-farm corn drying and storage facilities as influenced by the calendar days required for harvesting if on-farm drying and storage facilities are not available (Loewer et al., 1980).

and storage are available and that this number may change if the harvesting strategy changes. The total net return for a given system is the product of the net return per bushel and the number of "dry" bushels harvested.

Calendar Days for Harvesting with Drying. The relationship between the number of calendar days required to harvest the grain if farm storage is available and the expected net return to grain storage is shown in figure 9.17. As harvest time increases, net return to storage decreases, the average rate being approximately –0.258 ¢/bu/day of harvest delay. This assumes that for the total farm economic system no additional expense is incurred when the total harvest time is reduced.

Calendar Days for Harvesting without Drying. The relationship between the time required for harvesting if no grain facility is available and the net return to farm storage is shown in figure 9.18. As the time required for harvesting "without" a farm facility increases, harvest losses increase. This is reflected by an increase in the desirability of on-the-farm storage as shown in the general increase in expected net return. The net effort is approximately $0.0945/bu for each additional day that is required to harvest the grain when no facility is available.

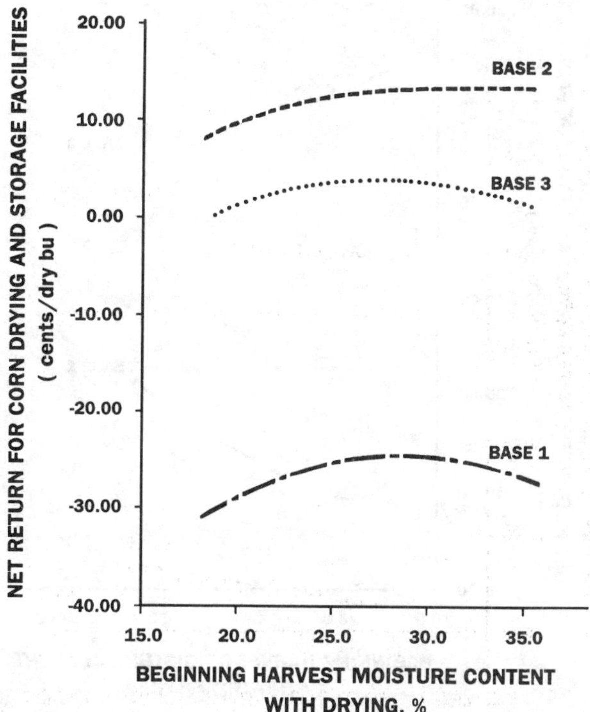

Figure 9.19–Net return to on-farm corn drying and storage facilities as influenced by the moisture content that harvest begins if on-farm drying and storage facilities are available (Loewer et al., 1980).

Beginning Harvest Moisture Content with Drying. The earlier that harvest begins, the less the harvest losses. However, more fuel is required for drying. The trade-off in these relationships in terms of net profit is shown in figure 9.19, with the optimum moisture to begin harvesting when on-the-farm storage is available being in the 26 to 28% range for the situations shown.

Beginning Harvest Moisture Content without Drying. One of the most common dockage rate schemes used by commercial elevators for excessive corn moisture is to reduce the price received per wet bushel by a fixed percentage of the dry bushel selling price for each point of moisture above the base moisture content levels (traditionally 15.5% for corn). This approach is used in the CACHE model using a value of "2" for the fixed percentage. As the moisture content at which harvest begins increases with the no-drying situation, the harvest losses decrease but the dockage increases. The trade-off between these relationships as defined by net profit is shown in figure 9.20. This indicates that if the farmer has no grain storage system and traditionally has begun

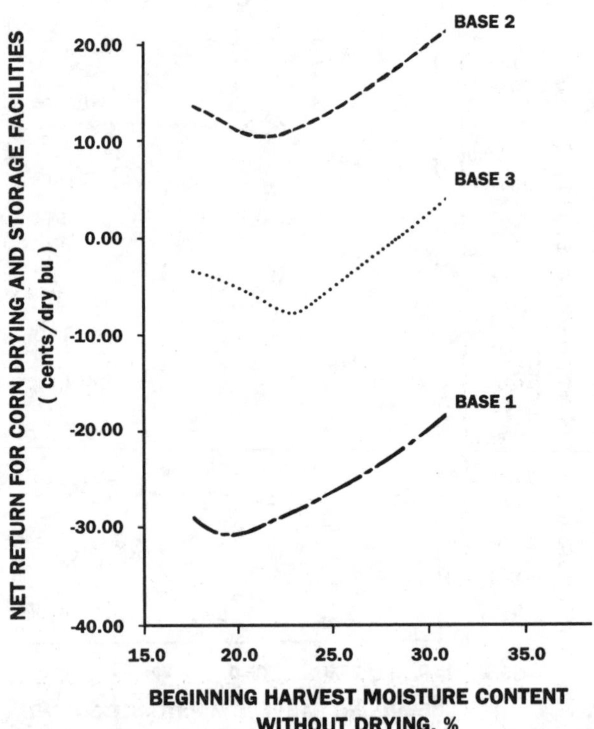

Figure 9.20–Net return to on-farm corn drying and storage facilities as influenced by the moisture content that harvest begins if on-farm drying and storage facilities are not available (Loewer et al., 1980).

harvesting in the 21 to 23% moisture content range, relatively less benefit will be received from the addition of a grain facility.

Harvesting Speed with Drying. The relationship between harvest speed when having a drying and storage system, and net economic return is shown in figure 9.21. As speed increases in the "with drying" option, harvest losses increase also which accounts for the reduction in net profit, the rate being approximately –5.168 ¢/bu/mph increase in harvest speed. This assumes that the increase in harvester speed will not reduce overall harvesting time in terms of calendar days.

Harvesting Speed without Drying. As harvesting speed increases when farm storage is not available, harvest losses increase also, resulting in a more favorable expected economic return to the "with drying" option (fig. 9.22). The rate of return is approximately $0.05649/bu for each additional mph in harvest speed.

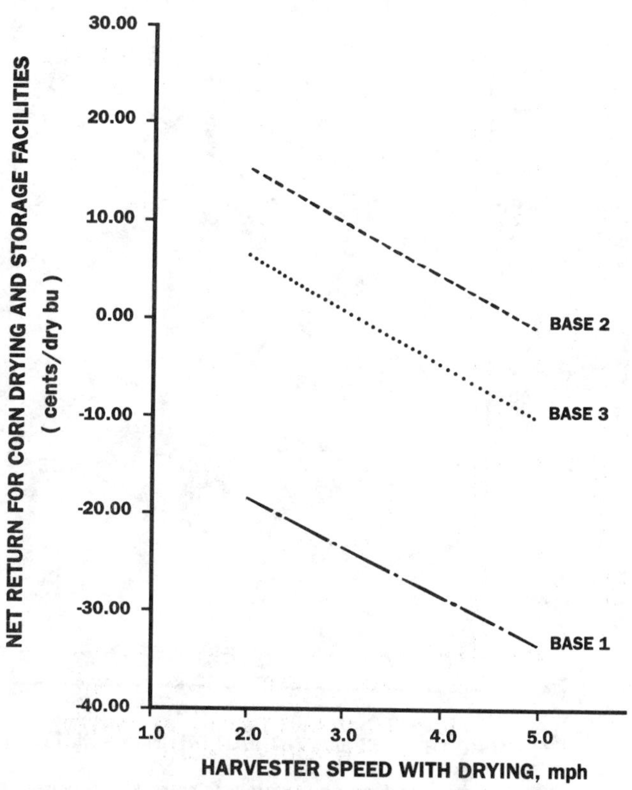

Figure 9.21–Net return to on-farm corn drying and storage facilities as influenced by the average speed of the harvester if on-farm drying and storage facilities are available (Loewer et al., 1980).

Facility Management Practices

Once the farmer has committed to the storage of farm-produced grain, decisions must be made concerning the management of on-farm drying and storage system.

Interest, Taxes, and Insurance. If grain were sold at harvest time, the money from this sale could be used to repay loans or placed in a savings account. Regardless, the interest charge to stored grain must be viewed as an opportunity cost. Likewise, taxes and insurance charges are a function of the value placed on the stored grain. The effects of a composite interest, taxes, and insurance charge on expected net returns are shown in figure 9.23. The slopes for the three base conditions are not parallel because of the differences in storage time and the value of the stored material. For purposes of this study, the effective percentage charged was based on the average value of the grain over the storage period. The interest cost on stored grain may be the single largest cost in many grain storage systems.

Number of Months that Grain is to be Stored. The number of months that grain is to be stored is one of the critical factors in determining the

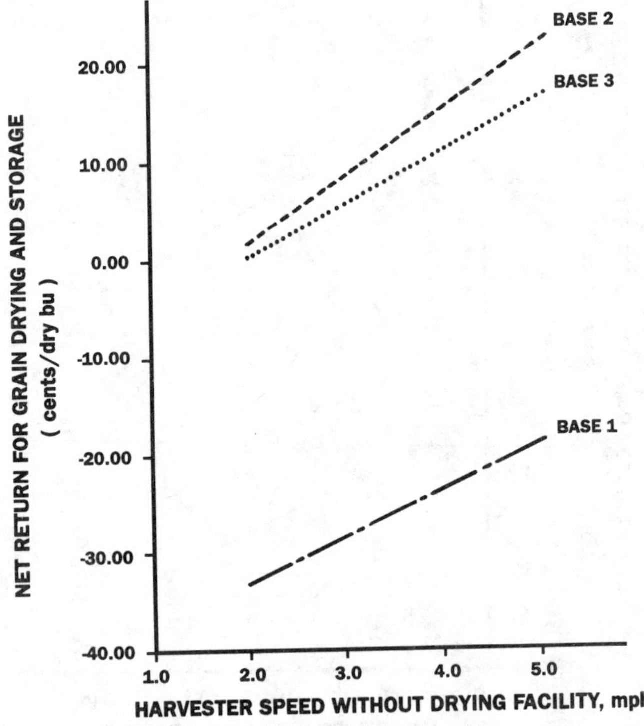

Figure 9.22–Net return to on-farm corn drying and storage facilities as influenced by the average speed of the harvester if on-farm drying and storage facilities are not available (Loewer et al., 1980).

effects of interest, taxes, and insurance on expected net return to grain storage (fig. 9.24). The variation in slopes of the three base conditions is because of differences in charges for interest, taxes, and insurance, and a difference in the average value of the stored grain over the storage period.

Moisture Content of Stored Corn. The moisture content of grain to be placed in storage depends on the expected temperatures during the storage period and the risk that the manager is willing to assume (Ross et al., 1973). Corn that is to be stored in the Midwest only during the winter months may retain its quality with a 15.5% storage moisture content if the manager is willing to assume some risk. This would be contrasted with a 12% storage moisture content in the South for the manager who plans to store grain until mid summer.

A storage moisture content below the base moisture standard for corn (given as 15.5% for purposes of example) results in economic loss to the farmer by (1) reducing the quantity of product the farmer has available for sale, and (2) increasing the cost of drying because of the

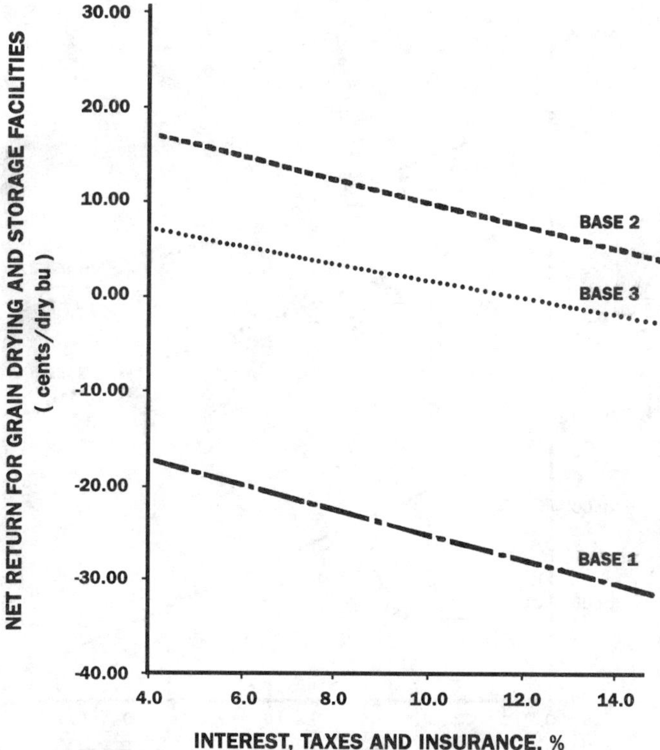

Figure 9.23–Net return to on-farm corn drying and storage facilities as influenced by the cumulative interest, tax, and insurance charges for the average value of the stored grain (Loewer et al., 1980).

extra moisture that must be removed. The effects of storage moisture content on expected net return is shown in figure 9.25.

If the grain is to be fed on the farm, the "overdrying" is less important in that the dry matter content of the corn remains essentially the same regardless of moisture content. Also, overdrying may be necessary for safe storage of the grain and, if so, it should be viewed as an essential cost of grain storage.

Market Conditions

One of the primary considerations in the purchase of grain drying and storage equipment is the expected economic benefits to be gained from holding the grain for future sale. Although it is impossible to predict the exact prices at a given time, the net return can be computed based on given market expectations.

Price of Corn at Harvest. The price of corn at harvest affects the profitability of on-the-farm grain storage in several ways. Given a constant percentage increase in price, the absolute price over the

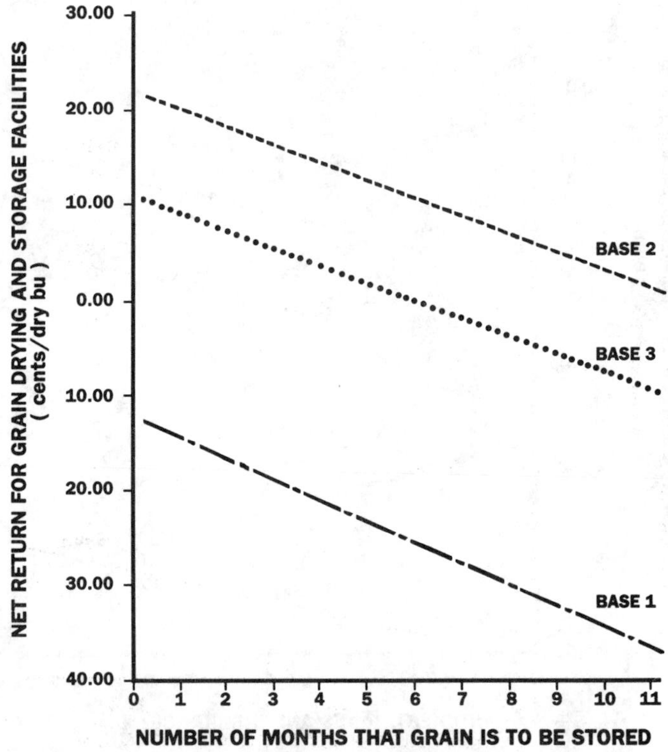

Figure 9.24–Net return to on-farm corn drying and storage facilities as influenced by the average number of months that the grain will be stored assuming the same price will be received regardless of storage time (Loewer et al., 1980).

storage period also increases. For example, if the farmer expects a 10% increase in the price of corn over the storage period, $2.00/bu corn will increase by $0.20/bu, while $3.00/bu corn increases $0.30/bu, a difference of $0.10/bu for the same expected percentage increase in value.

Based on the same approach, higher corn prices at harvest will reduce profitability somewhat because of an increased cost for interest, taxes, and insurance. Likewise, the cost of overdrying, because of a reduction of saleable product, increases also with an increase in the price of corn.

Probably the most significant factor of the increase in corn prices is related to harvest losses. The value of each unit lost increases directly with price because of the factors discussed in the section "Harvesting Strategy Factors". The net effects of these factors are shown in figure 9.26.

Expected Increase in the Value of Stored Grain. This is probably the single most important factor in determining expected net return to grain storage. If the expected increase in the value of stored grain is large

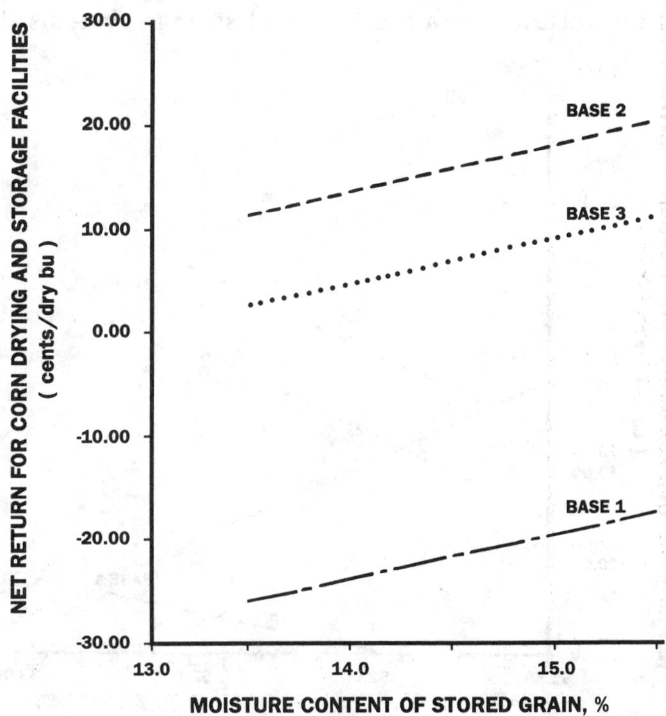

Figure 9.25–Net return to on-farm corn drying and storage facilities as influenced by the moisture content at which the grain will be stored assuming there are no in-storage losses related to excess moisture in the grain (Loewer et al., 1980).

enough, any grain system will be profitable. Likewise, it is difficult (but not impossible) for a system to show a profit if the expected increase in the value of the stored grain approaches zero.

The average value of the grain also increases as the expected increase in value becomes larger. Profitability is reduced in the same way as discussed in the previous section. The net effects of these factors are shown in figure 9.27.

Energy Considerations

When energy becomes less available and more expensive, increasing attention is given to fuel efficiency.

Amount of Fuel/Point of Moisture Removal. The grain system manager does have some flexibility in utilization of LP gas for drying. For example, using higher temperatures for drying or adopting the dryeration process may result in an energy saving (McKenzie et al., 1972; 1980).

Theoretically, it takes approximately 0.008 gal of LP gas/point of moisture removed. Conversion of LP gas to usable heat, when coupled with the drying process, increases this quantity by a factor of two to three times. Figure 9.28 relates the effects of energy utilization on the expected net return to grain drying and storage systems. Results

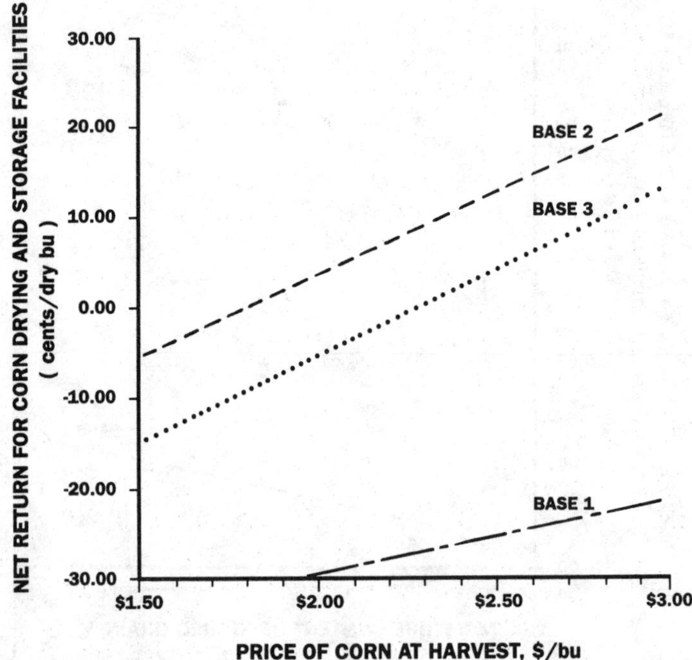

Figure 9.26–Net return to on-farm corn drying and storage facilities as influenced by the average price of corn at harvest (Loewer et al., 1980).

indicate that although energy utilization is important, only a relatively large change in drying and burning efficiencies would alter significantly expected net return.

Price of Fuel. The price of fuel for drying affects the expected net return to storage in much the same way as drying efficiency (fig. 9.29). Again, the cost of fuel is important, but if the expected net return to grain storage is sufficiently large, as with Base Condition 2, fuel prices alone will not be the deciding factor in the economic feasibility of on-the-farm grain storage. This point is very important in the evaluation of alternate sources of fuel for drying such as biomass conversion or solar. It may be possible to utilize these energy sources for drying at a much higher cost than the present cost of LP gas and still have a profitable total drying and storage system.

Moisture Drop/Day of Field Moisture Content after Harvesting Begins. The effects of varying this term are rather inconsistent indicating significant interactions among the associated input

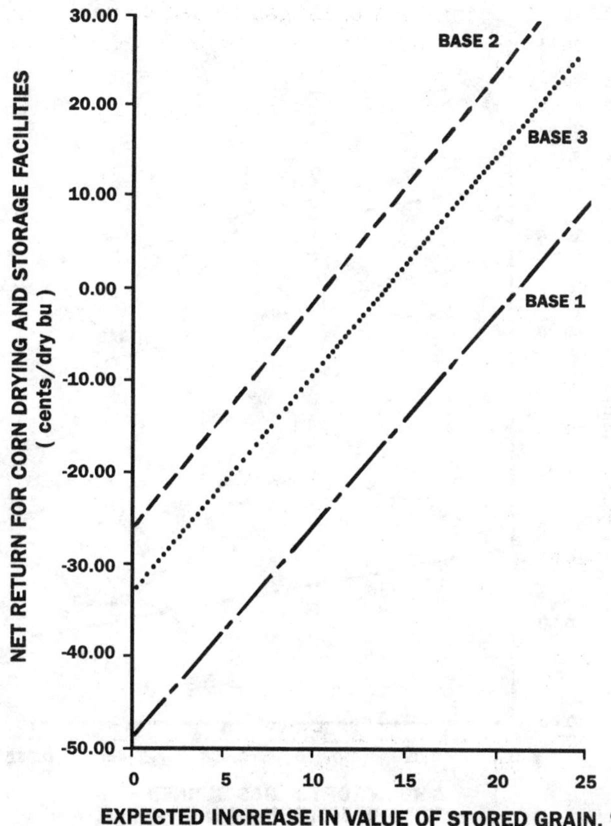

Figure 9.27–Net return to on-farm corn drying and storage facilities as influenced by the percentage increase in value of the stored corn over a given storage period (Loewer et al., 1980).

parameters (fig. 9.30). As the moisture content drops more rapidly in the field, the fuel required for on-farm drying decreases, but the benefits of the 2% dockage system favor the no-drying option. The CACHE model does not allow the average moisture content during the total harvesting season to fall below 18%, thus somewhat modifying the effects of relatively high rates of moisture drop in terms of expected net return.

Facility Component Costs

The cost of grain facility equipment on a per unit volume basis is largely a function of capacity, drying method, and degree of mechanization. The effects of these parameters, not considering tax savings associated with depreciation and investment credit, are shown in figures 9.31 to 9.33 (Loewer et al., 1976a). The following relationships influence facility size that, in turn, influences expected net return.

Potential Yield (bu/acre). Potential yield is defined as the yield if no harvest losses were encountered. As the potential yield increases, the

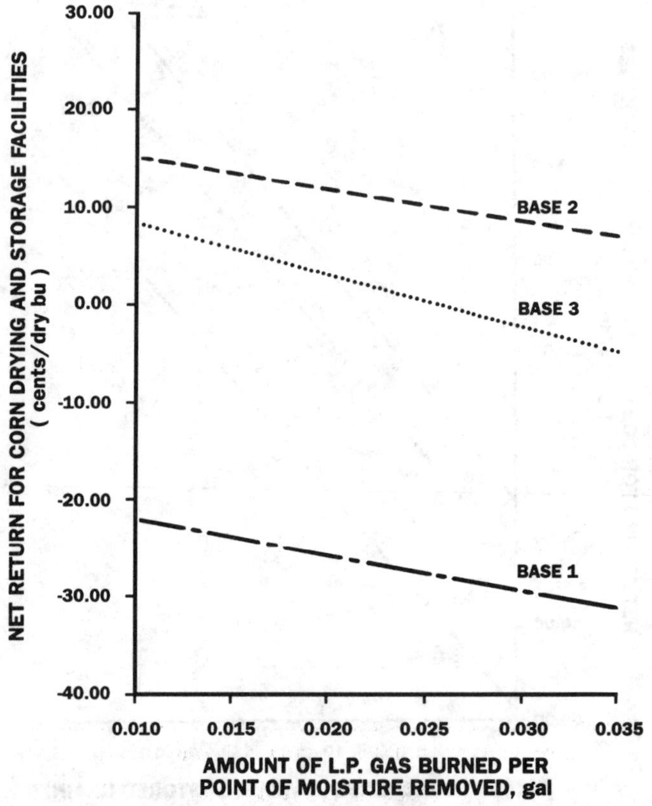

Figure 9.28–Net return to on-farm corn drying and storage facilities as influenced by the quantity of LP gas burned per point of moisture removed from the corn (Loewer et al., 1980).

required storage space also increases for the same harvesting strategy. The magnitude of harvest losses becomes more pronounced as do expenses and returns for drying. The effect of potential yield on expected net return is shown in figure 9.34.

Area of Corn to be Dried (acres). All the harvested corn in this study was dried and stored. The influence of this parameter is similar to potential yield although much more pronounced. The effect of acreage on expected net return (fig. 9.35) is very closely related to the curves presented in figures 9.31 – 9.33, indicating that much of the benefit of increased acreage is associated with reduced cost per bushel for storage facilities.

Summary

Expected net return to grain storage is a function of many parameters. A summary of the effects of several of these factors is shown in table 9.20. Even though prices change, a sensitivity analysis provides insight into the relative importance of the different parameters.

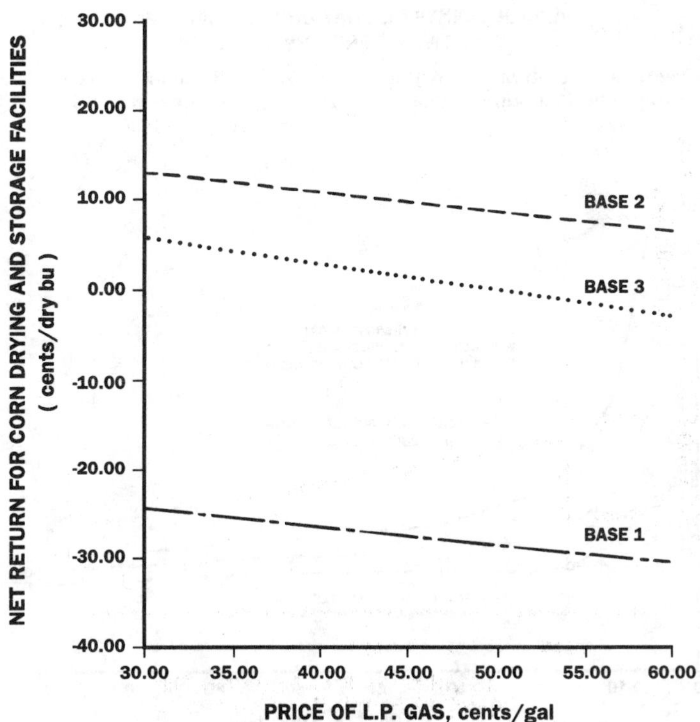

Figure 9.29–Net return to on-farm corn drying and storage facilities as influenced by the price of LP gas for drying (Loewer et al., 1980).

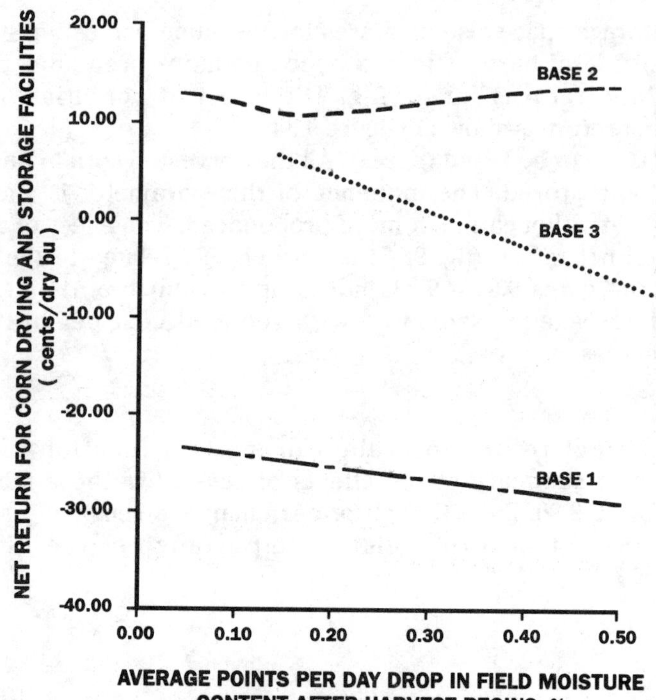

Figure 9.30–Net return to on-farm corn drying and storage facilities as influenced by the average point per day drop in field moisture content after harvest begins (Loewer et al., 1980).

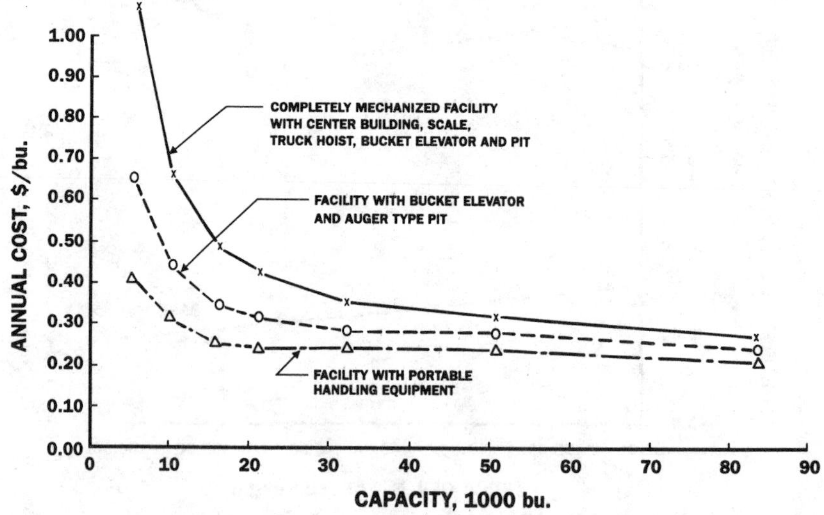

Figure 9.31–Annual cost to layer drying facility, 20-day harvest time (Loewer et al., 1976a).

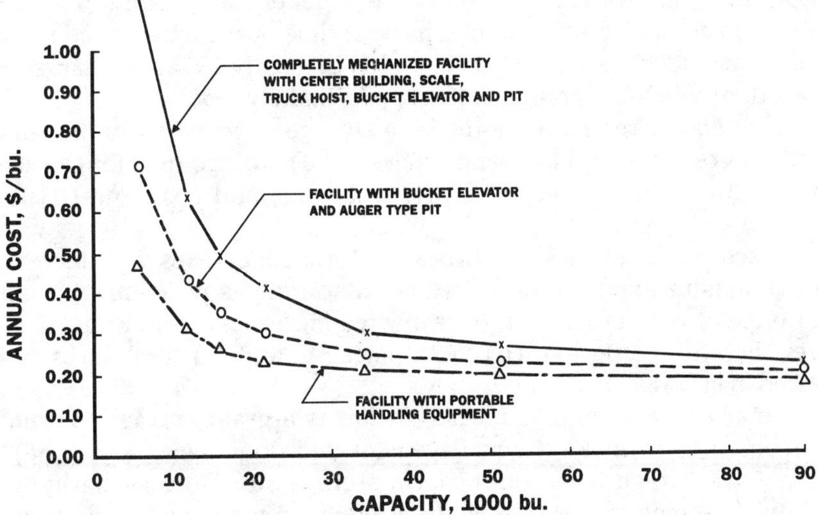

Figure 9.32–Annual cost to batch-in-bin drying facility, 20-day harvest time (Loewer et al., 1976a).

Generally, if on-the-farm storage is available, expected net return can be increased by reducing travel speed of the combine while keeping total harvesting time as short as possible. However, well-designed facilities free of bottlenecks may allow the farmer to accomplish both objectives (Benock et al., 1981).

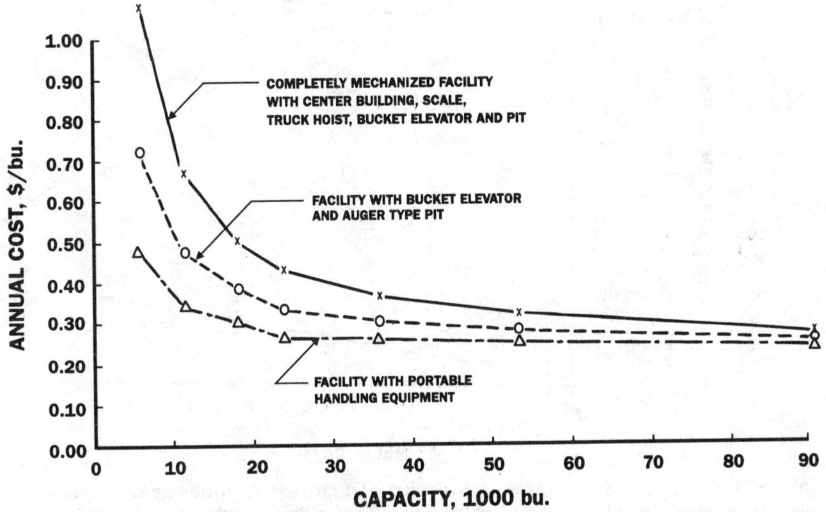

Figure 9.33–Annual cost to portable drying facility, 20-day harvest time (Loewer et al., 1976a).

On-the-farm storage is relatively less beneficial to the farmer who would not reduce individual total harvest time or combine speed if grain storage facilities were purchased, and is presently beginning harvesting operations when the grain reaches approximately 22%.

Expected net return to on-the-farm storage decreases with increases in (a) interest, tax and insurance charges, (b) storage time for the same expected rate of return, (c) drying fuel usage and cost, and (d) field drying rate.

Expected net return to on-the-farm storage increases with increases in (a) moisture content of stored grain so long as no damage occurs, (b) price of corn at harvest, (c) expected increases in the value of corn over the storage period, (d) yield per acre, and (e) total acres to be placed in storage.

The key to economic success, as far as a grain storage system is concerned, is to make the management decisions necessary to adjust from a no-storage to an on-the-farm storage situation. Primarily, this involves keeping harvest losses to a minimum with existing harvesting and delivery equipment, being aware of interest charges, building economical drying and storage facilities, and following correct marketing practices. Allowing the corn to dry in the field rather than

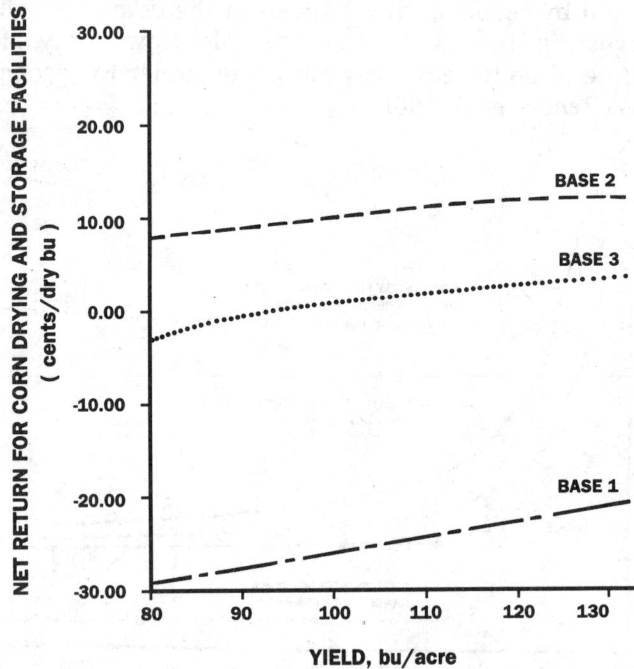

Figure 9.34–Net return to on-farm corn drying and storage facilities as influenced by the potential corn yield. Potential yield is the quantity of grain that would be harvested if there were no harvest losses (Loewer et al., 1980).

spending money for drying is false economy in most years. Holding corn for future sales must be justified with regard to interest charges on the stored grain.

The successful manager will be aware of the many trade-offs involved in grain facility management. This analysis provides some of the information needed to make insightful decisions.

Economics of Stirring Devices in Grain Drying

Stirring devices are used frequently with in-bin drying of grain. The advantages of stirring devices include reduced overdrying and an increased drying rate associated with the allowable use of a higher drying air temperature (Bridges et al., 1984). Also, airflow rates per unit of grain generally are considered to be somewhat higher when stirring devices are used (Bern et al., 1982; Williams et al., 1978). The primary disadvantages of using a stirring device are the additional capital investment and maintenance costs associated with the unit. The

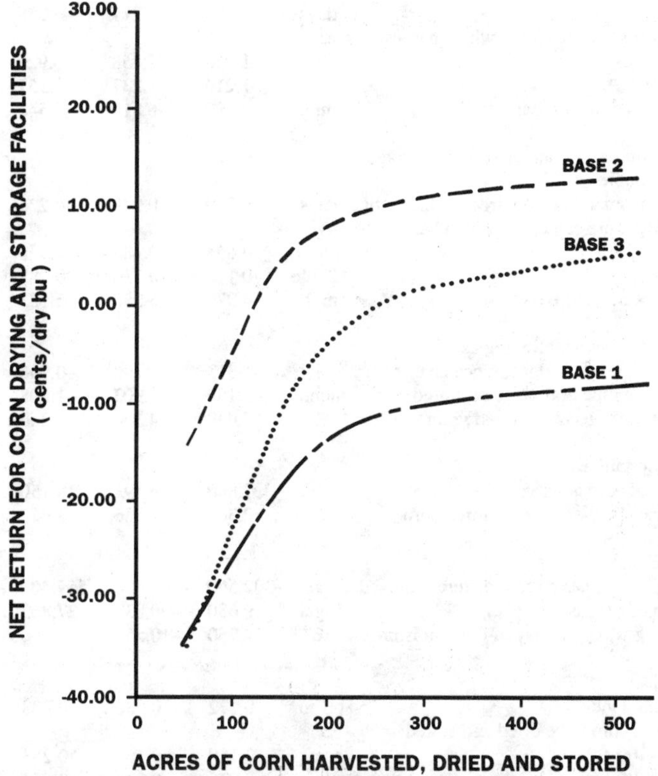

Figure 9.35–Net return to on-farm corn drying and storage facilities as influenced by the number of acres of corn that are harvested, dried and stored (Loewer et al., 1980).

primary question is whether stirring devices are economical and, if so, under what conditions. Therefore, the objectives of this chapter are to:

1. Develop an economic model of stirring device utilization.
2. Compute the expected economic return for representative model inputs including:
   a. In-bin Drying Method; natural, low temperature, and layer.

Table 9.20. Expected net return per unit volume as influenced by an increase in the stated system parameter assuming a linear relationship over the range tested

		Change/Unit Increase (cents/dry bu)			
System Parameter	Unit	Base 1	Base 2	Base 3	Average
I. Harvesting Strategies					
A. If on-the-farm grain drying and storage is not available					
1. Calendar days reuired for harvesting					
10-31	days	0.084	0.152	0.048	0.095
31-46	days	–0.050	0.211	–0.055	0.035
2. Moisture content at which harvest begins					
18-21%	%	–1.060	–1.158	–0.920	–1.046
21-30%	%	1.214	1.237	1.557	1.336
3. Speed of corn harvester	mph	4.590	6.810	5.548	5.649
B. If on-the-farm grain drying and storage is available					
1. Calendar days required for harvesting	days	–0.300	–0.248	–0.226	–0.258
2. Moisture content at which harvest begins					
18-28%	%	0.664	0.488	0.358	0.503
% percent %	–0.396	–0.120	–0.330	–0.282	
3. Speed of corn harvester	mph	–4.970	–5.350	–5.183	–5.168
II. Facility Management Strategies					
A. Interest, taxes, and insurance charge	%	–1.313	–1.190	–0.906	–1.136
B. Average time grain is to be stored	month	–2.187	–1.903	–1.808	–1.966
C. Moisture content of stored grain	%	4.042	4.340	4.260	4.214
III. Market Conditions					
A. Price of corn at harvest	$/bu	7.840	16.593	18.150	14.194
B. Increase in the value of stored corn	%	2.367	2.536	2.423	2.442
IV. Energy Considerations					
A. LP gas used per point moisture removed	gal	–392.500	–364.600	–557.400	–438.167
B. Price of LP gas	$/gal	–19.630	–20.833	–27.862	–22.775
C. Average drop per day in field moisture	%	–14.750	–10.555	–35.960	–20.422
V. Facility Design					
A. Potential yield	bu/ac	0.172	0.080	0.088	0.113
B. Area of corn to be dried and stored					
50 - 100 acres	acres	0.150	0.161	0.253	0.188
100 - 300 acres	acres	0.035	0.031	0.053	0.040
300 - 500 acres	acres	0.008	0.008	0.018	0.011

b. Price of Grain. $2.00 to $3.50/bu where a bushel is defined volumetrically.
c. Bin Size. 18-36 ft in diameter.

System Comparison

The stirring device economic model that is to be developed in the following discussion is called "STIR" (Loewer et al., 1984a). "STIR" is a static model that compares an existing in-bin drying system to that same system if a stirring device were added. Thus, no changes in bin structure or fan performance characteristics are considered. STIR assumes that 1 ft of bin wall will be displaced by the stirring device, and that the grain may be dried safely without economic differences in quality regardless of whether a stirring device is used. A 5-year life is assumed and the Accelerated Cost Recovery System (ACRS) depreciation schedule is used (15% of purchase price less one half of investment credit the first year, 22% the second year, and 21% for each of the next three years). It should be noted that tax regulations regarding such factors as investment credit and depreciation methods change from time to time. Thus, any economic analysis is also subject to modification over time. Along with the investment credit and depreciation assumptions, no salvage value is considered, and repair costs are assumed to occur uniformly over the life of the mechanism.

The model must know how drying system performance is altered by the addition of a stirring device; that is, if there is no change in performance, then there is no need to purchase the device. The set of questions must be addressed as if a stirring device were used and again as if a stirring device were not used. The common set of questions relates to:

- Airflow delivered per unit volume of grain.
- Temperature rise associated with drying air.
- Hours of operation of the fan.
- Power requirements for the fan.
- Average moisture content of the grain after drying.
- Bin diameter and eave height.

In addition, the following information must be known for the situation where a stirring device is to be used:

- Power requirements of the stirring device.
- Purchase price of the stirring device and associated bin reinforcement.
- Total repair cost of the stirring device over the next five years as a percentage of the purchase price.

For both the "with" and "without" stirring situations, the following information must be supplied:

- Individual tax bracket for each of the next five years.
- Cost of electricity per unit.
- Cost of liquid petroleum (LP) gas per unit.
- Value of money expressed as an annual interest rate.

STIR assumes that all fans and motors are powered by electricity and all heating of drying air is with LP gas. One may substitute electricity for LP gas by using the LP gas - electricity equivalent cost. Energy input for heating drying air may be estimated by the following equation:

$$\text{Btu} / \text{h} = 1.1 * \text{cfm} * (\text{temperature rise, } ^\circ\text{F}) \qquad (9.17)$$

where cfm (cubic feet per minute) represents the total air delivery of the fan.

Annual Stirring Costs

Drying Fan. The cost of electricity for the drying fan operation is:

$$\left(\begin{array}{c}\text{drying fan}\\ \text{cost with}\\ \text{stirring}\end{array}\right) = \left(\begin{array}{c}\text{power}\\ \text{requirements}\\ \text{for fan}\\ \text{operation}\\ [\text{units / hours}]\end{array}\right) * \left(\begin{array}{c}\text{hours to}\\ \text{dry if}\\ \text{stirrer}\\ \text{is used}\end{array}\right) * \left(\begin{array}{c}\text{cost per}\\ \text{unit of}\\ \text{electricity}\end{array}\right) \qquad (9.18)$$

Drying Fuel. The cost for drying fuel is:

$$\left(\begin{array}{c}\text{drying}\\ \text{fuel cost}\\ \text{with}\\ \text{stirring}\end{array}\right) = \left(\begin{array}{c}\text{rate of}\\ \text{energy}\\ \text{consumption}\\ \text{per hour}\\ [\text{equation 9.17}]\end{array}\right) * \left(\begin{array}{c}\text{hours}\\ \text{to dry}\\ \text{if stirrer}\\ \text{is used}\end{array}\right) * \left(\begin{array}{c}\text{cost per}\\ \text{unit of}\\ \text{LP gas}\end{array}\right) \qquad (9.19)$$

Stirring Motors. Stirring motor cost is as follows:

$$\left(\begin{array}{c}\text{stirring}\\ \text{motor}\\ \text{operationl}\\ \text{cost}\end{array}\right) = \left(\begin{array}{c}\text{power}\\ \text{requirements}\\ \text{of stirrer}\\ \text{motors}\\ [\text{units / h}]\end{array}\right) * \left(\begin{array}{c}\text{hours}\\ \text{to dry}\\ \text{if stirrer}\\ \text{is used}\end{array}\right) * \left(\begin{array}{c}\text{cost per}\\ \text{unit of}\\ \text{electricity}\end{array}\right) \qquad (9.20)$$

Repair Cost. Repair cost is equal to:

$$\begin{pmatrix}\text{repair}\\\text{cost}\\\text{per year}\end{pmatrix} = \begin{pmatrix}\text{total repair}\\\text{cost as a}\\\text{percentage}\\\text{of purchase}\\\text{price}\end{pmatrix} * \frac{(\text{purchase price of stirrer})}{(\text{life of stirrer [years]})} \tag{9.21}$$

Interest Cost. Interest cost is as follows:

$$\begin{pmatrix}\text{interest}\\\text{cost per}\\\text{year}\end{pmatrix} = \begin{pmatrix}\text{interest}\\\text{rate per}\\\text{year}\end{pmatrix} / 100 * \begin{pmatrix}\text{stirrer value less}\\\text{depreciation to date}\end{pmatrix} \tag{9.22}$$

Depreciation Cost. The depreciation costs for year N are based on the ACRS system and are equal to:

$$\begin{pmatrix}\text{depreciation}\\\text{for year N}\end{pmatrix} = \begin{pmatrix}\text{ACRS}\\\text{depreciation}\\\text{allowance}\\\text{for year N}\\\text{[\% / 100]}\end{pmatrix} * \begin{pmatrix}\text{purchase}\\\text{price of}\\\text{stirrer less}\\\text{1/2 of}\\\text{investment}\\\text{credit}\end{pmatrix} \tag{9.23}$$

Tax Savings. Tax savings are related to investment credit and total expenses.

(a) Investment credit is for the first year only and is computed as follows:

$$\begin{pmatrix}\text{investment}\\\text{credit}\\\text{for year 1}\end{pmatrix} = \begin{pmatrix}\text{five-year}\\\text{property}\\\text{factor,}\\\text{10\%}\end{pmatrix} * \begin{pmatrix}\text{purchase}\\\text{price}\end{pmatrix} \tag{9.24}$$

(b) Tax savings for year N without investment credit are equal to:

$$\begin{pmatrix}\text{tax savings}\\\text{for year N}\end{pmatrix} = \begin{pmatrix}\text{total for the}\\\text{annual cost of}\\\text{the drying fan,}\\\text{drying fuel,}\\\text{stirrer motors,}\\\text{repair, interest}\\\text{and depreciation}\\\text{for year N}\end{pmatrix} * \begin{pmatrix}\text{marginal tax}\\\text{bracket of the}\\\text{individual}\\\text{for year N}\end{pmatrix} \tag{9.25}$$

Total Annual Cost. The total annual cost for using a stirring device in year N is given as:

$$\begin{pmatrix}\text{total annual}\\ \text{cost for}\\ \text{adding a}\\ \text{stirrer for}\\ \text{year N}\end{pmatrix} = \begin{pmatrix}\text{the sum of the}\\ \text{annual costs for}\\ \text{the drying fan,}\\ \text{drying fuel,}\\ \text{stirrer motors,}\\ \text{repair, interest}\\ \text{and depreciation}\\ \text{year N)}\end{pmatrix} - \begin{pmatrix}\text{investment}\\ \text{credit}\\ [\text{for year 1 only}]\\ \text{plus tax savings}\\ \text{on expenses for}\\ \text{year N}\end{pmatrix} \quad (9.26)$$

Annual Conventional Costs

Drying Fan. Drying fan cost is equal to:

$$\begin{pmatrix}\text{drying}\\ \text{fan cost}\\ \text{without}\\ \text{stirring}\end{pmatrix} = \begin{pmatrix}\text{power}\\ \text{requirements}\\ \text{for fan}\\ \text{operation}\\ [\text{units / h}]\end{pmatrix} * \begin{pmatrix}\text{hours to}\\ \text{dry if a}\\ \text{stirrer}\\ \text{is not}\\ \text{used}\end{pmatrix} * \begin{pmatrix}\text{cost per}\\ \text{unit of}\\ \text{electricity}\end{pmatrix} \quad (9.27)$$

Drying Fuel. Drying fuel cost may be computed by the following equation:

$$\begin{pmatrix}\text{drying}\\ \text{fuel cost}\\ \text{without}\\ \text{stirring}\end{pmatrix} = \begin{pmatrix}\text{rate of energy}\\ \text{consumption}\\ \text{per hour}\\ [\text{equation 9.17}]\end{pmatrix} * \begin{pmatrix}\text{hours to}\\ \text{dry if}\\ \text{stirrer is}\\ \text{not used}\end{pmatrix} * \begin{pmatrix}\text{cost}\\ \text{per unit}\\ \text{of LP}\\ \text{gas}\end{pmatrix} \quad (9.28)$$

Overdrying Costs. Overdrying may be defined as the quantity of grain lost (measured as weight loss) based on the difference between the average grain moisture content when a stirrer is used as compared to when a stirrer is not used. All grain weight losses are considered real; that is, no consideration is given to the situation where nonstirred grain is fed on the farm, thus not reducing the real total of nutrients through overdrying. STIR assumes that a stirring device will mix all the grain in the bin to a uniform moisture content. The two options (to buy or not to buy a stirrer) imply that the stirred grain would not be lower in average moisture content than the non-stirred grain.

The cost equation for overdrying based entirely on user-defined parameters is as follows:

$$\begin{pmatrix}\text{cost of}\\\text{overdrying}\end{pmatrix} = \begin{pmatrix}\text{per unit value of grain at the}\\\text{final moisture content after}\\\text{drying with a stirrer}\end{pmatrix}$$

$$* \frac{\begin{pmatrix}\text{final \%}\\\text{moisture content}\\\text{if stirrer is used}\end{pmatrix} - \begin{pmatrix}\text{average \%}\\\text{moisture content}\\\text{if no stirrer is used}\end{pmatrix}}{\left[100 - (\text{average \% moisture content if no stirrer is used})\right]}$$

$$* \begin{pmatrix}\text{number of grain units dried}\\\text{when using a stirring device}\end{pmatrix} \qquad (9.29)$$

Tax Savings. The tax savings for the non-stirring option for year N are equal to:

$$\begin{pmatrix}\text{tax savings}\\\text{for year N}\end{pmatrix} = \begin{pmatrix}\text{total for the}\\\text{annual cost of}\\\text{the drying fan}\\\text{and drying fuel}\\\text{for year N}\end{pmatrix} * \begin{pmatrix}\text{marginal tax}\\\text{bracket of the}\\\text{individual for}\\\text{year N}\end{pmatrix} \qquad (9.30)$$

Total Annual Cost. The total annual cost if a stirring device is not used is:

$$\begin{pmatrix}\text{total annual}\\\text{cost without}\\\text{a stirrer}\\\text{for year N}\end{pmatrix} = \begin{pmatrix}\text{the sum of}\\\text{the annual}\\\text{cost for the}\\\text{drying fan}\\\text{and drying}\\\text{fuel for}\\\text{year N}\end{pmatrix} - \begin{pmatrix}\text{annual tax}\\\text{savings for}\\\text{year N}\end{pmatrix} \qquad (9.31)$$

Expected Net Return

At this point, the total annual cost considerations for the "with-stirrer" and "without stirrer" options have been presented. STIR takes these figures and makes the following computations:

$$\begin{pmatrix}\text{net annual cost}\\\text{difference}\\\text{for year N}\end{pmatrix} = \begin{pmatrix}\text{total annual}\\\text{cost of the}\\\text{stirrer for}\\\text{year N}\end{pmatrix} - \begin{pmatrix}\text{total annual}\\\text{cost if no}\\\text{stirrer is}\\\text{used in}\\\text{year N}\end{pmatrix} \qquad (9.32)$$

The "net annual difference" is converted to both a "present value" and a "future value" quantity using the following:

$$\begin{pmatrix}\text{total net}\\ \text{present}\\ \text{value for}\\ \text{N=1 to}\\ \text{5 years}\end{pmatrix} = \begin{matrix}\text{SUM}\\ \text{FOR}\\ \text{N=1,5}\end{matrix} \left| \begin{pmatrix}\text{net annual}\\ \text{cost}\\ \text{difference}\\ \text{for year N}\end{pmatrix} \right| * \frac{1}{(1 + ANR / 100)^N} \tag{9.33}$$

$$\begin{pmatrix}\text{total net}\\ \text{future}\\ \text{value for}\\ \text{N=1 to}\\ \text{5 years}\end{pmatrix} = \begin{matrix}\text{SUM}\\ \text{FOR}\\ \text{N=1,5}\end{matrix} \left| \begin{pmatrix}\text{net annual}\\ \text{cost}\\ \text{difference}\\ \text{for year N}\end{pmatrix} \right| * (1 + ANR / 100)^N \tag{9.34}$$

where
ANR = annual percent interest rate
N = year number

Table 9.21. Assumptions for the economic comparison between stirred and unstirred grain

Item	With Stirrer	Without Stirrer
Initial moisture content	20.5%	20.5%
Depth of grain in bin	15 ft	16 ft
Annual interest rate	15%	15%
LP Gas cost	$1.00/gal	$1.00/gal
Electricity cost	$0.06/kW-h	$0.06/kW-h
Air flow per unit of grain	1.1 cfm/bu	1.0 cfm/bu
Temperature rise:		
Natural	0° F	0° F
Low temperature	7° F	7° F
Layer	20° F	20° F
Hours to dry and final average moisture content		
Natural	647h, 15.5%	932h, 14.51%
Low temperature	356h, 15.5%	673h, 12.17%
Layer	184h, 15.5%	448h, 9.41%
Total repair cost for 5 years, % of purchase:	50%	0%
Individual tax bracket for each year	20%	20%
Stirrer cost by bin diameter:	(18 ft) $2232	$0
(List price, April 1983)	(21 ft) $2292	$0
(No bin wall stiffeners included)	(24 ft) $2339	$0
	(27 ft) $2458	$0
	(30 ft) $2561	$0
	(33 ft) $2715	$0
	(36 ft) $2841	$0

The present value figure is the net value of the stirrer and any associated equipment (such as bin stiffeners) at the time of purchase. The future value calculation is the value of the stirrer based on its projected performance five years from the time of purchase. Generally, present value analysis is the preferred method of making economic comparisons. Note in the equations 9.33 and 9.34 that when the cost of not using a stirrer is greater than having a stirrer, the present and future value statistics will be negative. Therefore, a stirrer should be purchased if in-bin drying is to be used and a negative present or future value is computed.

Example Analysis

The economics of using a stirring device for in-bin drying vary with each farm situation. The following analysis, however, is for a representative farm situation. Inputs that are common to all situations are given in table 9.21. The following inputs are varied:

Value of grain: $2.00 to $3.50/bu
Bin Diameter: 18 to 36 ft
Drying Method:
a. Natural air (no added heat)
b. Low temperature (7° F temperature rise)
c. Layer (20° F temperature rise)

Bins are considered full when drying begins. The natural drying air conditions are 53° F and 54% Rh. Low temperature drying conditions (60° F, 50% Rh) and layer drying conditions (73° F, 31% relative humidity) are based on the same absolute humidity as the natural air drying situation. Airflow per unit of grain is assumed to be 10% higher for the stirred grain. The hours required for drying, the final average moisture contents for the unstirred options, and fan power requirements are based on other simulation models (Thompson, 1975; Thompson et al., 1968). It is assumed that drying time required to obtain the average moisture content, as computed by Thompson's drying model, represents the stirred grain situation. Power requirements for the stirring devices are estimated to be the sum of the motor capacities. Electricity is considered to be the drying air heat source for low temperature drying with LP gas being used for layer drying. The prices of the stirrers were based on list values from a leading manufacturer for a two-auger device. Cost for associated equipment, such as bin wall stiffeners, was not considered.

Results of the Example Analysis

Results shown in figures 9.36 to 9.38 indicate that for the assumptions used, stirring devices are not economical for natural air in-bin drying

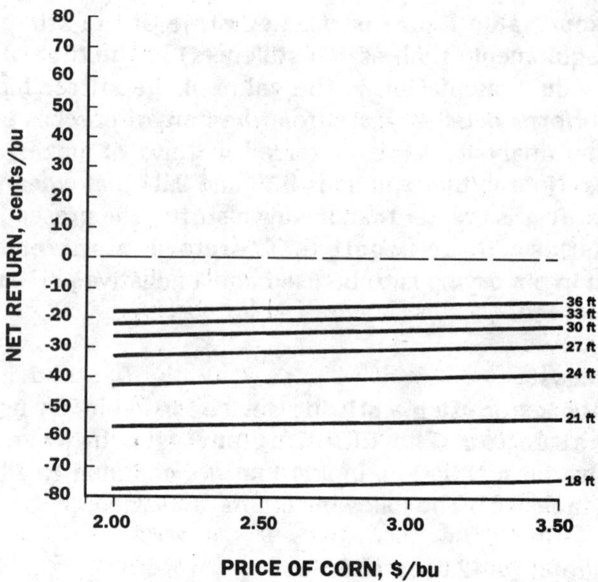

Figure 9.36–Net return to a stirring device when using natural air drying over a range of corn prices and bin diameters (Loewer et al., 1980).

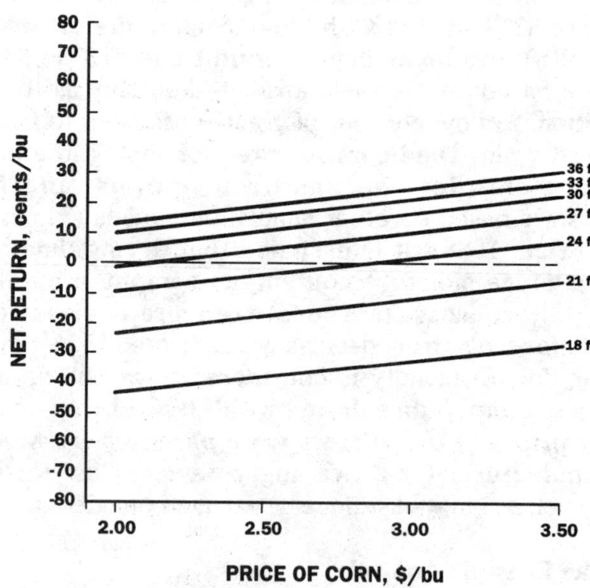

Figure 9.37–Net return to a stirring device when using low temperature drying over a range of corn prices and bin diameters (Loewer et al., 1980).

systems. Likewise, stirring devices are economically feasible for all but the smallest bin tested if layer drying is selected as the in-bin drying method.

The low-temperature in-bin drying choice offers mixed results. Generally, the three smallest diameters are not economical and the three largest are. Only the middle diameter bin (24 ft) crosses the zero present value line near the middle of the corn price range meaning that grain price is the determining factor insofar as economics were concerned. The influence of bin diameter is reflected in a significant decrease in investment cost per bushel for stirring devices as diameter increases (fig. 9.36).

What is the effect of altering list price values for the stirring device by ± 20% for the low temperature drying system? The 20% reduction represents a reasonable discount for the purchaser. The addition of 20% to investment cost is reflective of adding additional equipment (such as bin stiffeners), future price increases (other factors remaining the same) or of more costly stirring units than the one used in this analysis. The effects of altering purchase price by a ± 20% were of little importance when compared to the base assumptions, as is shown in figures 9.39 and 9.40. Basically, grain price is the determining factor within the range of purchase prices tested.

The effects of adding more heat are the most important consideration in that the avoidance of overdrying is the primary advantage of utilizing

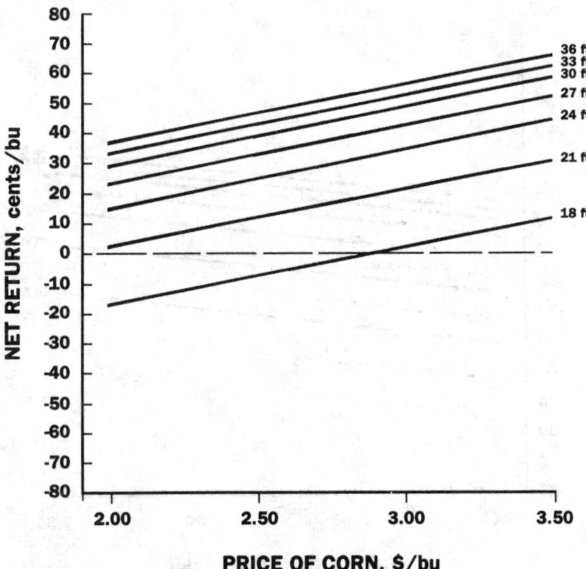

Figure 9.38–Net return to a stirring device when using layer drying over a range of corn prices and bin diameters (Loewer et al., 1980).

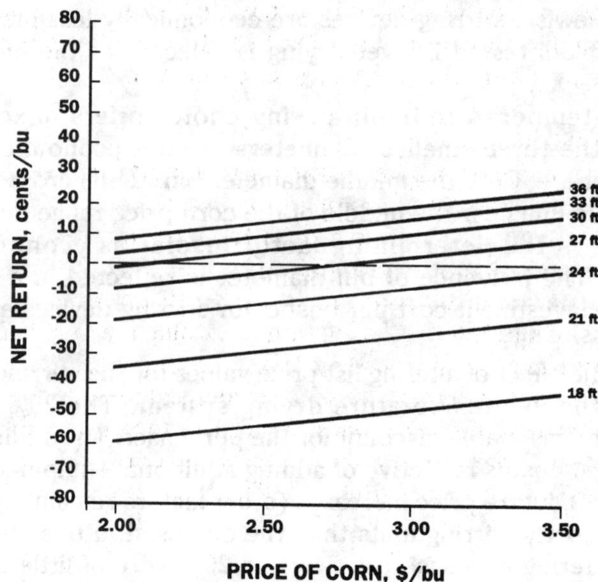

Figure 9.39–Net return to a stirring device when using low temperature drying and increasing the base price of the unit by 20% (Loewer et al., 1980).

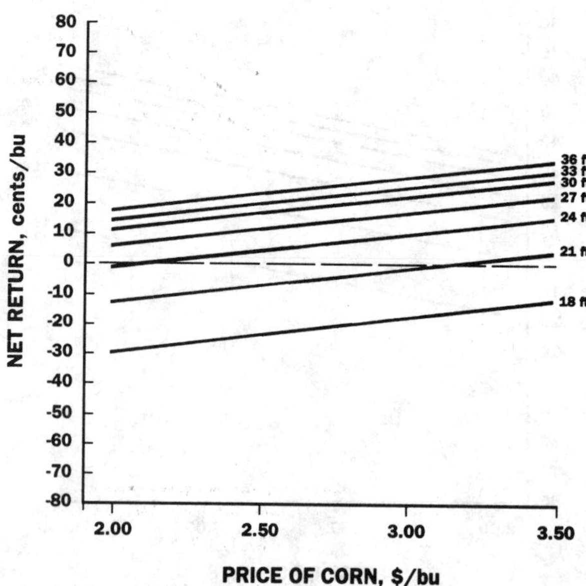

Figure 9.40–Net return to a stirring device when using low temperature drying and decreasing the base price of the unit by 20% (Loewer et al., 1984a).

stirring devices. The effects of heat addition as reflected in drying air temperature are shown in figure 9.41 for the low temperature drying situation using a 24-ft bin diameter over a range of grain prices.

Discussion of Assumptions

The above analysis assumes that corn will dry safely either with or without stirring. No cost is assigned for risk of spoilage or aflatoxin in either the stirred or unstirred situation. Dry matter is assumed to remain constant over the drying period.

No consideration is given to the overall economic or physical implications of altering the drying rate by adding stirring. For example, a farmer may have an undersized in-bin drying system. By adding a stirring device and increasing drying air temperature, it may be possible to harvest corn at higher moistures and reduce harvest losses. The effects of these types of decisions on total system economics have been evaluated previously (Loewer et al., 1980).

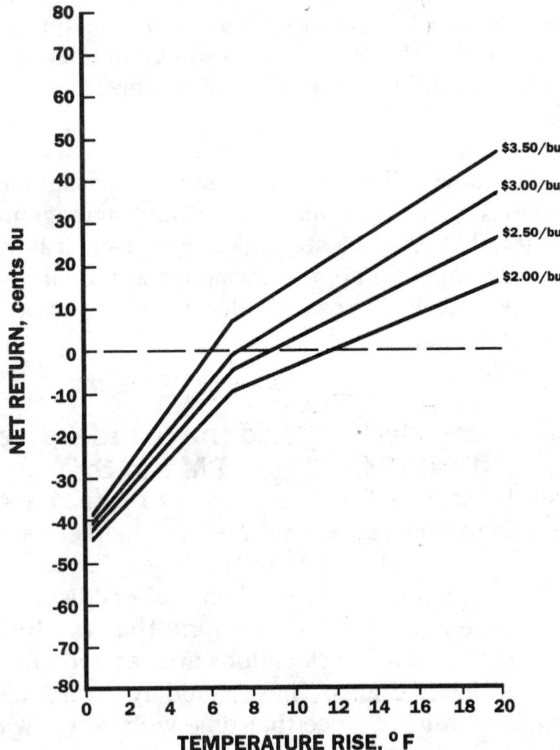

Figure 9.41–Net return to a stirring device in a 24-ft diameter bin as influenced by the temperature rise of the drying air (Loewer et al., 1984a).

The analysis assumes that stirring is equally effective for each of the bin diameters. In addition, the grain is assumed to be mixed uniformly to a final moisture content of 15.5%. If a lower final moisture content were selected, the difference in overdrying would be relatively less between the stirred and unstirred grain, thus reducing the economic advantage associated with using a stirring device.

Conclusions

With the assumptions of the analysis, the following conclusions may be drawn when evaluating the economics of the addition of a stirring device to an in-bin drying system for corn:

1. Stirring devices generally are not economical for natural air drying systems (no addition of heat).
2. Stirring devices generally are economical for layer type drying systems (increase in ambient air temperature of 20° F).
3. When using low temperature drying (increase in ambient air temperature of 7° F), stirring devices generally are not economical for bins less than or equal to 21 ft in diameter. Conversely, stirring devices generally are economical for bins greater than or equal to 30 ft in diameter. The price of corn will be the determining factor in evaluating the middle range of bin diameters.

Summary

An economic analysis is only as accurate as its assumptions. Changes in tax law, prices, physical components, and management affect the relative benefits of alternative strategies. However, it is very important that the decision maker be able to formalize and quantify assumptions in order to make the best possible decisions regardless of possible changes in basic inputs.

Dockage Methods and the Standardized Bushel Concept of Marketing

When grain is marketed, it is expected to satisfy a specific set of standards. If these standards are not met, the market value of the grain is reduced to reflect its true value to the buyer (Loewer et al., 1984c). The price reduction is referred to as "dockage" or "discounts".

Dockage may be accessed for anything that the buyer specifies. Usually, however, the grain specifications are based on the Official Grain Standards of the United States. These standards were adopted in 1916 and have remained in force since that time with very few changes (Hill, 1982). In 1987, moisture content was removed as a factor for determining grade. Examples of these standards for several types of grain are given in figure 9.44. The specifications differ for each grain

Grades and Grade Requirements

§ 810.404 Grades and grade requirements for corn.

Grade	Minimum test weight per bushel (pounds)	Maximum limits of— Damaged kernels: Heat damaged kernels (percent)	Damaged kernels: Total (percent)	Broken corn and foreign material (percent)
U.S. No. 1	56.0	0.1	3.0	2.0
U.S. No. 2	54.0	0.2	5.0	3.0
U.S. No. 3	52.0	0.5	7.0	4.0
U.S. No. 4	49.0	1.0	10.0	5.0
U.S. No. 5	46.0	3.0	15.0	7.0
U.S. Sample grade				

U.S. Sample grade is corn that:

(a) Does not meet the requirements for the grades U.S. Nos. 1, 2, 3, 4, or 5; or

(b) Contains 8 or more stones which have an aggregate weight in excess of 0.20 percent of the sample weight, 2 or more pieces of glass, 3 or more crotalaria seeds (Crotalaria spp.), 2 or more castor beans (Ricinus communis L.), 4 or more particles of an unknown foreign substance(s) or a commonly recognized harmful or toxic substance(s), 8 or more cocklebures (Xanthium spp.) or similar seeds singly or in combination, or animal filth in excess of 0.20 percent in 1000 grams; or

(c) Has a musty, sour, or commercially objectionable foreign odor; or

(d) Is heating or otherwise of distinctly low quality.

Figure 9.42–Excerpts from the Official Grain Standards of the United States. (a) corn. **(Continued on next page)**

Grades and Grade Requirements

§ 810.1404 Grades and grade requirements for sorghum

Grade	Minimum test weight per bushel (pounds)	Maximum limits of -- Damaged kernels: Heat damaged (percent)	Damaged kernels: Total (percent)	Broken kernels, foreign material and other grains (percent)
U.S. No. 1	57.0	0.2	2.0	4.0
U.S. No. 2	55.0	0.5	5.0	8.0
U.S. No. 3 1/	53.0	1.0	10.0	12.0
U.S. No. 4	51.0	3.0	15.0	15.0

U.S. Sample grade

U.S. Sample grade is sorghum that:

(a) Does not meet the requirements for the grades U.S. Nos. 1, 2, 3, or 4; or

(b) Contains 8 or more stones which have an aggregate weight in excess of 0.2 percent of the sample weight, 2 or more pieces of glass, 3 or more crotalaria seeds (*Crotalaria* spp.), 2 or more castor beans (*Ricinus communis* L.), 4 or more particles of an unknown foreign substance(s) or a commonly recognized harmful or toxic substance(s), 8 or more cocklebur (*Xanthium* spp.) or similar seeds singly or in combination, 10 or more rodent pellets, bird droppings, or equivalent quantity of other animal filth per 1,000 grams of sorghum; or

(c) Has a musty, sour, or commercially objectionable foreign odor (except smut odor); or

(d) Is badly weathered, heating, or distinctly low quality.

1/Sorghum which is distinctly discolored shall be graded not higher than U.S. No. 3.

Figure 9.43–Continued. (b) grain sorghum. **(Continued on next page)**

Grades and Grade Requirements

§ 810.1604 Grades and grade requirements for soybeans.

Grade	Minimum test weight per bushel	Maximum limits of --				
		Damaged kernels		Foreign material	Splits	Soybeans of other colors
		Heat damaged	Total			
	(pounds)	(percent)	(percent)	(percent)	(percent)	(percent)
U.S. No. 1	56.0	0.2	2.0	1.0	10.0	1.0
U.S. No. 2	54.0	0.5	3.0	2.0	20.0	2.0
U.S. No. 3 1/	52.0	1.0	5.0	3.0	30.0	5.0
U.S. No. 4 2/	49.0	3.0	8.0	5.0	40.0	10.0

U.S. Sample grade

U.S. Sample grade is soybeans that:

(a) Do not meet the requirements for U.S. Nos. 1, 2, 3, or 4; or

(b) Contain 8 or more stones which have an aggregate weight in excess of 0.2 percent of the sample weight, 2 or more pieces of glass, 3 or more Crotalaria seeds (*Crotalaria* spp.), 2 or more castor beans (*Ricinus communis* L.), 4 or more particles of an unknown foreign substance(s) or a commonly recognized harmful or toxic substance(s), 10 or more rodent pellets, bird droppings, or equivalent quantity of other animal filth per 1,000 grams of soybeans; or

(c) Have a musty, sour, or commercially objectionable foreign odor (except garlic odor); or

(d) Are heating or otherwise of distinctly low quality.

1/Soybeans that are purple mottled or stained are graded not higher than U.S. No. 3.
2/Soybeans that are materially weathered are graded not higher than U.S. No. 4.

Figure 9.44–Continued. (c) soybeans.

type. However, most types of grain are graded with reference to maximum allowable levels of damage and minimum test weights. The most limiting factor establishes the grade. The objective of the following discussion is to present the physical and economic significance of dockage from the perspective of both the farmer and the grain buyer, with special emphasis on the standardized bushel concept of marketing.

Marketing Principles

There are several market principles that should be considered by all members of the marketing system from producer to final user of the grain (Watson, 1983; Hill, 1982).

Principle 1: Farm prices of grain are set by the value of the final products and by competitive charges for services in the marketing channel.

Grain buying and selling is a competitive effort at all levels of pricing. Grain primarily is a food product for humans, either through direct processing of the material (e.g., bread, cereal, etc.) or indirect utilization (through animal feed). At each level of the marketing system, a service is administered which requires some adjustment in the price of grain because of the added value of the product. In theory, the competitive marketing system will not allow any single firm to reap excessive profits. This concept leads to the second principle.

Principle 2 : The division of income between farmers and elevator (buyers) is determined by competition and not by the grain standards.

The grain standards do not determine the value of the grain but rather provide information concerning quality. This makes the marketing process more efficient, thereby being of economic benefit to all members of the marketing system, from producers to consumers. The demand for grain, as reflected by its value in an end-product, establishes the price of grain, not the grain standard itself.

Principle 3: Water in grain has little intrinsic value.

Water is relatively plentiful and has little economic value in grain except as an aid in processing or in reducing handling damage. Excessive moisture adversely affects storability. However, water itself does not influence significantly the nutrient or energy value of the dry matter. Moisture will be addressed in more detail later in the discussion.

Principle 4 : Farmers can be expected to respond to the economic incentives created by the buyers in such a way as to maximize individual farm profits.

This statement is equally true at all levels of the marketing system. The dockage system employed by the buyer tells the farmer how to market the grain. It is only natural to expect each producer to try to maximize personal profits within the marketing rules as stated in the dockage specifications.

Principle 5 : Changes in the dockage method will not affect the profitability of the buyers or producers as a group over a long period of time, except that some benefits may accrue to all if the efficiency of the marketing system is increased.

The buyer (elevator) must obtain a sufficient price margin for the grain that passes through the facility in order to meet expenses and make an acceptable profit. The charges to the farmer are usually in proportion to services rendered. However, the objective of the elevator is to obtain a certain profit level rather than to distribute charges precisely to each of its customers. Thus, one farmer may pay a disproportionate share for certain services. The farmer, however, is under no obligation to use all or any of the services offered by the elevator although economic considerations may suggest certain management decisions. The key consideration for both the farmer and the elevator is to understand sufficiently the dockage system in order to make the best possible economic decisions, thus maximizing the profits to both and increasing the efficiency of the total marketing system. The following discussion will address the dockage system from the perspectives of both the farmer and the buyer, especially as related to moisture content.

Dockage for Moisture

The moisture contained in grain usually is of little economic benefit to either the farmer or the elevator. If the moisture content is too high, the grain will not store properly. If the moisture content is too low, excessive kernel breakage may occur during handling. These are common problems for both the farmer and the elevator.

Historically for most grains, the Grain Standards specified the base moisture content when grain is considered to be acceptably dry for marketing, or in essence, when a "wet bushel" becomes a "dry bushel". A change in the U.S. Grain Standards in 1987-1988 removed moisture content as a consideration for grade. Thus, local elevators presently establish the base moisture content that may vary depending on geographic location and type of grain.

Marketing of grain usually is conducted on a "per bushel" basis. However, there are significant variations in the definition of "bushel". Technically, a bushel is defined as the grain occupying a volume of 1.2445 ft^3. In market terminology, however, it is specified in terms of weight. For example, a bushel of corn is defined for marketing purposes as 56 lb of material. Similarly, soybean marketing is based on a weight of 60 lb/bu (table 9.22). The use of a weight measure to define a market bushel has the advantage of simplicity. However, the true value of grain depends on its dry matter content rather than its total weight. The relative amount of dry matter increases as moisture content decreases for a given weight of material. Thus, the base moisture content becomes the reference point for dry matter content. A bushel of corn at a base moisture content of 15.5% contains 47.32 lb of dry matter [(100% - 15.5%)/100 * 56 lb/bu]. Dockage rates are established, in part, to adjust for the excess water in 56 lb of corn when its moisture content is greater than 15.5%. The "extra water" (remember that even 15.5% corn has 8.68 lb of water/bu) above that associated with 15.5% moisture corn must be removed by drying to reach the base moisture content standard. The water that is removed is referred to as "shrink" because the initial weight of the grain has "shrunk" to a lower weight. Note, however, that the dry matter level in the original 56 lb of wet grain remains the same as before the grain was dried.

If grain is dried below the base moisture content, the value received decreases because fewer pounds of material (i.e., bu) remain than at the base even though there is the same amount of dry matter. This paradox resulted in much of the dissatisfaction associated with the U.S. Grain Standards in effect prior to 1987, especially in warmer and/or drier areas of the country where (1) grain must be dried to moisture contents lower than the base if it is to be stored safely, and/or (2) naturally occurring weather conditions result in the grain drying to moisture levels below the base moisture content.

Table 9.22. Bushel weights and base moisture contents for various grains (Hill, 1982)

Grain Type	Bushel Weight (lb)	Base Moisture Content (%)
Corn	56	15.5
Soybeans	60	13.0
Grain Sorghum	56	14.0
Wheat	60	13.5

For purposes of this discussion, a bushel is defined as an established weight regardless of moisture content. For example, a bushel of corn is defined to be 56 lb of grain at any moisture content. The term "market value" refers to the price per bushel of grain when at the elevator's base moisture content, For example, the market value of corn is defined as the value of 56 lb of material at 15.5% moisture. The extent to which shrinkage and the base moisture content affects market value is shown in figure 9.45 and table 9.23. Note that dockage is not assessed to grain that is below the base moisture content. Yet, farmers who deliver overdried grain penalize themselves because they provide more dry matter per unit of wet grain weight than is required by the standard. The elevator may blend the overdried grain with grain that is above its base moisture content in order to obtain a blend that is at the base moisture level and optimum insofar as marketing is concerned. This blending becomes part of the margin, so the elevator is indifferent as to which individual farmer brings in wet or overdried grain so long as an acceptable profit can be made. There is certainly a difference, however, to the particular farmer who delivers the overdried grain.

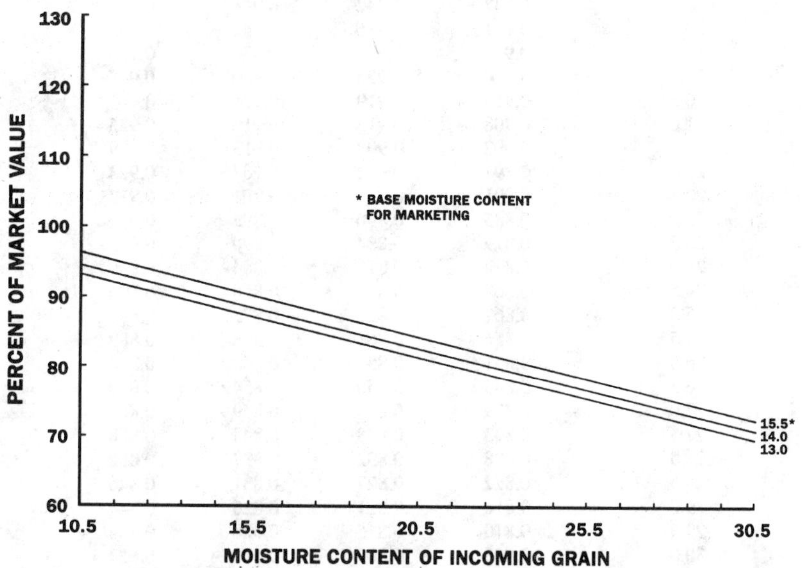

Figure 9.45–The relationship between the moisture content of the incoming grain and its true market value in terms of dry matter as influenced by the final moisture content of the grain. Market value is defined as the price per unit weight of grain at the base moisture content (Loewer et al., 1984c).

Table 9.23. The influence of grain moisture on the fraction of weight remaining after conditioning to the base moisture content, without allowance for handling losses

Field Moisture Content (%)	Fraction of Weight Remaining after Conditioning to the Following Base Moistures*			
	13.0%	13.5%	14.0%	15.5%
8.5	1.052	1.058	1.064	1.083
9.0	1.046	1.052	1.058	1.077
9.5	1.040	1.046	1.052	1.071
10.0	1.034	1.040	1.047	1.065
10.5	1.029	1.035	1.041	1.059
11.0	1.023	1.029	1.035	1.053
11.5	1.017	1.023	1.029	1.047
12.0	1.011	1.017	1.023	1.041
12.5	1.006	1.012	1.017	1.036
13.0	1.000	1.006	1.012	1.030
13.5	0.994	1.000	1.006	1.024
14.0	0.989	0.994	1.000	1.018
14.5	0.983	0.988	0.994	1.012
15.0	0.977	0.983	0.988	1.006
15.5	0.971	0.977	0.983	1.000
16.0	0.966	0.971	0.977	0.994
16.5	0.960	0.965	0.971	0.988
17.0	0.954	0.960	0.965	0.982
17.5	0.948	0.954	0.959	0.976
18.0	0.943	0.948	0.953	0.970
18.5	0.937	0.942	0.948	0.964
19.0	0.931	0.936	0.942	0.959
19.5	0.925	0.931	0.936	0.953
20.0	0.920	0.925	0.930	0.947
20.5	0.914	0.919	0.924	0.941
21.0	0.908	0.913	0.919	0.935
21.5	0.902	0.908	0.913	0.929
22.0	0.897	0.902	0.907	0.923
22.5	0.891	0.896	0.901	0.917
23.0	0.885	0.890	0.895	0.911
23.5	0.879	0.884	0.890	0.905
24.0	0.874	0.879	0.884	0.899
24.5	0.868	0.873	0.878	0.893
25.0	0.862	0.867	0.872	0.888
25.5	0.856	0.861	0.866	0.882
26.0	0.851	0.855	0.860	0.876
26.5	0.845	0.850	0.854	0.870
27.0	0.839	0.844	0.849	0.864
27.5	0.833	0.838	0.843	0.858
28.0	0.828	0.832	0.837	0.852
28.5	0.822	0.827	0.831	0.846
29.0	0.816	0.821	0.826	0.840
29.5	0.810	0.815	0.820	0.834
30.0	0.805	0.809	0.814	0.828
30.5	0.799	0.804	0.809	0.822

* (100 - initial MC) / (100 - base MC).

Dockage for Handling

When grain is handled, part of the weight of the marketed material will disappear in the form of dust, chaff, trash, etc. This loss will affect both the elevator and the farmer. A typical dockage assessment is 0.25 to 0.50% of the delivered weight of the wet material. In effect, this dockage charge represents a "handling shrink". It may be added as a constant to the moisture related shrinkage to obtain a total shrinkage charge.

Drying Cost

If grain moisture is above the base moisture content, it must be dried to at least the base value for marketing. Drying cost will be encountered by either the farmer or the elevator. The expense of drying is dependent on the speed and efficiency of the drying method.

The farmer and the elevator operator will, in all probability, have different drying costs because of different efficiencies in their drying operations. They are affected differently also with regard to overdrying. The elevator usually will not dry below the base moisture content, but the farmer may because of long-term storage requirements. If the elevator receives overdried grain, it may blend this grain with wet grain and avoid the drying cost associated with drying the wetter grain. Conversely, the farmer that overdries grain cannot recover overdrying cost except perhaps as part of the usual increase in the market value of grain that is stored later in the year. In addition, everything else considered equal, the farmer will use relatively greater quantities of energy per pound of moisture removed when drying to moisture contents lower than the base moisture.

The cost of drying can be expressed as a percentage of market value through the following equation:

$$\begin{matrix}\text{drying cost,}\\ \text{\% of}\\ \text{market value}\end{matrix} = \frac{\begin{pmatrix}\text{energy used}\\ \text{for drying}\end{pmatrix} * \begin{pmatrix}\text{price per}\\ \text{unit of energy}\end{pmatrix}}{\begin{pmatrix}\text{price of grain at}\\ \text{the base moisture content}\end{pmatrix}} * 100 \qquad (9.35)$$

For example, if 10,000 Btu of LP gas (92,000 Btu/gal) were required to dry a bushel of corn from 25.5 to 15.5% moisture content, and corn and LP gas were selling for \$3.00/bu (market value at 15.5% moisture) and \$1.00/gal, respectively, then:

$$\text{drying cost (\% of market value)} = \frac{(10{,}000\ \text{Btu/bu}) * [(\$1.00/\text{gal}) / (92.000\ \text{Btu/gal})] * 100}{\$3.00/\text{bu}}$$

$$= 3.62\%$$

Discount Schedules

Elevators assess discount charges to recover their expenses associated with moisture shrinkage, handling shrinkage, drying costs and other overhead charges and make a profit for services rendered. Discount schedules normally are related to the moisture content of the incoming grain because moisture shrinkage is such an important economic factor. Two types of schedules usually are used. One is based on a constant charge per point of moisture above the base moisture content. The other is established as a percentage of the market value of the grain to be charged for each point of moisture above the base. Both systems may appear "stair-stepped" because the dockage chart may indicate a constant charge over a range of moisture content values, usually one-half point.

The constant-charge-per-point system recognizes that much of the overhead associated with the elevator is not influenced by the moisture content of the grain. This method is satisfactory so long as grain prices do not change significantly. If, however, grain prices increase, the value of the shrink increases proportionally, and the elevator is penalized. Conversely, a decrease in grain price results in a decrease in the value of the shrink, and if no adjustment to the discount schedule is made, the farmer is penalized.

The fluctuations in market value and the corresponding change in the value of shrink led to the discounting of grain based on a percentage of its market value. The advantage of this method is that it accounts for changes in the value of the moisture shrinkage. However, it does not allow for an adjustment in margin related to other costs. For example, if the price of drying fuel increases, the elevator must either absorb the added cost or increase the discount percentage if its margin is to remain the same. Similarly, a decrease in grain prices also lowers the elevator's profit margin in that less gross income from the discount is received.

Both of these methods have their shortcomings. However, they are relatively simple. A comparison of the systems is shown in figure 9.46. The effects of these systems on profitability for both the farmer and the elevator will now be discussed.

To Dry or Not to Dry?

The elevator has little choice but to dry or blend delivered grain to the base moisture content level for marketing. The farmer, however, may choose to dry grain before delivering it to the elevator, thus avoiding the discounts for moisture. This person must pay for drying fuel and equipment, and absorb the moisture and handling shrinkage. This farmer's "value sold dry" is the value of a bushel of wet grain after making allowances for the fraction of the bushel that is lost because of moisture and handling shrinkage. The farmer's "value sold wet" is the value of one bushel of wet grain after elevator discounts for moisture and handling shrinkage have been made. The "gross-returns for farm drying" represent the difference between the "value-sold-wet" and the "value-sold-dry". The farmer must subtract all the on-farm drying and extra handling expenses from the "gross-returns for farm drying". This individual then must decide if any profit that remains is sufficient to warrant associated risk and management inputs. In addition, the farmer's drying and shrinkage expenses normally will be greater than that of the elevator if the producer chooses to dry the grain below the base moisture content. This overdrying must be viewed as a necessary part of storage expense.

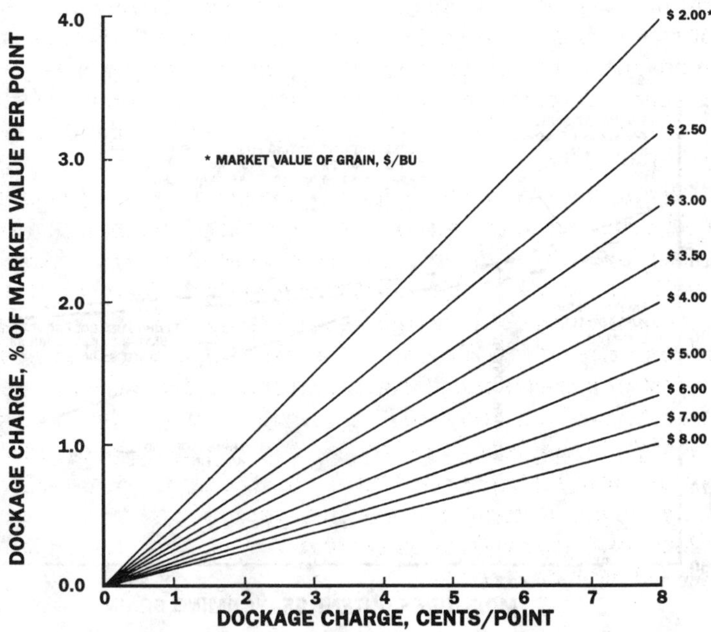

Figure 9.46–Comparison between the most common methods of determining dockage charges for excess moisture in grain (Loewer et al., 1984c).

Example Economic Comparisons

A comparison will now be made between the farmer and the elevator with regard to economic benefits. A series of illustrations will be presented to show the step-by-step evaluation that is made. Corn will be used as the example grain, but the same principles apply to other grain types. Two categories of delivered grain will be considered: corn that is greater than the base moisture content at the time of delivery to the elevator, and corn that is drier than the base moisture content when delivered.

Moisture Shrinkage. The percentage of moisture shrink may be determined by the following equation:

$$\text{moisture shrink (\%)} = \left[1.0 - \frac{[100 - (\text{moisture content of delivered grain, \%})]}{(100 - \text{base moisture content \%})}\right] * 100 \quad (9.36)$$

Both the elevator and the farmer will experience this weight loss.

Handling Shrink. Both the elevator and the farmer will experience a handling shrink. For purposes of this example, a constant 0.5% loss will

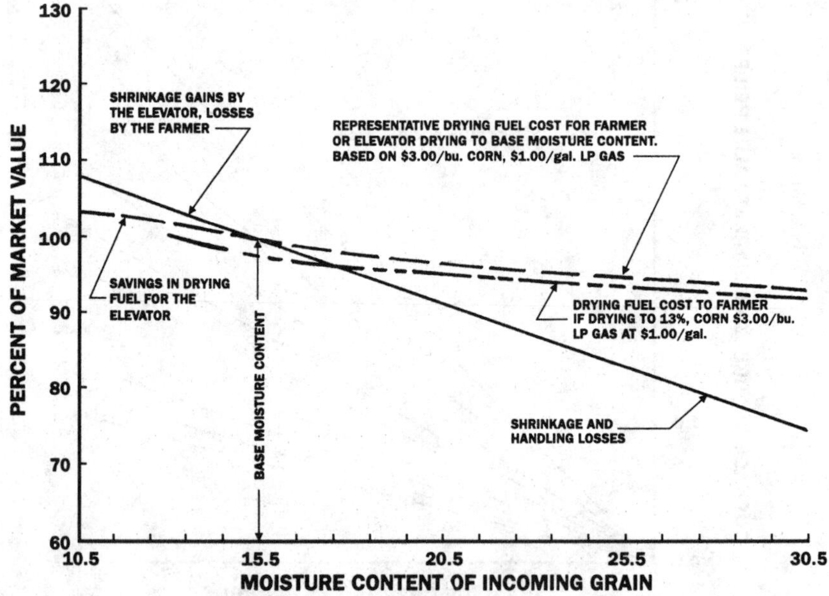

Figure 9.47–Influence of incoming grain moisture content on grain market value as independently influenced by shrinkage and drying fuel cost (based on corn and LP gas prices of $3.00/bu and $1.00/gal, respectively) (Loewer et al., 1984c).

be added to the above equation. The shrinkage and handling losses are added together and shown in figure 9.47.

Drying Fuel Cost. For simplicity, it will be assumed that the efficiencies of the dryers for the elevator and the farmer are the same. However, the farmer will use greater quantities of fuel if the grain is dried to moistures below that of the elevator's base moisture content, 15.5%. An example of fuel usage as a percentage of market value for drying to 15.5% and 13.0% is shown in figure 9.47. A $3.00/bu market value of U.S. No. 2 corn is assumed with LP gas costs being $1.00/gal. Equation 9.35 is used to make the conversion. Note that the elevator saves drying fuel from the delivery of the overdried corn that is in direct proportion to its drying value in blending. Hence, the elevator's drying fuel savings is a mirror image of its costs for removing the same quantity of moisture from wet grain. The farmer, however, cannot recover on-farm overdrying costs unless the on-farm overdried grain is also blended with wetter grain prior to the time of delivery.

Total Costs. The total costs for moisture shrinkage, handling shrinkage and drying fuel are shown in figure 9.48 for the elevator and the farmer who delivers grain at the base moisture content, and in figure 9.49 for the farmer who delivers 13.0% moisture grain.

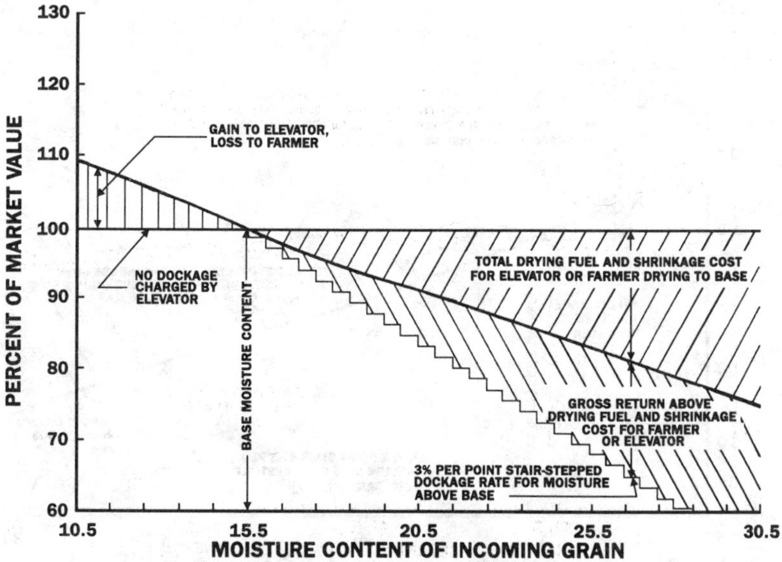

Figure 9.48–Influence of incoming grain moisture content on $3.00/bu corn when it is to be dried to the base moisture content using LP gas priced at $1.00/gal (applies to both farmer and elevator) (Loewer et al., 1984c).

Dockage Rates. A 3% per point stair-stepped dockage rate is shown in figures 9.48 and 9.49. This represents an equivalent per bushel charge per point of moisture of $0.09 for $3.00/bu corn (fig. 9.46).

Gross Return to Management, Labor, and Investment. The relative advantages to the elevator for receiving different moisture content grain are shown in figure 9.48 for a single dockage schedule. Note that the value of overdried corn to the elevator includes both shrinkage and drying fuel savings if the elevator can blend successfully to the base moisture content. To obtain this benefit, the elevator must supply adequate facilities for blending and absorb the extra breakage associated with the drier grain.

The individual farmer who delivers overdried grain to the elevator loses both the dry matter value and the extra expense for drying to a lower moisture content (fig. 9.49). Assuming the elevator to be a competitive firm, the losses incurred by this particular individual are considered part of the firm's profit margin so that a more attractive market price may be offered to all farmers. This individual farmer is, in essence, subsidizing the market price for other farmers who deliver grain above the base moisture content level.

Farmer Options. The farmer who normally would deliver overdried material may recover part of the dry matter value losses by adding water to the grain. The quantities of water required are relatively low

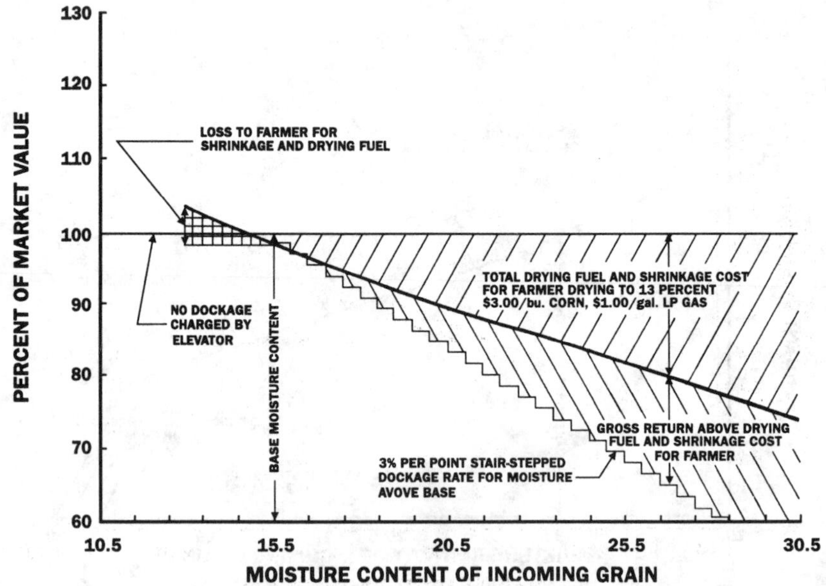

Figure 9.49–Influence of incoming grain moisture content on $3.00/bu corn when it is to be dried to 13% by the farmer using LP gas priced at $1.00/gal (Loewer et al., 1984c).

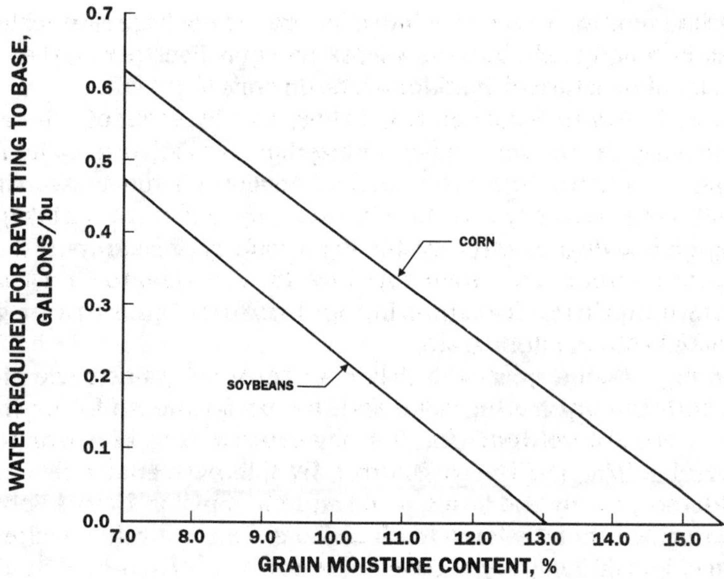

Figure 9.50–Water requirements needed to raise corn and soybeans to the base moisture content level (Loewer et al., 1984c).

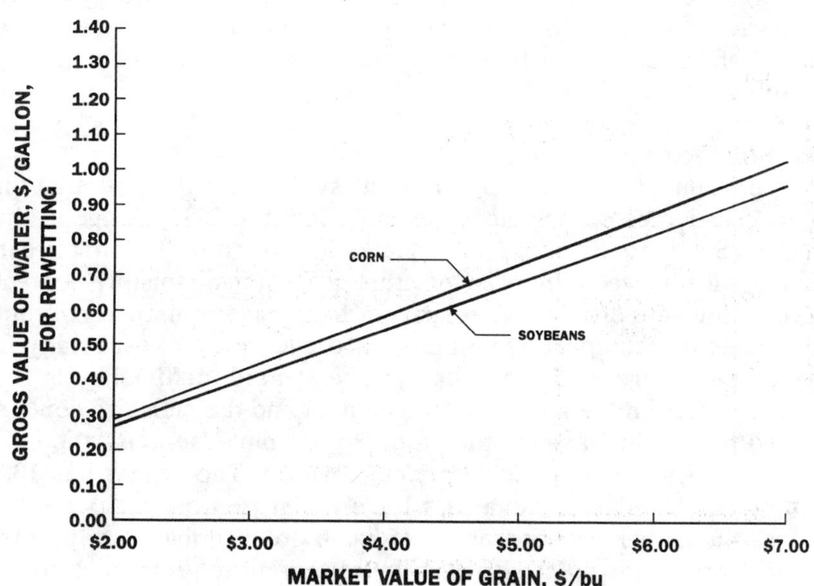

Figure 9.51–The gross value of adding water to grain below the base moisture content (Loewer et al., 1984c)

(fig. 9.50), and the value of the water is relatively high, approaching the value of LP gas on a per gallon basis, as shown in figure 9.51.

The farmer's options for adding water include:

1. Blending overdried grain with wetter grain.
2. Aerating corn with relatively wet air to increase the moisture levels within the grain.
3. Adding moisture directly, usually through mixing augers.

Blending requires that the farmer have sufficiently wet grain, adequate storage facilities, and the necessary materials handling equipment. Aerating corn to increase its moisture is a lengthy process resulting in relatively high expense to the farmer and generally is considered impractical. Adding water directly to grain can be done effectively and inexpensively (Bloome, 1983; Bloome et al., 1982). These practices, especially the latter, resulted in an opinion by the Food and Drug Administration that any rewetting of grain constituted adulteration of food (Bloome, 1983).

Elevator Options. The elevator operates on a price margin. Thus, if all farmers who deliver overdried corn normally began to rewet their grain to the base moisture level, the elevator would have to adjust its market price accordingly.

Grain may dry naturally to moisture contents below the base. Thus, the elevator may not have any wet grain to blend in with the overdried grain. In this case, there are no savings of drying fuel, and the options to the elevator are either to (1) rewet the grain themselves before selling to another buyer, or (2) build this consideration into their pricing structure.

System Inefficiencies

The present grain marketing discount system has the potential for two major sources of inefficiency even though the 1987 change in the U.S. Grain Standard did improve the marketing structure. The first lies in the difference between the base moisture content and moisture content required for safe storage. A relatively high base moisture content encourages the comparatively unsafe storage of grain, especially in warmer wetter areas of the country. The second inefficiency is the difference between the base moisture content and the moisture content required by the final user of the grain. For example, suppose the final processor wishes the corn to be 13% moisture. The farmer has 13% grain for sale, but the elevator's base moisture content is 15.5%. Suppose the farmer rewets grain to 15.5% in order to maximize profits. The final processor must also pay for redrying the grain back to its original 13%. This inefficiency associated with an extra drying results in losses to all except the gas industry. The two sources of inefficiency plus the inequity of elevator charges among individual farmers were contributing factors that led to the revisions in the U.S. Grain Standards for moisture.

Standardized Bushel

One alternative to present practices is to market grain on a dry-matter basis; that is, the total price received would be based on the quantity of dry matter sold rather than the total weight of material. One version of this alternative is the concept of a "standardized bushel" (Hill, 1982). The standardized bushel would retain the base moisture content previously used as a dry matter content reference. For example, a U.S. No. 2 bushel of corn would weigh 56 lb (47.32 lb are dry matter based on a 15.5 base moisture content). Moisture shrinkage would be applied to any moisture content grain delivered to the elevator in order to adjust the material's dry matter to that contained in the standardized bushel. In essence, grain moisture would cease to be a quality consideration. All delivered grain would be converted to the equivalent number of bushels in its commonly used market base moisture content. For corn, this would mean converting any delivered grain to its equivalent in U.S. No. 2 bushels before any discounts are assessed.

Changing to the "standardized bushel" would require that accurate moisture content readings be made over the entire range of moisture contents delivered to the elevator rather than only moisture levels that are above the base. The moisture content of the delivered grain would be entered into a table or calculator to make the conversion to the standardized bushel. Elevators would have to lower their market price to reflect the loss of profits they would normally receive from the delivery of overdried grain. The "average" farmer would receive the same price as with the existing dockage rates. The farmer who, under the present discount system, delivers overdried grain would receive a higher price with the standardized bushel concept. The farmer who delivers grain with moisture contents above the base would receive less because the elevator would have to adjust the market price downward. The elevator would receive the same profit as would the processor. The consumer would benefit because the marketing system is more efficient. This increase in efficiency also could increase the profits for all in the marketing system through increased demand for the product, although changes would probably be slight.

A Standardized Bushel Discount Schedule

The current dockage system is simple. However, it could be made more functional by separating the discounts that vary with the per bushel market value of grain from those that don't. Moisture shrinkage adjustments always affect the percentage of the per bushel market value in the same way. However, drying fuel costs are not directly influenced by the value of grain. The total shrinkage and drying cost to the elevator varies with both grain value and drying energy cost (if any) plus other overhead charges. A possible approach for the elevator is to first "shrink" the incoming grain based on its moisture content, add a constant percentage for handling losses, charge a discount based on the quantity of moisture that is actually removed by elevator and add a constant margin for overhead and profit. Discounts for other factors would be added to the total.

The suggested dockage schedule is based on the following equation:

$$\begin{aligned}\text{discount}\ \$/\text{wet bu} = {} & \left(\begin{matrix}\text{market value}\\ \text{of grain at \%}\\ \text{base moisture}\\ \text{content, \$ / bu}\end{matrix}\right) * 100 \\ & * \left(1.0 - \frac{(100.0 - \text{moisture content of received grain, \%})}{(100.0 - \text{base moisture content, \%})} + \frac{HL}{100.0}\right) \\ & + \left[\left(\begin{matrix}\text{moisture content}\\ \text{of received grain, \%}\end{matrix}\right) - \left(\begin{matrix}\text{base moisture}\\ \text{content, \%}\end{matrix}\right)\right] * D \\ & + (O + P + M) * \frac{(100 - \text{moisture content of delivered grain, \%})}{(100 - \text{base moisture content, \%})}\end{aligned} \tag{9.37}$$

where

discount = charge per "wet" (or delivered) bushel based on weight
HL = handling shrinkage (%)
D = drying charge (¢ /point of moisture removed)
O = constant overhead charge (¢ /standardized bushel)
P = profit margin (¢ /standardized bushel)
M = miscellaneous discounts for other quality factors (¢ /standardized bushel)

In the above equation, a bushel is defined by the weight of grain with no allowance for moisture. For example, a bushel of corn is defined as 56 lb of material regardless of its moisture content. The elevator's overhead, profit margin and miscellaneous discounts are based on the "standardized bushel". However the discounts are based on the number of harvested ("combined" or "56-lb") bushels (for corn) received by the elevator. Note that in this equation, the farmer receives credit for the dry matter in the overdried grain and for the drying fuel based on the elevator's drying expense. This "return" to the farmer must be covered by adjustment of the overhead charge or profit margin, or the elevator may add a "blending charge" for overdried grain as part of its discount schedule. Another possibility is to disregard completely the drying "bonus" for overdried grain below a certain moisture and assume that the fuel savings to the elevator would cover the blending charges and added expenses associated with grain breakage.

The effects of the above discount schedule are shown in figure 9.52. Note that the returns to the elevator are constant over the entire range of incoming grain moisture contents. If this type of dockage system were adopted by the industry, the elevator's margin, market price per bushel and drying charge would provide a simple means for comparing elevators as to the real prices offered for grain. The suggested procedure actually is much simpler than the collection of other systems presently in use in that much of the "mystery" is removed from the marketing system. The new system would encourage more competition and efficiency in the market. Also, the elevator would be ensured of a constant rather than variable price margin regardless of fluctuations in the prices of grain or drying fuel.

Other Discounts

Grain may be discounted for factors other than moisture. These include test weight, broken or damaged kernels, foreign material, and contamination by other types of seed. These factors usually relate to some measure of quality and storability. However, test weight may be a

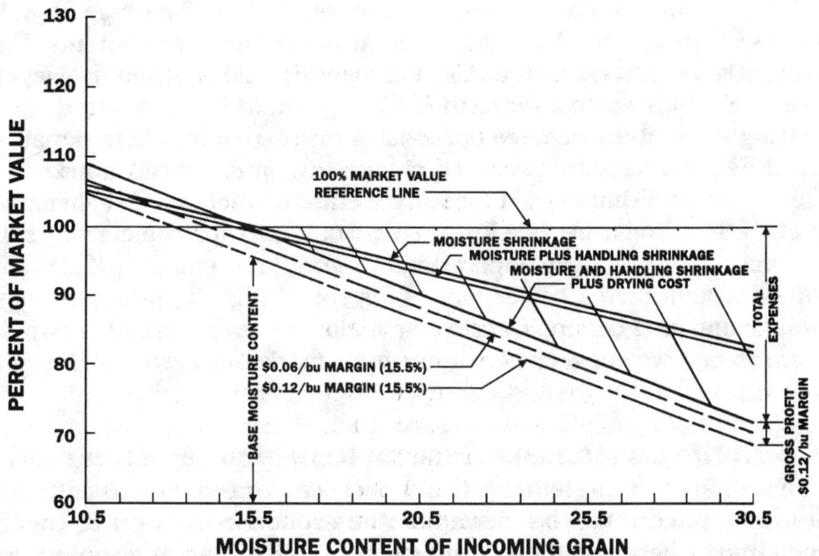

Figure 9.52–Influence of incoming grain moisture content on grain market value using a standardized bushel dockage approach with corn and LP gas priced at $3.00/bu and $1.00/gal, respectively (Loewer et al., 1984c).

misleading index of quality. For example, no close relationship has been found between test weight variations of mature corn and its quality for the major uses (Hall and Hill, 1974). Low test weights may also occur at very high or very low moisture contents (Hall and Hill, 1972) which have nothing to do with the chemical composition of the grain. For this situation, the farmer could effectively be "docked" twice for moisture.

The maximum allowable levels for broken or damaged kernels and foreign material allow for these materials to be blended profitably back into what otherwise would be cleaner grain. When this occurs, the effective grain quality decreases while the total value to the seller increases. As with moisture, the seller should be expected to maximize personal profits within the limits of the standard. The buyer will pay for the added quality needed, either in an increased price to the seller, or in equipment needed to process the grain to desirable quality levels. Again, the process of adding and then removing undesirable material from grain results in marketing inefficiency.

Summary

Grain is marketed according to certain standards. Presently, the variation in base moisture content among local elevators is, in effect, a determining factor in establishing the price received by farmers, often making price comparisons among elevators difficult. If the standardized bushel concept were utilized, the profit margin for the elevators (buyers) would remain the same. There would be more equitable distribution of elevator costs and profits among farmers (sellers). Grain quality probably would increase because relatively safe storage of grain could be more economically achieved at lower moisture contents. The drier grain might result in some increase in grain damage. However, added handling associated with blending would be eliminated, thus reducing some of the damage potential. Moisture content determination would become more critical to marketing, and overall marketing efficiency would improve. Discount methods, such as that given by equation 9.37, could be developed that would provide the elevators an economic hedge against fluctuations in grain and energy prices, and farmers would have a better means for comparing the price offerings among competing buyers. Overall, a revision in current practices would appear to be a welcome change in the grain marketing system.

Optimum Moisture Content to Begin Harvesting

The process of selecting the moisture content to initiate the harvesting of corn can be viewed as the economic trade-off of energy expenditures between the energy required for drying and the energy associated with harvest losses (Loewer et al., 1982; 1984d). The optimum moisture content to begin harvesting is based on the economic values of drying fuel energy and corn grain energy, and the drying method and harvesting rate.

The primary energy expenditures for grain drying are associated with fuel for drying (primarily LP gas) and energy in the harvest losses. Harvest loss energy results from field losses and machine losses. Typically, there is a trade-off between drying energy and harvest loss energy. That is, harvest losses may be reduced by earlier harvesting that in turn requires a relatively larger expenditure of drying energy. Drying energy also is influenced by drying rate, harvest rate, and grain moisture content. The economic value of grain energy as compared to drying fuel energy is the determining factor in deciding when to begin harvesting. With these factors in mind, the objectives of this section are to:

1. Compare the magnitude of energy expenditures from drying fuel and corn harvest losses as influenced by starting harvest moisture content, harvesting capacity, and drying method.
2. Compute the economic optimum moisture content to begin harvesting as influenced by harvesting capacity, drying method, and the relative prices of grain and drying fuel energy for one geographic location.

Procedure

The procedure used involves both physical considerations (such as drying energy input, weather, harvest losses) and economic considerations (value of a drying fuel and corn grain).

Physical Considerations. The evaluation of a dynamic harvesting system was accomplished by the development of a simulation model called HARVEST (Harvesting Analysis Requiring Various Estimates of

Table 9.24. Mean number of days with precipitation greater than or equal to 0.01 inch for Owensboro, Kentucky*

Month	Days
January	9
February	8
March	9
April	10
May	10
June	8
July	8
August	7
September	7
October	6
November	8
December	9

* Hill, 1976.

System Traits) (Loewer et al., 1984d). The model contains inputs relating to economics, weather, harvesting capacity, drying energy, machine losses, and field losses. Any grain type may be evaluated although only corn was used in this example.

In HARVEST, weather data may be supplied from either observed values or generated using probability distributions. The latter has been selected for this example. Average Owensboro, Kentucky rainfall data (Hill, 1976) was used in determining if harvesting occurred (table 9.24). No harvesting was scheduled if any rainfall occurred according to randomly generated precipitation events using table 9.24 inputs assuming a uniformly distributed probability of rain on a given day. The rainfall of a previous day was not considered. No harvesting was scheduled for Sundays, and no allowance was made for machine breakdown. Three runs were made of each specific harvesting strategy to reflect differences in weather during each harvesting season. The average of the three resulting energy expenditure values was used in the economic analysis.

Harvesting capacity as a function of grain moisture content is site specific. That is, it depends on the individual harvesting, delivery, handling-drying-storage equipment set. For this example, it was assumed that grain moisture content has no effect on harvesting capacity.

Energy for drying, machine losses and field losses were all considered to be functions of grain moisture content. The field moisture content variation over the harvest season was computed using a relationship given by Morey et al. (1971):

$$M(t) = M_e(t) + \left[\left(s/b\right) * \left(b*t - 1\right)\right] + \left[M_o + \left(s/b\right) - r - s*t_e\right] * e^{-(b*t)} \qquad (9.38)$$

where

M_o = initial moisture content specified at a time $t = 0$ (% wet basis)

$M_e(t)$ = equilibrium moisture content of grain in the field ($r + s*t_e$)

r = grain equilibrium moisture content at time t'

s = rate of daily change in $M_e(t)$ over the harvesting season

t = elapsed time from the date when a moisture content is known (days)

t_e = $t' - t$ (relates to a defined seasonal time axis t')

b = drying constant [$time^{-1}$ (0.06)]

It was assumed for this example that the equilibrium moisture content $M_e(t)$ on September 1 ($t' = 0$) was 15% and on November 30 ($t' = 91$) was 18%. With these assumptions, the parameters r and s in the expression for equilibrium moisture content are 15 and 0.033, respectively (Benock et al., 1979). Therefore, grain moisture content can be expressed as a function of elapsed time from a date when moisture content is known (t = 0). For this example, grain was at 30% moisture on October 1 and approached 18% on November 30.

Energy for drying varies with drying method, and ambient temperature and relative humidity, as well as initial and final moisture contents of the corn. Energy inputs for three representative drying systems (or drying rates) were determined for several initial moisture contents when drying to a final moisture content of 15.5% wet basis by using the crossflow drying model developed by Thompson et al. (1968). Changes in grain moisture content between these points were determined by linear interpolation. The three drying systems are representative of automatic, batch continuous-flow drying (high speed drying), batch-in-bin drying (medium speed drying) and layer drying (low speed drying). See table 9.25 for the energy comparison among the systems.

Daily machine and field losses (table 9.7) were expressed as a function of moisture content (Johnson and Lamp, 1966). Expressing losses in this fashion tends to reflect long-term averages rather than specific weather related events. The following terminology is used:

DE = Average energy required to dry the grain while not reflecting any inefficiencies of combustion.

ME = Average energy losses associated with harvesting machine inefficiency.

FE = Average energy losses associated with field losses not related to harvesting machinery.

Table 9.25. Energy values used to establish drying energy as a function of moisture content and drying rate when drying to a final moisture content of 15.5%

	Drying Energy*					
	Low Drying Rate		Medium Drying Rate		High Drying Rate	
Moisture Content (%)	mJ/kg water	Btu/lb water	mJ/kg water	Btu/lb water	mJ/kg water	Btu/lb water
18	5.07	2180	5.26	2263	8.25	3548
22	4.07	1748	4.38	1881	7.35	3159
26	3.99	1715	4.15	1783	7.26	3121
30	3.96	1702	4.08	1756	7.22	3104

* Low, medium, and high corresponds to typical layer, batch-in-bin and automatic-batch continuous-flow dryers. Energy values are reflective of total energy input comparisons.

Both ME and FE were assumed to be non-recoverable by livestock or other harvesting systems. The energy value used for corn was 18.43 MJ/kg of dry matter (7,926 Btu/lb) (Crampton and Harris, 1969).

In all instances, the same quantity of grain dry matter was assumed to exist on October 1. Harvesting continued until all of this material was either harvested or lost in the form of field or machine losses.

The HARVEST model was run for several different harvesting rates expressed as the number of work days required to harvest the crop if harvesting was not interrupted for any reason (1, 10, 15, 20, 25, 30, and for some analyses 40, 50 days). By using work days to express harvesting rates, the need to consider actual harvest time was eliminated. Harvesting was initiated when grain moisture reached the stated level of either 30, 28, 26, 24, 22, 20, or 18% wet basis.

Economic Considerations. One of the more important considerations for a corn farmer with an on-farm drying system is when to initiate harvesting to minimize the combined cost of drying fuel and harvest losses. By definition, the economic optimum is the point where the marginal value of the product being produced is equal to the marginal costs of the inputs used in production assuming that production is at a level where diminishing returns are occurring. For this example, the product is "saved grain" and the input is "drying fuel". The objective is to determine when the value of one additional unit of drying energy is equal to the value of the additional grain energy saved in the form of reduced harvest losses. An assumption is that the final equilibrium moisture content approaches 18%. HARVEST computed the total grain energy losses if harvesting was not initiated until the grain reached 18%. This energy value was established by definition as the greatest loss in grain energy (although certainly the grain could be left in the field until it was all lost). The relationship between additional drying energy and grain energy savings was developed for a constant harvesting rate by comparing drying energy expended and reductions in harvest losses as starting moisture content for harvesting was increased in stages from 18 to 30%. This procedure was repeated using HARVEST for each harvesting rate and method of drying. Using this data, an expression of the following form was found adequate to relate grain energy and drying fuel energy:

$$GE_{saved} = a_o + a_1 * DE_{exp} - a_2 * DE_{exp}^2 \quad (9.39)$$

where

GE_{saved} = grain energy saved

DE_{exp} = drying energy expended for a particular moisture content, drying method, and harvesting rate

a_o, a_1, a_2 = constants

The above equation is a physical relationship that may be converted to an economic expression of the following form:

$$P_{GE} * GE_{saved} = \left(a_o + a_1 * DE_{exp} - a_2 * DE_{exp}^2\right) * P_{GE} \qquad (9.40)$$

where
P_{GE} = price (or value) per unit of grain energy or the marginal value of each unit of grain energy

It was assumed for this example that P_{GE} and P_{DE} (price per unit of utilizable drying fuel energy or the marginal cost of drying fuel energy) are constant insofar as an individual grain producer is concerned. Therefore, the optimum level of DE for any price level of grain and drying fuel may be found, by those who are oriented mathematically, by taking the first derivative of equation 9.40 (marginal value of GE_{saved} as a function of DE_{exp}) and equating it to the marginal cost of DE_{exp} (equal to P_{DE}) as shown below:

$$DE_{opt} = \left(a_1 * P_{GE} - P_{DE}\right) / \left(2.0 * a_2 * P_{GE}\right) \qquad (9.41)$$

where
DE_{opt} = the optimum level of DE to be used

DE_{opt}, however, depends on drying method, harvesting rate, and moisture content. The more important question concerns the moisture content for beginning harvest to obtain DE_{opt}. The following relationship was found to relate moisture adequately at the beginning of harvest and DE_{exp} for each harvesting rate:

$$MC = b_0 + b_1 * DE_{exp} - b_2 * DE_{exp}^2 \qquad (9.42)$$

where
MC = moisture content at the beginning of harvest

By substituting the value of DE_{opt} into equation 9.42 for DE_{exp}, the optimum moisture content to begin harvesting (MC_{opt}) can be determined for a grain drying method and harvesting rate.

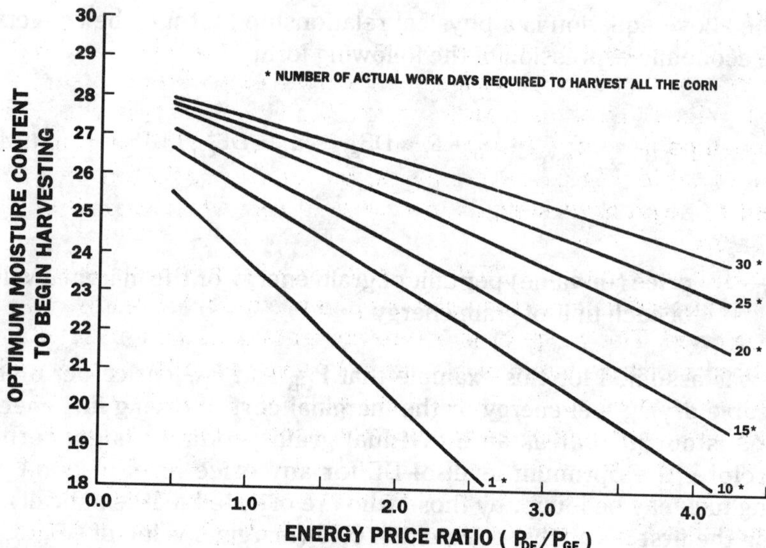

Figure 9.53–Influence of the energy price ratio of drying fuel energy and corn grain energy as it affects the moisture content to begin harvesting, for automatic-batch continuous-flow dryers. [The price of energy ratio is the price per unit of drying fuel energy divided by the price per unit of grain energy, and is also equal to 5.0959*($/gal of LP gas)/($/bu of corn); (Loewer et al., 1982; 1984d).]

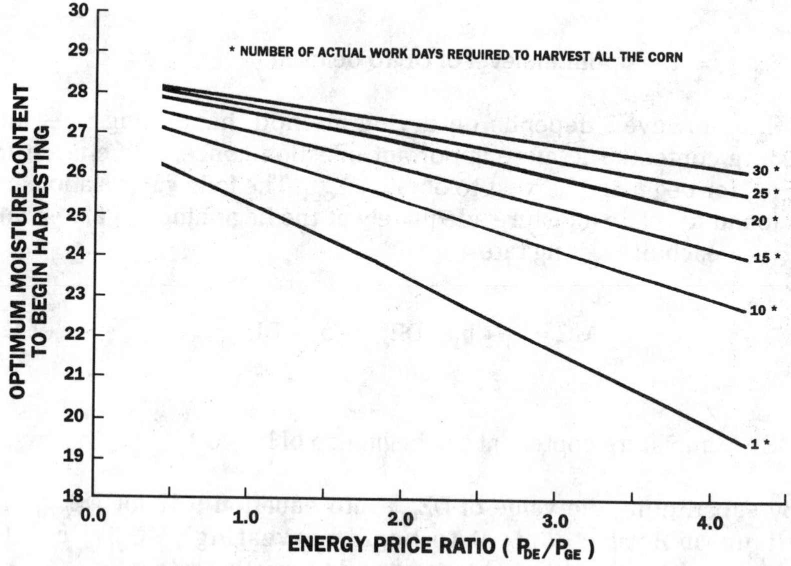

Figure 9.54–Influence of the energy price ratio of drying fuel energy and corn grain energy as it affects the moisture content to begin harvesting, for batch-in-bin dryers. [The price of energy ratio is the price per unit of drying fuel energy divided by the price per unit of grain energy, and is also equal to 5.0959*($/gal of LP gas)/($/bu of corn); (Loewer et al., 1982; 1984d).]

Results

P_{DE} and P_{GE} may be expressed in the form of a ratio to give the optimum moisture content to begin harvesting regardless of changes in actual prices. Table 9.7 and figures 9.53 to 9.55 represent the relationships developed from equations 9.39 to 9.42. The equations shown in table 9.26 allow computation for the optimum moisture content to begin harvesting over a range of harvesting capacities and price ratios.

The equations in table 9.26 do not state whether it is profitable to dry grain but instead indicate the optimum point to begin harvesting if grain is to be dried. The range of moisture contents used was 18 to 30%. A calculated moisture value of less than 18% would indicate that grain should be allowed to dry in the field as long as possible or until it reaches 18%. Remember, an assumption used in this example is that grain will not dry under field conditions to less than 18%, thereby requiring some additional drying if grain is to be stored safely. It is also assumed that the price of drying fuel energy (P_{DE}) is not sufficiently greater than the price per unit of grain energy (P_{GE}) to prevent economically any harvesting of grain that required drying energy before storage.

It is possible that starting moisture content calculations could result in values greater than 30%. The upper boundary of starting moisture

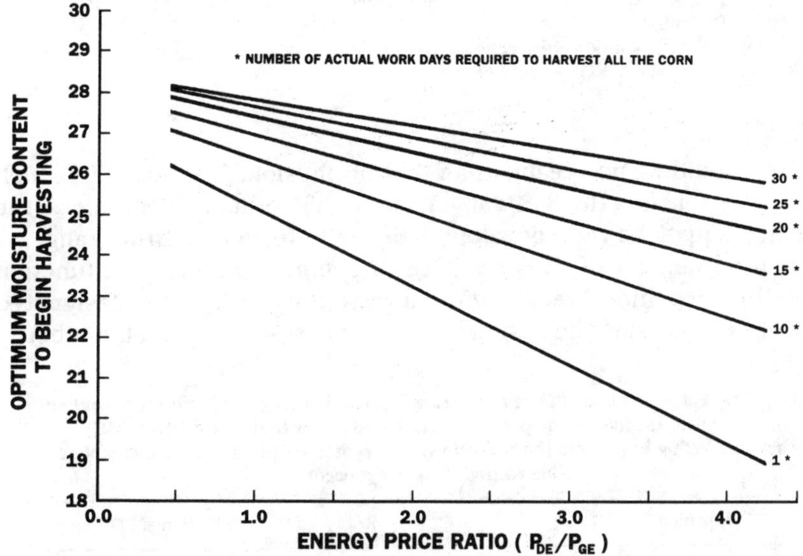

Figure 9.55–Influence of the energy price ratio of drying fuel energy and corn grain energy as it affects the moisture content to begin harvesting, for layer dryers. [The price of energy ratio is the price per unit of drying fuel energy divided by the price per unit of grain energy, and is also equal to 5.0959*($/gal of LP gas)/($/bu of corn); (Loewer et al., 1982; 1984d).]

Table 9.26. Optimum moisture (MC_{opt}) to initiate harvesting as a function of harvesting rate expressed as days to harvest if no delays occur (D) and the price ratios for grain energy (P_{GE}) and drying fuel energy (P_{DE})*

Drying Method	Equation	R^2	MC RGE CV	Drying RGE, %‡	Days‡
High	$MC_{opt} = 26.86 + 0.2216*D - 3.4916*(P_{DE}/P_{GE}) - 0.006119*D^2 + 0.08783*D*(P_{DE}/P_{GE})$	0.988	1.72	18-30	1-30
	or				
	$MC_{opt}† = 26.86 + 0.2216*D - 17.7982*(P_{LP}/P_c) - 0.006119*D^2 + 0.4476*D*(P_{LP}/P_c)$				
Medium	$MC_{opt} = 26.73 + 0.1716*D - 1.8284*(P_{DE}/P_{GE}) - 0.004063*D^2 + 0.04437*D*(P_{DE}/P_{GE})$	0.991	0.80	18-30	1-30
	or				
	$MC_{opt}† = 26.73 + 0.1716*D - 9.3173*(P_{LP}/P_c) - 0.004063*D^2 + 0.2261*D*(P_{LP}/P_c)$				
Low	$MC_{opt} = 26.75 + 0.1702*D - 1.7225*(P_{DE}/P_{GE}) - 0.004149*D^2 + 0.04330*D*(P_{DE}/P_{GE})$	0.990	0.79	18-30	1-30
	or				
	$MC_{opt}† = 26.75 + 0.1702*D - 8.7777*(P_{LP}/P_c) - 0.004149*D^2 + 0.2207*D*(P_{LP}/P_c)$				

* Also, included are expressions for MC_{opt} as a function of market prices for LP gas (P_{LP}, \$/gal) and corn (P_c, \$/bu).
† Reflects an 80% combustion efficiency.
‡ MC = moisture content; RGE = range of values.

content should be no greater than that of physiological maturity which may be greater than the 30% value used in this example. Regardless, the table 9.26 equations were developed for the 18 to 30% moisture range.

The total number of days required to complete harvest is a function of weather, machine breakdown, harvesting capacity, and designated non-work days. Using the assumptions of this study, the total number of

Table 9.27. Total days to complete harvesting (D_{av}) and average grain moisture content (MC_{av}) over the harvesting period as influenced by the moisture content (MC) when harvesting begins and the harvest rate (D) expressed as work days to complete harvesting if no delays occur

Equation	R^2	DV	MC-Range	D-Range
$D_{av} = 4.9155 + 1.37*D + 0.1896*MC$	0.997	3.32	18-30	1-50
$MC_{av} = 10.66 - 0.2745*D_{av} + 0.5514*MC + 0.003838*D_{av}^2$	0.756	7.33	18-30	1-50

days required to complete harvesting may be estimated from the first equation given in table 9.27. Note for this example that the drying method has no influence because it was assumed not to limit harvesting rate.

The moisture content of grain is subject also to weather in addition to possible varietal differences. The average moisture content over the harvesting period for this study may be estimated using the second equation given in table 9.27.

Summary

The relationships between the energy used in corn drying and that lost in the form of field and machine losses were simulated as a function of harvesting capacity, drying method, and the moisture content at which harvesting begins. From this information, the optimum economic moisture to begin harvesting has been related to harvesting capacity, drying method and the price ratio of drying energy to grain energy. This analysis does not consider any economic aspects of the total drying system other than artificial drying as compared to field drying. A static analysis of these other factors is presented earlier in this chapter. The assumptions used within the study also are very important to consider when using this information to plan harvesting strategy. Especially important are the following:

1. Generally, drying capacity increases with decreasing grain moisture content. This only becomes important to this analysis if grain drying capacity is the limiting factor in determining daily harvesting capacity.
2. The above analysis considered only one field moisture dry-down relationship regardless of weather or varietal differences.
3. All grain was dried to 15.5%. A different final moisture content would alter the energy consumption but not affect the economic optimum beginning moisture content for harvesting so long as the final field equilibrium moisture was 18%.
4. This analysis does not address the economic desirability of altering harvesting capacity to achieve greater profitability in drying. Rather, this example examines only the desirability of allowing corn to field dry.
5. Machine and field loss data in this study are somewhat dated and represent average daily losses.
6. Weather varies considerably over the corn growing area of the U.S. whereas this study used only one geographic location.

The above mentioned factors may alter the economic optimum point to initiate harvesting. However, the results obtained are reasonable and the procedures used are correct for any geographic location and type of grain. The intent is to define the computed values better as more information becomes available. Regardless, the computed optimum moisture content to begin harvesting as a function of harvesting capacity and the energy price ratio is of great interest to grain farmers.

Problems

9.1 Assume that soybean yield for an entire field is related to fertilizer applied by the relationship:

Y = 0.005*p**2 - 0.000006*p**3

where Y = bushels of soybeans produced, all other factors of soybean production besides fertilizer remain equal and p = pounds of fertilizer applied.

Using a spreadsheet, generate the expected soybean yield, average physical product, and marginal physical product (MPP = 0.01*P-0.000018*P**2) over a range of fertilizer application levels extending in 10-lb increments to the point where no soybeans are produced if additional fertilizer is applied.

Identify the fertilizer level(s) where:
(a) Soybean yields equal zero.
(b) Stage 2 production begins.
(c) Stage 3 production begins.

9.2 Given the above situation, assume that soybeans are priced at $6.00/bu and fertilizer at $0.25/lb. Compute the following over the range of fertilizer inputs used above:
(a) Marginal value of the fertilizer in terms of the extra soybeans that are produced.
(b) Average value of fertilizer in terms of the extra soybeans that are produced. Using the spreadsheet value, at what level of fertilizer input is soybean value at a maximum?

Repeat the analysis assuming that soybeans increase in value to $8.00/bu. How did the change in the value of soybeans affect the optimum level of fertilizer to be applied? How do fixed changes in the prices of inputs (fertilizer) and outputs (soybeans) impact the optimum level of fertilizer to be applied?

9.3 Compute the effective annual interest rate if the monthly interest rate is stated to be 1.5%.

9.4 A bank loan of $200,000 is to be obtained to finance the purchase of a grain storage facility over a 15-year period. The annual compound interest rate is 10%.
(a) Compute the size of the payment if the entire loan were to be repaid at the end of the 15-year period?
(b) How large would the annual payments be if they were of equal size?

9.5 An individual anticipates that he will have to add a continuous-flow grain dryer in seven years at a cost of $25,000. He believes that the prevailing interest rates over that period will be 12%.
(a) How much money would have to be invested today at this interest rate to have sufficient funds to purchase the dryer?
(b) What will be the payment size if he decides to make a series of uniform payments over the next five years for the same purchase?

9.6 A purchase of a grain facility is being considered by an individual under the condition that he may continue the annual loan payments of the previous owner which are $20,000/year for the next eight years. Upon further investigation, the potential new owner finds that the interest rate on the loan is 9%. What would be the size of the payment required to pay off the loan in a single payment at:
(a) The time of the purchase.
(b) At the end of eight years?

9.7 An individual is investigating the purchase of a materials handling system. Two used systems are located which are considered to be of equal value. In both cases, the systems may be purchased by repaying the bank loan. The first system has nine annual payments of $2,000 remaining at 10% interest. The second system has six annual payments of $3,000 remaining at an interest rate of 12%. Assume a zero salvage value at the end of the payment period. Using present value analysis:
(a) Which system is the least expensive and by how much?
(b) Repeat the problem if the annual operating cost for systems 1 and 2 are $500 and $700, respectively. Assume that the interest rates are the same as those used when making the annual payments.
(c) Repeat Part b of the problem given that anticipated repair cost at the end of the loan periods for system 1 and 2 are $1,000 and $3,500, respectively, with an annual interest rate of 11%.

9.8 A farmer has 300 acres of corn with an expected maximum potential yield (no harvest losses) of 140 bu/acre at a value of \$3.00/harvested bu. The time required to complete the harvest is estimated to be 33 days given the harvesting capacity, beginning when the corn has reached approximately 27% moisture content, and operating at a combine speed of 2 mph. Compute the change in the cost of harvest losses per acre and in total for the following conditions:

(a) Cool humid conditions exist (table 9.4) as compared to a warm dry season (table 9.5).

(b) Given cool dry conditions (table 9.6), harvester speed is increased from 2 to 5 mph.

(c) Given cool humid conditions (table 9.4), total harvesting time is decreased from 33 to 21 days. Increased from 33 to 45 days?

(d) Given a warm dry season (table 9.5), starting moisture content at harvest changes from 27% to 21.3%. From 27% to 33%?

9.9 One farmer reports the corn yield to be 130 bu/acre while another reports a yield of 140 bu/acre. The moisture contents at harvest time are 21% and 26%, respectively.

(a) Given a base moisture content of 15.5%, which farmer had the greatest "dry" bushels yield and by how much.

(b) Repeat the situation given a base moisture content of 14.0%.

9.10 A farmer harvests corn from a 1,000 acres and delivers it to a commercial elevator. If 125,000 bu of U.S. No. 2 corn (15.5% base) were delivered at a field moisture content of 18%, what is the "wet bushel" yield/acre? At 24%? At 30%.

9.11 A farmer is considering delivering corn to one of two commercial elevators. Elevator No. 1 has dockage charges of 2% of the selling price for each point of moisture above the U.S. No. 2 base (15.5%). Elevator No. 2 charges 6¢/point of moisture above the same base. Both elevators consider "invisible losses" to be 0.5%. Compute the gross return received by the farmer for each situation given the following:

(a) Corn is delivered at 20.5% moisture at a price of \$2.75/U.S. No. 2 bushel. At 25.5%? At 30.5%?

(b) Corn is delivered at 20.5% moisture at a price of \$2.25/U.S. No. 2 bushel. At 25.%? At 30.5%?

9.12 Given the situation in the previous problem, at what price will the dockage charge be the same at both elevators given a delivery moisture content of 20.5%? 25.5%? 30.5%?

9.13 A farmer estimates that his/her grain yield in "wet" bushels will be 150 bu/acre based on an average harvesting moisture content of 28% and a selling price of 3.00/bu at the point of delivery. The dockage rate used for purposes of discounts is 2% of the selling price/point of moisture removed including a 0.5% charge for "invisible" losses. Compute the following:

(a) Dockage charge per bushel delivered.
(b) Value of grain if sold wet (28%) ($/bu).
(c) Value of grain if sold dry (15.5%) ($/bu).
(d) Gross return available for on-farm drying ($/bu).

9.14 Given the previous problem situation, the farmer decides to purchase an on-farm drying system that will dry 2,000 bu of corn (at 56 lb/bu) to 15.5% moisture in 15 h. Assume electrical cost to be $0.06/kW-h and to be used at 100% efficiency. A 10-hp drying fan is used (multiply hp by 0.746 to obtain kW). LP gas cost is $0.80/gal with a burning efficiency of 80%. Drying energy requirements are 1,800 Btu/lb of water removed. Compute the following:

(a) Electrical cost ($/bu).
(b) LP gas cost ($/bu).
(c) Total energy cost ($/bu).
(c) Total fuel cost per day.

9.15 A farmer has 400 acres of corn with an average yield of 140 bu/acre. After all variable expenses are considered, a net return of $0.05/bu is expected if an on-farm drying system is purchased. The drying system has a useful life of 10 years with no salvage value. The interest rate associated with a loan is 11%.

(a) What is the maximum amount that can be spent each year on the system without experiencing a net loss?
(b) What is the net present value of this facility?

9.16 Referring to figures 9.7 to 9.9 (a) to (c), determine the following:

(a) The annual cost differential, per bushel and in total, for 20,000 bu of storage if selecting between a 1- or 4-bin layer drying facility.
(b) The per bushel annual cost differential for a 4-bin facility using batch-in-drying when comparing 10,000 bu of storage capacity to 50,000 bu of storage capacity.

(c) The annual cost differential, per bushel and in total, for a 40,000 four-bin bushel facility, if selecting between a layer drying and portable drying methods.

9.17 Referring to figures 9.10 to 9.12(a) to (c), compute the annual cost differential, per bushel and in total, for a 50,000 bu layer drying facility when comparing the ability to dry the entire capacity in 5 days as compared to 30 days. Repeat for batch-in-bin drying and portable drying.

9.18 For the conditions given in figure 9.13, compute the percentages of both the purchase and annual costs associated with the storage bin component for 10,000-bu layer drying facility. Repeat for an 80,000-bu facility.

9.19 Suppose two farmers were located in significantly different geographic areas in terms of climate. For the same level of risk, one could store a given type of grain at 14% moisture while the other had to store a similar type of grain at 13% moisture. Given that the grain base moisture content at the time of sale is 15.5% moisture, compute the over-drying cost for each farmer given:
(a) grain valued at $2.00/bu.
(b) $3.00/bu.
(c) $4.00/bu.

9.20 In terms of interest expense (opportunity cost), compute the total cost associated with storing 50,000 bu of grain given that grain is valued at $3.00/bu, the annual interest rate is 10%, and the storage period is three months. Repeat for storage times of 6, 9, and 12 months.

9.21 How much will the price of corn have to increase, both in absolute and in percentage, given the following conditions? A 3-bin, 50,000-bu portable drying facility is to be used exclusively for this corn [fig. 9.9 (c)]. The price of corn at the time of harvest (October 1) is $3.00/bu. The corn is to be sold as U.S. No. 2 (15.5% base moisture content) on March 1 of the following year. The annual interest rate is 11%, and taxes and insurance charges total 1% of the October 1 grain value. For simplicity, assume that the interest cost is based on the October 1 price rather than the average value of the grain over the storage period.

9.22 Compute the average annual fixed cost for a stationary mill based on a purchase price of $12,000, a usable life of 10 years with no salvage value, an annual interest rate of 9%, an annual tax rate of

1%, an annual insurance cost of 2%, and a repair and maintenance cost totaling 120% of the purchase cost of the mill over its life.

9.23 Using the previous situation, compute the operating cost per ton of processed material given that the mill capacity is 3 ton/h, the cost of labor is $5.00/h, 60% of the cost of labor actually is associated with operating and maintaining the mill, 10 hp is required to operate the electrically powered mill, and the cost of electricity is $0.07/kW-h.

9.24 Using the results generated from the above situations, generate the total annual costs per ton for processing 50, 100, 200, 400, 700, and 1,000 tons of feed/year.

9.25 Given that the commercial cost per ton for the processing of feed is $20/ton, compute the break-even point for both the stationary and portable mill systems in tables 9.17 and 9.18 and equation 9.16.

9.26 Using the CACHE computer program "base 3" conditions for each comparison, determine the net return differences for on-farm corn drying and storage facilities under the following independent comparative situations (return to the original base data in CACHE before making each comparison):

(a) operating a harvester at 5 mph as opposed to 2.5 mph in order to decrease the calendar days associated with harvesting from 45 to 35 if on-the-farm grain drying and storage is not available.

(b) corn is to be stored for three months at 14% moisture as compared to being stored for four months at 13.5% moisture content.

(c) the initial price of corn is $3.00/bu to increase in value by 7% in two months as compared to $2.50/bu corn that increases in value by 15% in 4 months.

(d) the LP gas used per point of moisture removed is 0.018 gal/point if the average moisture content to begin harvesting is 28% with on-the-farm drying as compared to 0.02 gal/point if the starting moisture content is 30%.

9.27 A farmer wishes to provide 2 ft^3/min/bu of grain in a bin that will contain 12,000 bu when filled to capacity.

(a) What heater capacity (Btu/h) is required to generate an 8° F temperature rise?

(b) What will be the temperature rise if the heater is installed and later on the bin capacity is increased to 15,000 bu by adding an additional ring? [Assume the fan delivers the same air volume per bushel as in (a)].

9.28 Given the following: 10-hp drying fan attached to a 10,000-bu grain bin; 250 h for drying with a stirrer; electricity cost of $0.07/kW-h; 100,000 Btu/h heater, LP gas cost of $0.75/gal; 92,000 Btu/gal of LP gas; 80% burning efficiency for LP gas burner; 1 kW stirring motor power requirements; stirrer purchase cost of $3,000; stirrer repair cost equal to 60% of the purchase price over an expected stirrer life of five years; annual interest rate of 10%. Compute the annual cost in total and per bushel for the following:
(a) Drying fan (to convert hp to kW, multiply by 0.746).
(b) Drying fuel cost.
(c) Operation of stirring motor.
(d) Stirring motor repair cost.
(e) Interest cost based on the average value of the stirring device assuming zero salvage value.

9.29 The after-drying average grain moisture content is 14% if a stirring device is used. Conversely, the after-drying average moisture content is 12% if a stirrer is not used. If the grain may be stored safely at 14%, compute the total and per bushel overdrying costs for 50,000 bu given a grain value of $3.00/bu at 14% moisture content.

9.30 Given figure 9.37, what are the break-even grain prices for 24 and 27-ft. diameter bins when using stirring devices?

9.31 Given figure 9.41, how much will the net return for stirring change for $3.50/bu corn (U.S. No. 2) if the drying air temperature increases from 6 to 12° F?

9.32 Determine the relative economics of using a stirring devise by running the STIREC program with default values given in table 9.21 except as follows: for both with and without stirring, the initial moisture content is 22%; annual interest rate is 10%; 15-hp drying fan; 1.5 hp total for stirring motors; 0% tax bracket for all years; and a 24-ft diameter bin. Economic feasibility is to be determined for corn prices ranging from $2.50 to $3.00/bu in $0.25 intervals and for natural, low and high temperature systems. (Note: A negative value in STIREC indicates that using stirring devices is feasible economically.)

9.33 Using figure 9.45 or equation 9.36, determine the market value of corn being harvested at 23% moisture if the market price is $3.25/bu at the base moisture content of:
(a) 15.5%.
(b) 14%.

9.34 Rice is to be dried from 20% to a base moisture content of 14%. Given that rice sells for $3.00/bu at 20% moisture, 5,000 Btu/bu is required for drying, and LP gas contains 92,000 Btu/gallon at a cost of $0.80/gal, compute the percentage of the moisture value that is associated with drying.

9.35 What percentage of weight will 28% grain lose if dried to a base moisture content of 15.5%?

9.36 Compute the equivalent dockage charge in cents per point for a dockage rate based on 2%/point of moisture with grain selling for:
(a) $2.50/bu.
(b) $3.50/bu.

9.37 Fifty-thousand bushels of corn are to be harvested at 28% moisture. The selling price for corn is $2.80/bu for U.S. No. 2 (15.5% moisture). The dockage rate being used by the local elevator is 2% of the selling price per point of moisture removed. Invisible losses associated with handling are not considered. Compute in total and on a per bushel basis:
(a) The value of the grain if sold directly to the elevator (value sold wet).
(b) The gross value of the grain if dried on the farm to the base moisture content.
(c) The gross return available to the farmer to pay for drying.

9.38 Given the above situation, the farmer estimates that 0.02 gal of LP gas will be required per point of moisture removed. LP gas may be purchased for $0.80/gallon. In addition, the cost of management and maintenance of the drying system is estimated to be $0.03/bu, and the life of the dryer is 10 years. What is the maximum that the farmer can pay for the drying system without losing money?

9.39 Repeat the above two problems given, for whatever reasons, that the farmer dries the grain to 13% before delivering it to the elevator.

9.40 How many gallons of water are lost per bushel if soybeans are dried to 10% below a base moisture content of 13%? What is the gross value of the water (in $/gal) given that soybeans are selling for $6.00/bu (see figs. 9.50 and 9.51)?

9.41 Assume that corn is selling for $2.50/bu, and the cost of LP gas for drying is $0.80/gal. Compute the optimum moisture content for beginning harvest (given the assumptions inherent to figs. 9.53 to9.55) if 25 working days are required to harvest the crop and the following drying systems are used:
(a) Automatic-batch/continuous-flow drying;
(b) Batch-in-bin drying?
(c) Layer drying?

9.42 Given the conditions in the above problem, compute for each of the drying systems (a) the total days required for harvesting, and (b) the average grain moisture content over the harvesting period (see table 9.27).

9.43 Using the computer program OPTMC, repeat the above two problems for the situation where 18 working days are required to harvest the corn; the final storage moisture content is 13.5%; the grain is stored 4 months; a 10% annual interest rate is in effect; and the cost for handling and delivery is $0.05/bu.

References for all chapters begin on page 541.

Appendix A

ASAE D271.2 DEC92

PSYCHROMETRIC DATA

Reviewed by ASAE's Structures and Environment Division and the Food Engineering Division Standards Committees; approved by the ASAE Electric Power and Processing Division Standards Committee; adopted by ASAE December 1963; reconfirmed December 1968; revised April 1974, April 1979; reconfirmed December 1983, December 1988, December 1989, December 1990, December 1991; reaffirmed for one year December 1992.

SECTION 1—PURPOSE AND SCOPE

1.1 The purpose of this Data is to assemble psychrometric data in chart and equation form in both SI and English units.

1.2 Psychrometric charts are presented that give data for dry bulb temperature ranges of -35 to 600 °F in English units and -10 to 120 °C in SI units.

1.3 Many analyses of psychrometric data are made on computers. The equations given in Sections 2 and 3 enable the calculation of all psychrometric data if any two independent psychrometric properties of an air-water vapor mixture are known in addition to the atmospheric pressure. In some cases, iteration procedures are necessary. In some instances, the range of data covered by the equation has been extended beyond that given in the original source. The equations yield results that agree closely with values given by Keenan and Keyes (1936) and existing psychrometric charts.

TABLE 1—SYMBOLS

Symbol	Definition
h	Enthalpy of air-vapor mixture, J/kg dry air or Btu/lb dry air
h_{fg}	Latent heat of vaporization of water at saturation, J/kg or Btu/lb
h'_{fg}	Latent heat of vaporization of water at T_{wb}, J/kg or Btu/lb
h''_{fg}	Latent heat of vaporization of water at T_{dp}, J/kg or Btu/lb
h_{ig}	Heat of sublimation of ice, J/kg or Btu/lb
h'_{ig}	Heat of sublimation of ice at T_{wb}, J/kg or Btu/lb
h''_{ig}	Heat of sublimation of ice at T_{dp}, J/kg or Btu/lb
H	Humidity ratio, kg water/kg dry air or lb water/lb dry air
ln	Natural logarithm (base e)
P_{atm}	Atmospheric pressure, Pa or psi
P_s	Saturation vapor pressure at T, Pa or psi
P_{swb}	Saturation vapor pressure at T_{wb}, Pa or psi
P_v	Vapor pressure, Pa or psi
rh	Relative humidity, decimal
T	Dry-bulb temperature, kelvin or rankine
T_{dp}	Dew-point temperature, kelvin or rankine
T_{wb}	Wet-bulb temperature, kelvin or rankine
V_{sa}	Air specific volume, m^3/kg dry air or ft^3/lb dry air

SECTION 2—PSYCHROMETRIC DATA IN SI UNITS

2.1 Psychrometric charts; two presented. One for a temperature range of -10 to 55 °C and one for a temperature range of 20 to 120 °C.

2.2 Psychrometric equations, SI units. Symbols are defined in Table 1.

2.2.1 Saturation line. P_s as a function of T

$$\ln P_s = 31.9602 - \frac{6270.3605}{T} - 0.46057 \ln T$$

Brooker (1967)

$255.38 \leq T \leq 273.16$

and

$$\ln(P_s/R) = \frac{A + BT + CT^2 + DT^3 + ET^4}{FT - GT^2}$$

Adapted from Keenan and Keyes (1936)

$273.16 \leq T \leq 533.16$

where

R = 22,105,649.25		D = 0.12558×10^{-3}
A = -27,405.526		E = -0.48502×10^{-7}
B = 97.5413		F = 4.34903
C = -0.146244		G = 0.39381×10^{-2}

2.2.2 Saturation line. T as a function of P_s

$$T - 255.38 = \sum_{i=0}^{i=8} A_i\,[\ln(0.00145 P_s)]^i$$

$620.52 < P_s < 4,688,396.00$ Steltz and Silvestri (1958)

A_0 = 19.5322
A_1 = 13.6626
A_2 = 1.17678
A_3 = -0.189693
A_4 = 0.087453
A_5 = -0.0174053
A_6 = 0.00214768
A_7 = -0.138343×10^{-3}
A_8 = 0.38×10^{-5}

2.2.3 Latent heat of sublimation at saturation

$$h_{ig} = 2,839,683.144 - 212.56384\,(T - 255.38)$$

$255.38 \leq T \leq 273.16$ Brooker (1967)

2.2.4 Latent heat of vaporization at saturation

$$h_{fg} = 2,502,535.259 - 2,385.76424\,(T - 273.16)$$

$273.16 \leq T \leq 338.72$ Brooker (1967)

$$h_{fg} = (7,329,155,978,000 - 15,995,964.08\,T^2)^{1/2}$$

$338.72 \leq T \leq 533.16$ Brooker (Unpublished)

2.2.5 Wet bulb line

$$P_{swb} - P_v = B'(T_{wb} - T)$$

Brunt (1941)

where

$$B' = \frac{1006.9254(P_{swb} - P_{atm})\left(1 + 0.15577\,\frac{P_v}{P_{atm}}\right)}{0.62194\,h'_{fg}}$$

Substitute h'_{ig} for h'_{fg} where $T_{wb} \leq 273.16$

$255.38 \leq T \leq 533.16$

2.2.6 Humidity ratio

$$H = \frac{0.6219\,P_v}{P_{atm} - P_v}$$

$255.38 \leq T \leq 533.16$

$P_v < P_{atm}$

2.2.7 Specific volume

$$V_{sa} = \frac{287\,T}{P_{atm} - P_v}$$

$255.38 \leq T \leq 533.16$

$P_v < P_{atm}$

2.2.8 Enthalpy

Enthalpy = enthalpy of air + enthalpy of water (or ice) at dew-point temperature + enthalpy of evaporation (or sublimation) at dew-point temperature + enthalpy added to the water vapor (super-heat) after vaporization.

$$h = 1006.92540\,(T - 273.16) - H[333{,}432.1 + 2030.5980(273.16 - T_{dp})] + h''_{ig}\,H + 1875.6864\,H(T - T_{dp})$$

$255.38 \leq T_{dp} \leq 273.16$

and

$$h = 1006.92540(T - 273.16) + 4186.8\,H\,(T_{dp} - 273.16) + h''_{fg}\,H + 1875.6864\,H\,(T - T_{dp})$$

$273.16 \leq T_{dp} \leq 373.16$

2.2.9 Relative humidity

$rh = P_v/P_s$

SECTION 3—PSYCHROMETRIC DATA IN ENGLISH UNITS

3.1 Three psychrometric charts, are presented with temperature ranges of -35 to 50 °F, 32 to 120 °F and 32 to 600 °F, respectively.

3.2 Psychrometric equations, English Units. Symbols are defined in Table 1.

3.2.1 Saturation line. P_s as a function of T

$$\ln P_s = 23.3924 - \frac{11286.6489}{T} - 0.46057 \ln T$$

Brooker (1967)

$459.69 \leq T \leq 491.69$

$$\ln(P_s/R) = \frac{A + BT + CT^2 + DT^3 + ET^4}{FT - GT^2}$$

Adapted from Keenan and Keyes (1936)

$491.69 \leq T \leq 959.69$

where

R = 3206.18
A = -27405.5
B = 54.1896
C = -0.045137
D = 0.215321×10^{-4}
E = -0.462027×10^{-8}
F = 2.41613
G = 0.00121547

3.2.2 Saturation line. T as a function of P_s

$$T - 459.69 = \sum_{i=0}^{i=8} A_i\,[\ln(10P_s)]^i$$ Steltz and Silvestri (1958)

$0.09 \leq P_s \leq 680$

where

A_0 = 35.1579
A_1 = 24.5926
A_2 = 2.11821
A_3 = -0.341447
A_4 = 0.157416
A_5 = -0.0313296
A_6 = 0.00386583
A_7 = -0.249018×10^{-3}
A_8 = 0.684016×10^{-5}

3.2.3 Latent heat of sublimation at saturation

$h_{ig} = 1220.844 - 0.05077\,(T - 459.69)$

$459.69 \leq T \leq 491.69$ Brooker (1967)

3.2.4 Latent heat of vaporization at saturation

$h_{fg} = 1075.8965 - 0.56983\,(T - 491.69)$ Brooker (1967)

$491.69 \leq T \leq 609.69$

$h_{fg} = (1354673.214 - 0.9125275587\,T^2)^{1/2}$

Brooker (Unpublished)

$609.69 \leq T \leq 959.69$

3.2.5 Wet bulb line

$P_{swb} - P_v = B'(T_{wb} - T)$ Brunt (1941)

where

$$B' = \frac{0.2405\,(P_{swb} - P_{atm})\,(1 + 0.15577 P_v/P_{atm})}{0.62194\,h'_{fg}}$$

Substitute h'_{ig} for h'_{fg} when $T_{wb} \leq 491.69$

$459.69 \leq T \leq 959.69$

3.2.6 Absolute humidity (humidity ratio)

$$H = \frac{0.6219\,P_v}{P_{atm} - P_v}$$

$459.69 \leq T \leq 959.69$

$P_v < P_{atm}$

3.2.7 Specific volume

$$V_{sa} = \frac{53.35 \times T}{144\,(P_{atm} - P_v)}$$

$459.69 \leq T \leq 959.69$

$P_v < P_{atm}$

3.2.8 Enthalpy

Enthalpy = enthalpy of air + enthalpy of water (or ice) at dew-point temperature + enthalpy of evaporation (or sublimation) at dew-point temperature + enthalpy added to the water vapor (super-heat) after vaporization.

$$h = 0.2405\,(T - 459.69) - H\,[143.35 + 0.485(491.69 - T_{dp})] + h''_{ig}\,H + 0.448\,H\,(T - T_{dp})$$

$$459.69 \leqslant T_{dp} \leqslant 491.69$$

$$h = 0.2405\,(T - 459.69) + H\,(T_{dp} - 491.69) + h''_{fg}\,H + 0.448H\,(T - T_{dp})$$

$$491.69 \leqslant T_{dp} \leqslant 671.69$$

3.2.9 Relative humidity

$$rh = P_v/P_s$$

Note: Psychrometric charts are printed with permission from the American Society of Heating, Refrigerating and Airconditioning Engineers, Inc., 345 E. 47th St., New York, NY; Proctor & Schwartz, Inc., 7th St. and Tabor Rd., Philadelphia, PA; and Carrier Corp., Carrier Parkway, Syracuse, NY.

References: Last printed in 1981 AGRICULTURAL ENGINEERS YEARBOOK; list available from ASAE Headquarters.

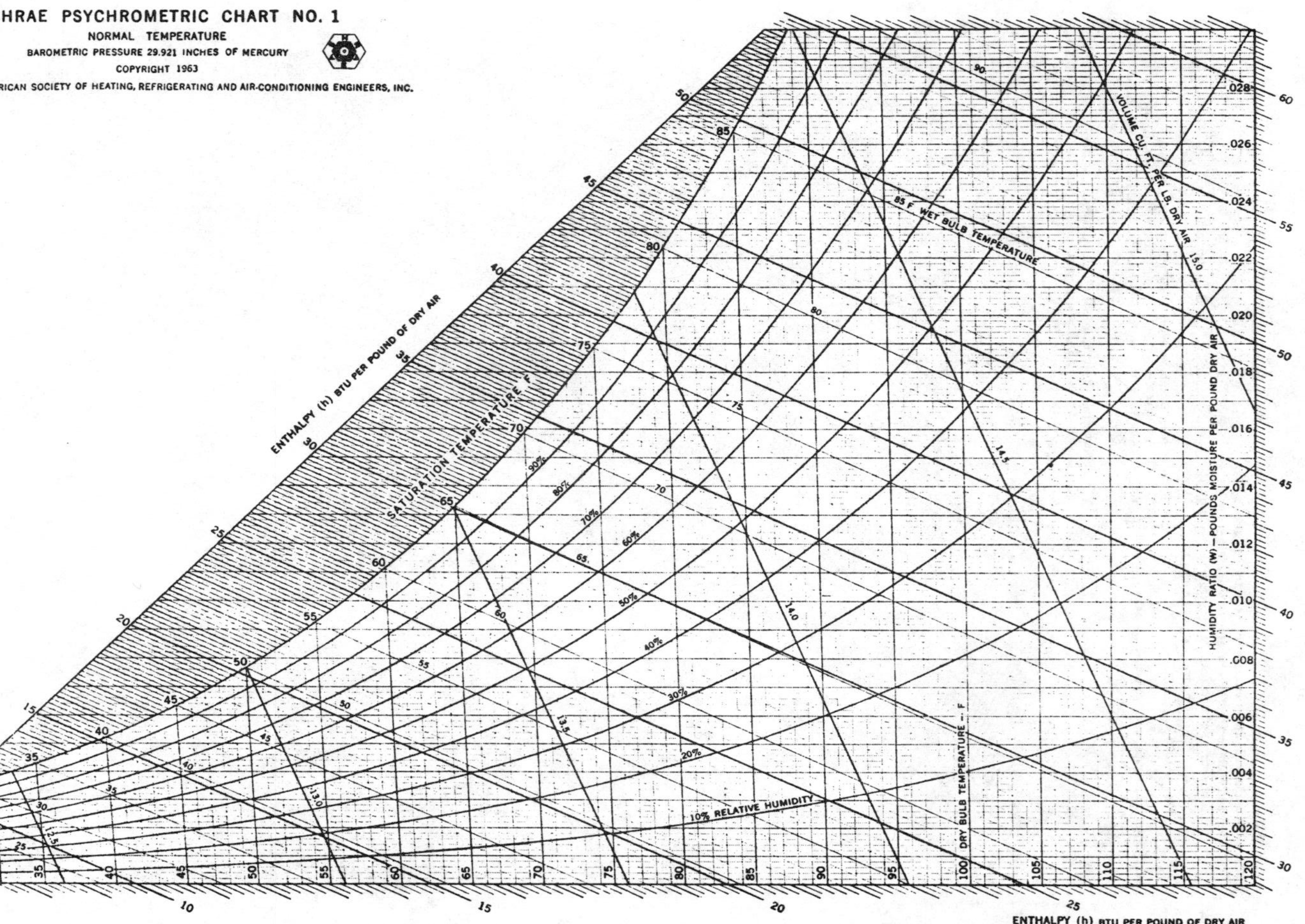
ASHRAE PSYCHROMETRIC CHART NO. 1
NORMAL TEMPERATURE
BAROMETRIC PRESSURE 29.921 INCHES OF MERCURY
COPYRIGHT 1963
AMERICAN SOCIETY OF HEATING, REFRIGERATING AND AIR-CONDITIONING ENGINEERS, INC.
ENTHALPY (h) BTU PER POUND OF DRY AIR
SATURATION TEMPERATURE F
85 F WET BULB TEMPERATURE
VOLUME CU. FT. PER LB. DRY AIR
HUMIDITY RATIO (W) — POUNDS MOISTURE PER POUND DRY AIR
DRY BULB TEMPERATURE — F
10% RELATIVE HUMIDITY
ENTHALPY (h) BTU PER POUND OF DRY AIR

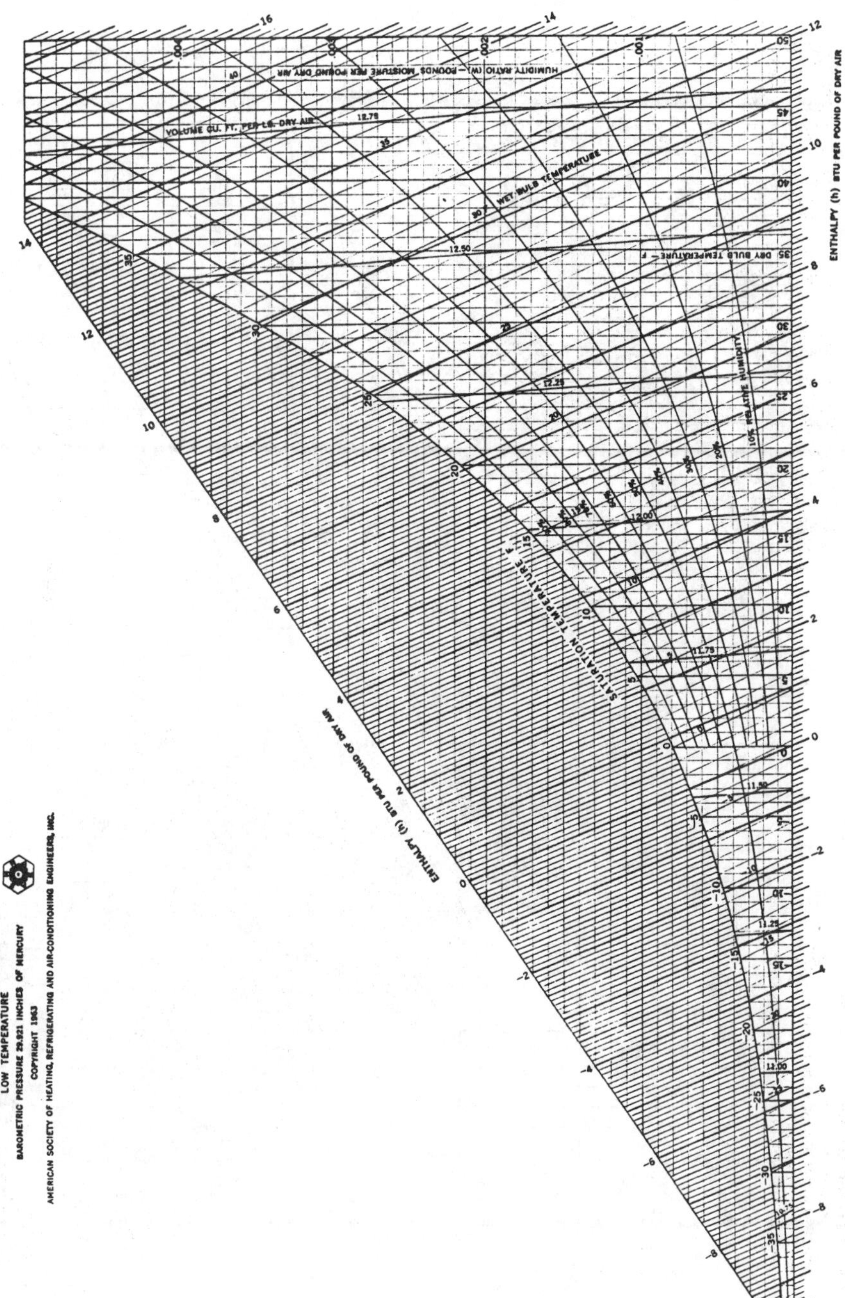
ASHRAE PSYCHROMETRIC CHART NO. 2
LOW TEMPERATURE
BAROMETRIC PRESSURE 29.921 INCHES OF MERCURY
COPYRIGHT 1963
AMERICAN SOCIETY OF HEATING, REFRIGERATING AND AIR-CONDITIONING ENGINEERS, INC.
ENTHALPY (h) BTU PER POUND OF DRY AIR
HUMIDITY RATIO (W) — POUNDS MOISTURE PER POUND DRY AIR
VOLUME CU. FT. PER LB. DRY AIR
WET BULB TEMPERATURE
DRY BULB TEMPERATURE — F
SATURATION TEMPERATURE F
10% RELATIVE HUMIDITY
ENTHALPY (h) BTU PER POUND OF DRY AIR

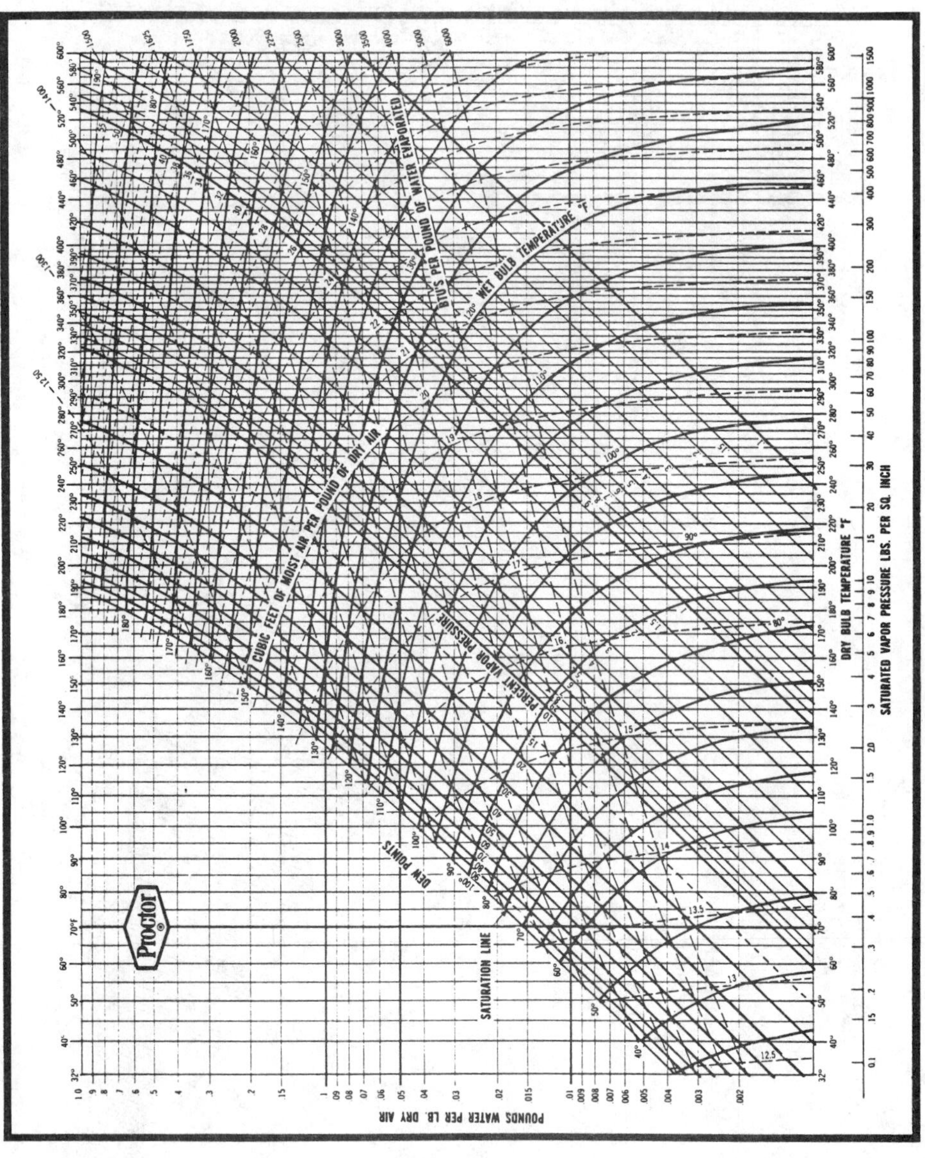
Proctor
CUBIC FEET OF MOIST AIR PER POUND OF DRY AIR
BTU'S PER POUND OF WATER EVAPORATED
WET BULB TEMPERATURE °F
PERCENT VAPOR PRESSURE
DEW POINTS
SATURATION LINE
DRY BULB TEMPERATURE °F
SATURATED VAPOR PRESSURE LBS. PER SQ. INCH
POUNDS WATER PER LB. DRY AIR

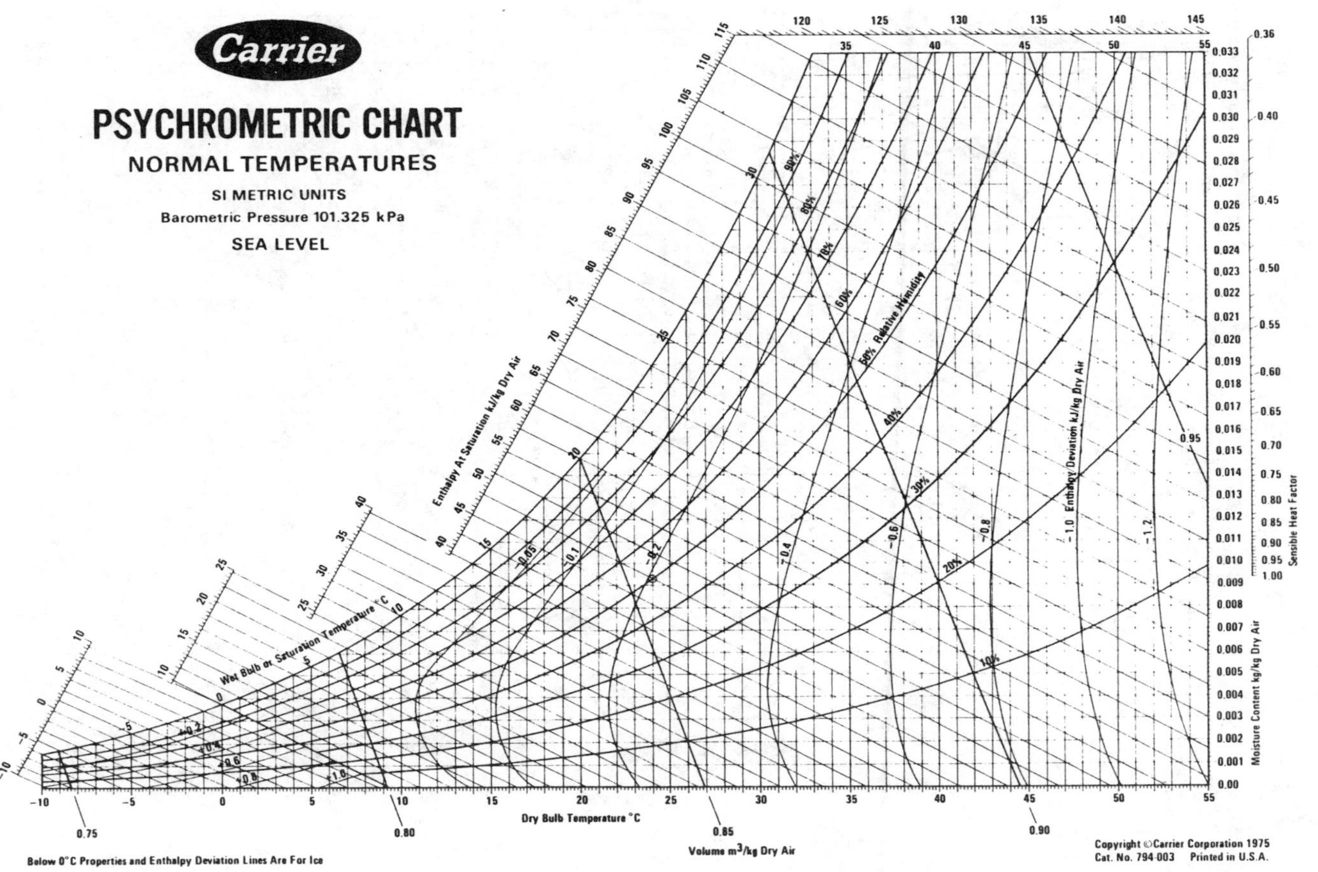
Carrier
PSYCHROMETRIC CHART
NORMAL TEMPERATURES
SI METRIC UNITS
Barometric Pressure 101.325 kPa
SEA LEVEL
Enthalpy At Saturation kJ/kg Dry Air
Wet Bulb or Saturation Temperature °C
Relative Humidity
Enthalpy Deviation kJ/kg Dry Air
Sensible Heat Factor
Moisture Content kg/kg Dry Air
Dry Bulb Temperature °C
Volume m3/kg Dry Air
Below 0°C Properties and Enthalpy Deviation Lines Are For Ice
Copyright © Carrier Corporation 1975
Cat. No. 794-003 Printed in U.S.A.

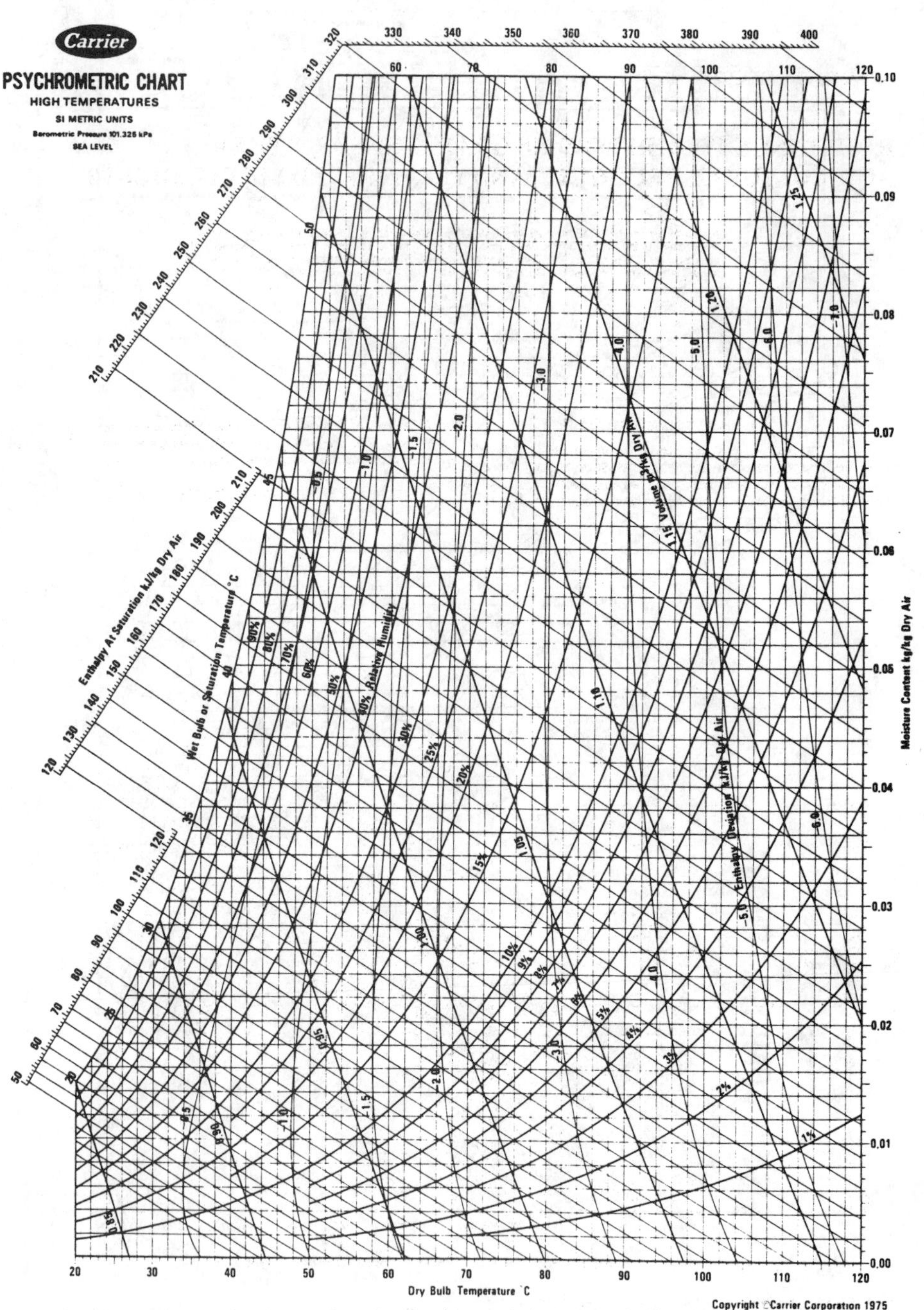
Carrier
PSYCHROMETRIC CHART
HIGH TEMPERATURES
SI METRIC UNITS
Barometric Pressure 101.325 kPa
SEA LEVEL
Enthalpy At Saturation kJ/kg Dry Air
Wet Bulb or Saturation Temperature °C
Relative Humidity
Volume m³/kg Dry Air
Enthalpy Deviation kJ/kg Dry Air
Moisture Content kg/kg Dry Air
Dry Bulb Temperature °C
Copyright ©Carrier Corporation 1975
Cat. No. 794-005 Printed in U.S.A.

Appendix B

ASAE D272.2 DEC92

RESISTANCE TO AIRFLOW OF GRAINS, SEEDS, OTHER AGRICULTURAL PRODUCTS, AND PERFORATED METAL SHEETS

Approved by the ASAE Committee on Technical Data; adopted by ASAE 1948; revised 1954, 1962; reconfirmed by the ASAE Electric Power and Processing Division Technical Committee December 1968, December 1973, December 1978, December 1979; revised December 1980; reconfirmed December 1985; revised March 1987; reconfirmed December 1991; reaffirmed for one year December 1992.

SECTION 1—PURPOSE AND SCOPE

1.1 These data can be used to estimate the resistance to airflow of beds of grain, seeds, and other agricultural products, and of perforated metal sheets. An estimate of this airflow resistance is the basis for the design of systems to dry or aerate agricultural products.

1.2 Data are included for common grains, seeds, other agricultural products, and for perforated metal sheets, over the airflow range common for aeration and drying systems.

SECTION 2—EMPIRICAL CURVES

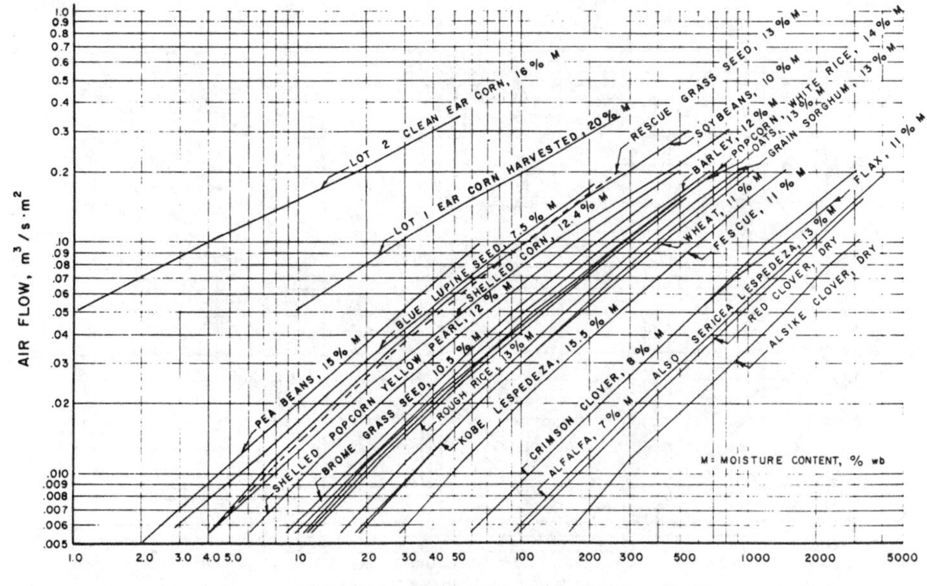

NOTES: This chart gives values for a loose fill (not packed) of clean, relatively dry grain.
For a loose fill of clean grain having high moisture content (in equilibrium with relative humidities exceeding 85 percent), use only 80 percent of the indicated pressure drop for a given rate of air flow.
Packing of the grain in a bin may cause 50 percent higher resistance to air flow than the values shown.
White rice is a variety of popcorn.

FIG. 1—RESISTANCE TO AIRFLOW OF GRAINS AND SEEDS (SI Units) (Shedd's data)

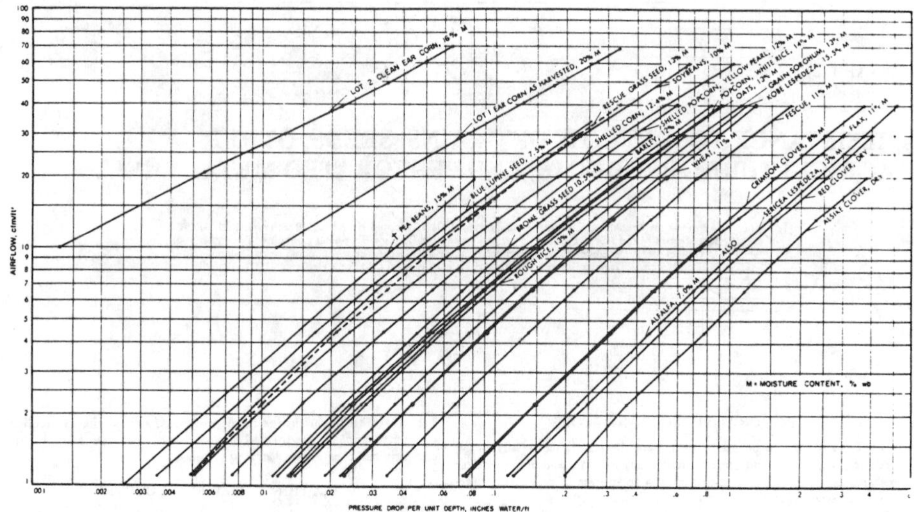

NOTES: This chart gives values for a loose fill (not packed) of clean, relatively dry grain.
For a loose fill of clean grain having high moisture content (in equilibrium with relative humidities exceeding 85 percent), use only 80 percent of the indicated pressure drop for a given rate of air flow.
Packing of the grain in a bin may cause 50 percent higher resistance to air flow than the values shown.
When foreign material is mixed with grain no specific correction can be recommended. However, it should be noted that resistance to air flow is increased if the foreign material is finer than the grain, and resistance to air flow is decreased if the foreign material is coarser than the grain.
White rice is a variety of popcorn.

FIG. 2—RESISTANCE TO AIRFLOW OF GRAINS AND SEEDS (Inch-pound units) (Shedd's data)

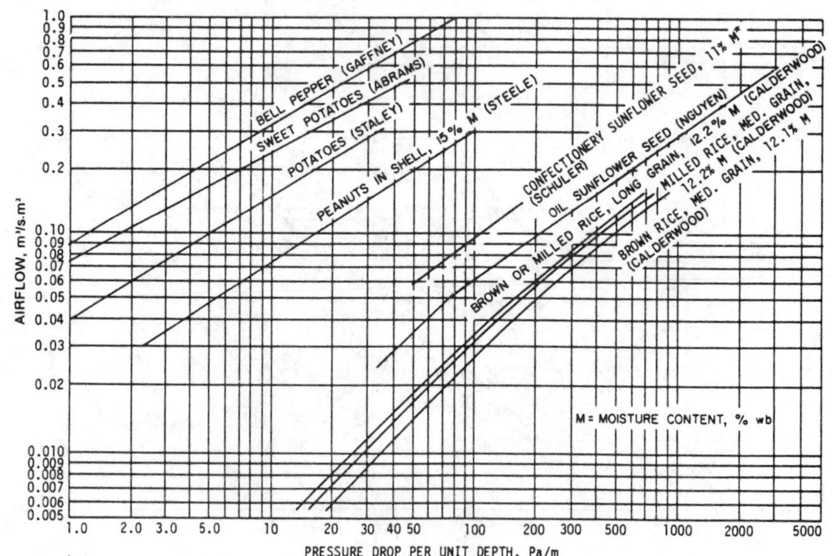

NOTES: Rice: Clean, loose-fill. A packing operation which raised the bulk density by 14-17 percent resulted in pressures 2.3 to 3.4 times those for loose fill.

FIG. 3—RESISTANCE TO AIRFLOW FOR OTHER AGRICULTURAL PRODUCTS (SI units)

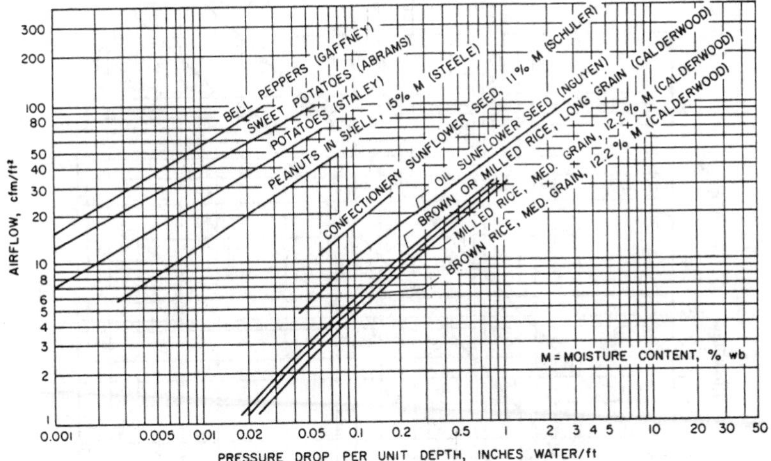

NOTES: Rice: Clean, loose-fill. A packing operation which raised the bulk density by 14 to 17 percent resulted in pressures 2.3 to 3.4 times those for loose fill.

FIG. 4—RESISTANCE TO AIRFLOW OF OTHER AGRICULTURAL PRODUCTS (Inch-pound units)

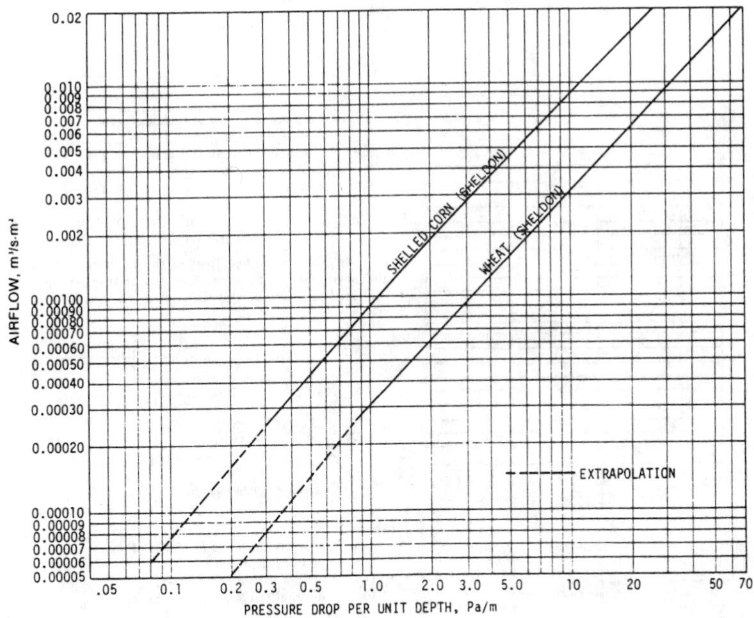

FIG. 5—RESISTANCE TO AIRFLOW OF SHELLED CORN AND WHEAT AT LOW AIRFLOWS (SI units)

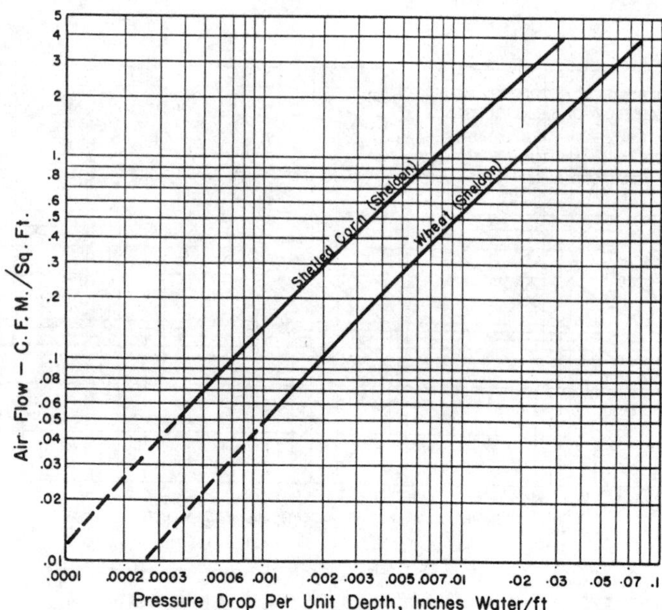

FIG. 6—RESISTANCE TO AIRFLOW OF SHELLED CORN AND WHEAT AT LOW AIRFLOWS (Inch-pound units)

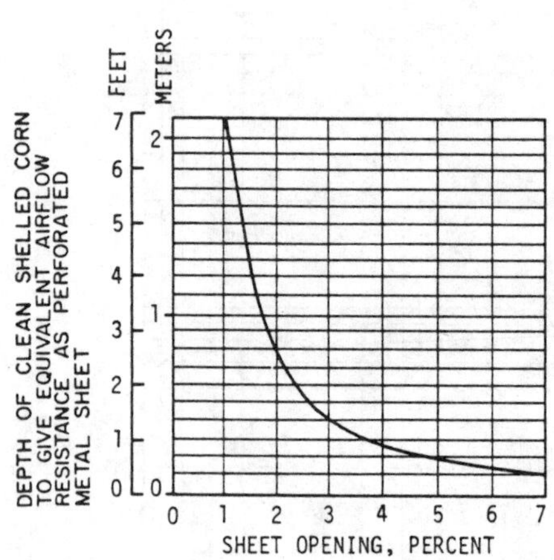

NOTES: When sheet openings amount to 20 percent, no additional resistance to airflow is produced.
A large number of small perforations is preferred to a smaller number of large perforations for the same amount of opening.
The curve shown is based on tests of sheets having width of perforations from 1 to 3.3 mm (0.04 to 0.13 in.).

FIG. 7—RESISTANCE TO AIRFLOW OF PERFORATED METAL SHEETS WHEN SUPPORTING GRAIN (Henderson)

SECTION 3—AIRFLOW RESISTANCE EQUATION

$$\frac{\Delta P}{L} = \frac{a\,Q^2}{\log_e(1+bQ)}$$

where
- ΔP = pressure drop, Pa or inches of water
- L = bed depth, m or ft
- a = constant for particular grain (see Table 1)
- Q = airflow, $m^3/s \cdot m^2$ or cfm/ft^2
- b = constant for particular grain (see Table 1)

SECTION 4—EFFECT OF FINES ON RESISTANCE TO AIRFLOW OF SHELLED CORN

4.1 An effect of adding fines to shelled corn is an increase in the airflow resistance of the corn. The pressure drop per unit bed depth can be corrected to account for fines using this equation:

SI units: $$\left(\frac{\Delta P}{L}\right)_{corrected} = \left(\frac{\Delta P}{L}\right)_{clean} (1 + (14.5566 - 26.418Q)(fm))$$

Customary units: $$\left(\frac{\Delta P}{L}\right)_{corrected} = \left(\frac{\Delta P}{L}\right)_{clean} (1 + (14.5566 - 0.1342Q)(fm))$$

where
- ΔP = pressure drop, Pa or inches of water
- L = bed depth, m or ft
- Q = airflow, $m^3/s \cdot m^2$ or cfm/ft^2
- fm = decimal fraction of fines, by weight

NOTES: Range of applicability: 0.076 to 0.20 $m^3/s \cdot m^2$ (15 to 40 CFM/FT^2) and $0 \leq fm \leq 0.2$. Broken grain and other matter which passed through a 4.76-mm (12/64-in.) round-hole sieve are defined as fines. (Hague)

SECTION 5—EFFECT OF BULK DENSITY ON RESISTANCE TO AIRFLOW OF SHELLED CORN

5.1 An increase in bulk density causes an increase in the airflow resistance per unit bed depth of the corn. The pressure drop per unit bed depth can be predicted as a function of airflow rate and corn bulk density by use of this empirical equation:

$$\frac{\Delta P}{L} = X_1 + X_2 \frac{\left(\frac{\rho_b}{\rho_k}\right)^2 Q}{\left(1 - \frac{\rho_b}{\rho_k}\right)^3} + X_3 \frac{\left(\frac{\rho_b}{\rho_k}\right) Q^2}{\left(1 - \frac{\rho_b}{\rho_k}\right)^3}$$

where

ΔP = pressure drop, Pa or inches of water
L = bed depth, m or ft
ρ_b = corn bulk density, kg/m^3 or lb/ft^3
ρ_k = corn kernel density, kg/m^3 or lb/ft^3
Q = airflow, $m^3/s \cdot m^2$ or cfm/ft^2
X_1, X_2, X_3 = constants (see Table 2 or Table 3.)

TABLE 2—VALUE FOR CONSTANTS (SI UNITS) FOR EQUATION IN PARAGRAPH 5.1

Airflow Range, $m^3/s \cdot m^2$	X_1	X_2	X_3
$0.027 \leq Q \leq 0.13$	-0.998	88.8	511
$0.13 \leq Q \leq 0.27$	-10.9	111	439
$0.27 \leq Q \leq 0.60$	-76.5	163	389

TABLE 3—VALUES FOR CONSTANTS (INCH-POUND UNITS) FOR EQUATION IN PARAGRAPH 5.1

Airflow range, cfm/ft^3	X_1	X_2	X_3
$5.3 \leq Q \leq 26.3$	-0.0012	5.53×10^{-4}	1.62×10^{-5}
$26.3 < Q \leq 52.5$	-0.013	6.94×10^{-4}	1.39×10^{-5}
$52.5 < Q \leq 117$	-0.094	10.2×10^{-4}	1.23×10^{-5}

NOTES: Range of applicability: 732 to 799 kg/m^3 (45.7 to 49.9 lb/ft^3) (corn bulk density) 0.027 to 0.60 $m^3/s \cdot m^2$ (5.3 to 117 cfm/ft^2). (Bern)

TABLE 1—VALUES FOR CONSTANTS IN AIRFLOW RESISTANCE EQUATION

Material	Value of a ($Pa \cdot s^2/m^3$)	Value of b ($m^2 \cdot s/m^3$)	Range of Q ($m^3/m^2 s$)	Reference
Alfalfa	6.40×10^4	3.99	0.0056 — 0.152	Shedd
Barley	2.14×10^4	13.2	0.0056 — 0.203	Shedd
Brome grass	1.35×10^4	8.88	0.0056 — 0.152	Shedd
Clover, alsike	6.11×10^4	2.24	0.0056 — 0.101	Shedd
Clover, crimson	5.32×10^4	5.12	0.0056 — 0.203	Shedd
Clover, red	6.24×10^4	3.55	0.0056 — 0.152	Shedd
Corn, ear (lot 1)	1.04×10^4	325.	0.051 — 0.353	Shedd
Corn, shelled	2.07×10^4	30.4	0.0056 — 0.304	Shedd
Corn, shelled (low airflow)	9.77×10^3	8.55	0.00025 — 0.0203	Sheldon
Fescue	3.15×10^4	6.70	0.0056 — 0.203	Shedd
Flax	8.63×10^4	8.29	0.0056 — 0.152	Shedd
Lespedeza, Kobe	1.95×10^4	6.30	0.0056 — 0.203	Shedd
Lespedeza, Sericea	6.40×10^4	3.99	0.0056 — 0.152	Shedd
Lupine, blue	1.07×10^4	21.1	0.0056 — 0.152	Shedd
Oats	2.41×10^4	13.9	0.0056 — 0.203	Shedd
Peanuts	3.80×10^3	111.	0.030 — 0.304	Steele
Peppers, bell	5.44×10^2	868.	0.030 — 1.00	Gaffney
Popcorn, white	2.19×10^4	11.8	0.0056 — 0.203	Shedd
Popcorn, yellow	1.78×10^4	17.6	0.0056 — 0.203	Shedd
Potatoes	2.18×10^3	824.	0.030 — 0.300	Staley
Rescue	8.11×10^3	11.7	0.0056 — 0.203	Shedd
Rice, rough	2.57×10^4	13.2	0.0056 — 0.152	Shedd
Rice, long brown	2.05×10^4	7.74	0.0055 — 0.164	Calderwood
Rice, long milled	2.18×10^4	8.34	0.0055 — 0.164	Calderwood
Rice, medium brown	3.49×10^4	10.9	0.0055 — 0.164	Calderwood
Rice, medium milled	2.90×10^4	10.6	0.0055 — 0.164	Calderwood
Sorghum	2.12×10^4	8.06	0.0056 — 0.203	Shedd
Soybeans	1.02×10^4	16.0	0.0056 — 0.304	Shedd
Sunflower, confectionery	1.10×10^4	18.1	0.055 — 0.178	Shuler
Sunflower, oil	2.49×10^4	23.7	0.025 — 0.570	Nguyen
Sweet Potatoes	3.40×10^3	6.10×10^8	0.050 — 0.499	Abrams
Wheat	2.70×10^4	8.77	0.0056 — 0.203	Shedd
Wheat (low airflow)	8.41×10^3	2.72	0.00025 — 0.0203	Sheldon

NOTES: The parameters given were determined by a least square fit of the data in Fig. 1-6. To obtain the corresponding values of (a) in inch-pound units (in.H_2O min^2/ft^3) divide the above a-values by 31635726. To obtain corresponding values of (b) in inch-pound units (ft^2/cfm) divide the above b-values by 196.85. Parameters for the Lot 2 Ear Corn data are not given since the above equation will not fit the data.

Although the parameters listed in this table were developed from data at moderate airflows, extrapolations of the curves for shelled corn, wheat, and sorghum agree well with available data (Stark) at airflows up to 1.0 $m^3/s \cdot m^2$.

References: Last printed in 1988 STANDARDS; list available from ASAE Headquarters.

Bibliography

Chapter 1

Benock, G., O. J. Loewer, T. C. Bridges, and D. H. Loewer. 1981. Grain flow restrictions in harvesting-delivery-drying systems. *Transactions of the ASAE* 24(5):1151-1161.

Bridges, T. C., O. J. Loewer, Jr., J. N. Walker, and D. G. Overhults. 1979. A computer model for evaluating corn harvesting, handling, drying and storage systems. *Transactions of the ASAE* 22(3):618-629.

Bridges, T. C., G. M. White, I. J. Ross, and O. J. Loewer. 1982. A computer aid for management of on-farm layer drying systems. *Transactions of the ASAE* 25(3):811-815.

Bridges, T. C., O. J. Loewer, G. M. White, and I. J. Ross. 1983. A management tool for predicting performance of continuous in-bin shelled corn drying systems. *Transactions of the ASAE* 26(5):1528-1532.

Bridges, T. C., D. G. Colliver, G. M. White, and O. J. Loewer. 1984. A computer aid for evaluation of on-farm stir drying systems. *Transactions of the ASAE* 27(5):1549-1555.

Bridges, T. C., D. G. Overhults, S. G. McNeill, and G. M. White. 1988. An aeration duct design model for flat grain storage. *Transactions of the ASAE* 31(4):1283-1288.

Brooker, D. B., F. Bakker-Arkema, and C. W. Hall. 1992. *Drying and Storage of Grains and Oilseeds*. Westport, CT: AVI Publishing Co., Inc.

Bucklin, R., T. Breeden, O. J. Loewer, T. C. Bridges, and G. Benock. 1982. Optimization of equipment and labor for seed processing and bagging systems. *Transactions of the ASAE* 25(4):1041-1044.

Bucklin, R., O. J. Loewer, T. Breeden, T. C. Bridges, and G. Benock. 1989. Simulation of the materials handling system in a seed bagging and processing plant. *Applied Engineering in Agriculture* 5(3):419-424.

EXSYS. 1985. Expert System Development Software. EXSYS, Inc., P.O. Box 75158, Contr. Sta. 14, Albuquerque, NM 87194.

Henderson, S. M., and R. L. Perry. 1976. *Agricultural Process Engineering*, 3rd Ed. Westport, CT: AVI Publishing Co., Inc.

Johnson, W. H., and B. J. Lamp. 1966. *Corn Harvesting*. Westport, CT: AVI Publishing Co., Inc.

Loewer, O. J., T. C. Bridges, and D. G. Overhults. 1976a. Computer layout and design of grain storage and handling facilities. *Transactions of the ASAE* 19(6):1130-1137.

Loewer, O. J., T. C. Bridges, and D. G. Overhults. 1976b. CACHE: Computer model for the analysis of the economics of corn harvesting and processing systems. Technical Series No. 10. Agr. Eng. Dept., University of Kentucky, Lexington.

Loewer, O. J., T. C. Bridges, and D. G. Overhults. 1977. Using the computer to analyze grain storage facilities. *Agricultural Engineering* 58(1):42-43.

Loewer, O. J., T. C. Bridges, G. M. White, and R. B. Razor. 1984a. Optimum moisture content to begin harvesting corn as influenced by energy cost. *Transactions of the ASAE* 27(2):362-365.

Loewer, O. J., T. C. Bridges, D. G. Colliver, and G. M. White. 1984b. Economics of stirring devices in grain drying. *Transactions of the ASAE* 27(2):603-608.

Loewer, O. J., and T. C. Bridges. 1986a. Utilizing computers in a grain drying and storage workshop. *Proceedings of the International Conference on Computers in Agricultural Extension Programs*. Feb. 5-6. Lake Buena Vista, Florida.

Loewer, O. J., T. J. Siebenmorgen, and I. L. Berry. 1986b. Geometric considerations in the design of circular grain storage systems. *Applied Engineering in Agriculture* 2(1):114-122.

Loewer, O. J., I. L. Berry, and T. J. Siebenmorgen. 1989. Geometric and economic considerations in the design of flat storage systems. *Applied Engineering in Agriculture* 5(2):259-264.

Loewer, O. J., M. K. Kocher, and J. Solamanian. 1990. An expert system for determining bottlenecks in on-farm grain processing systems. *Applied Engineering in Agriculture* 6(1):69-72.

Pritsker, A., and B. Alan. 1974. *The GASP IV Simulation Language*. New York: John Wiley & Sons.

Thompson, T. L., R. M. Peart, and G. H. Foster. 1968. Mathematical simulation of corn drying - A new model. *Transactions of the ASAE* 11(4):582-586.

Thompson, T. L. 1975a. FANMATCH: Fan performance computer programs. Presented at the 1975 Grain and Storage Computer Workshop, University of Kentucky, Lexington.

Thompson, T. L. 1975b. NATAIR: Natural air drying program. Presented at the 1975 Grain Drying and Storage Computer Workshop, University of Kentucky, Lexington.

Chapter 2

ASAE Standards, 32nd Ed. 1985. Data: D245.4. Moisture relationships of grains. St. Joseph, MI: ASAE.

Hurburgh, C. R., Jr., C. J. Bern, and S. N. Grama. 1981. Improvements in the accuracy of corn moisture measurement in Iowa. ASAE Paper No. 81-3515. St. Joseph, MI: ASAE.

Lorenzen, R. T. 1958. Effect of moisture on weight-volume relationships of small grain. ASAE Paper No. 58-111. St. Joseph, MI: ASAE.

Nelson, S. O. 1980. Moisture-dependent kernel- and bulk-density relationship. *Transactions of the ASAE* 23(1):139-143.

Paulsen, M. R., L. D. Hill, and B. L. Dixon. 1983. Moisture meter-to-oven comparisons for Illinois corn. *Transactions of the ASAE* 26(2):576-583.

USDA Grain Research Laboratory

Chapter 3

Bridges, T. C., O. J. Loewer, G. M. White, and I. J. Ross. 1983. A management tool for predicting performance of continuous in-bin shelled corn drying systems. *Transactions of the ASAE* 26(5):1528-1532.

Bridges, T. C., G. M. White, I. J. Ross, and O. J. Loewer. 1982. A computer aid for management of on-farm layer drying systems. *Transactions of the ASAE* 25(3): 811-815.

Bridges, T. C., D. G. Colliver, G. M. White, and O. J. Loewer. 1984. A computer aid for evaluation of on-farm stir drying systems. *Transactions of the ASAE* 27(5):1549-1555.

Brooker, D. B., F. W. Bakker-Arkema, and C. W. Hall. 1974. *Drying Cereal Grains*. Westport, CN: AVI Publishing Co., Inc.

Loewer, O. J., T. C. Bridges, D. G. Colliver, and G. M. White. 1984. Economics of stirring devices in grain drying. *Transactions of the ASAE* 27(2):603-608.

McKenzie, B. A., G. H. Foster, and S. S. DeForest. 1972. Dryeration: Better corn quality with high speed drying. Cooperative Extension Service Publication AE-72. West Lafayette, IN: Purdue University.

McKenzie, B. A., G. H. Foster, and S. S. DeForest. 1980. Dryeration & bin cooling systems for grain. Cooperative Extension Service Publication AE-107. West Lafayette, IN: Purdue University.

Siebenmorgen, T. J., and V. K. Jindal. 1986. Effects of moisture adsorption on the head rice yields of long-grain rough rice. *Transactions of the ASAE* 29(6):1767-1771.

Steffe, J. F., and R. P. Singh. 1980. Theoretical and practical aspects of rough rice tempering. *Transactions of the ASAE* 23(3):775-782.

Thompson, R. A., and G. H. Foster. 1967. Dryeration - High speed drying with delayed aeration cooling. ASAE Paper No. 67-843. St. Joseph, MI: ASAE.

Thompson, T. L. 1975a. NATAIR: Natural air drying program. Presented at the 1975 Grain Drying and Storage Computer Workshop, University of Kentucky, Lexington.

Thompson, T. L. 1975b. FANMATCH: Fan performance computer program. Presented at the 1975 Grain Storage and Computer Workshop. Agr. Eng. Dept. University of Kentucky, Lexington.

Thompson, T. L., R. M. Peart, and G. H. Foster. 1968. Mathematical simulation of corn drying - A new model. *Transactions of the ASAE* 11(4):582-586.

White, G. M., and I. J. Ross. 1972. Discoloration and stress-cracking in white corn as affected by drying temperature and cooling rate . *Transactions of the ASAE* 15(3):504-507.

White, G. M., O. J. Loewer, Jr., I. J. Ross, and D. B. Egli. 1976. Storage characteristics of soybean dried with heated air. *Transactions of the ASAE* 19(2):306-310.

White, G. M., O. J. Loewer, I. J. Ross, and D. B. Egli. 1980. Storage characteristics of soybeans dried with heated air. *Transactions of the ASAE* 19(2):306-310.

Chapter 4

ASAE Standards, 32nd Ed. 1985. ASAE D272.1. St. Joseph, MI: ASAE.

Stephens, L. E., and G. H. Foster. 1976. Grain bulk properties as affected by mechanical grain spreaders. *Transactions of the ASAE* 19(2):354-358.

Stephens, L. E., and G. H. Foster. 1978. Bulk properties of wheat and grain sorghum as affected by a mechanical grain spreader. *Transactions of the ASAE* 21(6):1217-1221.

Thompson, T. L. 1975. FANMATCH: Fan performance computer program. Presented at the 1975 Grain Storage and Computer Workshop. Agr. Eng. Dept. University of Kentucky, Lexington.

Chapter 5

Brekke, O. L., A. J. Peplinski, and E. B. Lancaster. 1975. Aflatoxin inactivation in corn by aqua ammonia. ASAE Paper No. 73-3507. St. Joseph, MI: ASAE.

Bridges, T. C., D. G. Overhults, S. G. McNeill, and G. M. White. 1988. An aeration duct design model for flat grain storage. *Transactions of the ASAE* 31(4):1283-1288.

Converse, H. H. 1973. Aerating corn for short-term storage. USDA-Agricultural Research. July.

Frus, J. D. 1967. Aeration of stored grain. Cooperative Extension Service Publication Pm-407. Iowa State University, Ames.

Gregory, W. W. 1973. Controlling insects in stored grain. Cooperative Extension Publication ENT-19. University of Kentucky, Lexington.

Hammond, C. 1977. Detoxifying corn using ammonia. Proceedings and recommendations resulting from the Corn Mycotoxin In-Service Training Meeting held in Atlanta, GA, Oct. 18-19.

Hellevang, K. J. 1984. Crop storage management. AE-791. Agr.Eng. Dept., North Dakota State University, Fargo.

Holman, L. E., ed. 1960. Aeration of grain in commercial storages. USDA, Marketing Research Report No. 178. Revised.

Kline, G. L., and H. H. Converse. 1961. Operating grain aeration systems in the hard winter wheat area. USDA, Marketing Research Report No. 480.

Loewer, O. J., and D. H. Loewer. 1975. Suffocation hazards in grain bins. Cooperative Extension Publication AEN-39. University of Kentucky, Lexington.

Loewer, O. J., I. J. Ross, and G. M. White. 1979. Aeration, inspection and sampling of grain in storage bins. Cooperative Extension Publication AEN-45. University of Kentucky, Lexington.

McKenzie, B. A. n.d. Suffocation hazards in flowing grain. Mimeograph, Agr. Eng. Dept. West Lafayette, IN: Purdue University.

Noyes, R. T. 1967. Aeration for safe grain storage. Cooperative Extension Service Publication AE-71. West Lafayette, IN: Purdue University.

Ross, I. J., H. E. Hamilton, and G. M. White. 1973. Principles of grain storage. Cooperative Extension Publication AEN-20. University of Kentucky, Lexington.

Schwab, C. V., I. J. Ross, and L. R. Piercy. 1982. Vertical pull and immersion velocity of a person trapped in enveloping grain flow. ASAE Paper No. 82-55-2. St. Joseph, MI: ASAE.

Spruill, D. G. 1977. Report of the Animal Science Sub-group Meeting of the Corn Mycotoxin Workshop. Proceedings and Recommendations resulting from the Corn Mycotoxin In-Service Training Meeting held in Atlanta, GA, Oct 18-19.

Thompson, T. L. 1975a. NATAIR: Natural air drying program. Presented at the 1975 Grain Drying and Storage Computer Workshop. University of Kentucky, Lexington.

Thompson, T.L. 1975b. FANMATCH: Fan performance computer program. Presented at the 1975 Grain Storage and Computer Workshop. Agr. Eng. Dept. University of Kentucky, Lexington.

White, G. M., I. J. Ross, and J. D. Klaiber. 1972. Moisture equilibrium in mixing of shelled corn. *Transactions of the ASAE* 15(3):508.

Willsey, F. R. 1972. Using hand signals for agriculture. Extension Publication S-68, West Lafayette, IN: Purdue University.

Chapter 6

Baker, K. D., R. L. Stroshine, G. H. Foster, and K. J. Magee. 1985. Performance of a pressure pneumatic grain conveying system. *Applied Engineering in Agriculture* 1(2):72-78.

Hellevang, K.J. 1985. Pneumatic grain conveyors. Cooperative Extension Service Paper No. 13ENG2-3. North Dakota State University, Fargo.

Henderson, S. M., and R. L. Perry. 1966. *Agricultural Process Engineering*, 2nd Ed. Ferguson Foundation.

Kentucky Agricultural Engineering Handbook. n.d. Agr. Eng. Dept. University of Kentucky, Lexington.

Link-Belt. 1968. Catalog 1000 - Materials handling and processing equipment; Catalog 1050 - Products and components for materials handling and power transmission. Link-Belt Division, FMC Corporation.

Magee, K. J., R. L. Stroshine, G. H. Foster, and K. D. Baker. 1983. Nature and extent of grain damage caused by pneumatic conveying systems. ASAE Paper No. 83-3508. St. Joseph, MI: ASAE.

Merkel, J. A. 1974. *Basic Engineering Principles*. Westport, CT: AVI Publishing Co.

Midwest Plan Service. 1968. Planning grain - Feed handling. MWPS-13. Iowa State University, Ames.

Norder, R., and S. Weiss. 1984. Bucket elevator design for farm grains. ASAE Paper No. 84-3512. St. Joseph, MI: ASAE.

Pierce, R. O., and B. A. McKenzie. 1984. Auger performance data summary for grain. ASAE Paper No. 84-3514. St. Joseph, MI: ASAE.

Segler, G. 1951. *Pneumatic Conveying*. Silsoe, Bedfordshire, England: National Institute of Agricultural Engineering.

Stephens, L. E., and G. H. Foster. 1976. Grain bulk properties as affected by mechanical grain spreaders. *Transactions of the ASAE* 19(2) 354-358.

Stephens, L. E., and G. H. Foster. 1978. Bulk properties of wheat and grain sorghum as affected by a mechanical grain spreader. *Transactions of the ASAE* 21(6):1217-1221.

Chapter 7

Benock, G., O. J. Loewer, Jr., T. C. Bridges, and D. H. Loewer. 1981. Grain flow restrictions in harvesting-delivery-drying systems. *Transactions of the ASAE* 24(5):1151-1161.

Bucklin, R., T. Breeden, O. J. Loewer, T. C. Bridges, and G. Benock. 1982. Optimization of equipment and labor for seed processing and bagging systems. *Transactions of the ASAE* 25(4):1041-1044.

Bucklin, R., O. J. Loewer, T. Breeden, T. C. Bridges, and G. Benock. 1989. Simulation of the materials handling system in a seed bagging and processing plant. *Applied Engineering in Agriculture* 5(3):419-424.

EXSYS. 1985. Expert System Development Software. EXSYS, Inc., P.O. Box 75158, Contr. Sta. 14, Albuquerque, NM 87194.

Johnson, W. H., and B. J. Lamp. 1966. *Corn Harvesting*. Westport, CT: AVI Publishing Co., Inc.

Khuri, R. S., W. D. Shoup, R. M. Peart, and R. L. Kilmer. 1988. Expert system for interpreting citrus harvest simulation. *Applied Engineering in Agriculture* 4(3):275-280.

Loewer, O. J., T. C. Bridges, and D. G. Overhults. 1977. Using the computer to analyze grain storage facilities. *Agricultural Engineering* 58(1):326-328.

Loewer, O. J., D. G. Overhults, T. C. Bridges, G. M. White, and I. J. Ross. 1978. Utilizing computers in a grain processing extension program. ASAE Paper No. 78-5004. St. Joseph, MI: ASAE.

Loewer, O. J., R. Bucklin, T. Breeden, T. C. Bridges, and G. Benock. 1979. Evaluation of the materials handling system in a seed bagging and processing plant. ASAE Paper No. 79-3513. St. Joseph, MI: ASAE.

Loewer, O. J., G. Benock, and T. C. Bridges. 1980. Effect of combine selection on grain drying and delivery system performance. *Transactions of the ASAE* 23(6):1548-1552.

Loewer, O. J., M. F. Kocher, and T. C. Bridges. 1988. An expert system for determining bottlenecks in on-farm grain processing systems. ASAE Paper No. 88-6073. St. Joseph, MI: ASAE.

Loewer, O. J., M. F. Kocher, and J. Solaimanian. 1990. An expert system for determining bottlenecks in on-farm grain processing systems. *Applied Engineering in Agriculture* 6(1):69-72.

Pritsker, A., and B. Alan. 1974. *The GASP IV Simulation Language*. New York: John Wiley & Sons.

Chapter 8

Benock, G., O. J. Loewer, T. C. Bridges, and D. H. Loewer. 1981. Grain flow restrictions in harvesting-delivery-drying systems. *Transactions of the ASAE* 24(6):1643-1646.

Bridges, T. C., O. J. Loewer, and D. G. Overhults. 1979a. The influence of harvest rate and drying time on grain drying and storage facility selection. *Transactions of the ASAE* 22(1):174-177.

Bridges, T. C., O. J. Loewer, J. N. Walker, and D. G. Overhults. 1979b. A computer model for evaluating corn harvesting, handling, drying and storage systems. *Transactions of the ASAE* 22(3):618-621, 629.

Loewer, O. J., T. C. Bridges, and D. G. Overhults. 1976a. Computer layout and design of grain storage facilities. *Transactions of the ASAE* 19(6):1130-1137.

Loewer, O. J., T. C. Bridges, and D. G. Overhults. 1976b. Facility costs of centralized grain storage systems utilizing computer design. *Transactions of the ASAE* 19(6):1163-1168.

Loewer, O. J., G. Benock, and T. C. Bridges. 1980. Effect of combine selection on grain drying and delivery system performance. *Transactions of the ASAE* 23(6):1548-1553.

Loewer, O. J., T. J. Siebenmorgen, and I. L. Berry. 1986. Geometric considerations in the design of circular grain storage systems. *Applied Engineering in Agriculture* 2(1):114-122.

Loewer, O. J., I. L. Berry, and T. J. Siebenmorgen. 1988. Geometric and economic considerations in the design of flat storage systems. ASAE Paper No. 88-6053. St. Joseph, MI: ASAE.

Loewer, O. J., I. L. Berry, and T. J. Siebenmorgen. 1989. Geometric and economic considerations in the design of flat storage systems. *Applied Engineering in Agriculture* 5(2):259-264.

Midwest Plan Service. 1983. Structures and environment handbook. Iowa State University, Ames.

Midwest Plan Service. 1986. Plan No. mwps-73210 (Aug). Iowa State University, Ames.

Siebenmorgen, T. J., M. W. Freer, R. C. Benz, and O. J. Loewer. 1986. Controlled atmosphere storage system for rice. ASAE Paper No. 86-6511. St. Joseph, MI: ASAE.

Thompson, S. A., S. G. McNeill, I. J. Ross, and T. C. Bridges. 1990. Computer model for predicting the packing factors of whole grains in flat-storage structures. *Applied Engineering in Agriculture* 6(4):465-470.

Thompson, S. A., I. J. Ross, and C. V. Schwab. 1992. Predicted packing factors for whole grains in metal grain bins. *Powder Handling & Processing* 4(3):265-269.

Chapter 9

Benock, G., O. J. Loewer, and T. C. Bridges. 1979. Influence of weather on the selection of harvesting, delivery and drying equipment. ASAE Paper No. 79-4043. University of Manitoba, Winnipeg, Canada.

Benock, G., O. J. Loewer, T. C. Bridges, and D. H. Loewer. 1981. Grain flow restrictions in harvesting-delivery-drying systems. *Transactions of the ASAE* 24(5):1151-1161.

Bern, C. J., M. E. Anderson, W. F. Wilcke, and C. R. Hurburgh. 1982. Auger-stirring of wet and dry corn-airflow resistance and bulk density effects. *Transactions of the ASAE* 25(1):217-220.

Bloome, P. D., G. H. Brusewitz, and D. C. Abbott. 1982. Moisture absorption by wheat. *Transactions of the ASAE* 25(4):1071-1075.

Bloome, P. D. 1983. Management implications of the market value of moisture in grain. ASAE Paper No. 83-3522. St. Joseph, MI: ASAE.

Bradford, L. A., and G. L. Johnson. 1953. *Farm Management Analysis*. New York: John Wiley & Son.

Bridges, T. C., D. G. Colliver, G. M. White, and O. J. Loewer. 1984. A computer aid for evaluation of on-farm stir drying systems. *Transactions of the ASAE* 27(5):1549-1555.

Byg, D. M., W. E. Gill, W. H. Johnson, and J. E. Henry. 1966. Machine losses in harvesting ear and shelled corn. ASAE Paper No. 66-611. St. Joseph, MI: ASAE.

Crampton, E. W., and L. E. Harris. 1969. *Applied Animal Nutrition*, 2nd Ed. San Francisco, CA: W. H. Freeman & Co.

Hall, G. E., and L. D. Hill. 1972. Test weight changes of shelled corn during drying. *Transactions of the ASAE* 15(2):320-323.

Hall, G. E., and L. D. Hill. 1974. Test weight adjustment based on moisture content and mechanical damage of corn kernels. *Transactions of the ASAE* 17(3):578-579.

Hill, J. 1976. Climate of Kentucky. Progress Report 221. Dept. of Agronomy, University of Kentucky, Lexington.

Hill, L. D. 1982. Evaluation of the issues in grain grades and optimum moistures. AE-4548. Dept. of Agr. Economics. Agricultural Experiment Station, College of Agriculture, University of Illinois, Urbana-Champaign.

Johnson, W. H., and B. J. Lamp. 1966. *Corn Harvesting*. Westport, CT: AVI Publishing Co., Inc.

Loewer, O. J., Jr., T. C. Bridges, and D. G. Overhults. 1976a. Facility costs of centralized grain storage systems utilizing computer design. *Transactions of the ASAE* 19(6):1163-1168.

Loewer, O. J., Jr., T. C. Bridges, and D. G. Overhults. 1976b. CACHE: Computer model for the analysis of the economics of corn harvesting and processing systems. Agricultural Engineering Technical Series, No. 10. University of Kentucky, Lexington.

Loewer, O. J., T. C. Bridges, G. M. White, and D. G. Overhults. 1980. The influence of harvesting strategies and economic constraints on the feasibility of farm grain drying and storage facilities. *Transactions of the ASAE* 23(2):468-476, 480.

Loewer, O. J., T. C. Bridges, G. M. White, and R. B. Razor. 1982. Energy expenditures for drying and harvesting losses as influenced by harvesting capacity and weather. ASAE Paper No. 82-3515. St. Joseph, MI: ASAE.

Loewer, O. J., T. C. Bridges, D. G. Colliver, and G. M. White. 1984a. Economics of stirring devices in grain drying. *Transactions of the ASAE* 27(2):603-608.

Loewer, O. J., T. C. Bridges, G. M. White, and D. G. Overhults. 1984b. The influence of harvesting strategies and economic constraints on the feasibility of farm grain drying and storage facilities. *Transactions of the ASAE* 27(2):468-476, 480.

Loewer, O. J., G. M. White, T. C. Bridges, I. J. Ross, and S. G. McNeill. 1984c. Comparison between current dockage methods and the standardized bushel concept of marketing. ASAE Paper No. 84-3518. St. Joseph, MI: ASAE.

Loewer, O. J., T. C. Bridges, G. M. White, and R. B. Razor. 1984d. Optimum moisture content to begin harvesting corn as influenced by energy cost. *Transactions of the ASAE* 27(2):362-365.

McKenzie, B. A., G. H. Foster, and S. S. DeForest. 1972. Dryeration: Better corn quality with high speed drying. Cooperative Extension Service Publication AE-72. West Lafayette, IN: Purdue University.

McKenzie, B. A., G. H. Foster, and S. S. DeForest. 1980. Dryeration & bin cooling systems for grain. Cooperative Extension Service Publication AE-107. West Lafayette, IN: Purdue University.

Morey, R. V., G. L. Zachariah, and R. M. Peart. 1971. Optimum policies for corn harvesting. *Transactions of the ASAE* 14(5):787-792.

Official United States Grain Standards. 1988. 7 CFR PART 810. Federal Grain Inspection Service. U.S. Dept. of Agriculture. P.O. Box 96454. Washington, D.C. 20090-6454.

Ross, I. J., H. E. Hamilton, and G. M. White. 1973. Principles of grain storage. Cooperative Extension Publication AEN-20. University of Kentucky, Lexington.

Thompson, T. L., R. M. Peart, and G. H. Foster. 1968. Mathematical simulation of corn drying - A new model. *Transactions of the ASAE* 11(4):582-586.

Thompson, T. L. 1975. NATAIR: Natural air drying program. Presented at the 1975 Grain Drying and Storage Computer Workshop, University of Kentucky, Lexington.

Watson, S. A. 1983. Grain Quality Newsletter - NC-151 participants and supporters. Vol. 5: 3. OARDC - Ohio State University, Wooster.

Williams, E. E., M. Fortes, D. G. Colliver, and M. R. Okas. 1978. Simulation of stirred bin low temperature corn drying. ASAE Paper No. 78-3012. St. Joseph, MI: ASAE.

Index